*

PANOPTICON

of

FASCINATING THINGS
AND
EVENTS

Mathias Scholz

Impressum

© 2024 Mathias Scholz

mathias.scholz@t-online.de

For my tomcat Humpel

1. Topics

Nichts im Leben, außer Gesundheit und Tugend, ist schätzenswerter als Kenntnis und Wissen; auch ist nichts so leicht zu erreichen und so wohlfeil zu erhandeln: die ganze Arbeit ist Ruhigsein und die Ausgabe Zeit, die wir nicht retten, ohne sie auszugeben.

Johann Wolfgang von Goethe

This booklet is an experiment. It deals with subjects and issues that are not entirely uninteresting, at least for the author - and ultimately, hopefully, also for you, the reader - in one form or another, because the topics addressed may be new to you in the context provided or connections might emerge that seem surprising at first glance. In short, the aim of this booklet is what in educated circles would be called "broadening horizons." It conveys knowledge for knowledge's sake in an entertaining way and under the premise that "any knowledge" (as opposed to "ignorance") is something useful and desirable, even if you can perhaps only "apply" it in cultivated small talk. And believe me - if you haven't already experienced it yourself - the more you know about our world, the more interesting and amazing it will seem to you. In this sense, this booklet is also intended as a small tribute to general education, the neglect of which is unfortunately becoming increasingly evident in a world where only narrowly defined specialist knowledge seems to be of any value.

Perhaps one or two of the topics will encourage you to engage with the subject matter in a different way - even if it's just an additional look at Wikipedia (to indulge in the zeitgeist) - or the booklet encourages you to skip the TV in the evening, contrary to your usual habit, and pick up a good book instead. After all, writing, books, and libraries are by far the most important inventions that humankind has ever produced - not television, the internet, the automobile, or the smartphone, as some might think. Books preserve knowledge beyond the lifetimes of those who collected, compiled, or even developed that knowledge themselves. They are the real memory of humankind and, therefore, the foundation of all scientific, technical, and cultural progress. So don't undervalue books! Without books and libraries, we certainly would not have all the things that make our lives so comfortable and worth living in modern society. The great Moravian pedagogue Johann Amos Comenius (1592-1670) wrote in 1650:

"If there were no books, we would all be completely crude and illiterate, for we would have no knowledge of the past, none of divine or human things. Even if we had any knowledge, it would be like legends that have been changed a thousand times by the fluidity of oral tradition. What a divine gift books are for the human spirit! No greater gift could be wished for a life of memory and judgment. Not to love them is not to love wisdom. But not to love wisdom is to be a fool. That is an insult to the divine Creator, who wants us to become his image."

2.   Cabinets of curiosities

Closely related to libraries (where books and writings of all kinds are collected) is the "cabinet of curiosities," also known as a "Panopticon," where "things" - primarily curiosities, natural history objects, and works of art - have been collected and stored since the late Middle Ages and early modern times, usually in anything but a systematic manner. A world-famous example of this is the "Green Vault" in Dresden, which you should definitely visit if you ever find yourself there. It is the treasure and curiosity chamber of the Wettin princes, who have made their art treasures accessible to the public since 1724. Although it is more of an art collection, it can hardly hide its origins in the desire to possess "strange and curious things," which was particularly widespread among secular rulers in the Baroque period. In fact, Augustus the Strong (1670-1733) himself employed skilled artists and craftsmen of his time to produce exhibits according to his wishes. I am thinking here of the court goldsmith Johann Melchior Dinglinger (1664-1731), whose works (for example, "The Golden Coffee Set" (1701) or the "Bath of Diana" (1704)) are still among the outstanding exhibits in the Green Vault today.

As far as I know, the term "Wunderkammer" was first used in the Zimmerische Chronik - the family chronicle of the Counts of Zimmern from Meßkirch in Baden-Württemberg (written between 1564 and 1566) - to describe a non-specific collection of art, natural objects, and curiosities, which were particularly common among sovereigns and wealthy citizens in the Baroque era. It was a way of showing off, proving one's high education, and investing one's wealth "wisely." There was certainly no shortage of exhibits. The voyages of discovery and trade to distant lands brought a constant supply of animal skins, trophies, and ethnographic curiosities, which then ended up in these cabinets of curiosities.

However, there were also "municipal cabinets of curiosities" which, like the one in Zittau in eastern Saxony, can be traced back to 1564. It was precisely in this year that its foundation stone was laid with the donation of a Viennese sundial. Over the following centuries, this rich collection of books developed into an extremely valuable library, parts of which are still preserved today in the old collection of the Christian Weise Library and are accessible to interested parties. Some of the art objects from it (such as the "Engelmann Celestial Globe" from 1690 or the armillary sphere from 1790) can be viewed in the Zittau City Museum.

In earlier times, such a "cabinet of curiosities" was usually referred to as a "Panopticon." Today, this term is typically used as a synonym for a "wax museum," which, if you think about it, is in a certain sense also just a special kind of cabinet of curiosities, especially when you consider how the extremely lifelike figures on display are made.

In line with the title of this book, we would prefer to imagine the "Panopticon" as a "Baroque cabinet of curiosities" that contains a multitude of different "wondrous things." The term "wondrous things" is defined broadly and should include not only physical objects but also incidents, life stories, scientific and philosophical discoveries, historical events, as well as technical and cultural achievements. And if these things are not only remarkable but also interesting (i.e., they evoke cognitive involvement in you as a reader or viewer even without an obviously higher purpose), then perhaps they are even the subject of this book...

Fig. 1: Pretiosa Hall in the "Green Vault" around 1930 (Wikimedia Commons)

3. Nothing new under the sun

And that's all there is to the preface. For as the Bible says in Ecclesiastes 1:9: *"There is nothing new under the sun."* Because, if you look at it very closely, everything that you will read in the following, or even everything that you have already read in your life to date or will read in the future, is contained as text in a coded form in a number whose first three digits (one digit before the decimal point, two digits after the decimal point) are guaranteed to be familiar to you. And if you also remember the following, somewhat bumpy German saying and recognize its deeper meaning, then you will know the circle number Pi - and that is what we are talking about here - to 31 decimal places:

„Sei e hoch i reell bezwecket, es klappt nicht mit jenem Exponent. Imaginäre Untiefe verdecket, wer Pi mit Primzahl zwei dezent zu dieser Zahl mit Mut verrührt, bis so Einheit Erweckung spürt."

4.   The mysterious number Pi

That is quite something. After all, who knows this number to 32 decimal places? Incidentally, the unofficial world record for memorizing Pi is 83,431 decimal places, and the official record is 70,030 - less than half the number of words in this book. Or, if you name one digit per second, it would take just 19 hours and 27 minutes to recite all 70,030 digits for the official world record... (but first, you have to memorize them!). It's amazing what the human brain can remember (we'll come back to that)...

If you want to try it yourself: Just start by memorizing the first 100 decimal places of PI. It's really not difficult...

The first 100 decimal places of the circle number Pi (π) are:

3,1415926535 8979323846 2643383279 5028841971 6939937510 5820974944 5923078164 0628620899 8628034825 3421170679 ...

Pi, the ratio of the circumference of a (Euclidean) circle to its diameter, is a very special number for a reason: it is an irrational real number (i.e., it cannot be written as a fraction of two whole numbers) and a transcendental number, which means that it has an infinite number of digits that do not repeat periodically, as is the case with rational numbers. Incidentally, the German mathematician Ferdinand von Lindemann (1852–1939) was the first to prove this with mathematical precision in 1882. Since Pi is also known to correspond to the area of a unit circle, it necessarily follows that it is impossible to square the circle using only a compass and ruler - a task that even the mathematicians of ancient Greece struggled with.

So, what is so special about such a transcendental number? Mathematicians would say that it is, firstly, irrational, and secondly, that it can never be the zero of any polynomial with integer coefficients - which mathematically justifies its transcendence. In practical terms, this means that such a number is a sequence of digits that does not exhibit any kind of regularity. You usually have to calculate each individual digit separately (which is called a "drip algorithm"), and it is never possible to predict exactly what the next digit in the sequence will be - unless you calculate it.[1] More than 105 trillion decimal places of pi are already known today (as of February 2024, calculation time 75 days). And the following applies:

"Any finite approximation of (no matter how large) is small compared to the infinite number of digits this number represents."

An approximate calculation of pi digit by digit is possible because this transcendental number can be represented in various forms as a mathematical series. One such series was developed in 1682 by the famous German polymath Gottfried Wilhelm Leibniz (1846-1716). Because it is so beautiful, it should also be reproduced here:

---

[1] Incidentally, the trillionth decimal place of pi is 0

$$\pi = 4 \sum_{k=0}^{\infty} \frac{(-1)^k}{2k+1}$$

Unfortunately, it is practically useless for calculating Pi, as the series converges extremely slowly (to obtain the first two decimal places, you have to calculate approximately 50 terms of this series). Therefore, efforts have been made to find faster-converging series. One of these came from the extraordinary and enigmatic Indian mathematician Srinivasa Ramanujan Aiyangar (1887–1920, called "S. Ramanujan" for short - if you know of him, you are certainly a mathematician), who at some point wrote the following series in his notebook, which remains enigmatic in many parts today:

$$\frac{1}{\pi} = \frac{\sqrt{8}}{9801} \sum_{k=0}^{\infty} \frac{(4k)! \, (1103 - 26390k)}{(k!)^4 \, 396^{4k}}$$

It converges incredibly quickly, as the sum of nine terms already provides 70 correct decimal places of Pi. However, even this series is no longer suitable for achieving world records in Pi calculation. For this purpose, so-called spigot algorithms are used, where one digit is calculated after the other. But that is a different topic, which we will only mention here without going into further detail. Instead, we would like to draw your attention to another well-founded but as yet unproven fact regarding Pi, which leads to astonishing consequences. If, in addition to Pi's irrationality and transcendence, it could also be proven that the circle number is a so-called "normal" number, then this would mean that every book in the world (strictly speaking, every finite document formulated in letters and digits) is encoded somewhere in the numerical sequence of Pi.

Fig. 2: Srinivasa Ramanujan had a habit of writing perfectly correct mathematical formulas in his notebook without proof, which sometimes took mathematicians many years to prove...

5.　　Every book in the world is included in Pi

But what is a "normal number" in the first place? In mathematics, a normal number is a real number among whose decimal digits all possible k-digit blocks of digits occur with the same asymptotic relative frequencies for each k > 1. Or to put it another way, each of the digits "0", "1", "2" ... "9" occurs statistically equally often in it.[2] And this is precisely what leads to the profoundly logical and inescapable consequence already mentioned:

This book you are reading here is contained somewhere in Pi in encrypted form. And also in all conceivable combinations of printing errors etc. pp.

"Encoded" here means nothing more than, for example, encoding any letter or character according to the ASCII code (as an example of a frequently used encoding) by three consecutive digits and in this way translating any sequence of numbers in Pi back into a sequence of letters. In this respect, "pi" (and any other transcendent "normal" number) can be understood as a "library of Babel" in the sense of the Argentinian writer Jorge Luis Borges (1899-1986), which really does contain all conceivable books.[3] It is pointless to find a specific book in it, because most of the books it contains are gibberish and incomprehensible. And it is also impossible to search for a specific book (e.g. this one). Even in the 105 trillion passages known to date, you will hardly find any longer coherent strings that make any deeper sense.

6.　　ISBN and Pi

But now a few enthusiasts thought they could take a relaxed approach and see which books are contained in the first 50 million decimal places of Pi - based on their ISBN numbers. So they quickly wrote a program and checked all ISBN numbers to see if their 13 digits appear anywhere in this tiny portion of the sequence of digits that make up Pi. In this case, as expected, they quickly found what they were looking for.

7.　　Snow White and Rose Red

On the first attempt, three books were clearly identified in Pi in this way, including the Dutch edition of the fairy tale "Snow White and Rose Red" by the Brothers Grimm. This fairy tale, which has unfortunately been somewhat forgotten in modern times, has recently experienced a certain renaissance through a song by the German rock group Rammstein (in the form of the title "Rosenrot" on the album of the same name from

---

[2] The question of whether pi is "normal" in this sense has not yet been proven. However, there is the assumption that at least every irrational algebraic number is normal (Bailey & Crandall, 2001), but pi is not one of them because it is transcendental. Whether pi is "normal" is an open problem in mathematics.

[3] Borges, (1944), Ficciones (Fictions), collection of stories

2005) and can quickly be found on YouTube. Originally, this fairy tale itself was an adaptation of another fairy tale, which goes back to the pedagogue Karoline Stahl (1776–1837) and which was included in the famous collection "Kinder- und Hausmärchen" by the Brothers Grimm (Jacob (1785–1863) and Wilhelm Grimm (1786–1859)) in 1837. However, neither Karoline Stahl, who conceived this fairy tale, nor the Brothers Grimm, who included it in their collection, knew that it had a particularly profound psychological significance. Only the well-known theologian and psychoanalyst Eugen Drewermann, who incidentally wrote a very readable biography of Giordano Bruno (1548–1600) (ISBN 3-423-30747-1), made this truly fundamental discovery...

8.   The Brothers Grimm and the "German Dictionary

But back to the Brothers Grimm. Their real achievement as Germanists lies in the publication of the "German Dictionary", a work of 34,824 pages that was only completed in 1961, 123 years after it was started. Today, it is available to everyone free of charge on the Internet but can also be purchased as a 33-volume printed edition in leather for a whopping €4,000 as an ornament for the bookcase at home.

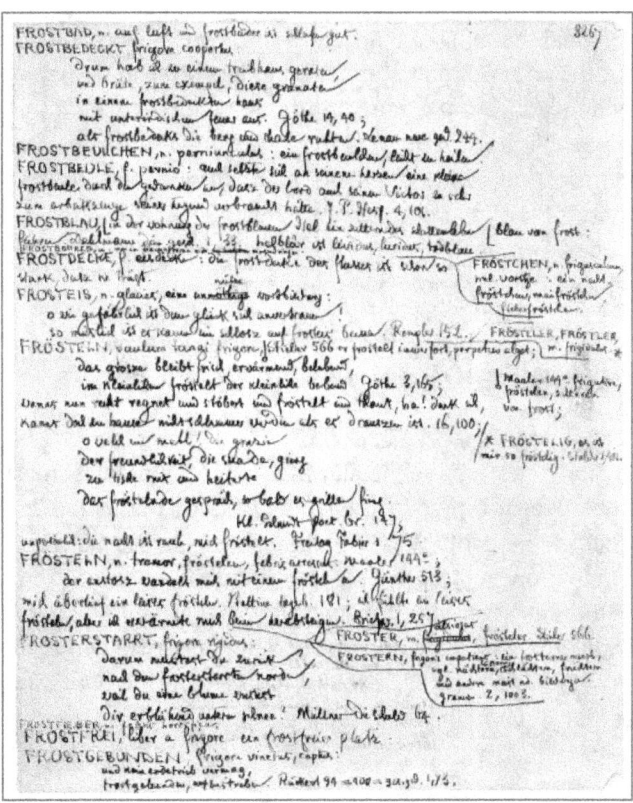

Fig. 3: Original manuscript page for the "German Dictionary" by Jacob Grimm (source Wikimedia)

So if you want to know the exact meaning of a certain German word that may not yet have made it into Wikipedia, Grimm's dictionary is undoubtedly the first choice for finding out. Or do you know right away what the verb "abblatten" means? "Defoliate", yes, but "flake off"? If so, then you are either a doctor or a hunter. Because the deer "defoliates" when it plucks the green leaves from young trees, or "the shoulder joint" when the suprascapular nerve is paralyzed for some reason. Words, therefore, have a meaning and an origin. Jacob and Wilhelm Grimm tried to determine both for every German word they could somehow get hold of.

9.   Who or what is an idiot?

Of course, some words originally slipped through their fingers, as they had not yet found their way into the normal vocabulary during their lifetime. For example, the beautiful and frequently used word "idiot", which comes from ancient Greek (ἰδιώτης) and actually refers to nothing other than a "private person" who was unwilling to hold any public office in the polis. Later, the meaning of the term shifted so that the now Latin word "idiōta" was used to refer to a "layman" or "incompetent". From the Renaissance onwards, the word began to be used to describe people with a certain "educational deficit", which eventually even became a medical term in modern times.

But back to the original meaning of the word. Today, we know that a private person can be an "idiot" under certain circumstances, namely when they behave like a fool or moron (probably derived from the word "trotteln", which is used in Austria for the leisurely gait of a horse).[4] The word "idiotes" has therefore undergone a certain change in meaning since Homer. That's why people feel insulted when they are called "private person" (or more often "official") in this way. As you can see, we all use a huge number of words on a daily basis whose meaning we are fully aware of as native speakers, but about whose origin we usually know next to nothing.

10.   Etymological dictionary

To prevent this "ignorance" from being cultivated, the "Etymologisches Wörterbuch der Deutschen Sprache" was invented in 1883 and has been published 25 times to date. It has since been incorporated into the "Digital Dictionary of the German Language"[5] and can be a lot of fun to browse through. It is always instructive. For example, you usually know that many illnesses are treated in a polyclinic ("Poliklinik", GDR term for a medical center) and that "polytheism" is the doctrine of many gods. But why is "Poli" spelled with an "i" one time and "Poly" with a "y" another time? Have you ever thought about this? The answer is actually quite simple. The Greek word "πολύς" (polys) means "many" -

---

[4] Incidentally, the most famous "idiot" to have found his way into world literature was the 26-year-old Prince Lev Myshkin, who was particularly notable at the time for his childishly naive behavior...

[5] See https://www.dwds.de/

hence "polyp" if you have many arms or "polyhistor" if you are a "polyhedron" like Gottfried Wilhelm Leibniz (1646-1716) or, among physicists, Lev Landau (1908-1968), or yourself if you have the time and courage to read this book to the end. The "police", on the other hand, like the "polyclinic", is nothing more than a municipal institution. And the Greeks called a city "πόλις" (polis) - just as the Americans logically called their big city in Indiana "Indianapolis" or their big city in Minnesota "Minneapolis". The fact that in some shady circles in Prussian Berlin, police officers were contemptuously referred to as "polyps" (and this was before the beautiful word "cops" entered common parlance, especially among some left-wing alternative citizens) must have been due either to their rudimentary knowledge of Greek or the ability of the police to apprehend lawbreakers like a "polyp", with "many arms" as it were.

11. Hans Albers as "Der Greifer"

From this, I believe, the quasi re-Germanized form "Der Greifer" (the Grabber) was derived for a very specific police officer from Essen (more precisely Otto Friedrich Dennert, "Kriminaloberkommissar"), who was played by Hans Albers (1891-1960) in the 1958 film of the same name. In GDR times, Hans Albers films were often seen as "guests" in Willi Schwabe's (1915-1991) "Rumpelkammer". Today you don't see him on television that often, which is a shame...

12. Munchausen's ride on the cannonball

But his ride on the cannonball in the 1943 film "Münchhausen" remains legendary. As a layman, you might ask: Is such a thing even possible? As a physicist, I would like to answer: Yes, of course. If the cannonball is lying on the ground, just sit on it. Otherwise, in flight, it's not feasible. The firing speed of a cannonball during the Russo-Austrian Turkish War (the movie is set during that period, 1736-1739) was probably around 150 m/s. It's not just the ascent that presents a serious problem. Even "holding on" would likely be quite difficult in real life. The feat that Hieronymus Carl Friedrich von Münchhausen (1720-1797) allegedly accomplished in front of the "Ochakov" fortress on the Black Sea, by switching to an oncoming cannonball in mid-flight, seems somewhat unrealistic at a relative speed of, let's say, 300 m/s, even in terms of feasibility. I think the Baron was clearly exaggerating here. The same also applies to "Münchhausen's theorem, horse, swamp, and mop" (Hans Magnus Enzensberger). Because if it were true and the Baron had really pulled himself and his horse out of the swamp by his own bootstraps, then Isaac Newton (1642-1726) and his theorem "actio = reactio" would really have been disproved. His "Newtonian mechanics" would then be empirically invalidated. The world would then behave much more strangely and would probably be unrecognizable. So, if you ever want to read the stories attributed to Baron von Münchhausen (which I can only recommend, e.g., on Google Books "Des Freiherrn von Münchhausen wunderbare Reisen und Abentheuer zu Wasser und zu Land"), always keep in mind: These are tall tales.

Fig. 4: This is how it is said to have looked when Hieronymus Carl Friedrich Freiherr von Münchhausen used the cannonball as a flying machine (Otto Lilienthal (1848-1896) was not yet born at the time). (Wikimedia, engraving by August von Wille (1828-1887)).

13. Lies, logic, paradoxes

As we all know, people have always lied. That is a truism. Small lies help to make life more pleasant, big lies tend to achieve the opposite. And lies can also lead to insurmountable logical problems, for example if you firmly claim *"This sentence is wrong!"*. As we all know, people have always lied. This is a truism. Small lies help to make life more pleasant, while big lies tend to achieve the opposite. Lies can also lead to insurmountable logical problems, for example, if you firmly claim, "This sentence is wrong!" This is a modern form of the statement *"All Cretans are liars"*[6] - as soon as it is articulated by a Cretan. Here, a statement collides with propositional logic. This is known as a "logical paradox." If the statement about Cretans is true (as Epimenides the Cretan once claimed), then it follows from self-reference that it is false - and vice versa. Incidentally, this direct self-reference can easily be canceled by two consecutive statements - but without resolving the issue: *"The next sentence is false. The previous sentence is true."* This also makes your head spin. If the statement is true, then it is false - so it cannot be true; and if the statement is false, then it is true - so it cannot be false. This paradox cannot be resolved logically.

It is similar with the statement, *"Nothing, absolutely nothing in this book is true,"* in the foreword of a book (not this one - it has none!). In the case of a novel, this is just a statement that certainly applies to the entire book (just think of the novel character James Bond by Ian Fleming (1908-1964)), but not to the foreword itself, which may have been written by the author. However, if the preface is part of the book, then the statement implies that the preface, including this sentence, is also untrue - meaning the novel itself is a fact (assuming the preface only consists of the remark, *"At least one statement in this book is false"*). Then the main part of the book must contain at least one error. Suppose, however, that the main part of the book is completely free of errors. Then the preface is true if it is false, and false if it is true.

14. Mr. Shanks has made a mistake

In a certain abstract sense, this paradox also reflects the tragedy of the mathematician William Shanks (1812-1882), who, of course, did not have access to the website www.pibel.de during his lifetime.[7] This is why he took the trouble to calculate the number pi, which we have already discussed in point 4, digit by digit by hand using a special series expansion. His tragedy was that he made a mistake when he reached the 528th decimal place (corresponding to the main part of his treatise, while in the preface he claimed that all digits were correct) and did not realize it. This mistake meant that all subsequent digits (he calculated pi up to the 707th decimal place without a calculator!) were also wrong. Fortunately, this error had no effect on Mr. Shanks' physical and mental well-being, as the mistake was only discovered in 1945, 63 years after his death. If he had known at the time that he had miscalculated at some point, the search for the incorrect digit would have been an ordeal. Because if the last digit he calculated is wrong, then it is highly likely that the digit before it is also wrong. But not just this one, but probably a whole series of digits (in this case, 176). To find the error, you would obviously have to start again from the place where you know exactly that it is correct.

---

[6] In the original Greek and in hexameter, it reads "εἶπέν τις ἐξ αὐτῶν ἴδιος αὐτῶν προφήτης- "Κρῆτες ἀεὶ ψεῦσται, κακὰ θηρία, γαστέρες ἀργαί."

[7] See www.pibel.de

15. Loose and fallacies

Just as you don't know where the error is in the sequence of numbers mentioned, you also don't know which ticket will win in a lottery pot with, say, 10,000 tickets. If you draw a ticket from this lottery pot, then with a probability of 1/10,000, you naturally don't expect to draw exactly the right one. With each subsequent draw, the likelihood of landing a winning ticket does not significantly increase. Interestingly, the expectation that you will not draw the right ticket contrasts with the fact that someone will win. The lottery industry relies on the belief of a ticket buyer: *"After all, someone has to win; why not me?"* At the same time, believing that none of the tickets you draw will win while accepting that a lucky winner will be chosen seems contradictory. It is therefore not rational to believe in both statements simultaneously. The only strategy that remains is to buy as many tickets as possible to significantly increase the probability of winning. However, this strategy is rendered ineffective if the prize for the winning ticket is, for example, 5,000 euros for 10,000 tickets, but each ticket costs 10 euros. The statement *"After all, someone has to win; why not me?"* is ultimately a fallacy.

Now, consider the "practitioner" who regularly plays the lottery (6 out of 49) with the hope of becoming a millionaire. A simple calculation using combinatorics shows that there are exactly 13,983,816 different ways to choose 6 numbers from 49. Let's assume that 32 lottery tickets, each with a different combination, stacked on top of each other, measure exactly one centimeter in thickness. Thus, 13,983,816 lottery tickets would form a stack approximately 4.37 kilometers high. Exactly one of these tickets must have the correct combination of six numbers to win. Therefore, "luck of the draw" in the sense of getting six correct numbers means drawing the right ticket from a 4.37-kilometer-high stack. If the super number is also correct, the stack would be 43.7 kilometers high. As you can see, even if you make 10 or 100 picks, the probability of becoming a millionaire this way is practically non-existent.[8]

16. Gourmet tip: Herring with whipped cream

Fallacies (paralogisms) are found quite frequently, for example, in gastronomy *("Whipped cream is good; herring is good. How much better must herring with whipped cream be!")* or in gynecology *("Storks return in spring. Birth rates increase in spring. The return of the storks evidently promotes the birth rate.")*. However, one is not immune to fallacies in general, as they do not necessarily have to be logically contradictory. Recognizing fallacies is often challenging, especially when they are based on apparent correlations for which there is no genuine causal relationship.

Here is an example from the field of climatology. At a specific location, temperature measurements have been continuously recorded for over 100 years and are now being

---

[8] unless you are extremely lucky...

statistically analyzed. There is a noticeable trend of increasing annual average temperatures over the years, which is often attributed to a general shift towards higher temperatures ("global warming"). However, the measurement station was originally located on the outskirts of a city, whose boundary gradually expanded outward over the century and eventually incorporated the measurement site. Cities, due to their construction, are known as "heat islands," meaning that the "concrete" of buildings heats up particularly strongly in the sun during the day. The observed "warming" at this specific measurement station could thus be attributed to the heat island effect (the "heat island" city expanding over time due to continuous development of the outskirts), potentially leading to incorrect conclusions if one is not careful.

17. Risk assessments

Many false conclusions are relatively easy to recognize. However, this does not apply to all situations. The interrelationships in nature and society are often so complex that inaccurate conclusions can arise from uncertain or incomplete information. Pure logic cannot always compensate for a lack of information or address situations where data is fuzzy or suggests correlations that do not exist. For example, there are often serious differences between objective and subjective risk assessments in everyday life. (What is objectively more risky: driving to the airport by car, or flying from one airport to another?)

Objective risk assessment is typically based on statistical data and scientific analysis, while subjective risk assessment is often strongly influenced by personal experience, emotional reactions, and psychological factors. For example, statistics clearly show that flying is safer than driving when measured by the number of accidents and deaths per mile traveled. Nevertheless, many people perceive flying as riskier because airplane crashes are more spectacular and receive more media attention, whereas car accidents seem commonplace and attract less attention.

Another example of the discrepancy between objective and subjective risk assessment is people's perception of natural disasters compared to everyday health risks. Although heart disease and strokes are among the most common causes of death worldwide, the risk they pose is often underestimated. In contrast, people often overestimate the risk of dying from an earthquake or volcanic eruption, despite the statistically low probability, because such events appear more dramatic and frightening.

These differences in risk perception can lead to misjudgments and inefficient decision-making processes. For example, individuals may overallocate resources to protecting themselves against rare but spectacular risks, while neglecting everyday but more dangerous risks. Politicians and society are therefore called upon to develop strategies that promote balanced and rational risk assessments. This requires not only providing scientifically sound information, but also understanding the psychological mechanisms that influence our perception of risk. Only by doing so can we make informed decisions that improve both our safety and quality of life.

18. About Knaffs, Plautze and Hunkies as well as Hemputis, which are predominantly red

Logic can also be applied to somewhat unfamiliar objects. For example, do you know what "green hunkis," Plautzes, Hemputis, or Knaffs are? With a bit of logical reasoning, you can certainly solve the following simple problem:

- All Knaffs have the same shape and size.
- All green Hunkis also have the same shape and size.
- Twenty Knaffs fit exactly into one Plautz.
- All Hemputis contain green Hunkis.
- A green Hunki is ten percent larger than a Knaff.
- A Hemputi is smaller than a Plautz.
- The contents of all Plautzes and Hemputis are predominantly red.
- Given these conditions, what is the maximum number of green Hunkis that can fit into a Hemputi?

Logic, which derives from the concept of reasoning, is a crucial tool in mathematics, especially in propositional logic, used to prove mathematical theorems. This is why mathematics students must study it. It is also helpful in everyday life.

In October 2010, the then Federal President of Germany, who is now largely forgotten, stated, *"Islam belongs to Germany."* This statement was later reaffirmed in 2015 by then Federal Chancellor Angela Merkel. This led to significant controversy. For instance, the then Prime Minister of Saxony, Mr. Tillich, countered with the statement, *"Islam does not belong to Saxony."* If this statement is affirmed, then, purely logically, Saxony does not belong to Germany. This raises the question: Is a Minister President allowed to make such a public statement?

19. Fuzzy logic

Logic is a sharp sword, as the ancient Greeks recognized. However, not everything in life is clear-cut and precise. Consider the eBook reader, tablet, or smartphone that you might be using to read this. All these devices have batteries that provide the electrical energy necessary for their operation. Unlike an on/off switch, which either establishes or interrupts the flow of energy, a battery is either empty or in a state somewhere between "empty" and "full." These intermediate states are called fuzzy. They cannot be neatly categorized into a "true" or "false" or "0" or "1" framework. To describe such states mathematically, "fuzzy logic" was developed around 1965. The term "fuzzy logic" was coined by American electrical engineering professor Lotfi A. Zadeh (1921-2017) of the University of California, Berkeley.

As digital technology, based on "sharp logic," became more established, its limitations were quickly recognized. This led to the development of multi-valued logic, which eventually evolved into modern fuzzy logic, with its vast range of applications, particularly in control and regulation technology.

For instance, if you have a translation app installed on your smartphone that translates the text of a document previously photographed with its camera into another language, this is made possible by fuzzy logic. In the future, when cars drive completely autonomously, allowing passengers to make calls or surf the internet without concern for traffic violations, this will also rely on complex control systems based on fuzzy logic.

In addition to propositional logic, predicate logic, and fuzzy logic (to name a few), there are also other forms of "logic," such as "women's logic" and "men's logic," which are often difficult for individuals of the opposite gender to understand. However, these are less a topic of mathematics and more a subject of the expanding literature on advice.

20. Advice literature: Ask Mummy!

This type of literature is a product of the Enlightenment and is currently reaching its peak on internet platforms such as "Frag Mutti".[9] In the past, it was often said that good advice was expensive. Today, there are both free and paid apps available for such advice. As the author, I can honestly say that I view this development quite positively. For instance, without the help of a corresponding YouTube video, I would never have been able to open the notebook I'm currently using and replace the annoyingly loud fan at a reasonable price. When I "screwed it back together," one screw remained. However, a service technician once explained to me with a smile that this is quite normal...

21. Screws and threads

Although the screw is a fairly old invention - consider Archimedes' screw - it was only with the technical revolution of the late 18th and early 19th centuries that it became the ultimate technical fastener. After the screwdriver was invented in 1744, an English inventor and mechanical engineer succeeded, just 53 years later, in creating the indispensable "thread cutter." This invention truly revolutionized screw manufacturing, making the effective use of a screwdriver possible and giving it a deeper significance. The unstoppable rise of the screw, known to be more effective than the long-established rivet for joining metal parts, was thus assured.

Today, screws with standardized metric threads are used almost exclusively. However, there are exceptions, particularly those that are British-made, such as the right-hand drive versions of various cars.

---

[9] Serious advice: If you can't ask your "mom" (or "grandma") for any reason, then go to https://www.frag-mutti.de/ . They will help you there...

One such example is the screw on the head of your photo tripod (if you have one), which is typically used to attach your digital camera to the tripod. This screw features a ¼ inch thread.

22. The fight against insufficient depth of field - focus stacking

A tripod is particularly useful when using lenses that are either not very fast or have a very long focal length. This is especially true for hobbies like macro photography, as the author indulges in. In macro photography, camera shake can be almost catastrophic if you aim for high image quality. This is due to the large reproduction scale, which is typically a maximum of 1:1 for good macro work. Additionally, the shallow depth of field is problematic, meaning only a very thin strip of the image is actually in focus when photographing close objects. This strip is often less than a millimeter wide.

To address this, you can attach a photo slide to the tripod head and then mount the camera. The photo slide allows you to move the camera closer to the subject, such as the head of a blowfly, and then take a series of shots. The slide is adjusted approximately 1/10 mm closer to the object between shots. Each image captures a different sharp sectional plane of the subject. The merging of these shots into a single image with a particularly high depth of field - something that cannot be achieved by simply stopping down the lens - can be done automatically by a computer program. This technique is known as "focus stacking."

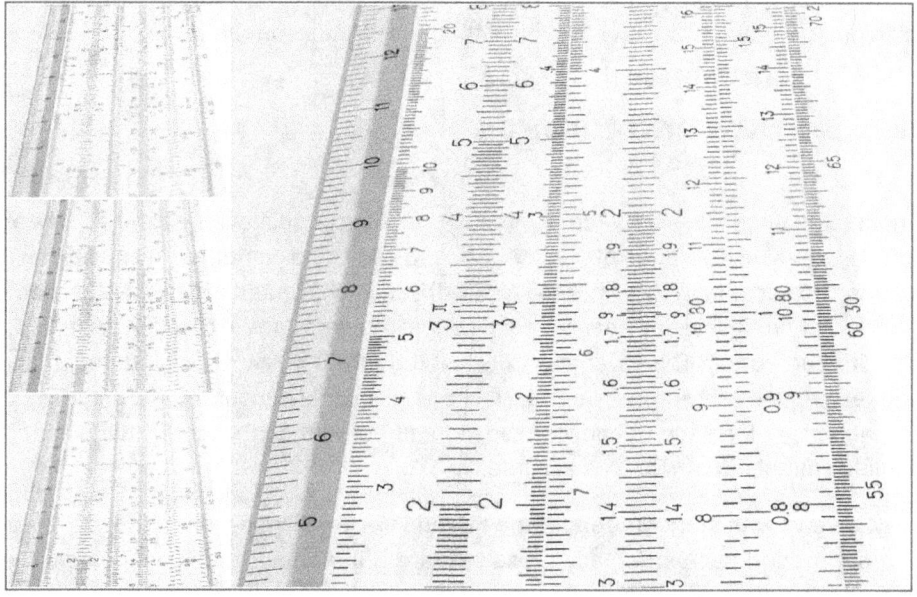

Fig. 5: The three images on the left have the focus point in the background, center and foreground. The large image on the right is obtained by increasing the depth of field (Wikimedia)

This method allows for extremely detailed pictures of small objects, like insects. While the process itself is not particularly difficult, the real challenge lies in getting the "blowfly" to stay still and "smile" at the camera throughout the 15 to 20-step approach - without moving...

23. Blowflies (Schmeißfliegen)

The german word "schmeißen," found in the term "Schmeißfliege," originates from Old High German and can best be translated as "to smear" or "to defile." In the past, before antioxidants were introduced into butter production and when refrigerators were still rare, butter often went rancid in the summer. Rancid butter, in which the fats and lipids have decomposed through oxidation, develops a distinctive odor that tends to attract Calliphoridae flies in large numbers. Once these flies land on the foul-smelling butter, they inevitably soil themselves with the rancid fat, which is why they were known as "Schmeißfliegen."

Since rancid butter has become relatively rare due to the addition of antioxidants and the widespread use of refrigerators, the fruiting body of the stinkhorn (Phallus impudicus), a foul-smelling edible mushroom, is now a better option for observing blowflies. The most well-known of these is the shiny green-gold "goldfly" (Lucilia sericata), which many people are familiar with.

Fig. 6: Goldfly in detail (Author)

24. Gourmet tip: Witch's eggs

This mushroom, unmistakable due to its phallic shape, is often so densely populated with blowflies and red-necked silphae (Oiceoptoma thoracicum, a carrion beetle) that its olive-colored, foul-smelling fruiting mass, called "gleba," can become almost completely obscured. This observation signals to the experienced mycologist that there is nothing left worth collecting. However, a few "witches' eggs" may still be found in the area, which some gourmet circles are not entirely averse to eating.[10] To prepare them, simply remove their outer, somewhat leathery skin and, if necessary, peel off the gelatinous layer underneath. The remaining part can be sliced and prepared like roast potatoes. They are reputed to be quite tasty.

One final tip: witches' eggs often break in the collection basket or bag, which can make them look unappetizing - especially when the greenish jelly oozes out. It helps immensely to use an egg carton, preferably one with a hinged lid, to keep them intact during transport.

Fig. 7: According to the German saying „Die Morchel mit dem ‚M' wie ‚Magen' kann jeder Mensch ganz gut vertragen. Die Lorchel mit dem ‚L' wie ‚Luder' ist dagegen ein meist giftiger Bruder" is clearly an edible mushroom. (Author)

---

[10] Tip: simply enter "Hexeneier zubereiten" on YouTube and trust Snokri's expertise...

25. Egg cartons and breakfast eggs

Such egg cartons with their contents are available in every supermarket. They hold brown or white eggs, which can be used for scrambling, frying, poaching, or making a breakfast dish. These eggs usually come from ecologically sound cage-free or barn-raised hens. Breakfast eggs are especially popular on weekends and are typically boiled for five minutes. (Note that "boiling water" does not truly exist because, unlike soup, water cannot reach a state of boiling while contained in an egg.) It is advisable to pierce the egg to prevent it from bursting when it comes into contact with hot water. Without this step, the air bubble trapped inside the egg expands as it heats up, and the resulting increase in pressure can force the shell to crack. This phenomenon is a direct consequence of Avogadro's Law, established by Amedeo Avogadro (1776-1856) in 1811, which states that the pressure of a gas in a constant volume (such as the inside of a chicken egg) increases with temperature. If the internal pressure exceeds the strength of the eggshell, the egg will burst. This can be easily identified by the egg white leaking out of the crack and forming unappetizing white threads as it coagulates in the cooking water.

26. Non-Newtonian liquids

Egg white is known to be liquid and is therefore considered a fluid. But it is not an ordinary (also known as "Newtonian") liquid, as you have no doubt often noticed yourself when trying to pick a small piece of eggshell out of a freshly beaten chicken egg with your fingers.[11] This is not so easy, because "egg white" is not an ordinary liquid like petrol or diesel (water only pretends to be an "ordinary" liquid, which is not true. Water is anything but an "ordinary" liquid). The egg yolk, on the other hand, can be easily separated from the egg white, preferably with the help of an empty plastic cola or juice bottle. Just give it a try...

But let's return to "liquids", or more precisely "fluids", which do not behave like "normal" liquids or fluids (even fine sand is a "fluid," as a glance at a classic egg timer proves).

According to "pure" theory, non-Newtonian fluids, for example, are "liquids" that exhibit strikingly non-linear viscous flow behavior, i.e., they change their viscosity ("flow behavior") when the shear forces acting on them change - something not observed in "ordinary" liquids. You can easily produce such a substance yourself by making a slightly mushy liquid from cornstarch and water. "Dilatancy" is achieved when the liquid can only be stirred very slowly with a spoon. If you increase the stirring speed, the mixture becomes increasingly difficult to stir, as the liquid behaves more and more like a solid in relation to the shear forces. You can even observe that cracks suddenly form in the medium, which is quite unusual for a pulpy mass. However, you can also take an ordinary hammer and place the "flat side" of its head on the cornstarch slurry. It will sink in immediately. However, if you hit the liquid surface quickly, it is as if you were hitting a board:

---

[11] The trick here is to first moisten the fingers with water...

in this case, the cornstarch slurry behaves like a solid. The sudden increase in viscosity is caused by a structural change in the fluid, which ensures that the individual fluid particles suddenly interact more strongly with each other. Osborne Reynolds (1842-1912) was able to explain this phenomenon in 1885 using the example of wet quicksand. This is known to behave "non-Newtonian" as well...

27. Sinking into quicksand?

"Quicksand" is, strictly speaking, a dispersion of fine sand and water that rarely occurs in nature (the edge of Lake Namak in Iran is notorious in this context – but the fact that peo-ple sink into its quicksand never to be seen again, as locals like to claim, is still just a fairy tale). The only time you see this is on TV or in the movies when you watch adventure films from the early 1960s. Back then, "sinking" into quicksand was a popular stylistic device to increase the suspense of the respective plot (because the more the victim moved to free himself, the faster he sank into the sand – a horrible idea!). Just think of "Lawrence of Arabia," which won seven Oscars and starred Peter O'Toole (1932–2013) and Omar Sharif (1932–2015). There is a very impressive scene in which one of T. E. Lawrence's servants (played masterfully by Peter O'Toole) sinks into quicksand in front of his master, who is unable to help him. However, this scene is not exactly physically accurate. Due to the high overall density of the water-sand dispersion, a person will not be able to sink any further into it than, say, up to their stomach, as the density of the human body is always lower than that of the quicksand. But without outside help, a person will still not be able to free themselves from this predicament in most cases - as they will be stuck in the "mud," so to speak.

Fig. 8: T.E. Lawrence's servant sinks into quicksand... (BBC)

And now a few words to explain the phenomenon of such a dilatant quicksand dispersion: at rest, the fine grains of sand are densely packed together, with the spaces between the grains completely filled with water. If this mixture is now subjected to shear stress, the sand particles slide past each other, with the water acting as a kind of lubricant and thus reducing the overall friction. A body (e.g., our hammer in the example of the cornstarch pulp) will then sink in at a speed that depends on the shear rate. If the shear load increas-es (hammer blow), the distance between the grains of sand becomes slightly greater and the water film connecting them tears, with the effect that the lubricating effect decreases and the frictional resistance increases accordingly.

By the way, if you fill a swimming pool with the aforementioned viscous cornstarch dispersion, you can do as Jesus did (see, for example, John 6:16–21) and walk across it with quick steps without sinking. But if you stop, you'll be like the famous diver in Arthur Schramm's two-line poem (we'll come back to this)...

At first glance, the fluid properties of normal liquids are very simple compared to those of dilatant liquids. Analogous to Newton's famous law of motion, according to which "force = (inertial) mass times acceleration," an equation of motion can also be established for them, which is able to describe all flow phenomena of "Newtonian liquids and gases" in a fundamental and conclusive manner. These equations are called the "Navier-Stokes equations" after their originators.

28. Navier-Stokes - a millennium problem

To scare you a little, I would like to briefly write down the equations (it is a system of non-linear partial differential equations) here, without wanting to or even being able to explain them properly (there is extensive specialist literature for this). This is just about "optics", so that you can see what mathematicians, physicists, meteorologists and engineers from the field of fluid mechanics have to deal with in their work:

Navier-Stokes equations

$$\frac{\partial}{\partial t} u_i + \sum_{j=1}^{n} u_j \frac{\partial u_i}{\partial x_j} = v \Delta u_i - \frac{\partial p}{\partial x_i} + f_i(x, t) \qquad (x \in \mathbb{R}^n, t \geq 0)$$

$$div\, u = \sum_{i=1}^{n} \frac{\partial u_i}{\partial x_i} = 0 \qquad\qquad (x \in \mathbb{R}^n, t \geq 0)$$

Initial conditions: $u(x, 0) = u^0(x)$ $\qquad\qquad (x \in \mathbb{R}^n)$

The applications of this equation system are enormous. It is used, for example, to optimize the streamlining of cars and to calculate the wing profiles and engine design of modern passenger and fighter aircraft. Astrophysicists use it to mathematically simulate "jets" that shoot out of the cores of active galaxies (quasars) in order to gain a better un-derstanding of their operating principles. They are just as indispensable for meteor-ologists who have to deal with the complex flow processes in the Earth's lower atmos-phere as they are for engineers whose task is to develop modern power plants or ship propulsion systems, for example. In short, the Navier-Stokes equations are a system of equations with outstanding practical relevance. And this is where the problem lies. It is still not known (apart from a few trivial individual cases) how to solve them exactly (i.e., analytically). However, finding approximate solutions for all possible applications "nu-merically" - i.e., with the help of computers - is no longer a problem. Nevertheless, it would be of the utmost importance to prove that there are exact solutions for the Navier-Stokes equations - regardless of the initial conditions. This would make it possible to investigate their tem-poral behavior in a quasi-determined way and thus gain new in-sights into complex dynam-ic systems, which are often associated with the term "deter-ministic chaos." Think of short-term weather and long-term climate prediction, the movement of planets and plan-etoids around stars (keyword "n-body problem"), com-plex life processes, or the develop-ment of stock market prices, which, as we know, can hardly be predicted in detail. For this reason, the question of exact solutions to the Na-vier-Stokes equations has also been included in the list of seven "millennium problems" in mathematics, each of which has a prize of one million US dollars. So if you have time and leisure and would also like to earn a million dollars... (item 429).

Incidentally, to date (2024), only one of these seven problems has been solved: the so-called "Poincaré conjecture." Several years of work by top-class specialists alone were invested in checking the solution until the community of mathematicians was certain that the proof presented by the flamboyant Russian mathematician Grigori Perelman (but never published by himself in a mathematical journal) was correct. The public paid less attention to the intellectual achievement of the now voluntarily unemployed math-emati-cian than to his refusal to accept both the Fields Medal (effectively the "Nobel Prize" for young mathematicians) and the prize money. By the way, if you want to read something more demanding than this text, the works of Grigori Perelman are easy to find on the Internet (ArXiv)...[12]

29. Why mathematics is important

Now a few brief remarks on how important a good knowledge of mathematics is if you want to become a scientist or engineer. Many a first-semester physics student has been anything but pleasantly surprised to find that the first two years consist mainly of mathe-matics lectures - and that university mathematics is presented very differently from what they were used to at school (in the first four semesters, it makes virtually no differ-ence whether you study mathematics or physics).

Doing physics professionally means that you first have to learn the "official language" of this highly demanding science, and that is mathematics. However, this does not mean that you have to be particularly good at "mental arithmetic" (at least in the past, many

---

[12] See https://arxiv.org/find/math/1/au:+Perelman_Grisha/0/1/0/all/0/1

sales assistants were often much better at this than some seasoned math or physics professors), but rather that you have to develop a mathematical understanding at the highest possible level in order to be able to use this "tool" successfully to solve problems. And you can only do this by practicing, practicing, and practicing again. Therefore, as a student, you shouldn't be surprised that you have to work on dozens of exercise sheets plus extensive practical protocols every week in addition to the obligatory lecture revision in order to get by reasonably successfully. Much more important than talent are diligence, a high frustration threshold, and the will to work on a problem until you have found "your" solution. And if you get it wrong, just sit down again and look for the mistake. If, on the other hand, you're just looking for a fun student life, then STEM subjects are definitely not suitable unless you're something of an "overachiever."

If you want to reduce the shock that the transition from secondary school mathematics to university mathematics causes for many students in advance, I recommend searching the internet for videos of mathematics lectures (e.g., on the Tübingen multimedia server TIMMS) and watching them. If you are comfortable with this, then you should really study something "math-heavy," because the intellectual reward that awaits you at the end is not to be sneezed at. Otherwise, you probably won't make it through the first few semesters...

According to the observations of the author of these lines, there is unfortunately a massive loss of mathematical education and skills in Germany as far as normal school mathematics is concerned. While in Asian countries the level is being raised higher and higher, here in Germany the focus on basic mathematical knowledge and skills has declined sharply since it has become all about "competencies" and equalization. The so-called "competence approach"[13], which aims to teach supposedly "soft skills" such as problem-solving strategies and teamwork, often means that the actual mathematical content and understanding of mathematical relationships are neglected in practice - with fatal consequences as soon as they are suddenly required in business practice or in university mathematics.

It is striking that in many German classrooms, mathematics is seen as a necessary evil to be mastered and not as a fascinating science that promotes creative thinking and analytical skills. Many students leave school with a poor understanding of basic math principles, which impairs their ability to succeed in technical or scientific careers.

In contrast, Asian education systems place great emphasis on a solid and comprehensive mathematical education. Mathematics is not just seen as necessary knowledge but as a foundation for the development of logical thinking and intellectual discipline. This is reflected in the impressive performance of Asian pupils in international comparative tests such as PISA and TIMSS, in which they regularly achieve top rankings.

The consequences of this trend are worrying. Weak mathematical education can impair Germany's competitiveness in global competition, as mathematics is a key skill in many

---

[13] It culminates in the "ticking competence" in written exams, which increasingly consist of multiple-choice questions...

industries of the future. In addition, the opportunity to give pupils the joy and opportunities of a sound mathematical education is being missed.

It is therefore urgently necessary for education policy and schools to recognize this alarm-ing trend and take countermeasures. Mathematics must once again be seen as a central part of the educational process, and lessons should aim to promote a deep understanding of mathematical concepts rather than just teaching superficial 'skills.' This could be achieved by placing a renewed emphasis on the fundamentals of mathematics, by enhancing the didactic skills of teachers, and by creating a positive attitude towards mathematics among students.

Let's return to the starting point of our profound reflections that led us to the Navier-Stokes equations: the Sunday breakfast egg. Most of us love it "soft-boiled," i.e., the yolk is still largely liquid in this case. The best way to enjoy it properly is to scoop it out. And to do this, you need a spoon or, even better and more appropriately, an "egg spoon."

30. Plastic and mother-of-pearl egg spoon

Such "egg spoons" are usually made of plastic (but not always). The reason for this is that such spoons are tasteless, unlike silver spoons, they do not tarnish unattractively, the egg mixture does not stick to them, and they do not draw heat from the egg, as is the case with metal spoons that conduct heat well.

The fact is, however, that "plastic" is a relatively new product of the chemical industry (spoon-compatible galalith was only invented around 1897 by Wilhelm Krische (1859–1909) and Adolf Spitteler (1846–1940)). For this reason, the "egg spoon" was first made of ivory in the Baroque period and then particularly popularly made of mother-of-pearl. And since the latter "mother-of-pearl spoons" are laboriously made from the shell of the mother-of-pearl oyster, one of them can quickly cost a hundred dollars or more on Amazon or eBay. That's why they are used for eating real caviar (a spoonful of which costs about the same as the aforementioned spoons) rather than for eating plain, more or less soft-boiled white or brown organic or non-organic hen's eggs..

31. Shells and mother-of-pearl

The special feature of these special spoons is not so much their intended use, but rather their mysterious iridescent sheen, which is typical of mussels and some sea snails. Even the large pond mussel (*Anodonta cygnea*) has it on its smooth inner surface. It is still present in large numbers in some shallow ponds here in Upper Lusatia (as long as they contain clean water). Their shells are relatively easy to find if you wade barefoot through the mud at the edge of the reed belt (in summer!) and pay attention to whether you step on something hard. According to binary logic, it is then either a pond mussel or no pond mussel (or pond mussel shell). But you can easily check this by reaching into the mud...

But back to the iridescent sheen on the inside of the mussel shell, which makes the egg spoons carved from it so sought-after and expensive. The reason for this is the mother-of-pearl, a composite material consisting of fine layers of calcium carbonate in the form of aragonite and various organic solids sandwiched in between. The soft body of the mussel deposits this aragonite in the form of approximately 10 µm wide and 0.5 µm thick more or less transparent platelets, which form slightly inclined stacks in both horizontal and vertical directions, the spaces between which are filled with organic substances (essentially as glue, e.g. in the form of chitin) (see Fig. 9).

0.5 µm is a layer thickness that corresponds almost exactly to the wavelength of visible light. As part of the incident white light is reflected on the upper side of each layer and another part is reflected on the lower side after passing through the layer, a path difference occurs, which is known to lead to interference. This causes certain parts of the visible spectrum to be extinguished and others to be enhanced, leaving different shades of color depending on the viewing angle. This is the reason why mother-of-pearl is iridescent - and why pearls made from shells such as the river pearl mussel or the oyster are so sought-after and expensive.

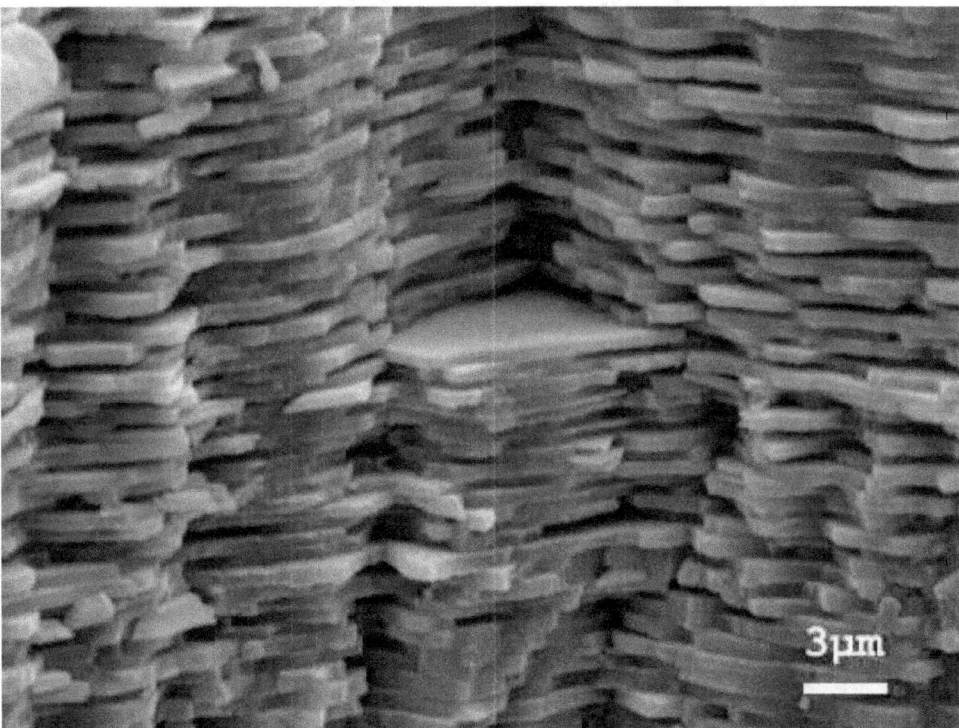

Fig. 9: View of the fracture surface of a mother-of-pearl mussel shell fragment under a scanning electron microscope (Wikimedia)

This is why jewelry made from pearls of the freshwater pearl mussel (and other mussel species) is so sought-after and expensive, as they also have an unmistakable sheen. Since the freshwater pearl mussel is virtually extinct in Central Europe, artificial "pearls" are often used instead of "freshwater pearl mussel pearls" to make pearl necklaces.

32.   Knight Runkel and the black pearl

Incidentally, black pearls are particularly valuable. The famous knight Runkel von Rübenstein (whose first name was Heino, by the way) found one of these by chance in the Gulf of Ormuz, in an oyster on a treasure chest that he had discovered during a dive to the wreck of the Poseidon. According to Hannes Hegen (1925–2014), this chest contained the golden armor of Alexander the Great.

Fig. 10: Knight Runkel (Heino) von Rübenstein, his two squires Dig and Dag and his faithful warhorse "Türkenschreck"... (Hannes Hegen)

33. Captain Nearchos and the Diadochi

The famous navigator Nearchus (360–312 BC), an admiral of Alexander the Great and a native of Crete, is said to have been shipwrecked on this ship just off the "Island of the Damned" near Ormuz, which is, of course, complete nonsense. What is true is that after the early death of Alexander (323 BC in Babylon), Nearchus entered the service of Antigonus Monophthalmus (i.e., the "one-eyed man," around 382–301 BC) and took an active part in the so-called Diadochi Wars. Monophthalmus is regarded as one of the most important Diadochi, i.e., the generals who divided up the vast Alexander Empire among themselves. He thus founded the Antigonid dynasty, which ruled large parts of Greece (Macedonia) until 168 BC.

34. Perseus, the last Antigonide

Their last king was Perseus (213–168 BC), who ascended to the Macedonian throne in 179 BC. While his predecessor was still on good terms with the Romans, Perseus made himself rather suspicious by secretly arming himself. The rising Roman Empire could not and would not tolerate this. In line with the policy of a "clean slate" (tabula rasa), legions were marched to Macedonia in 168 BC, where Perseus had entrenched himself above the river Elpeus in northeastern Greece. The Greek phalanx was crushed there on June 22 at Pydna by the Romans under the leadership of the Roman consul Lucius Aemilius Paullus Macedonicus (around 229–160 BC). Both armies were roughly equal in strength: the Romans had about 38,000 men, and the Macedonians had 44,000, of which around 4,000 fought on horseback on both sides. The battle was quickly decided thanks to the ingenious tactics of the "excellent" commander Paullus (according to Theodor Mommsen (1817–1903)). It is reported that the entire battle lasted no more than an hour and a half. Afterwards, 25,000 Macedonians were dead, wounded, or captured (unfortunately, there is no figure available for the Romans). The rest, including Perseus, sought their salvation in flight.

35. Perseus and 6000 talents of gold

The Macedonian king's 6,000 talents of gold, which he allegedly had with him when he fled, proved to be his undoing. If you consider that a talent was equivalent to a weight of between 25 kg and 30 kg, 6,000 talents seems a bit excessive. After all, this would correspond to a compact cube of gold with a side length of just over 2 meters, which is about half the weight of Britain's gold reserves today. But be that as it may. Perseus was betrayed by his escape companions after he had killed one of them. Thus, he was given the unintended opportunity to make a brief visit to Rome, where he was forced to participate in the obligatory triumphal procession before being relocated just as involuntarily to a damp dungeon on Lake Fucine. He died there a few years later.

The Roman victory at Pydna led to the abolition of the Macedonian monarchy and the division of the former state into four "free regions," which were not allowed to enter into relations with each other and also had to mint their own coins under Roman supervision. It goes without saying that the Romans pocketed the alleged 6,000 talents of gold. And if you're wondering where it went, well, we don't know. After all, it's been half an eternity since then...

36. Greek historiography

We know all this from the Greek historian Polybius of Megalopolis (c. 200–120 BC). He was taken hostage to Rome at the end of the Third Macedonian War, where he was responsible for the education of the two sons of the consul Lucius Aemilius Paullus (died 216 BC), Lucius Aemilius Paullus Macedonicus (c. 229–160 BC). There, he developed his interest in historiography, which culminated in his 40-volume work "Histories", written in ancient Greek, a kind of universal history of the Roman Empire.

His method of depicting historical events and placing them in their respective political contexts is characterized by a sobriety and a methodology that is always committed to the truth and includes source criticism. Alongside Herodotus (c. 490–430 BC) and Thucydides (c. 454–c. 399 BC), he is considered one of the three most important ancient historians whose pragmatic approach to historiography still influences modern historical writing. What was called "universal history" in Polybius' time would be called "world history" today.

37. Why it's good to study universal history

Over 225 years ago, a certain Friedrich Schiller (1759–1805) chose as the topic for his inaugural lecture at the University of Jena why it is still valuable and useful to study "universal history" today. In it, he posed the question that remains relevant today:

"What states did man pass through until he rose from that extremity to this extremity, from the unsociable cave-dweller to the witty thinker and educated man of the world?"

He viewed it as the task of "universal history" to make this journey tangible through a chain of human events linked by a teleological bond. It is not merely the enumeration of events in the form of a pure chronicle that is essential, but the recognition of the connections between individual events, their interdependencies, their social and political contexts, the description of the "zeitgeist," and what a "philosopher's mind" derives from this for his own actions.

This lecture caused a veritable storm of enthusiasm among the students of the time - it was the era of Kant and the Enlightenment. Unfortunately, his professorship was unpaid, so he had to continue earning a living through poetry. He was aided in this by the significant resonance his drama "Die Räuber" (The Robbers) received, not only in Germany.

While his second major historical work on the "History of the Thirty Years' War" was rather mediocre (his first, which earned him the professorship in Jena, was the "History of the Secession of the United Netherlands"), his dramas, poems, and ballads enriched German literature immensely and enduringly.

Fig. 11: Friedrich Schiller

38. Weimar Classics

In this context, his deep friendship with Johann Wolfgang von Goethe (1749–1832), with whom he, along with Johann Gottfried Herder (1744–1803) and Christoph Martin Wieland (1733–1813), founded an era that has gone down in literary history as "Weimar Classicism," should be particularly mentioned. Since then, many of his dramas and stage plays, as well as his poems and ballads, have become obligatory reading material for every generation of schoolchildren. The study of Schiller's works has increasingly shifted from memorizing famous ballads to interpreting them (poetry interpretation - *"What is the poet trying to tell us?"*).

39. The Bell (by Schiller)

A hundred years ago, pupils at humanistic grammar schools knew such mammoth poems as "Die Glocke" by heart without any mistakes. After all, the "Lied von der Glocke" consists of 425 lines of poetry. Today, people often ask in vain which poem the sentence *"It is dangerous to wake the leu, the tiger's tooth is pernicious; but the most terrible of horrors is man in his madness"* comes from. Young people often have problems interpreting the word "Leu".

But thank God, there were also gifted poets beyond Goethe and Schiller for whom the "Lied von der Glocke" itself seemed too long and who therefore worked intensively on a shorter, more pupil-friendly version.

40. Goethe, Schiller, Arthur Schramm are the best we have.

One of them was Arthur Schramm (1895-1994), who has remained largely unknown outside his home country not only as a gifted poet of memorable two- and four-liners, but also as a great inventor. His version of Schiller's "Glocke" is probably the shortest and most compact ever composed: *"Loch in Erde, Bronze rin, Glocke fertig, imm, bimm, bimm"* (Hole in the ground, bronze rinn, bell ready, bimm, bimm, bimm*)*. But even he was unable to dispel a fundamental criticism of Schiller's ballad. It is well known that a bell needs a clapper in order to go "bimm, bimm, bimm". But there is no mention of this in either Schiller or Schramm.

But back to Arthur Schramm, who was only known as the "Klaane Getu" in his home town of Annaberg (Erzgebirge) because of his size. It's worth reading a few more of his more or less well-known songs *("Rumpeldipumpel, weg war'n die Kumpel! Schippe drauf, Glück auf." „Gluck, gluck – weg war er"* (The diver)) or unknown *("Die Sonne scheint ins Kellerloch. Ach lass sie doch - ach lass sie doch!")* or politically incorrect at the time *("Der Kumpel aus dem Bergloch kriecht. Hurra, der Sozialismus siecht!")* or even timelessly beautiful sayings *(„Sommer, Sonne, Wellenpracht, Badehose, Sowjetmacht".)* from oblivion. And as far as his inventions are concerned, unfortunately only the "MIRAMM laundry tongs" in the form of the "wooden barbecue tongs" have survived the test of time.

41. A zeppelin for catching flies

And, of course, his greatest invention, the "Zeppelin flycatcher," which was made of cardboard, modeled after an airship, and lined on the inside with sweet-smelling strips of glue that were irresistibly fragrant to flies, must also be mentioned. The flies entered the interior of the cardboard zeppelin through small openings and quickly stuck to the glue strips due to the limited freedom of flight provided by the zeppelin envelope. This was definitely an advantage over the classic insect glue strip, which can still be purchased today as a hygiene product (invented in 1909 by the cough drop manufacturer Theodor Keyser and developed to mass production) and which proves its usefulness when trying to rid oneself of a summer fly infestation without using "chemicals." Arthur Schramm's groundbreaking invention also spared people from the distressing sight of flies dying on the glue strip, which, as we all know, can be quite unpleasant for sensitive individuals...

It should also be noted that before Theodor Keyser invented the insect glue strip, the so-called "fly swatter" was usually the device of choice for dealing with flies. Alternatively, housewives might attempt to poison the bothersome "fly," as the infamous Marie-Madeleine Marguerite d'Aubray, Marquise de Brinvilliers (1630–1676), did with her father and two of her brothers

42. Fly agaric and fly death

However, the summer should not be too dry, because you need at least one fly agaric (Amanita muscaria). The application is quite simple: Cut up the fly agaric, place the shreds on a plate that is easily accessible to flies, moisten the mushroom slightly, and sprinkle plenty of sugar over it. After a short time, the flies will be attracted to the sweet substance and, shortly afterward, fall from the ceiling or lamp (and if you're unlucky, into the soup) with symptoms of acute fly agaric poisoning. For humans, however, the fly agaric, considered "the poisonous mushroom" par excellence, is relatively harmless.

In contrast to the related death cap (*Amanita phalloides*), which is easily distinguishable by its coloration, no fatal poisoning has been reported from the fly agaric. You would have to consume more than a kilogram of fresh fly agarics to suffer a fate similar to that of the aforementioned Musca domestica species (stomach ailments occur with far less mushroom substance, thanks to the abundant ibotenic acid and its decomposition product, muscimol). However, fly agarics don't taste very good. During wartime, when food in general and edible mushrooms in particular were scarce (due to high levels of collecting activity), fly agarics were occasionally used as food supplements after being boiled several times and the cooking water discarded. What remained after this process was largely free of toxins, but certainly anything but tasty. As my grandmother used to say, "Hunger drives it in."

43. Caesar's mushroom

Many people think that the fly agaric is unmistakable because of its familiar shape and color. However, that's not entirely accurate. In fact, there is a similar relative, a mushroom so esteemed that in ancient Rome it was reserved primarily for the "emperor" (famously known as "Caesar") and his entourage. This mushroom is the Caesar's mushroom (Amanita caesarea). Although it is quite rare in Central Europe and especially in Germany, it is increasingly being observed to establish itself in southern Germany due to "general global warming." However, its primary distribution area is south of the Alps, where it is found exclusively (like the truffle) in deciduous forests.

The Caesar's mushroom, known in ancient Rome as "*fungus suillus*," along with the porcini mushroom, was a highly prized edible mushroom, sometimes even eaten raw by gourmets. As the distinction between "edible mushrooms" and "poisonous mushrooms" was not as clear-cut as it is today, an emetic was always kept on hand during a "mushroom meal" to ensure that the stomach could be emptied easily in the event of any indisposition. Galen of Pergamon (often called "Galen" for short, 129–216 AD), a famous physician, was familiar with a range of such remedies. They were typically used to clear the stomach after a protracted binge...

Fig. 12: The Caesar's mushroom - an excellent edible mushroom (Wikimedia)

44. Money doesn't stink

This is why such establishments were also equipped with a special room, the vomitorium, where Romans paying homage to Lucullus could take their emetics (also known as "vomitorium") undisturbed... The less well-off Roman had to explicitly visit the "latrina" for this purpose instead.

Latrines were extremely important facilities at the time and the separate urinals were even subject to a special tax (*Pecunia non olet!*) at times (for example under the Emperor

Vespasian). The reason for this was that the emperor, well, needed money. And it was convenient for his argument that some trades, e.g. cloth dyers and leather tanners and especially the laundries of the time, relied on the urine that accumulated in them, as the Romans did not yet know "Perwoll©" or "Ariel©".

45. Urinals and fine blue fabrics

As far as fabric dyeing is concerned, it is not the urine itself but the ammonia formed from it that is the essential reagent. It is needed, for example, to dye fabrics blue with the help of woad (*Isatis tinctoria*).

Fig. 13: Indigo dyer in Morocco (Wikimedia)

In this special dyeing process, known as indigo dyeing, the glycoside indican contained in the woad plant is first fermented. The fermented woad plants, which have been ground into a pulp, are formed into woad balls and then left to dry in the air. During this process, the indican is converted into the yellow indoxyl, which is insoluble in water. To dissolve it, these balls are soaked in stale urine (ammonia!) and then reduced with pot-ash (calcium carbonate). The now water-soluble dye can be easily transferred to fabrics such as linen, which initially turns bright yellow and then, under the influence of atmospheric oxygen, deep blue as the indoxyl oxidizes to form indigo.

Since indigo blue substances were in high demand in Rome (and elsewhere) and there-fore expensive, it is understandable why the Roman state imposed taxes on the latrine operators who sold the metabolic product they collected. After all, tanners, dyers, and launderers were left in a difficult position if they ran out of this essential "latrine raw material"...

But where does the ammonia come from that made the clarified urine a valuable economic asset in those days? The answer is urea (Latin: urea), a carbonic acid diamide, which is produced during the breakdown of amino acids in living organisms and, unlike ammonia - which is extremely toxic to the organism - urea is relatively non-toxic. This is why, a hundred years ago, it was still common practice for doctors to taste urine to determine whether a patient was diabetic. If the urine tasted sweet, the patient was likely diabetic; if not, they were not. For a long time, it was believed that urea was a substance that could only be produced in organisms through their inherent "life force.

46. Friedrich Wöhler and the urea

But then one day Friedrich Wöhler (1800-1882) mixed silver cyanate with ammonium chloride and, to his surprise, obtained urea. This experiment, conducted in 1828, was groundbreaking and is now considered a milestone in the history of chemistry. It demonstrated for the first time that organic compounds, which had previously been thought to be products of a mysterious "life force" or "*vis vitalis*," could actually be synthesized by purely chemical means. With this experiment, Wöhler refuted the widely held theory that organic substances could only be produced by living organisms.

Fig. 14: Friedrich Wöhler, the founder of "organic chemistry" (Wikimedia)

Wöhler not only received scientific acclaim for this achievement but also a professorship, which provided him with a secure income and allowed him to continue his research.

Today, 1828 is regarded as the birth year of "organic chemistry," and the concept of "vis vitalis" became obsolete - no more "vitality."

47.   What is "life"?

Biologists, therefore, had to develop new theories to explain the phenomenon of "life." As we know, this is no easy task. At NASA's second astrobiology conference in 2002, over 100 characteristics of living beings were compiled that are absent in inanimate nature. The challenge is to determine which of these characteristics must be present for an entity to be recognized as "life." A definition of "life" must encompass a water flea, a cat like my cat Humpel, a mold, a dandelion, and, of course, a human being. Crucially, it must also apply to a "minimalist" living being, a primitive microorganism. This definition should ideally include every conceivable form of "life," not just those on Earth. Experience shows that this is where the difficulties arise.

48.   The riding primordial dwarf

Even the most primitive earthly life form, *Nanoarchaeum equitans* (commonly known as the "riding primordial dwarf"), is a creature that is far more complex than the earliest forms of life that appeared on Earth at the end of chemical evolution (abiogenesis). It is unlikely that this early life form simply emerged with a "beep" from something "inorganic" becoming "living." Instead, the transition to life must have involved a process of many individual, successive stages. It is also conceivable that "life" on Earth may have originated multiple times in parallel, but only one form of life - namely "the one we know" - survived and evolved into the diverse and complex biosphere that makes our planet unique.

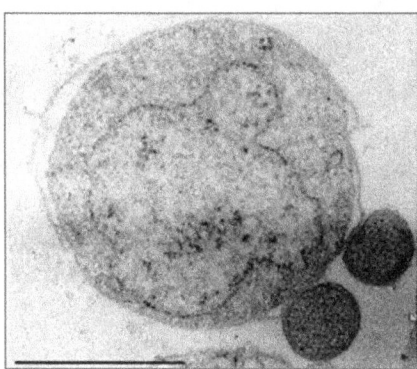

Fig. 15: Two Nanoarchaeum equitans, riding on the cell of Ignicoccus hospitalis (a thermophilic archaea) (Wikimedia)

49. When is life life at all?

Such a multi-stage process makes it difficult to attempt a universally valid definition of life, because it is necessary to explain precisely and justifiably at which stage the degree of organization of a living organism is reached - and even today's science cannot achieve this. This can be observed, for example, when attempting to formulate clear criteria for "minimalist life," such as Characteristic A = self-sustaining, Characteristic B = energy metabolism, Characteristic C = compartmentalization through a semi-permeable membrane, Characteristic D = ability to produce components (of the organism) itself, and so on. (Although there is no consensus on these features among the scientists working on this problem - and even less on their importance). If A+B+C+D defines "minimalist life," then A+B+C must still be "non-life" and A+B+C+D+E must already be "beyond minimalist life." An additional complicating factor is that the characteristics in the total can of course be permuted across all characteristics used for the decision... It is therefore sensible to approach the problem pragmatically. Admittedly, this means that a large number of "definitions" will continue to exist side by side, as they do today. However, there are various scientific and methodological ways to approach this astonishing phenomenon: chemically, biologically, in terms of systems theory, philosophically, and even physically, as the famous quantum theorist Erwin Schrödinger (1887-1961) once tried to do, for example.

Many "definitions" of "life" emphasize in particular the necessity of the existence of molecular information memories, which explicitly contain the blueprint of a living organism and which are able to pass on the information stored in them stably, i.e., quasi from generation to generation, over very long periods of time. The well-known evolutionary biologist Richard Dawkins expressed this in the most extreme terms when he emphasized that a living being is in principle only a vehicle for saving its genetic information in the form of DNA over time. Or to put it casually, *"the chicken is the method that nature has devised to transform an egg into an egg..."* Other definitions again emphasize the reproductive capabilities of living entities in connection with the above.

50. The mule paradox

And here's where things get complicated: The donkey is considered a living being, but a mule is not. This contentious insight is the crux of the "Mule Paradox," which has occupied biologists for some time. Mules, as is commonly known, are generally infertile. But who would dare to deny such a sympathetic and resilient animal the attribute of "life"? This leads to a broader problem when trying to encapsulate the phenomenon of "life" within a universally valid definition. To go beyond tautologies like *"life is an attribute of living systems,"* one must use terms that are just as difficult to define, such as complexity, information, and order, each with its own specific context. In summary, all attempts to define "life" remain highly contentious. It's not even clear whether there is a definitive definition that can adequately capture and represent all aspects of living matter.

But perhaps this is similar to the attempt to define, once and for all, what "music" is, for example. At least for the author, the answer is subjective yet pragmatic - "I know it when I hear it..." In fact, the question of when "sounds" become music and when "music" becomes noise is just as difficult to answer as the question of the transition between non-living and living nature at the end of chemical evolution around 3.8 billion years ago. This latter question might even be more difficult because the assessment of when sounds become "music" is only possible subjectively, while the second question can at least theoretically be answered objectively. I can certainly imagine that some modern orchestral works would have sounded like terrible noise to someone like Johann Sebastian Bach (1685–1750). But this also shows that music is a reflection of its time, a "Zeitgeist," as Johann Gottfried Herder (1744–1803) might say, reflecting the mentality of an era. If one wants to understand the world, one cannot avoid analyzing it historically.

51. Stupidity as a World Power

As Karl Jaspers (1883–1969) aptly put it: *"Anyone who only knows the present is bound to become stupid."* This brings us to the topic of intellectual poverty and a world power like no other: stupidity. Its opposite is, in a sense, intelligence. And nothing in the world is more equally distributed than intelligence, for everyone believes they have enough of it. Both stupidity and intelligence form a dialectical opposition, and both are universal human phenomena that can change the world - for better or worse. *"O sancta simplicitas!"* are said to have been the last words of Jan Hus (c. 1370–1415) as he was burned at the stake in Konstanz on July 6, 1415. If the severe consequences of this politically foolish act had been foreseen at that time, much suffering in the following decades could have been prevented. But on the other hand, if the Hussite Wars hadn't happened, other wars would likely have occurred instead. Ah, human stupidity...

Stupidity is, first and foremost, nothing more than a lack of judgment due to limited knowledge, where inadequate intelligence might be a factor, but it doesn't have to be. Therefore, in what follows, "stupidity" should not be understood as a pathological impairment, like idiocy or feeblemindedness, for which the affected individuals bear no blame. Instead, we are dealing with self-inflicted stupidity and (mostly unconscious) stupidity as a widespread way of life. This also includes the observation that many people exhibit extraordinary wisdom in one area while being shockingly stupid in another. But this is due to the high degree of specialization, as specialized knowledge is naturally more in demand in the workplace than general knowledge. Daily or general stupidity, on the other hand, manifests in a pathological disinterest in things that a wise person seeks out of an internal drive and for which they often make considerable efforts to intellectually engage. Incidentally, this was already noticed by Arthur Schopenhauer (1788–1860), who wrote in his collection of aphorisms, Parerga und Paralipomena:

"...most people, although not always consciously aware, fundamentally have in their hearts the supreme maxim and guiding principle of getting by with the least amount of thinking possible because thinking is a burden and a bother to them. Accordingly, they think just enough to get by in their occupations and then only as much as their various

pastimes, like conversations or games, require - both of which must be structured so that they can be engaged with a minimum of thought. If, during leisure hours, there is a lack of such distractions, they would rather spend hours at the window staring at the most insignificant events than pick up a book, because reading would require mental effort."

Notice anything? Nothing significant has changed since Schopenhauer's time. The bench in front of the window has simply been replaced by the armchair in front of the TV or the chair in front of the gaming console. Even with a smartphone, one can spend hours without any intellectual benefit. This brings us to a second profound insight, already articulated by Erasmus of Rotterdam (1466?–1536) in his Praise of Folly: *"Stupidity is timeless."* And as Christ famously said in his Sermon on the Mount (quoted here in Church Latin): *Beati pauperes spiritu quoniam ipsorum est regnum caelorum*, which, with a bit of cheek, can be loosely translated as "Dumb and happy" as an enviable state of mind (though admittedly, this is a somewhat mischievous interpretation...).

A relevant point to consider was raised by Robert Musil (1880–1942) in 1937 when he noted, *"Whoever speaks of stupidity naturally assumes they are not affected by it - or at least not to the extent that would prevent them from judging it clearly. For considering oneself clever is often seen as a sign of stupidity."*

And here is a keen observation, put into rhyme by Wilhelm Busch (1832–1908), which reflects deep insight into human nature and behavior:

> *Oftmals paaret im Gemüte*
> *Dummheit sich mit Herzensgüte*
> *während höh're Geistesgaben*
> *meistens böse Menschen haben.*

Stupidity is, first and foremost, an individual phenomenon, which is either learned (by refusing to "learn" - i.e., to acquire knowledge) or "indoctrinated." In the latter case, we speak of "dumbing down" - and if the aim is to intellectually suppress entire groups of people, of "mass stupefaction."

52. Mass stupefaction

Let's take a look at some examples of how "dumbing down" works - specifically as a citizen of the (former) country of "poets and thinkers." After all, we have an undeniably long and glorious tradition in this country when it comes to the art of mass stupefaction. So let's dig out some exquisite examples from our own cultural treasure chest:

Modern Media Landscape

When you watch most German-speaking private TV channels, you might conclude as an educated and neutral observer that the program directors have practically devised a secret plan to systematically undermine the intellectual abilities of their viewers. How else can you explain the seemingly endless flood of shows like "*Deutschland sucht den Superstar,*" "*Ich bin ein Star – Holt mich hier raus!*" (*The Jungle Camp*), and "*Berlin – Tag & Nacht*"? Here, a platform has been created where pseudo-celebrities and wannabe stars can showcase their incredible lack of talent and moral flexibility while the audience comfortably dulls their minds on the couch. Even the public broadcasters ("*Öffentlich-Rechtliche*") are no exception, actively contributing to the infantilization - especially in the realm of politics - beyond their responsibilities outlined in the Broadcasting Treaty. For many years, there has been a noticeable trend where intellectually free individuals increasingly avoid consuming ARD and ZDF, and if they do, they prefer channels like ARTE and (to a lesser extent) PHÖNIX...

German Education System

We live in a time when education is more important than ever. It says a lot about a so-called "nation of education" when universities and colleges - especially in non-liberal arts subjects - have to offer remedial courses at the start of studies to teach new students basic skills in math or German spelling, skills that should be a given after passing the Abitur (high school diploma). When it comes to STEM subjects, compare the admission requirements of a German university with those of universities in India or China, where additional tests must be passed to prove competence...

The once-leading German educational landscape (especially during the German Empire) now suffers from three major issues, all of which are political in nature. The first problem is a combination of federal fragmentation, chronic underfunding (evident in dilapidated school buildings, overcrowded classrooms, and a severe teacher shortage), and (mostly pointless) over-bureaucratization.

Due to the specific problems in large cities (such as the significant migrant communities with language barriers), the second issue is growing educational inequality, where the educational opportunities for all children are no longer ensured. This inequality manifests on multiple levels and leads to a deep division within society.

On one side, there are children from wealthier families living in good neighborhoods with access to high-quality schools; on the other, there are children from socially disadvantaged families, often with a migration background, growing up in poorer districts. The latter often attend poorly equipped schools with overwhelmed teachers and insufficient support to compensate for language deficits or other educational disadvantages.

The German system of early selection after the fourth grade also contributes to educational inequality. Children are divided into different school types - Hauptschule, Realschule, Gymnasium - often exacerbating social differences. Children from academically oriented families are more likely to attend Gymnasium, while children from less educated or migrant families are often relegated to Hauptschule or Realschule. This early separation cements social inequalities and hinders equal opportunities.

Politically driven educational inequality is thus the second major issue of the German education system. It not only leads to individual disadvantages but also threatens social cohesion and peace. A society that fails to provide equal educational opportunities for all children squanders its future and the potential of its younger generation. The dumbing down of the populace, here in the form of systematic discrimination and exclusion of entire population groups, is the sad result of misguided policies.

The last issue is an educational policy aspect, where long-standing, proven didactic methods and content were suddenly replaced by hasty reforms and "innovative" approaches, without sufficient testing or scientific basis. The (German) school system increasingly became an experimental field for eccentric educators at university pedagogical departments, exemplified by concepts like "writing by ear" (thankfully a complete failure) or "competency-based learning" (instead of acquiring fundamental skills in arithmetic and writing). Just take a look at "modern" math textbooks with their many colorful pictures and often childish examples and exercises that convey anything but the beauty of an exciting and exact science.

In short, German schools have been turned into experimental fields through uncoordinated and often ideologically driven reforms, with students bearing the brunt. Proven didactic methods were recklessly discarded to make way for questionable innovations. The consequences of this reform frenzy are unstable learning environments, insecure teachers and students, and an increasingly fragmented and inefficient education system. Dumbing down, in this case through the uncritical adoption and implementation of untested pedagogical theories, is the sad result of a misguided education policy. But maybe that's even intentional. After all, smart citizens are always a threat to certain political circles due to their right to vote...

Fig. 16: It says on the stamp...

Politics

Politics actively promotes the dumbing down of the populace in Germany through a series of actions and failures, often hidden behind pleasant-sounding terms and popular buzzwords. A prominent example is the so-called "double whammy" ("Doppelwumms"), a term for massive government spending programs launched in response to crises like the COVID-19 pandemic or the energy crisis. The use of terms like "double whammy" or "bazooka" in political discourse deliberately contributes to the infantilization of public debate. Such childish and catchy terms are chosen to present complex political measures in a simple and memorable way. While this may be effective in the short term for attracting attention, it leads to a simplified and often distorted view of serious and multifaceted issues. This prevents a deep engagement with the actual content and consequences of political actions, making it easier for politicians to push through their ideological fantasies, regardless of reality. Just think of the energy transition, climate policy, or the (highly controversial) COVID-19 measures (keyword: "RKI Files").

53.   Is Stupidity Harmful?

The question of whether stupidity is harmful is certainly an interesting one. It's not a question that can be answered with a simple yes or no. Even in today's civilized and so-called "highly cultured nations," a certain degree of stupidity might actually be more beneficial than detrimental to the survival of the individual (Horst Geyer, 1907–1958). Additionally, one must distinguish between the personal harm it causes (which the individual usually doesn't even notice) and the harm that foolish decisions can inflict on others. Ignorance and the lack of critical thinking - something Kant encapsulated in the phrase "Have the courage to use your own understanding!" - are still shamelessly exploited in politics today. These shortcomings are used not for the common good but to push decisions that serve particular interests or various ideologies. Sometimes it seems that the old saying, "Keep them stupid, and I'll keep them poor," remains a guiding principle of political influence.

"Stupidity" becomes dangerous to society when "stupid" people ascend to decision-making positions for which they are intellectually unqualified. They often no longer base their decisions on objective realities *("We are surrounded by reality!")* but on ideological wishful thinking. In doing so, they have the real potential to destroy a country or economy from within, as one can currently observe with a sense of amazement in the case of "Germany." And a dumbed-down populace is the perfect enabler for this process (thus completing the circle)

54.   Then Why Don't You Move to Fukushima!

Take, for instance, the so-called "Energiewende" (energy transition), a topic that could warrant an entire treatise on its own. For those who are experts in the field - by which I mean people who understand how energy generation and distribution systems work and what it means that at any given moment, exactly as much electrical energy must be supplied as is being "consumed" at that same moment - there is a glaring disconnect between the technical reality and the opinions and perceptions held by the majority of the public. The confusion between installed capacity (nominal power) and available capacity, particularly concerning "renewables," is just one of the many "cognitive errors" that complicate a factual discussion about the energy transition. This often causes such discussions to quickly devolve into quasi-religious declarations of belief.

Consider the following observation: When reading forums where the (very much debatable) pros and cons of nuclear energy are argued, it's easy to spot the participants who are woefully uninformed. These individuals often respond to objections with the overly "clever" phrase, *"Then why don't you move to Fukushima!"* - and contribute nothing else of substance to the discussion. And they are not few in number. In topics like this, rational arguments generally don't get very far because they require too much effort to understand. People prefer to fall back on vague "gut feelings" or resort to personal attacks.

What can be observed on this smaller scale certainly has larger implications when it comes to, for example, swaying majorities.

55. Democracy

And democracy, when you put it differently, really just means "the struggle for majorities," because in a democracy, it's ultimately the "majority" that decides the direction things will take. And if you're outside the majority, such as in the symbiosis of "lobbyist and politician," then you'll engage in "shaping opinions" in your favor, or have it done by your supporters. For this purpose, you have access to the press, radio, and, of course, television, as well as the internet - though the latter, with its uncontrollable variety of information, is often seen as a threat. But this threat is manageable. Those who get their opinions and knowledge solely from the big headlines of the tabloid press or manipulative TV news are unlikely to search the internet for alternative views or perspectives on a topic and form their own, still subjective but more informed and not entirely externally controlled opinion. And since this seems to be the majority, and politics is all about majorities, the masses, as sad as it may be, remain easily manipulated. As Andreas Tenzer once said: *"For a people that dares to use its own intellect, democracy is the best of all forms of government; for a dumbed-down people, it is the worst."*

56. Democracy and the Pitfalls of Electoral Law

But is democracy per se an ideal form of government? The term "democracy" etymologically means "rule of the dēmos" - the people of the state. And what is the best way to organize the rule of a state's people? By electing representatives endowed with corresponding powers who each represent a segment of this "people" both in personnel and programmatically. The majority forms the government - and the majority (or the "people's representatives") is determined through free and secret elections. So much for the theory.

Let's consider the pure form of the majoritarian electoral system, as it was introduced in South Korea in 1987 following the end of a brutal military dictatorship. At that time, there were three presidential candidates: Roh Tae Woo (Democratic Justice Party, 1932–

2021), Kim Dae Jung (1924–2009), who would later win the Nobel Peace Prize and become President of Korea, and Kim Young Sam (1927–2015), both representing the "New Korean Democratic Party" - two liberals and one staunch conservative militarist. The people yearned for freedom, and as a result, the two liberal candidates received almost two-thirds of the votes. However, the conservative military representative fell well below 50%, yet he still secured the presidency. This outcome clearly did not reflect the will of the people, but such a result is not uncommon under a majoritarian electoral system. What followed was predictably chaotic: the losing opposition denounced this blatant "electoral fraud" and demanded new elections. Radical students responded, as usual, with violent protests, which were, as usual, suppressed with the customary severity by the police. All this turmoil was caused by an electoral system that apparently only seemed fair. It has been known at least since 1785 that no electoral system can truly represent the will of the voters in a fully representative way. This is due to purely logical and mathematical reasons.

57. The Condorcet Paradox

Here, the "election researcher" refers to the "Condorcet Paradox," named after Marie Jean Antoine Nicolas Caritat, Marquis de Condorcet (1743–1794). He demonstrated during the French Revolution that the outcome of an election can depend not only on the votes cast but also on how they are counted. This is why electoral systems (and there are many!) in democratic and pseudo-democratic countries around the world are often extremely complex, and attempts are made to obscure or correct obvious injustices through systems of overhang mandates or by implementing second and third rounds of voting.

In many countries, including Germany, the focus is ostensibly on voting for parties rather than individual candidates, with each party presenting its specific election program. However, the contents of these programs often fade from memory quickly after the election. This leads to a problem that can potentially undermine the system. It is best explained with an example, assuming there are only two competing parties, each with a different election program as an offer to voters. For clarity, let's use the CDU and the SPD, with the following "assumptions" intended only as examples.

The main campaign issues are: 1. Attitude towards Russia; 2. Energy transition; and 3. Subsidy policy towards the economy. The CDU advocates 1. no support for Russia, 2. a reduction in energy costs, and 3. a reduction in subsidies. The SPD supports (in this example!) exactly the opposite. After the election, four voter domains can be identified: Domain A (20% of votes) – no support for Russia, increase in energy prices, maintenance of subsidies; Domain B (20% of votes) – support for Russia, reduction in energy prices, maintenance of subsidies; Domain C (20% of votes) – support for Russia, increase in energy prices, reduction in subsidies; Domain D (40% of votes) – no support for Russia, reduction in energy prices, reduction in subsidies. This results in the following distribution of voter support for the campaign issues: 60% of voters oppose support for Russia, 60% support a reduction in energy prices, and 60% favor a reduction in subsidies. On the other hand, Voter Group A voted for the SPD because they wanted higher energy costs and maintenance of subsidies. Voter Group B also voted for the SPD because they wanted to support Russia and were against subsidy cuts. Likewise, Voter Group C voted for the SPD because they were okay with higher energy costs and supporting Russia. Only Voter Group D, with 40% of the votes, chose the CDU, as they agreed with all the points in the CDU's program. Evidently, the SPD wins with 60% of the vote, even though 60% of voters oppose each of the SPD's individual campaign promises! In none of the issues does this party, in this example, represent the majority. No wonder voters feel deceived.

I will leave it at that for now, in accordance with the saying "Si tacuisses, philosophus mansisses" (If you had remained silent, you would have remained a philosopher). However, I must address a folly that is still treated like science in some circles today, following this brief excursion into the depths of democratic government: astrology.

58. Astrologers, Horoscopes, and Scientific Education

One of the greatest scientists of the German-speaking world (if not the greatest of all time!), Johannes Kepler (1571-1630), had to make a living from astrology despite not being entirely convinced by it. He is credited with the following quote:

„Es ist wol diese Astrologia ein närrisches Töchterlein … aber lieber Gott | wo wolt jhr Mutter die hochvernünftige Astronomia bleiben | wenn sie diese jhre närrische Tochter nur hette | ist doch die Welt viel närrischer | und so närrisch | daß deroselben zu jhrer selbst frommen diese alte verständige Mutter die Astronomia durch der Tochter Narrentaydung | … | nue eyngeschwatzt und eyngelogen werden muß | Auch sind sonsten der mathematicorum salaria so seltsam und gering, daß die Mutter gewißlich Hunger leiden müßte, wenn die Tochter nichts erwürbe."

It can be found in his 1610 publication in Frankfurt am Main, titled *Tertius interveniens, das ist die Warnung an etliche Theologos, Medicos und Philosophis, sonderlich D. Philippum Feselium, daß sie bey billicher Verwerffung der Sternguckerischen Ab-erglauben / nicht das Kindt mit dem Badt außschütten / und hiermit ihrer Profession vnwissendt zuwiderhandlen*. Anyone interested can read it in the original.[14]

Astrology is commonly understood as the esoteric belief that the positions of celestial bodies (especially the planets) at the time of a person's birth can predict their fate and personality traits. This "popular" and still very prevalent doctrine is far more complex in its "serious" form than one might assume from reading horoscopes in tabloids. It has a long history and tradition that dates back to the Babylonians and Assyrians and was only methodically separated from astronomy at the beginning of the modern era (at least in the Old World). Most astronomers of the late Middle Ages and early modern period were

---

[14] https://www.deutschestextarchiv.de/book/show/keppler_tertius_1610

also often recognized astrologers, as astronomy and astrology were almost always taught together in academic education at that time.

Fig. 17: Johannes Kepler

The first major codification of astrology, attempting to postulate physical transmission mechanisms affecting human fate, comes from Claudius Ptolemy (circa 100 to 175 AD). In his work Tetrabiblos (the "Four Books," around 150 AD), he provides a summary of Hellenistic astrological concepts, linking them with the prediction of astronomical phenomena such as specific planetary configurations, lunar phases, eclipses, etc. Like many of his successors, Ptolemy was aware of astrology's fatalistic nature, which, in its consistent form, denies free will. This dilemma of classical astrology could only be mitigated but never fully resolved and remains a major counterargument against an astrological worldview today (Why do the life paths and characters of twins born under the same "star signs" often differ fundamentally?).

Doubts about the declared functionality of astrology began to arise during the Renaissance. For example, Giovanni Pico della Mirandola (1463–1494) published statistical studies in his Disputationes adversus astrologiam comparing astrological weather predictions with "actual weather." Johannes Kepler, as mentioned, spent a significant part of his livelihood creating horoscopes and had some doubts about the fateful significance of the stars, although he did not completely reject astrology. The strongest arguments against astrology do not stem from its lack of empirical confirmation but from its internal logical structure, making it an epistemological issue. The question is whether it is logically possible to imbue astrology with content that translates the claimed influence of celestial bodies on human traits into a comprehensible and verifiable chain of effects, which can then be subjected to the criteria for scientific theories formulated by Karl Popper (1902–1994). From a scientific standpoint, this question must be answered in the negative. This is evident from the fact that astrology uses symbols and relationships between these symbols, which, upon closer inspection, prove to be artificial conventions treated like dogmas (such as the zodiac signs). For example, it is highly unclear why only the zodiac constellations, in conjunction with the planets, sun, and moon when located within them, are supposed to have effects on people, while other notable objects like the constellation Orion or the star Sirius are not. It is therefore accurate, despite the scientific methodology used by astrologers in creating horoscopes, to describe astrology as a pseudoscience (or "cargo cult science," which I will return to in Point 447). This assessment is warranted because astrology is inherently resistant to intersubjective verification and lacks openness to empirical falsifiability.

The strong interest in astrology today can be easily explained by the human desire to know about one's future fate. In contrast, the scientific response is rather sober. Complex systems, such as human society, in which every individual is embedded, are fundamentally unpredictable. Astrology offers an apparent life aid, which need not always be viewed critically. Where scientific education is underdeveloped or lacking, which applies to the majority of humanity, people will naturally be guided more by esoteric than scientific considerations in everyday life. And a horoscope is an ideal tool for this purpose. Its popularity among more simple-minded individuals partly stems from the fact that at least the horoscopes found in various tabloids (which have nothing to do with "real" horoscopes!) seemingly have a high success rate. This is not surprising, as their statements, produced en masse by professional (read "greedy") astrologers, are very general, subtly positive, and create a kind of recognition effect. This is based on the psychologically understandable tendency to perceive vague, general, and subtly positive statements about oneself as accurate.

59. Barnum Statements

Such statements are commonly referred to as Barnum statements. Their name comes from Phineas Taylor Barnum (1810–1891), the founder of Barnum's American Museum in New York. Barnum started his career as a lottery seller and newspaper publisher before turning to show business. In 1842, he established his "American Museum" in New York, a conglomeration of a curiosity cabinet, a zoo, waxworks, and a vaudeville theater. It was highly successful, but unfortunately, it was completely destroyed by a fire on July 13, 1865. The New York Times reported extensively on the fire, estimating the damage at over one million dollars - at a time when a dollar still had significant value! In short, the damage was immense, and it wasn't until the year 2000 that the museum could be reopened under the auspices of the City University of New York. Barnum, who had sought his fortune in politics after the fire, did not live to see this momentous event, having died 109 years before it. He would certainly have enjoyed it. But back to the great fire.

Fig. 18: Great Fire on July 13, 1865 - Barnum's American Museum in Flames!

60. Chang and Eng Bunker from Siam

For a short time, Chang Bunker and Eng Bunker (1811–1874) from Tambon in Siam (now part of Thailand) were out of work, as they regularly performed there. However, their unemployment was brief, as they were a genuine attraction and had learned to market themselves profitably. Chang and Eng were conjoined twins, which is not particularly unusual in itself. What made them special was their birthplace, Siam, and the fact that they were joined at the side. In short, Chang and Eng Bunker were Siamese twins. Due to this quirk of nature, they were destined to spend their lives literally together. At the age of 17, they came to the United States and performed in various shows, including Barnum's American Museum. Despite or perhaps because of their condition and their business acumen, they were well-regarded individuals. At 30, they married a pair of sisters and settled in North Carolina. They fathered a total of 22 children. In 1870, they visited Germany and met the renowned physician Rudolf Virchow (1821–1902), who proposed surgically separating the brothers. However, since it was not known at the time whether they might share some organs (which was not the case, as an autopsy after their deaths clarified), the idea was abandoned. Nowadays, such procedures are more feasible.

Fig. 19: Chang and Eng Bunker (1811-1874) from Siam

61. Separation of Siamese Twins

Siamese twins are indeed born from time to time, with a probability of about 1 in 1 million. The most common type of conjoined twins are those with chest fusion, like Chang and Eng. These can often be relatively easily separated surgically, just like twins joined at the hips. Nowadays, attempts are also made to surgically separate children joined at the head. A particularly notable case was the operation on the twin sisters Ladan and Laleh Bijani from Iran. For anatomical reasons, separation was long considered completely impossible. It was not until a team of doctors from Singapore saw potential for success that an attempt was made in 2003. Unfortunately, the attempt was a complete failure. Due to severe blood loss during the procedure conducted by 28 doctors, both sisters died of circulatory failure.

Siamese twins can also occur in animals. In particular, farm animals with two heads (which are not "true" Siamese twins but suffer from dicephaly) occasionally make headlines, as the press, which is very keen on such "sensations," tends to cover them extensively. They promise significant attention and thus high circulation, much like disaster reports and gossip stories.

62. Daily Newspapers and Journals

The world's first "true" daily newspaper (which already focused on such content) was published in Leipzig in 1650. The widespread adoption of printing, based on Johannes Gutenberg's (1400-1468) invention, made it possible to publish a "newspaper" every day of the week (except Sunday, as the Church was still opposed at that time). Before this, "Zeytungen" (which means "news") were either published irregularly or at most once a week. Their distribution was usually very local and limited to larger cities and residences. This is often reflected in their names, such as the "Wiener Zeitung," which has been published continuously since 1703, and the oldest newspaper in Germany, the "Hildesheimer Allgemeine Zeitung" (since 1705).

During the Enlightenment, newspapers - and even more so, the more specialized magazines ("Journals") - became increasingly popular, and reading them developed into a regular weekend activity among educated circles. In this way, the enlightened citizen was continually supplied with topics for refined conversations, as vividly illustrated in Goethe's "Faust" in the section about the "Easter Walk":

„Nichts Bessers weiß ich mir an Sonn- und Feiertagen, als ein Gespräch von Krieg und Kriegsgeschrei, wenn hinten, weit, in der Türkei, die Völker aufeinander schlagen. Man steht am Fenster, trinkt sein Gläschen aus und sieht den Fluss hinab die bunten Schiffe gleiten; Dann kehrt man abends froh nach Haus, und segnet Fried' und Friedenszeiten."[15]

And the attitude expressed in these lines is likely familiar to the discerning reader. From the very beginning, newspapers and magazines pursued two primary goals: to engage readers and to generate revenue through advertisements and inserts. Additionally, they have served as ideal platforms for propaganda. Depending on the intended audience, this led to the development of two main types of press: serious informational press, which focuses on topics related to politics and economics, and gossip press, which caters to entertainment purposes or the informational needs of less discerning minds.

---

[15] On Sundays, holidays, there's naught I take delight in,
Like gossiping of war, and war's array,
When down in Turkey, far away,
The foreign people are a-fighting.
One at the window sits, with glass and friends,
And sees all sorts of ships go down the river gliding:
And blesses then, as home he wends
At night, our times of peace abiding.

Fig. 20: Since 1705, the "Hildesheimer Allgemeine Zeitung" (originally titled "*Hildes-heimer Relations-Courier*") has been providing its readers with news, advertisements, and various pieces of information. As of 2024, following the cessation of the "Wiener Zeitung," it is now the oldest daily newspaper in the world.

63. Boulevard and Public Dumbing Down

Today, the term "boulevard" has come to denote a more refined category of media. It now includes not only print publications but also occupies a significant portion of tele-vision airtime. This complements the aim of "public dumbing down" (see Point 52) with a particularly effective approach: shaping ordinary people into a politically agreeable, contented, and ideally non-rebellious electorate. Boulevard media primarily focus on "money, boredom, and the disruption of social peace," often through the stimulation of fear, schadenfreude, and envy. They thrive on banality, trivialities, and private matters of little public interest - often only elevated due to the prominence of the individuals involved. This content, frequently marked by foolishness and second-hand embarrassment, guarantees high circulation for publishers and high viewership for television channels. Interestingly, one could even use these figures to create an index that might compare the mental health and societal health of different nations.

Fig. 21: The boulevard press has been increasingly satisfying less of the public's informational needs since the rise of alternative media on the internet (unless, of course, they are sitting at the hairdresser's).

64. „Quality Press"

The "quality press" stands apart from the "tabloid press" in its commitment to providing verifiable facts and serious journalism. Publications like the Washington Post, the Frankfurter Allgemeine, and the Weltwoche are examples of this "serious" press, which, in principle, offers factual information, clearly distinguishes between commentaries and opinions, and avoids excessive ideologizing or propaganda. It aims to respect the intelligence of its readers rather than treating them as naive.

However, this distinction is increasingly questioned, especially in the realm of political reporting. Many readers now have access to a variety of alternative information sources, such as the Internet, and are often proficient in foreign languages, which facilitates the verification of news and claims. This access has highlighted discrepancies between public opinion and what is reported in the media, leading to the term "lying press," which resonates particularly with politically aware and informed individuals.

In Germany, for instance, this issue has been evident in the coverage of the Euro crisis, the COVID-19 pandemic, the Ukraine conflict, the Greek financial crisis, the fight against the Islamic State, the refugee crisis, and the strained relationship with Russia. The term "quality press" has thus come to be seen by some as a dysphemism. With the decline in print media's influence, this shift is detrimental to the economic success of newspapers and magazines. Yet, this reality seems to be acknowledged only slowly by some editorial offices.

65. Sapere aude! - The Best Defense Against Propaganda

An opinion should be formed by the reader themselves, in the spirit of Enlightenment (sapere aude!), and not be presented ready-made for uncritical adoption, like feeding silage to a cow. These principles of good journalism are currently being undermined (with a few commendable exceptions[16]). This includes concealing vested interests, trimming or omitting information, and uncritically accepting information (as can be demonstrated by many examples during the Ukraine crisis, where some editorial offices didn't even check for plausibility). The demonization of individuals and, particularly in online media, the disabling of comment functions (of course, in the name of quality assurance) are also becoming more prevalent. Moreover, headline-grabbing false reports are increasingly sold as "truth" on the front page, and when part of the audience (and knowledgeable people) recognize these falsehoods (in modern terms, "fakes"), the editorial offices often don't consider it necessary to issue a correction (and if they do, it's often brief and buried in the editorial section - after all, as Plutarch knew: *Calumniare audacter, semper aliquid haeret*[17]).

Journalists are likely aware that political conflicts can indeed be written into existence. When an unchecked rumor or a false report propagated by interested parties is presented as "news" that the recipient (the reader) is expected to believe, and this creates a mood leading to corresponding reactions, things can quickly spiral out of control. After all, such rumors and falsehoods have been used to spark wars that could never have been sold to the public without appropriate journalistic and propagandistic preparation. I think readers can come up with some relevant examples from recent history themselves.

---

[16] I'm thinking specifically of the Swiss Weltwoche.

[17] "Accuse boldly; something always sticks!"

66. Self-Fulfilling Prophecies

Rumors or false reports, if circulated often enough, can sometimes evolve into self-fulfilling prophecies. This concept refers to "*an assumption or prediction that, by virtue of being made, causes the assumed, expected, or predicted outcome to become reality, thus confirming its own correctness*" (Paul Watzlawick, 1921-2007). The "most famous" example is the bank rumored to have financial difficulties. If this rumor spreads, many customers might feel compelled to withdraw their money as quickly as possible, which can then lead to the bank actually experiencing financial difficulties, and so on.

67. Oedipus and the Oracle of Delphi

A "classic" example is the story of Oedipus, which can be summarized as follows: The Oracle of Delphi once predicted to his father:

"If you ever have a son, he will kill you and marry your mother."

When Oedipus was born, he was abandoned as a child to die in the wilderness, so he never met his biological father or mother. However, he was rescued and grew up in a foreign family (the family of King Polybus of Corinth, for those who want the details), where he grew strong and became a young man.

Upon hearing a rumor that he was not actually the king's son, he was troubled and decided to travel to Delphi, the sanctuary of Apollo at the foot of Mount Parnassus, to consult the Oracle. The Pythia, however, did not provide a satisfactory answer and only told him that he would kill his father and marry his mother. To avoid this fate (believing his parents were in Corinth), he left his hometown for Daulis. Along the way, he had a quarrel with the driver of an oncoming carriage. The outcome, known for over 2,700 years, is that Oedipus killed the driver and his companion. Unfortunately, the companion was his father Laius, thus fulfilling the first part of the prophecy. The continuation of the story is well known, as otherwise, Gustav Schwab (1792-1850) would not have been able to recount it so compellingly.

Oedipus succeeds in solving the Sphinx's riddle and is rewarded with the queen, Jocasta, as his wife. As it turns out, she was previously Laius's wife and thus his mother. Of course, he was unaware of this at the time, as his belief in oracles was evidently quite limited compared to that of Croesus (see below). Consequently, Oedipus fathered twins with his mother (not Castor and Pollux, but Eteocles and Polynices).

Fig. 22: Die Pythia auf ihren dreibeinigen Schemel beim „Orakeln"... (Wikimedia)

However, disaster soon struck the land in the form of a devastating plague, and the Oracle of Delphi declared that it could only be averted if Laius's murderer was found. As was to be expected, the blind seer Tiresias revealed that Oedipus was the murderer. Jocasta subsequently hanged herself, and the details of what happened to Oedipus thereafter are not fully known. Various accounts exist about his later life and death, allowing everyone to choose their preferred version of the story.

The fate of Oedipus has been extensively explored in art and literature both in antiquity (by Sophocles, Aeschylus, Euripides) and in modern times.

68. Jim Morrison and the Oedipus Complex

A passage in Jim Morrison's (1943-1971) famous song "The End" ("*This is the end, my only friend...*") is associated with the Oedipus Complex. This term refers to the libidinous attachment of a boy to his mother, accompanied by jealousy and hostility towards his father, as an expression of childish sexuality. According to Sigmund Freud (1856-1939), this complex is believed to explain various neuroses in adulthood, though this theory is controversial.

69. At the Foot of Mount Parnassus...

In the story of Oedipus, the Oracle of Delphi plays a significant role at several points. Delphi was one of several oracle sites in ancient Greece but was considered the most important spiritual center around 700 BC. It can still be visited today, though all that remains are ruins and an actress mimicking the Pythia alongside the obligatory cashier at the entrance to the sanctuary - a faint echo of the time when Delphi was the challenging destination for many famous kings such as Laius, the King of Thebes, Oedipus, his son, King Midas of Phrygia, and, of course, Croesus, the last king of Lydia. I would now like to recount something I find quite interesting about the last two, as their fates were significantly influenced by the Pythia's pronouncements.

It's important to know that Delphi was a sanctuary of the god Apollo, which was established at the foot of the mighty Mount Parnassus (incidentally, the Apollo butterfly, Parnassius apollo, lives there). The famous historian Plutarch (45-125 AD), from a certain temporal distance, has left us records of how things went between the 6th and 5th centuries BC. Initially, the oracle operated only once a year, but as it became clear that "oracling" was highly profitable for the people of Delphi (they had to build extra treasuries to store the offerings from pilgrims), the high priestess, known as the Pythia, would dress in traditional white robes and enter the inner sanctum of the temple. There, she would sit on a highly uncomfortable three-legged stool, while being enveloped in a hallucinogenic mist rising from a fissure. In a trance, she would answer the questions shouted to her by the priests, with the belief that the answers were whispered directly to her by Apollo. These answers were often highly interpretable, if not completely incomprehensible.

70. The Misfortune of Croesus

One who was deeply convinced of the truthfulness of the Delphic Oracle - after having put the Pythia to the test - was the king and bon vivant Croesus (590-541 BC) from Lydia. Here's how it went. When he had the Pythia guess his favorite dish, she spoke the following words, cast in ancient Greek hexameters and rendered here in German:

„Wohl weiß ich, wieviel Sand im Meer, wie die Weite des Wassers, selbst den Stummen vernehm' ich und höre des Schweigenden Worte. In dem Sinne dringt mir der Geruch der gepanzerten Kröte, wie man sie kocht zusammen mit Lammfleisch in eherner Pfanne, Erz umschließt sie von unten, wie Erz auch darübergezogen."[18]

---

[18] "I know well how much sand is in the sea, how vast the water is; even the mute I hear and the silent I listen to. In this sense, the smell of the armored toad reaches me, how it is cooked together with lamb meat in a bronze pan, bronze encloses it from below, as bronze is also spread over it."

And indeed, lamb with turtle was Croesus's favorite dish. So it was not surprising that when he planned a military campaign against the Persian king Cyrus II, he first inquired about its outcome from the Delphic Pythia. And the answer was crystal clear to him:

„If you cross the Halys, you will destroy a great empire."

His misfortune was that he could not imagine that it was his own empire he was destroying by crossing the border river to Cappadocia, the then boundary with Persia. It seems like a avoidable misunderstanding today. The Pythia's wise sayings were indeed ambiguous…

71.   The Gold of Midas

King Midas of Phrygia, the legendary figure, is said to have consulted the Oracle of Delphi frequently and left many gifts there. However, he is more famously known for another event. According to legend, one day a few farmers found a friend of the god Dionysus lying on the ground, asleep from his drunken state, and they dragged him to King Midas. Midas received him warmly and invited him to a feast before returning him to Dionysus. The god, grateful for the hospitality shown to his friend, granted Midas a wish. In his naivety and greed, Midas wished that everything he touched would turn to gold. The wish was generously granted.

Soon, it became apparent that this wish led to certain, previously unconsidered problems - indeed, existential problems. For he could neither eat nor drink, as everything he touched turned to gold, just as he had wished. Desperate, he went back to Dionysus to beg him to lift the curse. Dionysus advised the king to wash himself in the river Pactolus, which, according to tradition, indeed worked. Since then, this river near the Aegean coast has been known for its abundant gold deposits. The story serves as a lesson to those who hear it: *"Gold (or money) cannot be eaten."* However, it should also be noted that gold can still be somewhat helpful with eating, as in the case of a gold crown in one's dentures…

72.   Programming with Delphi

While the oracle site of Delphi has been around for about 2800 years, the programming language "Delphi" has only existed since 1995. It is an object-oriented derivative of Pascal (the procedural programming language "Pascal" - named after the mathematician Blaise Pascal - was developed in 1971 by Niklaus Wirth at the Swiss Federal Institute of Technology in Zurich) with a particularly strong database connectivity, initially through the BDE (Borland Database Engine).

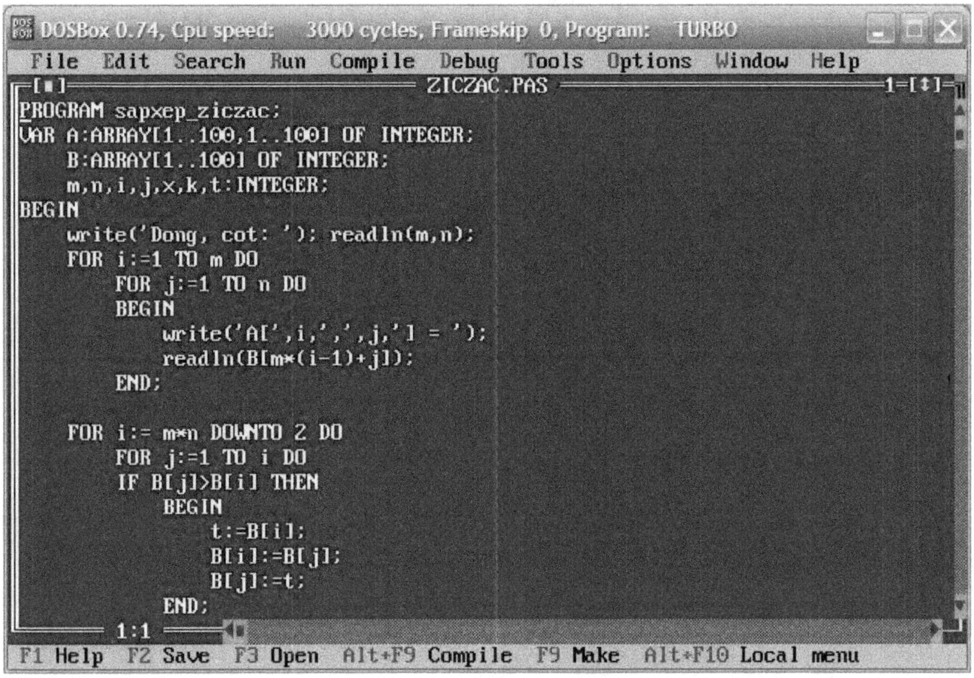

```
DOSBox 0.74, Cpu speed:    3000 cycles, Frameskip 0, Program:   TURBO
 File   Edit   Search   Run   Compile   Debug   Tools   Options   Window   Help
┌[■]═══════════════════════════ ZICZAC.PAS ═══════════════════════════1=[↕]═┐
│PROGRAM sapxep_ziczac;                                                      ▲
│VAR A:ARRAY[1..100,1..100] OF INTEGER;                                      ▓
│    B:ARRAY[1..100] OF INTEGER;                                             
│    m,n,i,j,x,k,t:INTEGER;                                                  
│BEGIN                                                                       
│    write('Dong, cot: '); readln(m,n);                                      
│    FOR i:=1 TO m DO                                                        
│        FOR j:=1 TO n DO                                                    
│        BEGIN                                                               
│            write('A[',i,',',j,'] = ');                                     
│            readln(B[m*(i-1)+j]);                                           
│        END;                                                                
│                                                                            
│    FOR i:= m*n DOWNTO 2 DO                                                 
│        FOR j:=1 TO i DO                                                    
│        IF B[j]>B[i] THEN                                                   
│            BEGIN                                                           
│                t:=B[i];                                                    
│                B[i]:=B[j];                                                 
│                B[j]:=t;                                                    
│            END;                                                            
│            END;                                                            ▼
└─── 1:1 ═══════◄█                                                     ►──────┘
 F1 Help  F2 Save  F3 Open  Alt+F9 Compile  F9 Make  Alt+F10 Local menu
```

Fig. 23: Here's how the program editor of the legendary "Turbo Pascal" once looked...

Since in the mid-1990s there was only one reasonably powerful database management system available, namely "Oracle," this new programming environment based on Borland's Object Pascal was aptly named "Delphi." It is still around today and can be used to write extremely powerful Windows and Linux programs. However, the company "Borland" has not existed since 2009.[19]

73.   Blaise Pascal and his Pascaline

Blaise Pascal (1623-1662) did not invent the computer, but he did invent a mechanical calculating machine. He created it as early as 1642. One of these machines can be viewed at the Mathematical-Physical Salon in the Dresden Zwinger.

Since building these mechanical calculators, known as "Pascalines," was very complex and people were generally better at mental arithmetic back then than they are today, the machines saw little widespread use. Moreover, the patent Blaise Pascal received for this invention did not make him wealthy.

---

[19] Today, "Delphi" is developed further by Embarcadero Technologies, Inc., based in Austin, Texas. There is also an open-source variant that is similarly powerful: Free Pascal.

Fig. 24: A Pascaline from 1652 (Wikimedia)

74. Invention of the Omnibus

Another invention by Blaise Pascal, which he managed to bring to fruition shortly before his death in 1662, has survived in a much more modern form to this day: the horse-drawn omnibus (shortened to "*vehicle for all*"), which operated on a fixed route with set times, making its arrival and departure times predictable (unlike, as is well known, with Deutsche Bahn). It was referred to as the "five-sou carriage" (*carrosses à cinq sols*) and, drawn by two horses, serviced several districts of Paris.

Fig. 25: Berlin Horse-Drawn Omnibus of Line 19[20] (Source Bundesarchiv)

---

[20] Today, Line 19 operates between Grunewald S-Bahn Station and Mehringdamm U-Bahn Station.

75. Why Paris is Called "Paris"

No, the name Paris for the city "Paris" has a different origin than most people believe (see Point 76) - and this can be precisely traced back to the well-known historical work of Gaius Julius Caesar (100 – 44 BC), *De Bello Gallico* (in English, "The Gallic War"), which you can read for yourself at any time.[21] The great Roman general and statesman Julius Caesar reports that the Gallic king Vercingetorix, along with the warriors of the Celtic tribe of the Parisii, fought against Caesar's legions. They left their native settlement Lutetia on an island in the Seine and, as expected, were defeated by the Romans. The Romans then took over their settlement, expanded it along the riverbank, renamed it "Lutetia Parisiorum," and developed it into a city that, for its time, was truly modern, and was later simply called "Paris."

A Roman who found himself there, whether privately or on official business, was considered to have struck it lucky. An aqueduct supplied the city with fresh water, baths invited people to bathe, and several theaters provided a cultural environment. The economic conditions in and around the settlement were also excellent for the time. This laid the foundation for the city's great future. However, there was no metal tower held together with 2,500,000 rivets back then...

After the fall of the Western Roman Empire, Frankish kings resided there starting in 508.

76. Trojan Troubles

Many people with at least a rudimentary classical education (Homer) believe that the city name "Paris" is connected to "Paris," the son of the Trojan king Priam, who, as is generally known among this group, demonstrated particularly good taste when he chose Aphrodite from among the three candidates (because she promised him the most beautiful woman among mortals).

When Paris later saw Helen, he believed he had found the promised gift of Aphrodite - so he took her to his homeland of Troy on the Lydian coast without much hesitation. Unfortunately, Helen was the wife of the powerful King Menelaus of Sparta, which led to the outbreak of the Trojan War. In the chaos of the war (which Homer recounts in detail in the Iliad), Paris shot an arrow into the heel of the hero Achilles, a wound from which Achilles, as is well known, did not survive. Thus, Achilles was lost to the ongoing siege of Troy (which was quite tragic for the Achaeans at the time), and the Greeks under their leader Agamemnon only regained their luck in the war when the hollow wooden horse devised by Odysseus was completed and secretly filled with soldiers. What happened next from the Trojan side can only be described as "extremely foolish" (similar to opening unknown email attachments today). The outcome is well known.

---

[21] In the "good old days," this used to be standard curriculum material in humanistic high schools.

Fig. 26: The hero Achilles, fatally wounded by an arrow to the heel, depicted in a sculpture by Ernst Herter from 1884.

77. The Hunchback of Notre-Dame

But back to the city of Paris. In 1163, the foundation stone for the Notre Dame de Paris Cathedral was laid, and in 1831, the famous novel by Victor Hugo (1802-1885) about the gypsy girl Esmeralda and the bell ringer Quasimodo, which is set precisely in and around this church, was published. Most people know this work of the great French novelist only through its film adaptations, which is actually a shame. The novel has an unmatched epic grandeur that vividly and opulently brings the life of late medieval Paris (1482) to life, with an exciting, intricately woven plot of diverse characters. One should really be familiar with both: the novel and the wonderful 1956 film adaptation with Anthony Quinn as Quasimodo, the incomparable Gina Lollobrigida as Esmeralda, and Alain Cuny as Claude Frollo.

Fig. 27: World literature – it's a must-read!

Life in a late medieval city, unless one belonged to a better-off class, was usually arduous and short. The average life expectancy was just 35 to 40 years, due in part to poor sanitary conditions, which fostered devastating epidemics (such as the plague) and also led to high child mortality rates. All these factors, combined with medical care that was barely affordable for most, hard physical labor, frequent military conflicts, and poor nutrition, contributed to an early death.

Fig. 28: Esmeralda (Gina Lollobrigida (1927-2023)) could have certainly competed with Helena (scene from *The Hunchback of Notre Dame* by director Jean Delannoy (1908-2008)).

78. Death in the Middle Ages

In the Middle Ages and early modern period, death lost its cheerful, Epicurean aspect and became a terror associated with damnation, purgatory, and the torments of hell. These concepts, stripped of their theological significance, became real and imaginable facts that one could only escape through a life pleasing to God. In short, the goal of medieval life was to die a good death (*bona mors*). Death confronted Christians directly with the question of salvation or damnation, with the medieval Church being strong in its descriptions of damnation (consider, for instance, Hieronymus Bosch's Last Judgment or Dante's Divine Comedy) but surprisingly weak in visions of heaven. The terrors were further fueled by the impressions left on people by sudden, massive epidemics (e.g., the plague waves of the late Middle Ages), the horrors of wars, or the very direct experiences from frequent heretic and witch burnings of the time.

Fig. 29: Hieronymus Bosch's *Triptych of the Last Judgment* (1450-1516) – it addresses the themes of "Paradise," "Last Judgment," and "Hell" in three panels, beginning with the creation of Eve and the Fall of Man on the left panel, followed by the sinful behavior of humanity and the resulting end of the world on the central panel, and concluding with the terrors of Hell on the right panel (Wikimedia).

The greatest fear for medieval people was dying too early, while still in a state of sin (mala mors), for "Mors peccatorum pessima" – the death of sinners is exceedingly bad. Therefore, in the Middle Ages, "living" meant preparing concretely for death, as nothing was worse than a sudden death without absolution, the last rites, and ecclesiastical support. Daily prayers, devout attendance at Mass, good deeds, etc., were meant to prevent a "bad death." This idea was supported by the principle of Augustine, "One cannot die badly who has lived well" (Non potest male mori, qui bene vixerit). To aid in this, since the 12th century, a form of divine intercession developed, based on Saint Christopher (who, according to legend, carried the Christ child across the river). His feast day in the Catholic calendar is July 24. The aim of this intercession was to avert "death without the means of grace." For this, one needed an image of the saint and a corresponding intercession formula.

79. Saint Christopher Sheets

With the rise of printing, a large number of prints, mostly woodcuts depicting Saint Christopher, were produced and distributed among the people. These were known as "Saint Christopher sheets" (Christophorusblätter). Despite their large quantity, relatively few have survived to this day.

With the onset of the Reformation, criticism of the cult of Saint Christopher grew increasingly prominent among educated circles. Erasmus of Rotterdam (c. 1466-1536) was particularly vocal in his criticism, notably in his work *In Praise of Folly*. For Luther, ultimately, the "viewing of the image of Saint Christopher" was merely a specific form of idolatry.

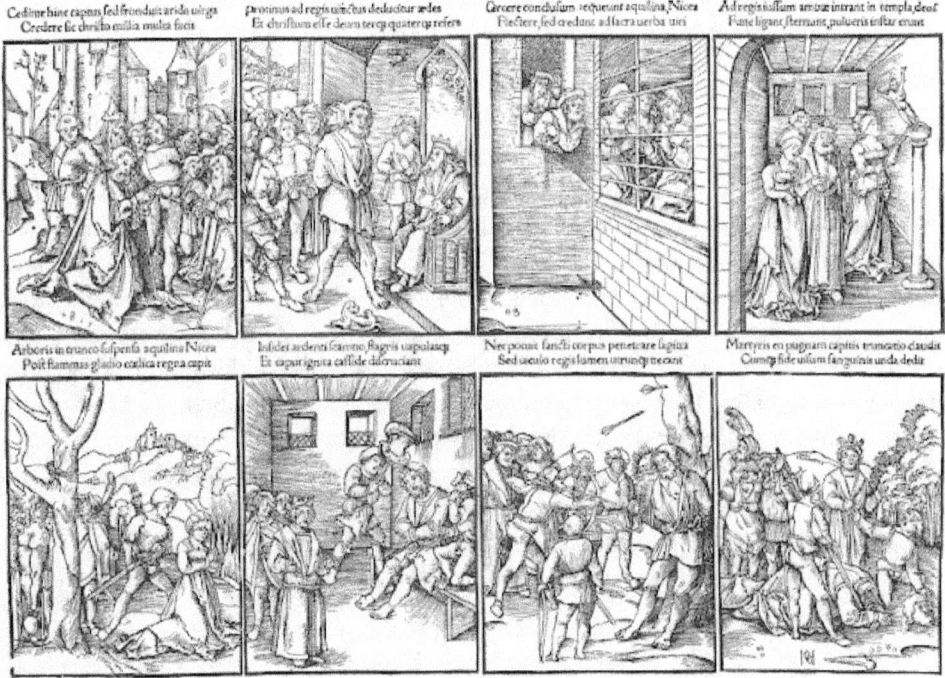

Fig. 30: Saint Christopher sheets were particularly popular during the Reformation period.

When it came to the reconstruction of St. Peter's Basilica in Rome, and it became clear that it was going to be more expensive than anticipated - much like the recent airport construction in Berlin - the Pope and his cardinals had to figure out where to find the necessary funds.[22] After realizing that the revenues from the Peter's Pence, a voluntary donation from the faithful to the papacy, were insufficient, they recalled the deep fears of late medieval people: namely, the high likelihood of ending up in hell or purgatory after death.

80. Indulgences

It is remarkable that church law provided the poor sinner with ways to avoid it or at least significantly shorten the time of penance in hell. Before the construction of St. Peter's Basilica, it was sufficient to undertake a pilgrimage (in the worst case, as far as Jerusalem) or, for smaller sins, to repeat a specific prayer a certain number of times after confession. However, the practice of indulgences became commercialized with the introduction of indulgence certificates, which anyone could buy for themselves or others for "cash" without any other requirements (confession, communion). And those who were planning a "sin" could even purchase a prophylactic exemption from temporal punishment. For *"When the coin in the coffer rings, the soul from purgatory springs"* was a common saying of the indulgence sellers at that time, with the Dominican friar Johann Tetzel (1460-1519) being one of the most prominent (and successful) among them.

Since only a part of the funds raised from indulgence sales actually made its way to Rome (the other part typically ended up in the treasury of the local bishop or the indulgence seller himself), the indulgence sellers received significant support from local church leaders (such as the Hohenzollern prince and three-time archbishop Albrecht of Brandenburg). The indulgence sellers made loud and exaggerated claims, painting the horrors of hell in even darker terms than Hieronymus Bosch (1450-1516) could have, thus extracting the last penny from the gullible people. However, note that the method behind indulgence trading is still effective today, though it has become more subtle. For example, one might recall "Telekom stocks" (perhaps remembering Manne Krug, "lawyer's favorite" and Ron Sommer?) or some forms of "Riester pension plans" (during the zero-interest policy!). But there was, unfortunately for the church leaders and the pope, a learned professor of biblical exegesis who found the excessive indulgence trade quite suspicious at his university in Wittenberg. He was less concerned with the spiritual intent behind it than with the fact that indulgences could also be bought for the deceased. Thus, he decided, as was customary at the time, to initiate a learned disputation on the subject.

---

[22] The so-called „Sondervermögen" ("special fund") had not yet been invented at that time...

Fig. 31: Such a note could be purchased for a small amount of money during Martin Luther's time to protect one's soul from potential harm after death.

81. Martin Luther and his 95 Theses

In 1517, this learned monk named Dr. Martin Luther (1483-1546) published his 95 Theses on the efficacy of indulgences, triggering a sequence of events that led to the enduring modern schism in the Church. The success of the Reformation and Luther's escape from the fate of Jan Hus were partly due to a rather average but very wise prince, namely Elector Frederick III "the Wise" of Saxony (1463-1525), under whom Luther was at the time. Despite being deeply Catholic, Frederick extended his protective hand over the professor from his newly founded University of Wittenberg, who had fallen under the Imperial ban by the Edict of Worms. And thanks to Frederick's arranged forced stay of Luther at Wartburg Castle near Eisenach, we Germans owe our first German Bible translation and thus the foundations of our written language. No one has shaped the German standard language more than Martin Luther. He paid close attention to the people's language, struggled with every word while translating the Holy Scriptures, coined memorable phrases that later entered everyday language, and laid the foundation for a unified written German for a nation divided into numerous dialects.

Fig. 32: Portrait of Martin Luther from 1528 by Lucas Cranach the Elder (1472-1553). (Wikimedia, Luther House Wittenberg)

82. Martin Luther and the German Language

Just one example: he begins the Christmas story not with a fairy-tale „Es war einmal..." ("Once upon a time...") but with the elevated and elegantly phrased „*Es begab sich aber zu der Zeit...*" ("And it came to pass in those days..."). This alone makes it worthwhile to read the Gospel of Luke once again, especially today, when sloppy language and gender nonsense have seemingly become the norm. If you don't believe it, just read the emails that come in every day at work. Sometimes, what is presented in the context of ordinary business correspondence makes one increasingly doubt the existence of a cultured nation...

But back to Martin Luther. During his lifetime, there were around 20 different languages or dialects spoken in the German-speaking world, which could be broadly categorized into Low German (in the north) and Upper German (in the south). The Electorate of Saxony was situated precisely on the language boundary, which explains why words from both linguistic regions were used in the Luther Bible, with the pleasant effect that today even Frisians and Bavarians can understand each other almost effortlessly.

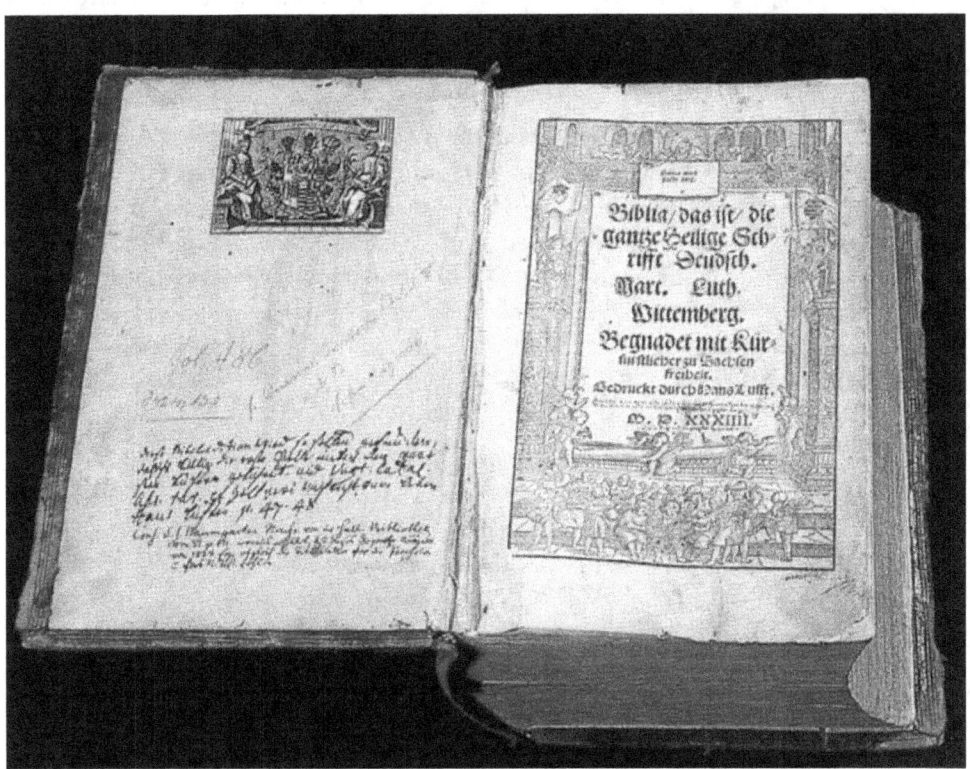

Fig. 33: One of the most important books ever written in the German language – the Luther Bible (Wikimedia)

Few people know that phrases such as wie „*Der Mensch lebt nicht nur vom Brot al-lein*", „*Niemand kann zwei Herren dienen*" ("Man does not live by bread alone," "No one can serve two masters,") or „*Sein Licht unter den Scheffel stellen*" ("Hide his light under a bushel") all come from Luther. This is because he put considerable thought into formulations. Literal translation was not necessarily his thing. For him, it was more important to grasp the meaning and derive a formulation that everyone could understand, rather than sticking rigidly to the words. Just try translating the Latin phrase "*Ex abundantia cordis os loquitur*" into German using Google and compare the result with Luther's version "*Wes das Herz voll ist, des gehet der Mund über*" – and you will recognize a familiar proverb. [23] This is also why the Luther Bible is not just a mere translation from the Latin and Greek originals, but a literary masterpiece and thus a cultural asset in its own right. Many words that were once part of a dialect and only spoken and understood locally

---

[23] „Aus der Fülle des Herzens spricht der Mund" ("Out of the abundance of the heart, the mouth speaks.")

found their way into modern German through Luther's Bible translation. For example, the word "ruchlos," which is known to be another term for "rücksichtslos" (ruthless), as well as "Feuereifer" (zeal), "Machtwort" (authoritative word), "Morgenland" (the Orient), or "Lästermaul" (slanderous mouth) were words that Luther used for the first time and are now part of the everyday vocabulary of every German.

Because the language of the Luther Bible did not change in the following centuries (it was virtually sacrosanct) and dialect-related comprehension issues were mitigated with "translation aids," it established a certain standard that ultimately even students learned to read and write by. The fact that God's word was now proclaimed from the pulpits in Bible German instead of the incomprehensible Church Latin was an important step toward the unification of the language. Although it took some time before the written language initiated by Luther also reflected in spoken High German (which even most Swiss speak), this created another foundation for a future unified German nation-state, which, as we know, became a reality in 1871. And one more thing might be interesting: The written form you see here – the capitalization of nouns typical of German – was preserved because of Luther, as he applied it consistently. This is actually a good reason to maintain it in the future.

Finally, a few words about the Reformation.

83. Martin Luther and the Reformation

Chronologically, the Reformation is situated between Luther's posting of the 95 Theses in 1517 and the end of the Thirty Years' War with the Peace of Westphalia in 1648. This period marks one of the most significant eras in European history, characterized by profound upheavals in both the Church and society. It is undoubtedly the most important ecclesiastical (and societal) renewal movement of the early modern period, originating in Germany and gradually extending to neighboring countries - first to Switzerland under Ulrich Zwingli (1484-1531) - transforming Europe into a new shape. This movement brought about not only religious but also political and cultural changes that continue to have an impact today. The cost of these upheavals was, however, steep. One need only consider the Thirty Years' War, which plunged Central Europe, and particularly the German lands, into unprecedented misery and devastated entire regions.

84. Prague in Bohemia

The starting point for the most devastating war of all wars that has ever afflicted our continent was Prague, the capital and heart of Bohemia. For many centuries, this city on the Vltava was itself the center of Europe. In St. Vitus Cathedral on the Hradčany, the castle hill high above the Vltava, are buried four emperors of the Holy Roman Empire as well as several Bohemian kings and queens. They reflect a significant part of the history of the Sacrum Imperium Romanum, from the Přemyslids in the 14th century (Ottokar I and II), through the Luxembourgs like Charles IV, to the Habsburgs Ferdinand I, Maximilian II, and Rudolf II, with effects still felt today.

Fig. 34: Prague, the "Golden City"

In Prague, Jan Hus (1370?-1415) worked at the Charles University, founded there in 1348 by Charles IV (the oldest university north of the Alps!). The first Prague Defenestration, which marked the beginning of the Hussite Wars, took place in 1419, and the second Prague Defenestration, which marked the start of the Thirty Years' War, occurred in 1618. Johannes Kepler served as the court mathematician to Emperor Rudolf II from 1600 to 1612 in Prague, where he developed his major work, *Astronomia Nova*, one of the most important and significant books in the long history of astronomy.

85.   The Prague St. Wenceslas Treaty

And not to be forgotten, it was here in Prague that the St. Wenceslas Treaty was signed in 1517, which granted noble landowners the right to brew beer on their estates. This privilege marked a turning point in Bohemian brewing culture and led to the establishment of a brewery around 1570 in the village of Kruschowitz (Krušovice), located about 50 kilometers northwest of Prague. This brewery quickly became an important economic factor in the region. Since it was acquired by Emperor Rudolf II in 1581, it has operated under the name "Imperial Brewery Kruschowitz" and was highly popular at the imperial court. So, if you ever travel through Bohemia, look out for the distinctive logo with the name "Krušovice" and the year 1581 on either side of the crown – a symbol of a centuries-old tradition and the legacy of truly imperial brewing craftsmanship.

86. Krušovice Dark

It's definitely worth stopping to enjoy a half-liter of "Krušovice Dark" (Krušovice Černé), just as Emperor Rudolf II often did. A beer connoisseur has described it as follows:

"The beer pours deep black with a beautiful, brownish foam and has aromas of rich roasted malt, coffee, and smoke. Surprisingly, the taste is quite different: initially, you get the expected pronounced roasted flavors, coffee, a hint of bitter chocolate, and grain, but immediately after, there's no heaviness. The beer flows smoothly down the throat and feels refreshingly mild. It finishes with a subtle bitterness and is reminiscent of a pilsner. A glance at the alcohol content explains this lightness."

Fig. 35: This trademark © of the Imperial Brewery in Krušovice should be a guidepost for every visitor to Bohemia - especially if they enjoy dark beer.

87. Braunschweig Mumme

"Black" beers are, by the way, a truly German invention. Their history can be traced with some uncertainty back to around 1390, in the form of "Braunschweig Mumme." This special, extremely durable dark beer has been brewed in Braunschweig since 1492 and enjoyed great popularity over the centuries. Even today, you can purchase Mumme, although it is usually found as a non-alcoholic malt beverage that serves as a strength and revitalization aid.

For a long time, Mumme was the beer of sailors. Its high alcohol and sugar content made it long-lasting even without refrigeration, protecting sailors from the dreaded scurvy. Mumme was not just a drink but an essential companion on long sea voyages. In Lower

Saxony, this beverage was considered a staple for centuries, never missing from any pantry. There was even a well-known proverb praising the virtues of this beer:

„Ein starker Sachse wird, wie alle Völker sagen - Nie schmal in Schulter seyn und schlappe Lenden tragen. - Fragt einer, welches denn die Ursach dessen sey? - Er isset Speck und Wurst und trinket Mumm' dabey".[24]

Despite numerous copies and more or less successful imitations, Braunschweig Mumme has preserved its uniqueness to this day. It is not only a symbol of brewing craftsmanship but also a living emblem of the city of Braunschweig, extending into the present and reflecting the long tradition of this unique beer.

Fig. 36: Mumme was a very popular beverage among sailors, which was brewed not only in Braunschweig.

88. Beer with a Head

A freshly tapped beer is not only a culinary delight but also an aesthetic pleasure when it has a "head" (also known as a "crown" or "cap" elsewhere). This refers to the fine, creamy foam that forms on top of the liquid when poured and is almost always white in color. Its formation is due to the carbon dioxide dissolved in the beer from the fermentation process.

When pouring, there is a slight reduction in pressure and a minor warming of the liquid, which causes carbon dioxide gas bubbles to form. According to the principle discovered by Archimedes some time ago, these bubbles rise to the top and continue to grow

---

[24] "A strong Saxon, as all peoples say,
Will never have narrow shoulders and weak loins.
If someone asks what the cause is,
He eats bacon and sausage and drinks Mumme."

due to the decreasing pressure. Special proteins are carried upward with the bubbles, forming foam bubbles when they reach the surface of the liquid. Smaller bubbles may combine to form larger ones, and the supply of bubbles from below pushes them upward along with their neighboring bubbles, creating a frothy crown - "the head" - on the beer.

The walls of the bubbles are transparent, but they can both refract and reflect light. When light falls on the beer foam, it scatters in all directions, resulting in a white foam regardless of the color of the beer. The uniqueness of this foam lies in its stability, which lasts for a longer time, even though no surfactants (like those in dishwashing liquid or bath foam) are involved in its formation. This property, often used to test the "freshness" of a beer, remained a mystery for centuries because such long-lasting foam formation is unusual for "normal" proteins.

It took 496 years since the introduction of the "German Beer Purity Law" (which explicitly bans dishwashing liquids in beer) for this mystery to be solved. And, as expected, the culprit was genetics. Specifically, a gene from the yeast Saccharomyces pastorianus named CFG1, where "CFG" stands for "Carlsbergensis foaming gene." When this gene is expressed in the yeast cell, it produces the shell protein Cfg1p at the ribosomes, the cell's protein factories, which stabilizes the beer foam. This appears to be particularly effective in the beers of the Danish Carlsberg Brewery.

"Carlsberg" and "Tuborg" were, by the way, the favorite beers of the members of the "Olsen Gang" Egon, Kjeld, and Benny. These two Danish beer brands were often consumed by the Olsen Gang members while they were planning a new heist, and Egon explained to his two companions what he needed for it...

Stable foams are not only desirable in beer.

89. Not Always Are Dreams Foam – Carrie Jane Everson

Around 1885, an American teacher from Chicago named Carrie Jane Everson (1843–1914) noticed something interesting while washing oil-soaked sacks that had previously contained crushed copper ore. She observed that when a mixture of ore and "gangue" (i.e., the non-valuable accompanying minerals) was slurried in a foaming oil emulsion, the ore particles adhered to the foam bubbles and could be easily skimmed off. The reason for this effect is that foam bubbles readily adhere to hydrophobic, or poorly wetted surfaces, causing them to float on top of the slurry of ore and gangue.

This serendipitous discovery led to the development of the flotation separation process, which is now considered one of the most important methods in ore processing. By systematically applying Everson's observations, ores could be processed more efficiently with reduced losses, significantly increasing the yield of valuable metals. Mrs. Everson's patent laid the foundation for this technique, which, in its modified form, continues to play a central role in the mining industry today. It is used not only for ore separation but also in other industries, such as wastewater treatment and raw material processing. Thus, Everson's discovery, which initially seemed like a simple everyday observation, has had a lasting impact on industrial development and remains an example of how innovative ideas often come from the most unexpected sources.

Fig. 37: Carrie Jane Everson discovered and developed the flotation process, which is crucial for metal extraction. As a result, the prestigious Mining Journal dedicated a posthumous article to her in 1916.

90. Gold Extraction

Gold dust, however, cannot be separated into "gold" and "sand" using this method. The preferred substance here is "mercury." Since mercury is well-known to be a hazardous environmental toxin, this process causes numerous environmental damages in various parts of the world, especially in the Amazon and its gold-rich tributaries. The easiest method is still searching for gold nuggets. However, most of the time, you only find gold flakes in your "gold pan," as so-called "placers"[25] are not very common. Extracting these "gold flakes" is also not a particularly lucrative activity (but it is an interesting hobby that you can successfully pursue in Germany, albeit without serious profit intentions).

[25] These are places where nuggets are just lying around...

Fig. 38: In Germany, there are also some places where you can indeed successfully "pan for gold." However, the operating result "before taxes" will always be negative. But it's fun (say the "gold panners")...

It would be better if you could bind the gold to something that can be more easily separated from the sand and from which the gold can later be recovered. One such substance is mercury. It is easy to handle because it is liquid and can be easily separated from the sand since many small mercury droplets can coalesce into larger ones due to their extremely high surface tension. However, the most important aspect is that mercury binds gold into an amalgam. This chemical process was a common method of gold extraction for centuries, especially in manual gold panning.

However, the use of mercury has severe drawbacks. Aside from its efficiency in gold extraction, mercury is a highly toxic metal that can severely poison both humans and animals. This fact has marred the enjoyment of many gold prospectors, as the health risks often outweigh the gains. Once the amalgam is formed, it only needs to be heated (for example, with a classic blowtorch) so that the mercury evaporates, leaving behind the pure gold.

The consequences of this method are devastating. It is estimated that throughout the history of gold mining, over 2,000 tons of mercury have been released into the environment in the Amazon alone. This massive influx of mercury represents an unparalleled environmental disaster due to its extreme toxicity. The long-term damage to ecosystems and human communities in these areas is enormous, clearly demonstrating that the practice of using mercury in gold extraction comes at a high price - not only for health but also for the environment.

91.   „Biogold" and the Golden Fleece

In contrast, the "predecessor method" used in antiquity in the Caucasus region is completely ecologically harmless, as it only uses natural substances ("Alles Bio oder was?"). Today, some circles would certainly say that it produced genuine and pure "organic gold." But let's take it step by step. This concerns the real background of the story of the fleece of Chrysomeles, a ram sired by Poseidon, whose body hair entered Greek mythology as "The Golden Fleece" because it shone with a "golden" luster. Not quite as golden in appearance, this ram can also be found in the night sky as the constellation representing the zodiac sign Aries.

Fig. 39: Ancient Gold Jewelry (Wikimedia)

92.   The Argonaut Saga

In the time of the Argonauts, the ram (by its own request, by the way) had already been sacrificed to the war god Ares, and its removed fleece was set up in Colchis in the grove of Ares. The Argonauts, led by Jason of Argos, were very eager to obtain this fleece. They sailed across the Black Sea in their ship "Argo," and after many adventures, they finally reached the mouth of the River Rioni (then called Phasis) in present-day Georgia. There, they demanded the fleece from the king, who made its release contingent upon completing a difficult task related somewhat to the activity of "plowing." Jason accomplished the task with distinction, thanks to tips from Medea, the king's daughter, who had fallen in love with him. She helped him by betraying a malicious plan devised by her father, Aeetes, to steal the golden fleece. The theft was successful. After a lively return journey, which has several versions, Jason and Medea eventually arrived in Corinth, where they lived together for a time until Jason abandoned her to marry Glauce, the daughter of King Creon...

93. Medea's Revenge

But Medea was resourceful (*nomen est omen*) and devised a terrible revenge plan, which resulted in the deaths of King Aeetes, his daughter Glauce, and also her own two children. Since then, the murder of one's own children has been inextricably linked to Medea, vividly portrayed in the eponymous tragedy by Euripides (480–406 BCE). Franz Grillparzer (1791–1872) also produced an interesting theatrical adaptation around 1820 as part of a trilogy. And, of course, the 1969 film directed by Pier Paolo Pasolini (1922–1975), featuring Maria Callas (1923–1977) in the lead role (who did not need to sing in this role as an opera diva), must not be forgotten. But what is the "true" background of the "Golden Fleece"?

Fig. 40: Maria Callas as "Medea" in the 1969 film adaptation of Euripides' famous ancient tragedy by Pier Paolo Pasolini

To extract gold from mountain streams, specifically fine gold flakes and gold particles, sheep pelts were placed in the current. The heavy gold dust would settle between the hairs of the pelt, allowing it to be concentrated. This method was used in antiquity to exploit gold deposits that were rich in gold dust but poor in nuggets. Historians believe that this technique is the basis for the legend of the golden fleece.

94.   Gold, Gold, Gold...

It is estimated that the total amount of gold ever mined on Earth is about 166,000 tons. This corresponds to a golden cube with an edge length of just 20 meters. Since the volume of a cube increases with the cube of its edge length, despite an annual production of about 3,000 tons, the increase in the size of this cube remains modest. Gold is simply not very abundant, which is why a ceramic crown for a defective tooth is usually less expensive than a gold crown.

95.   How is Gold Created in the Cosmos?

The truly interesting question is where and how gold is created in the cosmos and how it ultimately ended up on Earth. Elements are typically forged in stars, a process known in astrophysics as "burning" - a different term for "nuclear fusion." Only deuterium, lithium (found in our lithium-ion batteries), beryllium, and boron, along with omnipresent hydrogen and most of the helium, originated from the first minutes of the Big Bang nearly 13.79 billion years ago.

First-generation stars were composed only of hydrogen and helium, with trace amounts of lithium and beryllium (the latter two elements being negligible in abundance compared to H and He). These stars were massive and had short lifespans (only a few million years), ending their lives rapidly in a special type of supernova now known as a pair-instability supernova. A supernova can briefly outshine all the stars in a galaxy combined - and it's not advisable to witness such an event from close range (<100 light-years away).

The "explosion debris" from such stars gradually accumulated in interstellar matter, which then gave rise to subsequent generations of stars. This "debris" contained all the elements that were either fused in stars or formed through other processes (which astronomers refer to as s-processes for "slow" processes or r-processes for "rapid" processes).

It's important to note that a star can only generate energy through nuclear fusion up to the element iron (Fe) with atomic number 26. This energy is required to maintain the star's stability. A star remains stable as long as, at every point within it, the inward gravitational force is balanced by the gas pressure (and for massive stars, also the radiation pressure) at that point. When this condition is met, the star is in hydrostatic equilibrium. If this condition is not met, the star will attempt to compensate for the energy deficit by contracting its core. This process taps into the star's gravitational binding energy, with some of this energy heating the stellar material and the rest radiating into space. Thus, the life of a star is essentially a story of the contraction of its core. Beginning with the contraction of a gas and dust cloud, at the end of a star's life, only a compact remnant remains in the form of a white dwarf, a neutron star, or a black hole (and, in extremely rare cases, possibly nothing at all).

96. Life History of a Star with Eight Solar Masses

For a star to reach the fusion of iron during its lifetime, it must have an initial mass of at least about eight solar masses. Without going into extensive details (which are highly interesting!), the life of such a massive star generally unfolds as follows: After forming from cosmic gas (99% hydrogen and helium) and dust, and reaching the ignition temperature for "hydrogen burning" (about 20 million degrees Celsius) during its contraction phase, the star will gradually convert hydrogen into helium over the next million years. This phase is known as a "main sequence star." During this burning phase, the heavier helium accumulates in the star's core until the hydrogen in the core is depleted to the point where the energy required to maintain hydrostatic equilibrium can no longer be supported by hydrogen burning. At this moment, the star's core becomes unstable and collapses. This releases energy and increases the core temperature until it reaches approximately 200 million degrees Celsius, the ignition temperature for helium burning, and the core collapse ends.

With the fusion of helium into carbon and oxygen (the triple-alpha process, involving three helium nuclei), the star can sustain itself for the next 10,000 years. The "ash," i.e., the carbon and oxygen produced, accumulates again in the core. This continues as long as there is enough helium to undergo thermonuclear fusion. Eventually, this process ends, and the cycle of core contraction and temperature increase until the next burning phase ignites repeats. Each new burning phase produces less energy (a phenomenon easily explained by physicists), so the duration of these primary burning phases becomes progressively shorter. Thus, after helium burning at a core temperature of 800 million degrees Celsius, carbon burning occurs (about 600 years), followed by neon burning at 1.4 billion degrees Celsius for one year, then oxygen burning at 2 billion degrees Celsius for six months, and silicon burning at 3.5 billion degrees Celsius for one day (producing iron). Finally, the iron core collapses to form a neutron star with a diameter of about 20 kilometers, and the outer shell of the star explodes: We observe a hydrodynamic supernova. This is likely the birth moment of gold, which might eventually get trapped in the fine hairs of a fleece with some luck...

Gold is also produced in large amounts when two neutron stars collide - an extremely rare and colossal event in the universe. Such a collision, known as a "kilonova," releases enormous amounts of energy, generating heavy elements like gold and platinum. All the gold found on Earth originates from these supernova and kilonova catastrophes, making gold not only a symbol of wealth but also a relic of the most powerful processes observable in the universe (Point 98).

97. Heavy elements are built from neutrons

The nucleus of a stable gold atom is known to contain 79 protons and 118 neutrons. When the iron nucleus collapses in a supernova, nuclear processes occur that lead to a huge flow of neutrons. As neutrons have no electrical charge, they can, for example, attach themselves to existing iron nuclei, which become heavier and heavier as a result, i.e. heavier and heavier isotopes of this metal are produced within a very short time. These isotopes are not stable, but radioactive. This means that individual neutrons decay into protons, electrons and anti-electron neutrinos (the neutron is said to undergo beta-minus decay), which in turn causes the isotope to move up one step on the element ladder (periodic table). Only when a nucleus combines 79 protons and 118 neutrons in this way is a stable gold atom formed (albeit still without an electron shell). Of course, other heavy elements are also formed in this way, such as uranium with a proton number of 92 and even plutonium with a proton number of 94. Incidentally, plutonium is the last element in the periodic table that still occurs naturally on Earth. All elements in the periodic table with an even higher atomic number were created artificially. However, their lifespan (half-life) is quite limited due to their radioactivity. Since there are people with gold teeth, since there are people who can afford to eat from golden plates with golden knives and forks and, not to forget, also because there are "opponents of nuclear power" who have something against "nuclear power" (uranium) for reasons that are not always rational, it must be assumed that a supernova must somehow have been involved in the formation of the earth more than 4.56 billion years ago. And this fact alone tells us a lot about the early history of the sun and the planets. Before we go into this in more detail, however, it should be pointed out that most of the gold in the cosmos is probably produced by the collision of neutron stars (point 96).

98. When Neutron Stars Collide

Neutron stars are extremely dense remnants of massive stars that remain after a supernova explosion and can sometimes be observed as so-called "pulsars" for a period. They are composed almost entirely of degenerate neutron fluid and possess an astonishing density - one teaspoon of neutron star matter would weigh about a billion tons on Earth.

When two such neutron stars orbit each other in a binary system, they lose energy through the emission of gravitational waves and gradually spiral towards each other until they eventually collide. The process of "spiraling in" has already been observed in a specific double pulsar system (the Taylor-Hulse Pulsar, PSR J1915+1606), with the components predicted to collide in approximately 85 million years. During this collision, extreme conditions occur, which, as described earlier, enable the synthesis of particularly heavy elements. The colliding neutron stars release an enormous amount of neutrons, which are ejected into the surrounding space. These neutrons can be captured by atomic nuclei through the aforementioned r-process (rapid neutron capture process) before they decay radioactively. Subsequently, the captured neutrons transform into protons via beta decay, causing the newly formed element to shift to heavier elements on the periodic table. This process continues until stable or long-lived heavy elements are formed, including gold (Au).

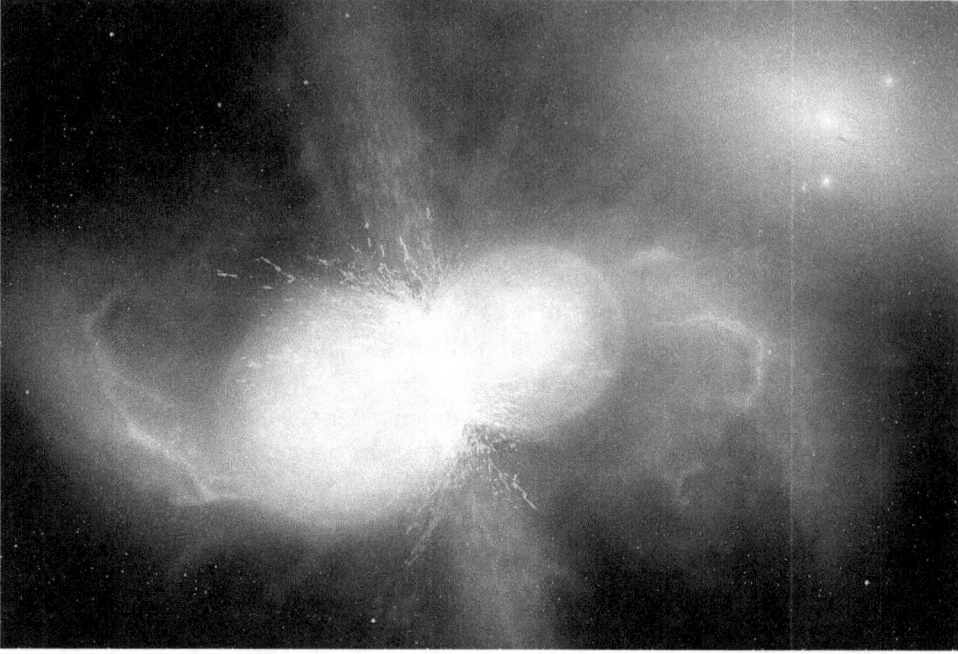

Fig. 41: Artistic representation of the merger of two neutron stars, during which enormous amounts of gold (and other heavy elements) are produced (Wikimedia)

Physicists are currently trying to understand what exactly happens during a neutron star collision through computer simulations. A key question is what fraction of the mass escapes into interstellar space and whether (and how) the remainder collapses into a black hole. However, all calculations in this area are hindered by the uncertainty of the properties of high-density nuclear matter, which is represented by a variety of hypothetical equations of state. These equations of state are largely based on theoretical considerations and are being tested through experiments in heavy-ion accelerators. "Soft" nuclear matter leads to the formation of a black hole even in lighter collisions, while "stiff" matter can stabilize larger masses. Thus, the maximum mass of neutron stars provides information about the properties of nuclear matter and could even clarify whether the building blocks of nuclei dissolve into quarks during a gravitational collapse.

Regarding the formation of gold, model calculations suggest that about 10^{-6} to 10^{-7} of the total mass of the colliding neutron stars is converted into gold. This means that in a typical neutron star collision, where the total mass of the two stars is about 2.8 times the mass of the Sun, approximately 0.000028 solar masses are transformed into gold and other heavy elements. Unfortunately, it is not possible to determine precisely how much of this ends up in a potential black hole and thus remains out of reach for future treasure hunters.

99.   A Supernova and the Early History of the Solar System

Let's return to the question of where the gold nuggets and gold flecks in our gold-bearing streams (including those found even in Saxon Upper Lusatia!) actually come from. The answer is that they were formed during a supernova explosion that occurred before the formation of the Sun and Earth.

The fundamental information in this context is as follows: The supernova must have originally been a star with more than 25 solar masses, whose lifespan lasted no more than five million years. Such massive stars are very rare. A star cluster formed from an interstellar gas and dust cloud (the cloud fragments into many individual clouds during its collapse, from which protostars then form through further contraction) must have about 2,000 members to ensure that at least one object of this caliber is among them. Since the formation of the Sun and its planets also took only a few million years, this star must have experienced its core collapse while still in the protostar phase at a not too great distance (estimated at ~1 light-year), and some of its explosion debris must have penetrated into the material from which the planets were later formed. Fortunately, there was also some gold and uranium among this material. It is even possible - and there are serious reasons to believe this - that without this supernova, we might not exist at all. The radioactive elements from the supernova contribute to maintaining the plasticity of the Earth's mantle through their decay. Without a convective mantle, there would be no plate tectonics; without plate tectonics, there would be no continents; without continents, there would be no long-term climate stabilization. And without all this, no humans. And without humans? Well, we don't know what would be, because if someone "does not exist," they cannot ask why...

Thus, we arrive at the non-trivial problem of whether everything we perceive with our senses is truly as it is.

100. Is Reality Real?

This is simply the age-old philosophical question about "reality," specifically whether there is a world outside of our individual consciousness, and if so, how it can be proven. As Edgar Allan Poe (1809-1849) once put it, "*All that we see or seem is but a dream within a dream...*"

But this question also has a historical context: In the 18th century, there was the profession known as the "scholar." These were usually more or less well-to-do individuals who, free from material concerns, could devote themselves to intellectual and scientific matters, thereby gaining a certain prestige in their society. I will briefly mention Johann Burckhard Mencke from Leipzig, who lived from 1674 to 1732. His greatest achievement was the publication of the three-volume "*Scriptores rerum Germanicarum praecipue Saxonicarum*," which still provides important source material for historians today. But that is not the main topic here. Instead, we focus on a sentence from his 1716 work "*Charlataneria eruditorum*" (yes, Burckhard Mencke significantly enriched the German language with the word "Scharlatan"). It says:

"*...That you are the only ones in the world; every other would only exist in their own thoughts...*"

This brings us to a paradox that I would like to briefly describe.

So, back to the beginning. Each of us owns a computer, and research institutes have supercomputers whose computational power has now surpassed the 1.1 exaflop threshold. And this computational power will continue to increase in the future (according to Moore's Law).[26] With their help, one can simulate highly complex processes based on known natural laws and advanced mathematical methods, processes that are difficult or even impossible to access experimentally. I am thinking here of weather forecasting, the simulation of atomic bomb explosions, or even the simulation of the "Big Bang" and the resulting cosmic structure formation processes.

Clearly, natural phenomena can be mathematically simulated in any complexity if the required computational power is available (this statement is not entirely accurate, as both mathematics (Gödel's Theorem) and physics (quantum mechanics) impose fundamental limits, but let's set that aside for now).

101. The Brain in the Nutrient Broth

Now, let's take another step. What defines us humans is our brain. Everything we perceive with our senses enters our brain in the form of electrical impulses, creating the world we are profoundly convinced exists independently of us. Sunlight reflecting off a rose passes through the lens of our eyes to the retina, where an upside-down, distorted, and reduced image of the rose forms. This image is converted into electrical signals by receptors in the form of rods and cones through complex biochemical processes. These signals travel along nerve pathways to the brain, where it rectifies the image of the rose and translates it into a sensory impression accessible to our consciousness. Simultaneously, electrical nerve impulses from the cells of the nasal mucosa arrive, allowing us to perceive the delicate scent of the rose. If our fingers are pricked by the rose's thorns, our brain also registers this through well-defined weak nerve impulses. All these electrical pulses can be measured today. It is only these electrical impulses that create a world in our brain, allowing us to "think" and "act," and determine our entire emotional experience.

Now, imagine that our brain is floating in a laboratory in a delicate nutrient broth (which the French might call "bouillon"), and all incoming nerves are connected to a supercom-

---

[26] Moore's Law originally stated that the number of transistors on integrated circuits would approximately double every two years, which has been a fundamental basis for the rapid development of computer technology and the digital revolution.

puter that simulates the "outside world." This supercomputer generates impulses exactly as our sensory organs would normally do. Imagine that this machine is simulating a rose with its typical shape, its rich red color, its pleasant scent, and the thorns on its stem, along with the hand holding it. And that this machine produces exactly the same weak impulses as the sensory organs would. And instead of sending these impulses to a monitor or speakers, it sends them through the connected nerve pathways to the brain in the nutrient broth. In this "retort brain," the distinct illusion of a rose with its rich red color, scent, and thorns on its stem would be created. And if you (i.e., your brain in the nutrient broth) also hear birdsong, see blue mountains in the background, and white clouds in a blue sky, these too are merely illusions and a result of the simulation. Even if you feel (and seemingly willfully cause) your hand to move through your hair, it is still just a simulation. The disturbing aspect is that there is no way for you (or me, etc.) to distinguish between our real existence and the existence of a "retort brain" in this simulation.

Fig. 42: Is the rose real or just a simulation? What do you think?

102. Is there an external world?

And the big philosophical question that comes with it (see the introductory quote by Johann Burckhard Mencke, point 100) is: *How can we be sure that something like an "external world" truly exists?* The essence of these considerations is that we cannot be certain that an external world exists, that our consciousness resides in a brain, and that this brain is in a body. This is a fundamental problem that science cannot resolve (any scientific knowledge we acquire could, in the sense described here, also be "simulated").

Do you think this is all nonsense? So do I. But think about it for a moment. And perhaps occasionally watch the film "The Matrix" again... By the way, there are two sequels to this quite interesting film, though they tend to be weaker ("The Matrix Reloaded" and "The Matrix Revolutions"). In these films, just like in the first part of the trilogy, there are many philosophical and historical references if you're familiar with these fields.

103. Solipsism

The philosophical idea that the "world" is merely a simulated illusion and that only the "self" is the sole existing subject is called "solipsism." In a broader sense, this term (from Latin solus meaning "alone" and ipse meaning "self") means that everything, i.e., the entire world, or the whole "being," is exhausted in the subjective contents of an individual's consciousness (specifically yours, dear reader). This is referred to as "metaphysical solipsism" - and this is what we will discuss here. A metaphysical solipsist claims the nonexistence of an external world (more precisely, a world outside of their own consciousness), with the awareness that this claim can never be proven. It is indeed possible to assume that all sensory impressions are merely products of those senses themselves, without necessitating the existence of anything objective outside of consciousness - see the metaphor of the brain in a vat. Or, to use Edgar Allan Poe's quote again, "*All that we see or seem is but a dream within a dream...*".

104. Claudia Brücken and Propaganda

The related and relatively obscure poem "*A Dream Within a Dream*" from 1849 gained some attention again in 1984 when it was recited by Claudia Brücken and set to music on the debut album of the Düsseldorf-based pop band "Propaganda." This debut album is titled "*A Secret Wish*" and remains quite popular among aficionados even today.

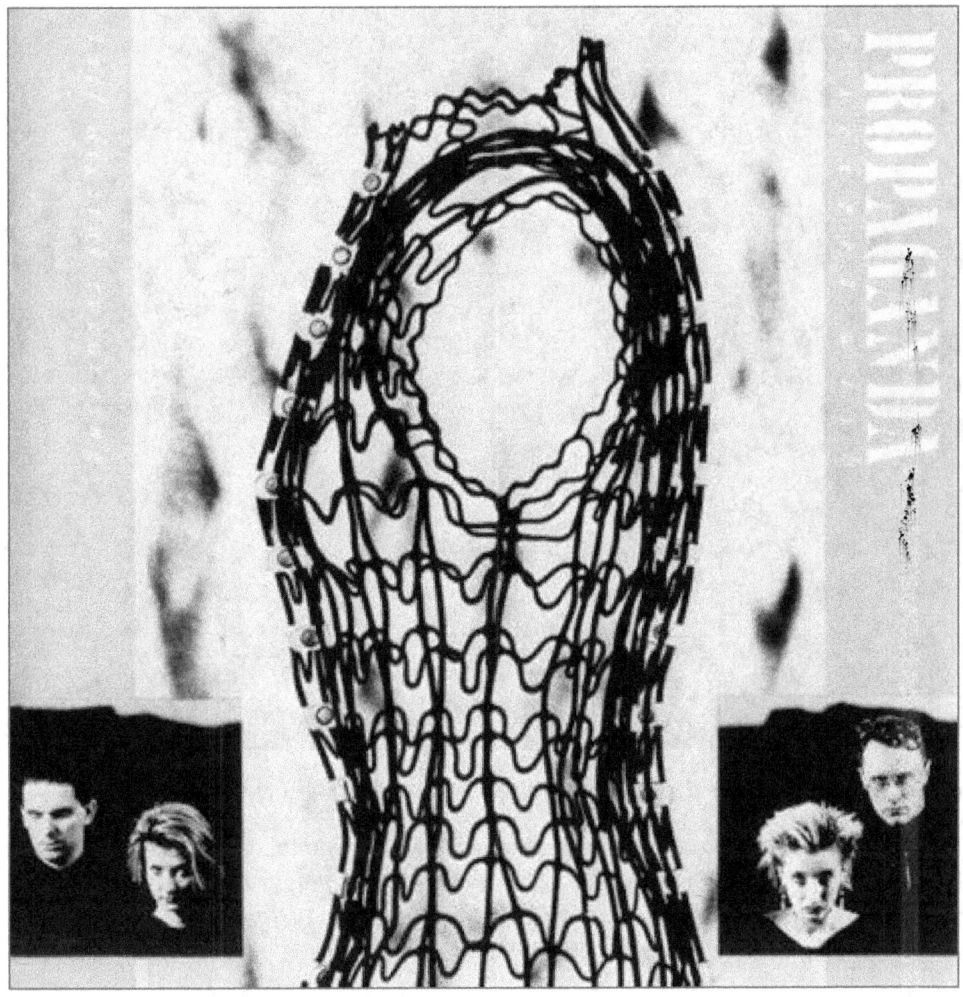

Fig. 43: Album cover of "The Secret Wish" by the German pop band "Propaganda"

105. Dr. Mabuse

This is partly due to the songs "Duell" and, of course, "Dr. Mabuse." This name represents a supervillain created in 1919 by the Luxembourgian writer Norbert Jaques (1880-1954). The character gained a certain popularity through a number of film adaptations in the early 1960s, competing with the contemporary Edgar Wallace mysteries. The titles speak for themselves: *„Im Stahlnetz des Dr. Mabuse"* (The Steel Net of Dr. Mabuse), *Das Testament des Dr. Mabuse"* (The Testament of Dr. Mabuse), *„Die unsichtbaren Krallen des Dr. Mabuse"* (The Invisible Claws of Dr. Mabuse), *„Scotland Yard jagt Dr. Mabuse"* (Scotland Yard Hunts Dr. Mabuse) and, of course, *„Die Todesstrahlen des Dr. Mabuse"* (The Death Rays of Dr. Mabuse.)

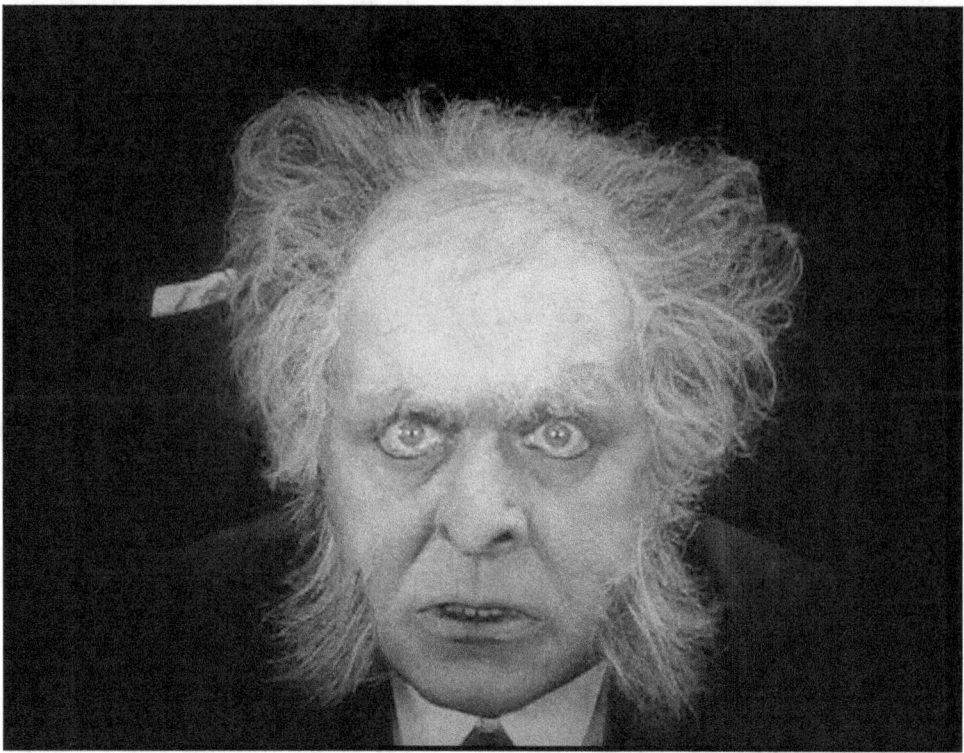

Fig. 44: Dr. Mabuse was certainly an intriguing figure (Film Archive).

To be precise, the black-and-white film series is a continuation or remake of two films by Fritz Lang (1890-1976) from 1922 („*Dr. Mabuse, der Spieler*" (Dr. Mabuse, the Gambler) - a two-part silent film) and 1933 („*Das Testament des Dr. Mabuse*" (The Testament of Dr. Mabuse), this time with sound and banned by the Nazis).

The fascination with Dr. Mabuse's character lies in the fact that he uses his brilliant intellect, his profession as a psychoanalyst (Freud!), and his extraordinary hypnotic abilities to lead a criminal organization and commit crimes himself, all while holding a vision of a better and fairer world. In these adaptations, one can admire the "who's who" of the German acting scene of that time, including Gert Fröbe (1913-1988), who played Commissioner Lohmann in two Mabuse films and became world-famous. And rightly so, in my opinion. The younger generation might only recognize him from James Bond's "Goldfinger" (1964), which still occasionally airs on television. But that was just one of more than 40 films in which he appeared, often as a character actor.

106. Those Magnificent Men in Their Flying Machines

In the English film from 1965, "*Those Magnificent Men in Their Flying Machines or How I Flew from London to Paris in 25 Hours 11 Minutes*," which was shortened for the German posters to the more concise and memorable title "*Die Tollkühnen Männer in ihren fliegenden Kisten*", was an amusing film for anyone interested in the history of aviation. Similarly, the film "*Schussfahrt nach San Remo*" (starring André Robert Raimbourg (1917-1970), better known as "Bourvil") was just as entertaining for those who are more interested in the history of road cycling.

Fig. 45: Colonel Manfred von Holstein (right in period bathing attire)

The "daring men" are pilots ("*Richtge Männer wie wir...*") who, with their contraptions cobbled together from plywood and piano wire, aim to compete in a flying race across the English Channel organized by a newspaper in 1910. Of course, a delegation from the German Empire (after all, our Emperor Wilhelm II was a grandson of British Queen Victoria!) must be part of the event, stylishly represented by Colonel Manfred von Holstein and his adjutant, Captain Rumpelstoß.

107. Gert (Karl-Gerhart) Fröbe

The former was played in an unmistakably Prussian manner by Gert Fröbe. Captain Rumpelstoß, on the other hand (played by Karl-Michael Vogler (1928-2009)), who was originally designated as the pilot of the imperial aircraft, unfortunately fell ill due to acute diarrhea, forcing Colonel Holstein to take over his role with the bold and memorable words, "*There is nothing a German officer cannot do!*" As one might expect, despite his intensive study of the aircraft's instruction manual (specifically the „Heeresdienstanweisung zur Bedienung eines Flugzeugs" (Army Service Directive for Operating an Airplane), he was unable to navigate it across the English Channel...

By 1910, the motorized airplane was already a practical flying machine. The year before, aviation pioneer Louis Blériot (1872-1936) had successfully crossed the English Channel for the first time.

108. Why Don't Ton-Heavy Airplanes Fall from the Sky?

Back then, people were still amazed that such a heavy machine as an airplane could stay in the air, given that everything that falls from one's hand inevitably lands on the ground. Today, no one is surprised anymore, although very few people can explain why this is the case. Well, the "secret" lies in the wings, specifically in their profile and their angle of attack relative to the airflow.

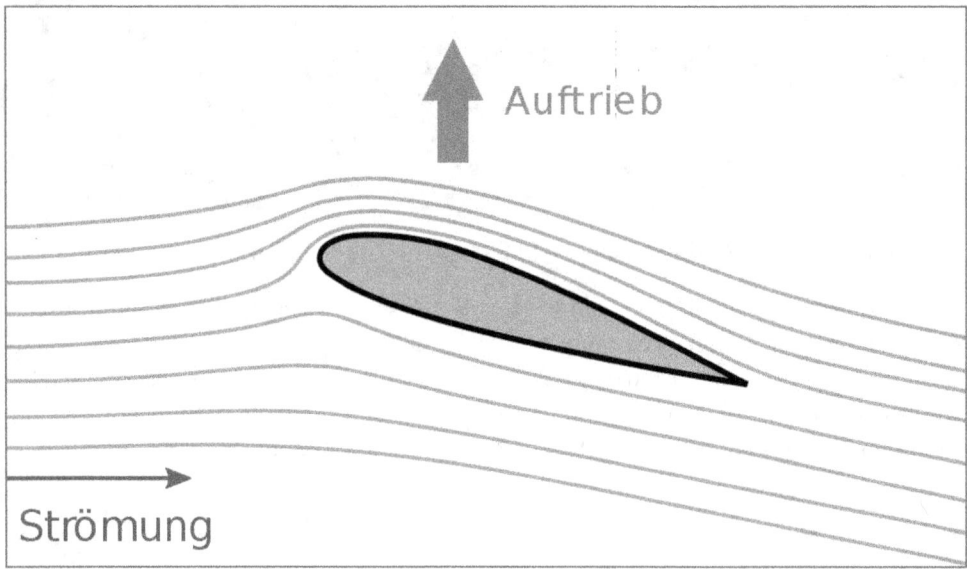

Fig. 46: Deflection of Laminar Airflow Around a Wing Profile and the Direction of Lift. (Wikimedia)

A wing profile is designed so that the upper surface is more curved than the lower surface. When it is slightly tilted against the airflow, the air travels faster over the top and slower underneath. According to the laws of aerodynamics, this creates an upward pressure force underneath (which accounts for about one-third of the lift) and a "tension force" above. Both forces point in the same direction and are thus capable of supporting the entire aircraft, unless the airflow speed decreases so much that it causes a stall. When that happens, the aircraft has only one direction to go, which is down, always following gravity. Therefore, it is crucial for the pilot to be aware of the airspeed in every flight situation. To achieve this, the airspeed must be measured accurately.

109. Airspeed Measurement with the Pitot Tube

This is typically done with a Pitot tube, which was invented by the German physicist and co-founder of fluid mechanics, Ludwig Prandtl (1875-1953). If it fails, for example, due to icing at high altitudes, it can cause an aircraft to crash. There are several historical examples of this. One of the most consequential was the crash of an Air France Airbus A330-203 over the central Atlantic on May 31, 2009, which resulted in the deaths of all 228 people on board.

Fluid mechanics is a very important branch of physics with an extremely broad range of applications (see the "Navier-Stokes Equations," Point 28). Nevertheless, there seem to be phenomena where it appears not to work as expected.

110. The Bumblebee Paradox

I am thinking of the bumblebee, which, according to the laws of aerodynamics, should actually not be able to fly, as Ludwig Prandtl (1875-1953) humorously claimed in his lectures. Yet it does fly, as you can observe with your own eyes in any clover field or meadow edge during the summer. The great German fluid dynamicist once jokingly suggested that this is because the laws of fluid mechanics mean nothing to it. This may be true, but it doesn't explain its more or less graceful flying ability. Bumblebee researchers, as well as fluid dynamicists, were not satisfied with this problem. It quickly became clear that this "bumblebee paradox" is not a paradox at all. When calculating the aerodynamic behavior of a bumblebee (weight, wing area, wingbeat frequency) with the correct value for a parameter known as the "Reynolds number" and considering the vortices occurring during flight, it becomes clear that a bumblebee can indeed fly. However, this was already proven by observation, so it was nothing new. Now, we finally understand why. All these studies greatly enhance our confidence in aerodynamics, and today you can board a modern aircraft with even greater reassurance - thanks to the little bumblebee. And one more thing, before I forget: a bumblebee can reach a flight speed of over 10 km/h, and even a bit more with a tailwind...

111. The Buff-tailed Bumblebee as a Rodent

The first buff-tailed bumblebees (*Bombus terrestris*) can be observed starting from late March, when they emerge from hibernation and visit the flowers of the false lupine at the edge of a deciduous forest. Unfortunately, their proboscis isn't long enough to reach the sweet nectar from the long corolla of these flowers. However, while a bumblebee may be small, it is far from being dumb. It simply chews a hole at the end of the flower's corolla, thereby easily accessing the sweet nectar. If you take a closer look at false lupine flowers (or even the flowers of the spring vetch) during a little spring walk, you will undoubtedly find flowers with a small hole at the base of the corolla.

112. Spring Aspect

Spring is indeed a great time for an educational forest walk. The deciduous trees are still without leaves, allowing sunlight to easily reach the forest floor. It is the time of spring bloomers. For a few weeks, the forest floor transforms into a sea of flowers, and the knowledgeable observer can distinguish between yellow and white wood anemones, liverworts, barrenwort, hazelwort, lesser celandine, goldstars, various species of violets, spring vetches, primroses, and, in moist areas, butterwort and golden-saxifrage, as well as the peculiar common toothwort.

113. Bee flies

A bit later, large white patches of starwort appear, where you can often see bee flies hovering in the air like tiny hummingbirds. These insects are worth a closer look. In flight, you can see their extremely long, straight proboscis, which almost matches the length of their brown, heavily hairy body (about 8 to 12 mm) (they cannot sting, so there's no need to worry). Additionally, their fairly long legs (six in total, as they are not spiders) are noticeable, with four extended backward and two forward during flight.

In Germany, there are 34 different species, among which the small bee-fly and the large bee-fly are the most common. Their flight behavior is so distinctive (hovering in the air, then rapidly changing position like summer hoverflies...) that they are almost impossible to overlook. Before they perform these aerial feats, these flies undergo an interesting developmental process.

Fig. 47: Photos of bee flies are relatively common, but photos of bee-fly shadows are quite rare, as a quick Google image search will show...

Take, for example, the large bee fly, which can be seen from early April on blooming willow trees (catkin). It depends on the presence of certain solitary bees and wasps, whose larvae are parasitized by "its" larvae for a while. The female bee-flies lay their eggs close to the entrances of these underground nests. Once the bee fly larva hatches, it instinctively migrates to the next brood nest of such a wasp and begins to consume a portion of its food supplies, uninvited. After reaching a certain size, it then also starts feeding on the wasp larvae themselves as a pure carnivore. After several molts, it pupates in a characteristic fly-shaped barrel-like pupa, overwinters, and emerges the following spring as a pure vegan, solely interested in the nectar of spring flowers.

As you can see, a spring walk through nature offers many interesting discoveries. This includes some colorful butterflies that have survived the winter in a state of torpor and now enhance the flowering meadows with their vibrant colors. Notable among these are the magnificent Peacock butterfly, the Small Tortoiseshell, the now rare Large Tortoiseshell, the Mourning Cloak, the Comma, and the Brimstone butterfly, all of which like to warm themselves in the sun with their wings spread wide. In April, the Aurora butterfly, various whites, the Swallowtail, and sometimes even the Scarce Swallowtail butterfly in warmer spots, which all survived the winter in the pupal stage, also make an appearance.

Fig. 48: The Scarce Swallowtail (*Iphiclides podalirius*) was, at the beginning of the 20th century, quite a common butterfly, similar to its close relative, the Swallowtail (*Papilio machaon*). After that, its population rapidly declined, and it disappeared from many places where it was once common. Currently, however, there is a noticeable recovery in the population of this largest native swallowtail. The photo was taken in the warm Elbe valley near the "Porta Bohemica," on the "Radebeule" mountain above the old bishopric town of Leitmeritz. (Author)

114. Why are butterflies called "Schmetterlinge" in German?

But have you ever wondered why butterflies are actually in german called "Schmetter-linge"? In English, butterflies are known as "butterflies," which literally translates to "butter-fly" or "butter-bird." This seemingly peculiar name does not immediately reveal its etymological connection to the German word "Schmetterling."

The colorful daytime butterflies, which were once far more numerous in past centuries than they are today, likely prompted people to give them names in their respective dialects, names whose meanings have largely been forgotten over time. In the novels of Theodor Fontane (1819-1898), for example, the term "Kalitten" is occasionally used for butterflies - an old name still common in the Havelland region in the 19th century, preserved in the children's rhyme: "*Kalitte, Kalitte setze dir, ich gebe dir auch Wein und Bier...*" (a call for the butterflies to settle on a flower). This term likely represents a distortion and generalization of the name "Kohlweißling" ("Kolwite") to include "white," "yellow," and "brown" butterflies. However, this is not entirely certain, as there was already a word "Kielitt" in Low German ("Plattdeutsch") for butterfly.

But back to "Schmetterling." In the Middle Ages, there was a widespread superstition that certain nocturnal butterflies would visit cows' udders in the dark to suckle milk. "Molkadiebe" (milk thieves) is a fitting name that has survived in some regions of Germany to this day. Additionally, it was widely believed that witches could transform into gray butterflies at night to steal butter, cream, and lard from farmers. In the Czech language, "*smetana*" means cream, and in East Central German, the word "Schmette" or "Schmetten" was used for cream. Through dissimilation, this led to the term "Schmetterling" for these "milk thieves," which we still use for this insect order today. According to another interpretation, "Schmette" might also mean "Seym" or honey. Therefore, butterflies would be "honey-people" that move from flower to flower to suck nectar.

115. Molkadiebschnopper, Mottakeenich and Tuud

In the „Glatzer Bergland" on the other side of the Adler Mountains in Silesia, the local people referred to a daytime butterfly as "Molkadiebschnopper" and a large nocturnal moth as "Mottakeenich" or "Tuud."

The origin of the word "Falter" is also difficult to reconstruct from the mists of the past. It likely derives from the ancient word "Byfaltera," which evolved into "Zwyfalter" and eventually the term "Falter." Today, it is not clear whether this term refers to the two pairs of wings that butterflies can fold together when at rest (similar to praying hands, as masterfully depicted by the Nuremberg artist Albrecht Dürer (1471-1528)).

In a German butterfly book from 1660, the Painted Lady (*Vanessa cardui*) is called "Brauner Weyfalter." In old writings, besides the term "Sommervogelin" (from the time of Walther von der Vogelweide), the name "Pfeifholter" (found as "Phiffholtir" in the Glossaria Augiensia of the Reichenau Monastery on Lake Constance from the 13th century) is also used for butterfly. Perhaps this can be traced back to the Latin word for butterfly, "Papilio"?

Individual butterfly names also show that our ancestors were keen observers. For some, the colors were decisive (such as Scarce Copper, Mourning Cloak, Blue), while for others, their distinctive behavior was key (e.g., "Bachantin" or "Tänzer" for *Lopinga achine*). The "Blutströpfchen" are named for their blood-red spots on a dark background. Some butterfly names refer to noticeable physical features, such as "Erpelschwanz" (*Clostera pigra*, Chocolate-Tip) or "Taubenschwänzchen" (*Macroglossum stellatarum*, Hummingbird hawk-moth). Nocturnal moths are often colloquially referred to as "Motten" (moths). However, I cannot provide much information about the origin of this term, except that it is related to the word "Made" (maggot).

116. The Book- and Clothes-Eating Moth

As early as the 13th century, writings mentioned the book-destroying impact of "moths." However, real "moths" were primarily seen as destroyers of textiles, which were stored in wardrobes or other closed containers for preservation and occasional use. At some point, the tiny clothes moth (*Tineola bisselliella*) chose such a "clothing container" as its new habitat - its larvae originally lived in bird nests - and thus became an enemy of humans. Even the Assyrians, over 2,800 years ago, had to contend with this pest. Since then, the history of fashion has also been intertwined with the history of moth control, which reached its first peak with the invention of the mothball in the late 19th century.

117. Naphthalene

The magic ingredient was called naphthalene, a white substance extracted from tar that - well, you guessed it - smells like mothballs. And that scent is something moth larvae simply cannot tolerate, causing them to perish.

The benefit of a wardrobe free from moth larvae holes used to come at the cost of wearing clothes that smelled like mothballs. However, the unpleasant odor of mothballs quickly led people to search for alternatives. The chemical industry was quite successful in this regard, and today's modern textiles often contain substances that deter clothes moths from the start. These advancements have largely made the tedious task of "mothballing" clothes for the winter or summer, depending on the season, a thing of the past. Instead, we can now safely store our clothes in the closet year-round without worrying about moth larvae damaging them.

Fig. 49: The Clothes Moth isn't so rare that it can no longer be photographed (Author)

The development of these protective textile treatments, however, has also impacted the population of clothes moths. As they find fewer suitable food sources, this little moth, once feared in many households, is becoming increasingly rare. While this is a relief for most people, there is a small group that views this decline with mixed feelings: lepidopterists, the butterfly researchers. For them, the decline of the clothes moth is a loss, as this unassuming moth plays an interesting role in the complex web of nature. Thus, while technological progress in the fight against moths has brought many benefits, it also comes at a cost to biodiversity.

118. The Skirt and What's Underneath...

In the Middle Ages, a "rock" typically referred to an outer garment with sleeves, worn by both men and women. This term has nothing to do with the English word "rock," meaning "stone," as an Anglicist might assume. Instead, it derives from the Old High German word "roc" (meaning "fabric") or "ruc" (meaning "to spin"). Before the advent of trousers for men or undergarments for women, the skirt, along with long stockings, was often the only piece of clothing worn by the less affluent. For men, it usually extended to about the knee, while for women, it reached down to the ankles. Typically, nothing was worn underneath the skirt, a fact that Scottish men in earlier times used as a psychological weapon in battles to intimidate their enemies.

Interestingly, the skirt - especially the women's skirt - gained a certain geopolitical significance two years after the storming of the Bastille, the event that triggered the French Revolution.

119. The Massacre on the Champ de Mars

I'm referring to the horrific massacre on the "Field of the Federation," commonly known as the Champ de Mars, on July 17, 1791, in Paris. On that exceptionally beautiful day, as historians report, a mass gathering of the Republican Party was planned. The people were called upon to sign a petition for the dethronement of the imprisoned King Louis XVI. To support this cause, a wooden platform was erected, where the petition was placed on the "Altar of the Fatherland" for signing.

A broad wooden staircase led up to the platform, which the citizens - and particularly the women with their long skirts (and wearing nothing underneath) - had to ascend to sign the petition. This "nothing underneath" detail caught the interest of two young men who, being French, had arrived early with provisions and a small keg of wine. They positioned themselves under the staircase, having drilled a few holes for better visibility. However, their venture went terribly wrong.

Fig. 50: Champ-de-Mars. On the day of the memorandum signing (July 17, 1791), a massacre took place among the participants...

120. Women turn into Hyenas

A few women discovered the voyeurs and began to beat them mercilessly, in line with Schiller's observation that in such situations, "...women turn into hyenas." Word quickly spread that two royalist spies had been found under the staircase with a barrel of gunpowder, supposedly planning to blow up the "Altar of the Fatherland." Furious men then joined the fray - not with fists and feet like the women, but with knives in hand. According to one version of events, the men stabbed the two voyeurs, decapitated them, placed their heads on pikes, and triumphantly paraded them around the square. Another, more plausible version claims that they were simply hanged from the nearest lamppost without much ado.

The ensuing uproar alarmed the royalist-leaning military under Marie-Joseph Motier, Marquis de La Fayette (1757-1834), who was then pelted with stones by the enraged crowd - all Republicans. At some point, the first shot was fired. Shortly thereafter, over a hundred people (though newer research suggests it was at most 50) lay dead around the "Altar of the Fatherland." The rest of the story is quickly told. La Fayette lost his reputation among the people, Danton fled to England, Desmoulins and Marat went underground, and many others were arrested. Finally, on January 21, 1793, Louis XVI was executed, followed by his wife, Marie Antoinette (the one with the cake), on October 16 of the same year. Both were beheaded by the famous and highly regarded Parisian executioner, Charles-Henry Sanson, using a machine invented by a doctor whose name remains well-known to this day. It's intriguing to imagine what might have happened if, back then, the women had already been wearing underwear and shorter skirts...

121. Decisive Moments in History

There are moments that, at first glance, seem insignificant yet have the potential to steer world history in an entirely new direction. These seemingly trivial instances can become pivotal turning points that define entire eras and forever alter the course of history. The Austrian writer Stefan Zweig (1881-1942) compiled a remarkable collection of such events under the title Sternstunden der Menschheit (translated as Decisive Moments in History). First published in 1927 and expanded in 1943, Zweig vividly recounts twelve historical moments when history held its breath. His book may not be lengthy, but the power and significance of the episodes he describes make it essential reading for anyone interested in the nuances and twists of human history.

Zweig masterfully transports the reader into the crucial moments of these "decisive hours," making the drama and fatefulness of the events palpable. Each chapter recounts a moment when the destiny of a nation - or even all of humanity - hung by a thread, and a single decision or tiny event tipped the scales one way or the other.

I would like to briefly mention three of these "decisive moments" that personally impressed me the most: The Fall of Constantinople in 1453 by Sultan Mehmed II, an event that marked the end of the Middle Ages and the beginning of a new era; the death of Leo Tolstoy, a farewell to one of the greatest writers in world literature, a moment of symbolic significance not just for Russia but for the entire world; and finally, the race to the South Pole, the tragic story of polar explorer Robert Falcon Scott (1868-1912).

122. "... if not, what remains of glory?" - Robert Falcon Scott

„Was bleibt, was bleibt nach dem Tode, wenn nicht bleibt, wenn nicht bleibt der Ruhm? – Große Tat, großes Menschentum!" is a lyric from the 1976 song by "Stern Meißen," who were inspired by this story (the text, which is easily found on the Internet, was written by Kurt Demmler (1943-2009)).

Fig. 51: In 1901, Scott (fifth from the right) was the captain of the wooden three-masted ship Discovery, with which he undertook his first expeditions to Antarctic waters (Wikimedia).

Robert Falcon Scott, like his rival Roald Amundsen (1872-1928), was no novice in polar exploration when he set out on November 1, 1911, for his ill-fated march to the South Pole. By that time, he had already attempted to reach the South Pole once before, along with Edward Adrian Wilson (1872-1912) and Ernest Shackleton (1874-1922). During the so-called Discovery Expedition (named after the ship that carried them to Antarctica), he managed to cross the 82nd southern latitude, a record at the time. If anyone could reach the South Pole, it was him. Thus, he and his companions, including the ornithologist Edward A. Wilson, prepared for a new expedition, which entered polar exploration history as the Terra Nova Expedition (again named after the ship). The expedition was successful to the extent that Scott and his three companions - Edward A. Wilson, Henry R. Bowers, and Edgar Evans - reached their goal, the South Pole, on January 17, 1912. However, Norwegian Roald Amundsen had already arrived nearly five weeks earlier, on December 14, 1911, as part of the Amundsen-Fram Expedition, and had hoisted the Norwegian flag to mark his success. The return journey of the British team across the Ross Ice Shelf turned disastrous and ended with the deaths of the last three expedition members around March 29, 1912, after Edgar Evans had already died earlier from injuries sustained in an accident (he fell into a crevasse and suffered a concussion, from which he could not recover under the harsh Antarctic conditions).

This return march of over 1,000 kilometers can be well reconstructed, as Robert Scott kept a regular diary, which was later found in November 1912 along with his mortal remains. Since then, this expedition has been romanticized as a tragic race between Scott and Amundsen and has been depicted in various literary forms. "Was bleibt nach dem Tode, wenn nicht bleibt, wenn nicht bleibt der Ruhm?" (What remains after death, if not glory?) as the song by Stern Meißen says. The bodies of Scott, Wilson, and Bowers, encased in the ice of the Antarctic Ross Ice Shelf, will likely be released into the Ross Sea around the year 2270 due to glacier drift. Their current position is unknown.

123. The Whaler Terra Nova

The story of the polar-capable sailing ship Terra Nova, used by Scott's expedition, is also fascinating. Originally a whaling ship, it spent nearly 60 years navigating the polar seas of both hemispheres before sinking in 1943 after a collision with an iceberg off Greenland. (The 24-man crew was rescued by the U.S. Coast Guard.) At that time, it was serving as a supply ship for strategically important stations in Greenland. Built in 1884 from thick oak in Dundee, Scotland, the three-masted vessel was chartered by the British Admiralty in 1903 to retrieve the stranded Discovery Expedition, in which Robert Scott also participated, as previously mentioned. Subsequently, it primarily operated in the high northern latitudes according to its original purpose until it was chosen for the 1911 expedition to the South Pole, following extensive modifications.

With a crew of 50, three motor sledges (which failed miserably), 19 Siberian ponies, and far too few sled dogs, the Terra Nova finally reached Cape Evans in January 1911, from where Robert Scott would set out on his ill-fated march ten months later. The wreck of the S.S. Terra Nova was discovered in 2012 at a depth of 300 meters in the Labrador Sea off Greenland.

124. Sonar

This accidental discovery occurred during the testing of a new specialized sonar system ("*Multi-beam Mapping Sonar*") on the flagship of the private Schmidt Ocean Institute in Palo Alto, California. Sonar is a typical invention from wartime. It emerged alongside Germany's declaration of unrestricted submarine warfare against Britain in 1917, which made the search for an effective method of locating submarines imperative. Consequently, many engineers in all the naval warfare nations worked on finding a practical solution. In Germany, the physicist Alexander Behm (1880-1952) from Mecklenburg was the one who filed a patent application with the Reich Patent Office in 1913 for a system that anticipated the fundamental operating principles of sonar.

Fig. 52: Submarine warfare during World War II increasingly became a suicide mission for German U-boat crews due to sonar technology (scene from "Das Boot" by director Wolfgang Petersen (1941-2022)).

To commercially exploit this invention, the Behm Sonar Company was founded in Kiel in 1920. Its most prominent client was the Graf Zeppelin shipyard, which showed significant interest in a sonar-based method for determining altitude over the seabed and conducted corresponding tests. However, it was still a long way from achieving a practically useful application that could accurately locate not just water depths but also the position of submarines. Between the world wars, several efforts were made in this regard, leading to various deployable sonar devices. The breakthrough came in World War II with the development of active sonar. Its search signal is identified by the characteristic "ping," often used in submarine films. This device allows for determining both the location and distance of objects that reflect sound waves. The first wartime-ready device was the British ASDIC, which, in combination with radar from 1943, allowed the Allies to increasingly turn the Battle of the Atlantic in their favor. As a result, submarines gradually lost their tactical advantages in the tonnage war, a fact that the German War Ministry under Grand Admiral Dönitz (1891-1980) was reluctant to acknowledge for a long time.

Submarine warfare was almost always a deadly profession during World War II. Anyone interested in learning more should definitely read Lothar Günther Buchheim's (1918-2007) book "Das Boot" (or alternatively, watch the film of the same name by Wolfgang Petersen, preferably the director's cut). It compellingly conveys the claustrophobic feeling of alternating between the debilitating boredom aboard during the approach to the combat area and the terror of responding to a torpedo attack, experienced by the crew in the oppressive confines of a steel tube.

125. Claustrophobia

"Claustrophobics" were evidently completely out of place here. It is now estimated that three to four percent of all people suffer to varying degrees from this psychological disorder, which is popularly but not entirely accurately referred to as "agoraphobia."[27]

126. Assassination of the President

Abraham Lincoln (1809-1865), the 16th President of the United States, is also believed to have suffered from this condition, though it had little impact on his administration. Lincoln apparently had a sufficiently large office, so this issue largely went unnoticed by outsiders and had no known effects on his presidency. However, his crucial term came to a sudden and tragic end on the evening of April 14, 1865. That night, he attended Ford's Theatre in Washington with his wife, Mary Todd, and some friends, including Major Henry Rathbone, to watch the comedy *Our American Cousin*. During the performance, actor and staunch Confederate supporter John Wilkes Booth (1838-1865) entered the presidential box and shot Lincoln in the back of the head at close range. The severe injury led to Lincoln's death the following day, plunging the nation into deep mourning.

Booth, who believed that by carrying out the assassination he was aiding the Southern cause, survived the President by only a few days. After a massive manhunt, he was tracked down by Union soldiers on April 26, 1865, in a barn in Virginia. During an attempt to arrest him, a shootout ensued in which Booth was killed. The assassination of Lincoln marked not only the end of a significant political career but also a turning point in American history that forever altered the course of the nation.

---

[27] Cases of claustrophobia are not well-documented in historical records. However, there are indications and hints that Napoleon Bonaparte and Nicholas II of Russia suffered from agoraphobia...

127. The Mary Surratt Justice Scandal

This was followed by what can only be termed a "judicial scandal" by today's standards. In a court proceeding, four individuals involved in or allegedly involved in the conspiracy were sentenced to death and publicly hanged, including Mary Surratt (1823-1865), who was highly likely to have been innocent. She was, incidentally, the first woman in the United States to be executed at the behest of federal authorities, a grim tradition that continues to this day.

128. Selbstmordattentäter, bomb ein bischen später...

Assassinations and assassins have been prevalent throughout human history, and in some cultural contexts, assassination - often carried out by a "suicide bomber" - remains the weapon of choice for terrorists. Observing the news, it seems that particularly the Sunni and Shia Muslim communities are heavily impacted by this form of terrorism. There is a reason for this.

("Selbstmordattentäter, bomb ein bisschen später" is the refrain of a satirical song by the cabaret artist Andreas Rebers).

129. Sunnis and Shiites

You may remember that in the 7th century, the direct descendants of the Prophet Muhammad conquered large parts of the Middle East, North Africa, and the Iberian Peninsula, giving rise to the Umayyad Caliphate. Its most important representative was Caliph Uthman ibn Affan (574-656), who ruled from 644 to 656. He would have likely ruled a little longer, but he was murdered, as was not uncommon at the time, by being lynched (to use the modern term) by his Egyptian opponents. Thus, Ali, the Prophet's son-in-law, suddenly became caliph of the vast empire that had been united. The Umayyad governor of Syria, Muawiya ibn Abi Sufyan (603-680), was not pleased with this development and declared himself "counter-caliph," leading to a civil war. An enmity developed that did not end with Ali's assassination just four years later. Ali's followers considered themselves the rightful heirs of the Prophet, while the Umayyad supporters insisted on representing the orthodox position of Islam, the Sunnah. They became Sunnis, while the followers of Ali became Shiites. One might cynically observe that their mutual animosity led to violent conflicts - a grim chapter in the long history of human folly. A particularly devastating form of such conflict, the suicide attack, has unfortunately emerged in the modern era.

The one ingredient has always existed: fanatical, uneducated, easily influenced, desperate, ideologically misled people, abused by third parties and, of course, fighting for their country for patriotic reasons.[28] Effective explosives carried invisibly on the body in special vests and vehicles in which, as is well known, particularly large quantities of them can be deposited are the ingredients that only modern times have provided.

Most suicide attacks today are clearly linked to Islamic fundamentalism, in which martyrdom and obscure promises of paradise form the ideological foundation that drives people, including highly educated individuals, to commit suicide with the fervent desire to take many others with them to their deaths. Consider the events of September 11, 2001, or the ongoing violence in Iraq today. The overriding aim is almost always to achieve the greatest possible media attention through terror in order to achieve political goals in the medium and long term (which, however, rarely succeeds in the long run). Examples of this can be found in both the offline and online press every day, but they are usually only mentioned briefly or as side comments (unless Europe is directly affected).

130. Tyrannicide

While today's suicide attacks aim to achieve a broad effect, i.e., as many deaths as possible among the "enemies" or "infidels," the classic "tyrannicide" was focused solely on one person - the "tyrant." However, this type of murder, which aims to eliminate an unjust or perceived unjust ruler, was rarely conceived as a suicide attack, although it almost always ended up being one. In more recent history, the Hitler assassination attempt of July 20, 1944 (by Colonel Claus Schenk Graf von Stauffenberg) and, in earlier history, the murder of Gaius Julius Caesar on the "Ides of March" in 44 BC are just two examples.

Since ancient times, legal scholars have debated at length the conditions under which the murder of a tyrant is legitimate and when it constitutes a crime. Ultimately, the view of Thomas Aquinas (1225-1274) prevailed in this respect, at least in the Christian West. He considered the violent removal of a person who had usurped state power to be legal and necessary under certain conditions. This view is reflected, for example, in Article 20, paragraph 4 of the German Basic Law.[29]

131. The older and younger tyranny

The term "tyrannis" did not have a negative connotation from the outset. In ancient Greece, it initially referred only to a specific form of rule over a polis (city-state), which can be described somewhat vaguely as "autocracy" (often referred to as "older tyranny"). The ruler was usually an influential and well-known man with a personal following who helped him to power by proclamation in agreement with the "city people" (but

---

[28] This is certainly a bit of a simplification, as suicide attacks are, strictly speaking, a complex phenomenon and cannot be reduced to a single cause or group.

[29] „Gegen jeden, der es unternimmt, diese Ordnung zu beseitigen, haben alle Deutschen das Recht zum Widerstand, wenn andere Abhilfe nicht möglich ist." (All Germans have the right to resist anyone who undertakes to remove this order if no other remedy is possible.)

not necessarily altruistically). Only rule usurped by a "tyrant" was considered objection-able even in antiquity and, as it was later perceived as the opposite of legal "kingship," was referred to as "lesser tyranny." This was always the case when this rule was based on violence and could only be maintained under the threat or exercise of violence. Such a "tyranny" often met the fate of being overthrown by force itself - whereupon the next tyrant usually followed...

According to the "*Hyginus Mythographus*," a collection of Latin excerpts from the 2nd century AD, Dionysius II of Syracuse in Sicily (396–337 BC) is said to have been a partic-ularly cruel tyrant.

132. „Die Bürgschaft" - by Schiller

The story it recounts inspired Friedrich Schiller (1759-1805) to write his famous ballad "Die Bürgschaft", the last lines of which, "*Ich sey, gewährt mir die Bitte, In eurem Bunde der dritte.*" ("I am, grant me the request, the third in your alliance"), have probably been heard by everyone at least once.

133. Jurassic Park

The rather negative meaning of "Tyrannis" is likely to have prompted Henry F. Osborne (1857-1935), president of the American Museum of Natural History, in 1905 to name the animal, of which a few particularly impressive fossilized bones had previously been found in Colorado and Wyoming, "*Tyrannosaurus rex*", meaning nothing more than "king lizard." The name is well chosen because, in real life, you really wouldn't want to meet this animal, which, thank God, is long extinct. The basic data derived from the bones alone are impressive: length approximately 12 meters; height 9 meters; weight between 5 and 6 tons; skull length up to 1.5 meters; carnivore (likely preyed on Triceratops, but probably also scavenged carrion); bipedal posture; maximum running speed 15 to 30 km/h; lived 67 to 65 million years ago in North America; and lived to be around 30 years old. It can be seen in action in various Hollywood films, where it makes "Jurassic Park" virtually unsafe. Unfortunately, there is no record of the gruesome sounds it made as it moved through the landscapes of the Maastrichtian.[30]. In this respect, therefore, the films mentioned are not necessarily to be trusted.

A particularly beautiful bony specimen of a king lizard is "Sue", whose (almost) complete skeleton was found in 1990 in South Dakota on the Cheyenne River Sioux Indian Reser-vation by Sue Hendrickson and carefully excavated. It is now fully re-stored in the Field Museum of Natural Historyin Chicago and can be viewed there if you ever find yourself

---

[30] This chronostratigraphic stage of the Upper Cretaceous covers the period from around 72 to around 66 million years ago and, as is well known, ended with a big bang in the area of the Gulf of Mexico, which also killed off the T-rex...

in Chicago. But anyone who believes that "*Tyrannosaurus rex*" was the largest land-dwelling carnivore of all time is mistaken...

134. The big brother of Sue

And then there is the Spinosaurus, a truly impressive dinosaur. While the famous *Tyrannosaurus rex* was only 12 meters long, the Spinosaurus was a true giant among predatory dinosaurs, reaching a stately 18 meters. This giant lizard, which was found fossilized in what is now North Africa, surpassed the T. rex not only in length but also in weight. Its maximum weight was probably around nine tons, while the T. rex weighed only around six tons. The huge teeth of the Spinosaurus, which adorned its approximately 1.75-meter-long skull, were particularly striking. With these huge teeth, it presumably not only hunted large fish but may also have preyed on other dinosaurs of its time - the early Upper Cretaceous.

Fig. 53: Exposed skeleton of a *Spinosaurus aegyptiacus* in the museum of the National Geographic Society in Washington D.C. It was probably, as recent studies seem to confirm, more of an aquatic animal,[31] but could also move quite well - similar to today's crocodiles - on land.

However, the most striking feature of Spinosaurus was undoubtedly its mighty dorsal sail. The bones of this sail protruded like spikes from the sedimentary rock where it was found in Morocco. This is why this dinosaur was given the name "spiny lizard" - an apt description for a predatory dinosaur of this superlative size, at least among carnivores.

---

[31] Henderson, D.M. (2018). A buoyancy, balance and stability challenge to the hypothesis of a semi-aquatic Spinosaurus Stromer, 1915 (Dinosauria: Theropoda).

In contrast to its American relatives such as the Tyrannosaurus rex, the Spinosaurus lived mainly in water and fed primarily on slightly larger fish. On land, it probably appeared rather clumsy and awkward. It is best imagined as an oversized mixture of crocodile and duck, mainly because of its paddle-like feet, which enabled it to live an amphibious life.

The discovery of its fossilized bones in 1975 not only provided new insights into this fascinating dinosaur but also answered a long-standing question in paleontology. Scientists had wondered whether there were carnivorous dinosaurs that lived not only on land but also in water. The example of Spinosaurus showed that there were even quite large specimens of these amphibious predatory dinosaurs in Africa.

135. How big can an Animal get?

Which raises the question: How big can an animal, especially a land dweller, actually get? Are there any limits, and if so, what are they? The fact that animals cannot grow to any size is a result of biomechanics. The balance between volume (weight), strength, the architecture of the internal skeleton, and mobility requirements plays an important role. After all, you don't want to collapse under your own weight - not even as a dinosaur. And the bigger you are, the bigger your appetite, of course. As is well known, a huge body size can only be sustained with a corresponding abundance of food.

Fig. 54: This is what it is said to have looked like, the *Mapusaurus roseae*, which once made Argentina dangerous for various herbivores (Wikimedia)

In the case of dinosaurs, however, this phenomenon, known in technical jargon as "gigantism," was not solely a matter of appetite. The reason lay in a "parallel development" between carnivores and herbivores, with herbivores being preyed upon by carnivores. The evolutionary drive, further supported by the physiology of reptiles (reproduction through eggs, bird-like lungs, high metabolic rate with high food intake), resulted from the fact that potential prey tried to reach a size that was no longer manageable for carnivores' mouths. Thus, their size and bodies grew with each evolutionary step. This allowed them to be preyed upon by even larger predators, which, in turn, led to herbivores evolving to become even larger. This evolutionary arms race culminated in the Argentine Puertasaurus, which during the Upper Cretaceous period was a harmless herbivore reaching up to 40 meters in length and the height of a multi-story building. Not even the fearsome Mapusaurus roseae, which at 12.5 meters long was slightly larger than its North American relative, the T. rex, dared to approach this towering giant of flesh.

Gigantism is an evolutionary concept that is not limited to dinosaurs. There are also many examples of it today. Just think of the giant squid (up to 14 meters long) or the giant oarfish, which can grow to a length of 17 meters, making it the longest bony fish in the world. Both examples live at great depths, which is why they are rarely seen. There are many stories about them, but most are dismissed as "sailor's yarn," although there is a grain of truth in some of them...

However, gigantism is not just a phenomenon of the animal world.

136. The Tower of Babel

It has also become particularly widespread in the human world. It started with the "Tower of Babel" and is far from ending with Erdogan's palace in Ankara, which our new Pope Francis (who certainly knows the story of Babel inside out) was allowed to visit some time ago. And the story of Babel and its tower, which Pieter Bruegel the Elder (1530?-1569) painted particularly impressively in 1563 and which, as Robert Koldewey (1855-1925) discovered,[32] was in reality a ziggurat, should be a lesson. More details can be found in the Bible, in the Old Testament, Genesis 11, 1-9, where you can read in detail:

"Now the whole world had one language and a common speech. As people moved eastward, they found a plain in Shinar and settled there.

They said to each other, "Come, let's make bricks and bake them thoroughly." They used brick instead of stone, and tar for mortar. Then they said, "Come, let us build ourselves a city, with a tower that reaches to the heavens, so that we may make a name for ourselves; otherwise we will be scattered over the face of the whole earth."

But the Lord came down to see the city and the tower the people were building. The Lord said, "If as one people speaking the same language they have begun to do this, then nothing they plan to do will be impossible for them. Come, let us go down and confuse their language so they will not understand each other."

---

[32] If you would like to find out more about Robert Koldewey, you should read the book "Götter, Gräber und Gelehrte" by Kurt Wilhelm Marek ("Ceram"), which is no longer up to date historically, but is a fabulous read...

So the Lord scattered them from there over all the earth, and they stopped building the city. That is why it was called Babel - because there the Lord confused the language of the whole world. From there the Lord scattered them over the face of the whole earth."

And that is the deeper reason why foreign language lessons are necessary today, why there are tools like Google Translate and DeepL, why Leo is not only understood as a useful computer application for students, and why you can earn real money with Babbel if you own it.

Fig. 55: The Tower of Babel as imagined by the Flemish painter Lucas van Valckenborch (1535-1597). The original oil painting can be admired in the Landesmuseum Mainz. It was probably painted in 1595 (Wikimedia)

But back to the tower and its discoverer, Robert Koldewey. Born in 1855, he studied architecture, archaeology, and art history in Berlin, Munich, and Vienna (though he did not obtain a degree). He then became a civil servant in Hamburg, a position that did not suit him. In 1882, he embarked on an expedition to Greece, where he met Francis H. Bacon from Boston (1856-1940), who later excavated Assos on the island of Lesbos, and his friend Clarke. Here, Koldewey developed an interest in archaeological research, which he pursued for the next few years on behalf of the German Archaeological Institute. His work took him first to Greece (Lesbos) and Sicily, and later, from 1892 to 1894 (with interruptions), to Mesopotamia, where he worked with Otto Puchstein (1856-1911). This was followed by a period as a teacher at a building trade school in Görlitz, where he visited both the Landeskrone and Mount Oybin, with its monastery and castle. He then returned to Mesopotamia at the age of 43 to excavate Babylon on behalf of Kaiser Wilhelm II. His excavations revealed many significant findings, some of which can now be admired in the original at the Pergamon Museum on Museum Island in Berlin. Among his discoveries are the famous Processional Way and the so-called "Hanging Gardens of

Semiramis," as well as the foundations of the Etemenanki, the main ziggurat of Babylon - known as the "Tower of Babel." Its base was a square with sides 90 meters long (which can still be seen on satellite images today). Above this base rose a stepped, truncated pyramid of probably seven stories, reaching an estimated total height of around 90 meters (not quite as high as the Great Pyramid of Giza).

The exact start of construction of this sanctuary is not known. However, literary sources written in cuneiform script estimate that it can be dated to the second millennium BC. The tower itself was destroyed several times in its history (most recently on the orders of Alexander the Great (356-323 BC)) and then rebuilt. All that remains in the memory of mankind, however, is the tale of the sacrilegious and unjustifiable desire to build a tower in the sky, with the aforementioned consequences.

It is interesting to note that the "Tower of Babel" - although undoubtedly a gigantomaniac structure - never found its way onto the list of the Seven Wonders of the Ancient World. The reason for this is probably quite banal. The Persian king Xerxes I (519?-465 BC) had it torn down as early as 480 BC, and Alexander finished it off around 323 BC after his original order to restore it was not followed. When the list of the Seven Wonders was drawn up, the tower probably no longer existed in a form that would have allowed people to remember its magnificence or see it for themselves. Experience shows that all buildings, as wonderful as they may be, have a tendency to collapse at some point (I mean, of course, with the exception of the Pyramids of Giza...) if nothing is done to preserve them.

137. The Entropie Theorem

From a physical point of view, this is a direct consequence of the so-called 2nd law of thermodynamics, which in its Faustian form reads: *"Alles, was entsteht, ist wert, dass es zugrunde geht"* (Everything that is created is worth destroying). In "scientific" terms, however, it sounds like this: *"The entropy of a closed system is either constant or increasing"* - which is why the 2nd law is usually referred to as the "entropy law" for short. However, in order to understand what "entropy" actually is, you first have to understand what "energy" is.

138. What is Energy?

"Energy" is a very frequently used word that appears in many different contexts. Someone who is bursting with energy has a lot of energy. Someone who wants to achieve a certain goal must expend a lot of energy to do so. And if you feel tired, "Red Bull," the energy drink, helps. All of this can certainly be linked to the Greek word "energeia," which means "acting force" and also leads directly to the physical concept of energy.

Energy is, in everyday terms, the ability of a body to perform work.

Admittedly, this is a very vague definition, since in addition to "mechanical" energy, there is also chemical energy, thermal energy, radiant energy, electrical energy, atomic energy, and so on, which seem to be completely different. This already shows that the physical concept of energy is obviously very abstract. It has taken a correspondingly long time to define this concept scientifically. Today we know that energy, or what this concept describes, is a conserved quantity resulting from a fundamental symmetry principle of our world, namely the temporal invariance of physical laws. In a world whose description does not depend on the point in time to which we refer (i.e., the starting time is freely selectable and does not influence the laws of nature), there must be a quantity (as stated in the famous Noether theorem, point 259) that remains constant under all conceivable circumstances. And this quantity is energy.

139. Energy cannot be renewed

Energy cannot be consumed or destroyed, or even renewed; it can only be converted into different forms of energy. If you want to remain conceptually precise, you should therefore never speak of "renewable" energy but at most "regenerable" energy if you wish to engage in the popular and mostly misguided energy debate. But that's just a side note. So, what actually happens when "energy is generated" or "energy is consumed"? Ultimately, only one form of energy is converted into another form. The total energy involved in these conversion processes does not change; it remains the same. This is precisely what the fundamental law of conservation of energy states - a law that cannot be repealed by any parliament in the world (not even by Brussels).

Here is an example that is easy to understand: I am referring to the gasoline engine that usually powers our car (unless it is a diesel or electric vehicle).

Gasoline is a mixture of different hydrocarbons whose molecular bonds contain chemical energy. During combustion in the engine, which chemically corresponds to oxidation, these bonds are broken, and new, lower-energy bonds are formed with the oxygen extracted from the air (producing, among other things, the well-known carbon dioxide and, at high temperatures, nitrogen oxides). However, the energy content of the reaction products (more precisely, their chemical energy content) after oxidation is lower than the energy content of the unburned gasoline. Since the inexorable law of conservation of energy watches over everything, the difference must be converted into another form of energy, in this case, thermal energy. This thermal energy rapidly heats the gas in the cylinder, which expands and moves the piston (mechanical energy or mechanical work output). However, not all of the thermal energy is transformed into mechanical energy as one might think. A significant amount escapes with the gas from the cylinder into the environment - through the exhaust (where, as every Formula 1 fan knows, some of the energy can still be converted into acoustic energy). Via the crankshaft and the gearbox, the piston drives the wheels that move the vehicle. Since friction occurs everywhere despite oiling, only a small part of the chemical energy actually reaches the wheels to give the vehicle the desired speed (expressed by its kinetic energy).

If you now take the total energy contained in the chemical bonds at the beginning and released during combustion and compare it with the sum of the forms of energy generated in the engine, you will find that they are exactly the same. This is exactly what the law of conservation of energy states. Such a conversion chain applies to every energetic process.

140. There is no energy consumption

In this sense, there is no energy consumption. This term describes something else, as I will explain later. But it goes even further. In every energy conversion process, as I mentioned, there are losses that reduce the proportion of useful energy. They usually manifest themselves in the form of thermal energy, which is released into the environment. This type of energy has the property that, in most cases, it can hardly be used for any practical purpose (well, you could fry an egg on the engine block). In our gasoline engine, for example, this includes the heat from the heated gases that leaves through the exhaust or heats the engine block itself. It therefore makes perfect sense to speak of a "value" (in the sense of utility value) of a form of energy.

141. The value of energy

The "value" of energy lies in driving physical processes. Or to put it more precisely, energy keeps the world in motion by converting "valuable energy" (free, concentrated, capable of doing work) into "less useful" energy (bound, depleted, diluted, so to speak). In the process, the energy is "devalued," which is usually referred to as "energy consumption." For example, electrical energy in a traditional light bulb can be very effectively converted into barely usable heat energy and a fraction of this (~5%) into useful light energy. This light energy could, in turn, be used to illuminate the photovoltaic panels on the roof of the house at night, which are useless at this time. After all, they would be able to convert ~20% of the incident light back into electrical energy. However, a cycle in which the electrical energy generated in this way is used to feed the light bulb will not be achieved in this manner. That would also be a so-called "perpetual motion machine," and many "inventors" have already tried to create one, but, predictably, without success.

As you can see, a light bulb is the ideal device for wasting energy very effectively. The ratio of useful power (light) to the power consumed is what is known as efficiency: A normal light bulb has an efficiency of just 5%.

Physically, the "devaluation" of energy is linked to the irreversible processes that always accompany it. Energy devaluation cannot be reversed by itself. However, other irreversible processes can certainly be used to reverse a given process that is irreversible in itself. And I, the author, must now[33] try this out for myself. My cup of coffee has become

---

[33] 24.02.2024, 17:30, study, together with Humpel the cat, who is licking up the piece of cream cake mentioned in the example from the floor

less valuable, as I have just realized with dismay, because it has cooled down while I am laboriously transferring my thoughts to the notebook keyboard. To reverse this process, I have no choice but to put it in the microwave for 30 seconds to consume valuable electrical energy (how much is measurable, e.g. via its price in euros). In any case, my coffee would never have warmed up again by itself...

142. What is Entropie?

And this brings us to another extremely abstract concept, the aforementioned concept of entropy. To put it simply, entropy is what prevents you from turning a scrambled egg back into an intact hen's egg, or that cold coffee does not become hot again on its own, or that my piece of cream cake, which has fallen on the floor, does not jump back onto my plate by itself so I can continue to enjoy it. And - I'm sure you can think of plenty more examples yourself...

As I have just explained, the reversal of a physical process that occurs on its own can always be influenced by the subsequent course of another physical process that takes place independently. This means that the "degree" of energy degradation can now be quantified. The physical process that can reverse the other, so to speak, is the one with the greater energy degradation. Here is where the concept of entropy comes into play. Etymologically, the Greek word "entropy" simply means "transformation".

As has just happened, the value of my cup of coffee has been greatly reduced by cooling. It therefore makes sense to take entropy as a measure of this devaluation, or, as physicists like to do, to formally consider the amount of devalued energy (= entropy) as proportional to the amount of heat released into the environment, i.e. as the entropy difference between hot and cold coffee: $\Delta S \sim \Delta E = \Delta Q$.

Now, it obviously makes a difference whether my hot coffee cools down in the garden in winter when it's freezing or in my 20 °C warm room (in winter it can even freeze if you're not careful - you could then quite rightly call it "iced coffee").

In the first case, the amount of heat released into the environment is obviously greater than in the second case (we assume that the coffee cools down completely, i.e. to the ambient temperature). In other words, the ambient temperature also plays a role in the quantification of entropy.

Formally, you can write $\Delta S = \Delta Q / T$ for it. So what does the total entropy change of my beloved cup of coffee look like? It obviously consists of two parts. Firstly, the entropy of the coffee decreases by $\Delta S_{Kaffee} = -\dfrac{\Delta Q}{T_{Kaffee}}$ as it cools down to the ambient temperature and secondly, the entropy of the environment increases by $\Delta S_{Umgebung} = \dfrac{\Delta Q}{T_{Umgebung}}$. The total entropy is therefore $\Delta S_{gesamt} = \Delta Q \left(\dfrac{1}{T_{Umgebung}} - \dfrac{1}{T_{Kaffee}} \right) > 0$.

It is only important to note here that the total entropy always increases to the same extent as the energy devaluation progresses. Energy devaluation is therefore equivalent to the generation of entropy. And the total entropy in a closed system can only ever increase: $\Delta S_{gesamt} \geq 0$. This truly fundamental theorem is the second law of thermodynamics. And it also has fundamental consequences for the generation of energy (= generation of high-quality usable energy) and for the consumption of energy (= the devaluation of energy by initiating useful physical processes, e.g. to enable a computer to calculate).

What I have just tried to explain is the thermodynamic aspect of entropy. However, entropy, like energy, is a very abstract concept that has other interpretations (which are, strictly speaking, equivalent to thermodynamic entropy, but this is usually not apparent at first glance). The actual (microscopic) justification for this special state variable is found in statistical mechanics. Additionally, entropy has a fundamental significance in information theory (in this context, we refer to it as "Shannon entropy").

Whenever something changes in our world, energy conversion is involved. However, according to the second law of thermodynamics, energy conversion always results in entropy production. This manifests itself in the even distribution of heat or substances in the environment (e.g., as waste).

143. Heat death of the universe

When all free energy has been converted and all substances are evenly distributed, a (closed) physical system eventually comes to rest. This is called its "heat death." An increase in entropy in a physical system is therefore associated with a reduction in its degree of order, i.e., every self-contained system tends to move toward a state of maximum disorder. This can be seen every day on the author's desk at home, for example: Three books there, the calculator here, the ballpoint pen next to it, a few scribbles, the coffee cup, the notebook, a pile of files on the left, and Humpel the tomcat in the middle of the writing surface, taking up all the remaining space. A formerly well-ordered desk inevitably turns into a disordered state over time (this is not just a fact of experience, but a law of nature!). You could also say that its entropy inevitably increases over time. Entropy therefore has something to do with the distribution of things. There are many more possibilities for an untidy desk (high entropy) than for a well-organized desk (low entropy). In statistical mechanics, this consideration is "refined" and we speak of microstates and macrostates. Those who are mentally healthy will find a lot about this in Landau's textbook on theoretical physics (according to the saying, *"Wer Landau liest und Sommerfeld, der hat ein Recht auf Krankengeld"* (Whoever reads Landau and Sommerfeld is entitled to sick pay). So what is the conclusion? Energy must be used to tidy up a desk. The pragmatic person knows that this is a Sisyphean task. That's why I recommend aiming for a desk of medium entropy.

Fig. 56: Desk with relatively high entropy (tomcat Humpel on the right)

Let us now return to energy, keeping in mind the key point from the entropy discussion: "You must use energy if entropy is to decrease." Additionally, remember that the production of heat is associated with high entropy production. Furthermore, it is important to note that the entropy of a closed system can only increase. Unlike energy, entropy is not conserved. Energy, on the other hand, is always conserved; it can only be converted into different forms through various processes, although this conversion is never complete because a portion of the energy is always converted into heat (which can be considered a synonym for entropy). The "quality" of such a conversion process is measured by its efficiency. Achieving 100% efficiency is impossible.

144. The "measured Variable" Energy

Now we come to the question of how energy is measured. Today, everyone knows the unit of measurement in the SI system: the joule. Occasionally, you might still encounter the calorie (cal) in cooking programs or discussions about "losing weight," as it is sometimes considered more descriptive than the joule. A calorie is defined as the energy required to heat exactly one gram of water from 14.5 °C to 15.5 °C at one atmosphere of pressure (101.3 kPa). On the other hand, one joule is the energy required to lift a mass of 0.102 kg (approximately a ham sandwich) by one meter. If you hit a soccer ball inelastically against a wall at a speed of 30 m/s (108 km/h) (with an energy of about 194 J), this energy is theoretically sufficient to heat a glass of water by one degree. Although this seems like a very small amount of energy, it can, in the worst case, lead to a visit to the hospital emergency room. However, as paying citizens, we are generally accustomed to a different unit of energy: the kilowatt-hour, or kWh for short. One joule is exactly one watt-second (Ws). Since one hour has 3600 seconds, one kilowatt-hour corresponds to exactly 3600 kJ. This means that a 60-watt light bulb can be operated for approximately 17 hours with one kilowatt-hour. With one joule, it would only last 1/60 of a second. The energy from the soccer ball would not even keep the light bulb lit for 3 seconds – these are the amounts of energy we deal with in the household.

On the energy production side, however, the prefix "kilo" doesn't get us very far. There, the common units for power usually start with "M" for mega or "G" for giga, making it easy to confuse megawatts (MW) with megabytes and gigawatts (GW) with gigabytes, as a contemporary documentary preserved on YouTube embarrassingly proves. A large lignite-fired power plant has an electrical output of 2200 MW. If it operates continuously at this output for a year (i.e., 86,400 x 365 = 31,536,000 seconds), it produces an energy of approximately 2200 x 31,536,000 = 69,385,200,000,000 J (69.4 TJ). This means that our light bulb could stay lit for approximately 37 million years if it did not burn out before then. To generate this amount of energy, you would need 20.4 million tons of brown coal. If shaped into a cube, this would form a cube with an edge length of about 280 meters (many times larger than the Tower of Babel). This implies that over 2000 tons of brown coal must be burned every hour. The same amount of energy could theoretically be generated from a cube of natural uranium (0.72% U235) with an edge length of 1.8 meters. Germany currently has eight lignite-fired power plants with a gross output of more than 1500 MW each (2023). Yet, a power plant with 2200 MW supplies just 0.5% of the country's primary energy consumption.

145. Primary energy and useful energy

This brings us to the important concepts of "primary energy" and "useful energy," which differ greatly due to losses (often related to the concept of "entropy") in the conversion chain. Primary energy is the energy available from naturally occurring sources (fossil fuels, wind, geothermal energy, sunlight, nuclear fuels) for initial energy production. Secondary energy is derived from this through conversion processes (primarily electrical energy, and sometimes heat energy for heating purposes), but is significantly reduced due to losses during energy recovery (for instance, German lignite-fired power plants have an average efficiency of about 38%). Moreover, this energy is further degraded during consumption, resulting in even less useful energy. It is unfortunate but true: ultimately, energy dissipates into heat, bringing the world ever closer to the concept of "heat death."[34]

---

[34] Of course, only if the world is thermodynamically a closed system. But who knows...

146. Primary Sources of Energy

o what are the primary sources of energy on Earth that are available to us for use? There are actually only three: sunlight, the tides, and the heat generated in the Earth's interior by the decay of radioactive isotopes. The latter are primarily derived from elements with which the matter from which our solar system was formed 4.53 billion years ago was interspersed. A supernova was not entirely innocent in this process. As we have already seen (see point 99), it made today's nuclear power plants with their opponents and supporters and their gold bars, gold teeth, and gold rings (if they should possess such jewels) possible in the first place.

All fossil energy sources - natural gas, oil, coal - have their origin in biomass, which in turn was primarily formed from solar energy over many millions of years through the ingenious process of photosynthesis. Wind energy and the energy of hydro-power also depend primarily on the sun, because the sun keeps the circulation of the atmosphere and the oceans going. Only tidal energy, which can be tapped by tidal power plants, comes from the gravitational effect of the sun and moon.

Our sun is an ordinary main sequence star of spectral class G2V with the easy-to-remember parameters of exactly one solar mass, one solar radius, and the luminosity of one solar luminosity (the bright shoulder star of Orion, Betelgeuse, on the other hand, has a mass of around 20 solar masses, a variable radius of around 660 solar radii, and a luminosity of around 55,000 solar luminosities, which is admittedly more difficult to remember).

So what is a solar luminosity? The luminosity of a star is the total energy that a star emits into the cosmos in a unit of time (i.e. a power). For the sun this is $3,845 \cdot 10^{26}$ W and for Betelgeuse 55,000 times as much. Of this, 1367 J per square meter and second arrive at the position of the Earth (i.e. at a distance of ~149 million kilometers). The quantity $S=1367$ W/m^2 is called the "solar constant". If you compare this with the current energy consumption of mankind ($\sim 1,74 \cdot 10^{17}$ W), then the solar power reaching the earth is around 12,000 times greater. Thank goodness this energy is not stored on earth but is radiated back into the cosmos as reflected or thermal radiation. This state is called "radiation equilibrium". It determines the average temperature over the entire earth. Under Central European conditions, Germany's entire electricity production ($6,21 \cdot 10^{14}$ W) could be achieved, at least in terms of figures, using solar cells installed over an area of approximately (140x140) km^2 (~2800 times the size of Lake Müggelsee) - er, of course, only if the sun is shining...[35]

While we're on the subject of "sunshine", let's apply what we've learned about entropy.

---

[35] A fact that is difficult to convey to many enthusiasts of the energy transition in Germany

147. Entropy export and structure formation

The earth is an ideal exporter of entropy. This is the only reason why structure formation is possible at all and the only reason why we exist. The reason for this is the temperature gradient between the solar photosphere (~5800 K) and the Earth's surface (~260 K at the atmospheric boundary). In the equilibrium case (~$1{,}74 \cdot 10^{17}$ W), incoming radiation is equal to outgoing radiation. However, there is a qualitative difference between the two radiations. High-energy photons from the sun are converted into low-energy photons (thermal radiation) by absorption and re-emission and radiated back into the cosmos. The entropy change per unit of time then results in

$$\approx 10^{17}\ \text{W}\ ((1/(5800\ \text{K}) - 1/(260\ \text{K}))) = -4 \cdot 10^{14}\ \text{W/K}.$$

This means that around $4 \cdot 10^{14}$ "entropy units" are released into the cosmos every second. This corresponds to an entropy export of ~ 1 W/(Km^2) in relation to the earth's surface. In this way, we are not, colloquially speaking, suffocating in waste, but have free capacity for self-organization and structure formation. The fact that so much happens

148. Primary energy consumption

Let us return to the concept of primary energy consumption. This is the "consumption" of energy (e.g. through combustion, through "alternative" energies, through nuclear power and photovoltaics, etc.) that is needed to produce "useful" high-value energy, usually electrical energy. The latter is usually used to generate useful energy (for example to cook coffee or semolina porridge), whereby $E_{primär} > E_{Elektro} > E_{Nutz}$. For physical reasons, there are losses in each process step, and the more process steps are required before utilization, the greater the losses.

149. Biogas power plants

Let's assume that biogas (methane) is produced using a biogas reactor. By burning it, heat for heating and electricity can be generated relatively efficiently. However, it can also be converted into diesel via several process steps (keyword "steam reforming") in order to be able to drive around in a car. However, the usable energy is then only a fraction of that which could have been achieved by direct combustion of the biogas. Just as the primary energy can never be converted into "final energy" without losses, the "final energy" (e.g. in the form of electrical energy) can only be converted into useful energy with losses. The difference between primary energy and final energy in Germany is around $2{,}6 \cdot 10^{12}$ kWh. This corresponds to around 80% of the annual electrical energy production of all nuclear power plants in the world combined. It should also be mentioned that energy must, of course, also be used to produce primary energy. After all, something has to drive the coal excavators and oil pumps. And even if you dig for coal with a spade, you have to eat something first in order to be able to do the necessary work.

Today, humanity satisfies most of its hunger for energy by burning the energy stored in fossil fuels over many millions of years: 37% oil, 25% coal, and 23% natural gas, i.e., 85% of global energy production is based on fossil fuels. The rest is provided by nuclear power (8%), biomass (4%), and the famous "alternative energies" (3.8%). This leaves photovoltaics and geothermal energy, although their significance is negligible in global terms. Unfortunately, fossil fuels are finite. This also applies to nuclear energy through nuclear fission (uranium reserves worth mining are also limited), but this could be countered by a sustainable nuclear industry, as nuclear fuel can be "incubated."[36] Since "fossil" energy sources cannot be regenerated and nuclear power is frowned upon by some people, at least in Germany, let's take a look at the next candidate in the list, "biomass," which is usually green and the favorite energy source of the "Greens" - because it is "sustainable." Every year, you can virtually watch the "growth" of biomass when, in spring and summer, crop-free fields are transformed into beautiful rapeseed and maize fields within a few months. And from this, the biofuel E10 can then be magnificently produced.

The process that makes rapeseed and maize grow is photosynthesis. It was "invented" by "life" more than 3.2 billion years ago and is able to fix carbon dioxide driven by light and build up plant biological substances (e.g., carbohydrates such as sugar, starch, and cellulose) in the process. The reaction chain is very complicated in detail, and I will not go into it here. Just this much: In a so-called light reaction, water is split using sunlight, oxygen is released, and the hydrogen is temporarily stored in a special enzyme, which is known (not quite exactly) to biology students under the pretty name "nicotinamide adenine dinucleotide phosphate," at least shortly before the exam. In addition, the extremely energy-rich compound ATP (adenosine triphosphate) is produced. In the subsequent dark reaction, carbon dioxide is extracted from the atmosphere and converted into carbohydrates with the help of these two substances.

What does the energy balance look like? How much primary energy can be produced on our fields in this way? Well, the balance is sobering. Only around 1% of the incident light energy can be used at all through photosynthesis. What's more, maize doesn't emerge until the end of May, and the growing season is already over in October. Incidentally, this is the best time for wild boars, which can live in the maize fields hidden from the eyes of hunters as if they were in a land of milk and honey. Of the approximately 1000 kWh of solar energy that falls on one square meter of land in Germany on an annual average, just 10 kWh is stored in the form of biomass. This makes you wonder whether it wouldn't be better for the farmer to fill his field with solar cells right away. There is no need for sowing and harvesting, and the primary energy yield is around 30 times higher.

---

[36] At least in Germany. The situation is quite different worldwide.

Fig. 57: Energy maize - the symbol of an ecological desert and a symbol of the German "Energiewende"

If you do some research, you will find, for example, that approximately 4600 m^3 of methane can be produced from one hectare (100x100 meters) of maize using modern biogas plants. Methane has an energy content of 50 MJ per kg. Therefore, 4600 m^3 has a calorific value of approximately 46,000 kWh. This, in turn, can be used to generate around 17,000 kWh of electrical energy (due to the poor efficiency of approximately one-third). Ultimately, 100% of light energy is converted to about 0.17% electrical energy. Fertilization and transport (which also require energy, or at least biodiesel) further reduce the result. Now, it should be clear why agricultural subsidies are necessary...

As for bioethanol and biodiesel, I'll avoid an assessment of their effectiveness. The use of biomass to generate energy clearly only makes sense if waste that is produced anyway is utilized in this way. It simply cannot be that rapeseed, maize, and wheat are used primarily to produce E10 while people around the world are starving.

150. "Energiewende" and castles in the air

Perhaps you now see the "energy transition" a little differently and listen more attentively when the media proclaim huge successes in this respect, which, on closer inspection, turn out to be castles in the air. It would have been better if the apologists for the "energy transition" had read Schopenhauer's "Parerga and Paralipomena." For there is the theorem:

„In allem, was unser Wohl und Wehe betrifft, sollen wir die Phantasie im Zügel halten, also zuvörderst keine Luftschlösser bauen, weil diese zu kostspielig sind."[37]

Or, to put it another way: it would have been much more sensible to address the problem of effective energy storage first and develop technical solutions beyond "castles in the air", rather than starting with the expansion of "alternative energies" that are volatile, disfigure old cultural landscapes, cause noise pollution, negatively impact wildlife, and prove to be completely useless when the wind isn't blowing or at night.

This brings us to the not exactly trivial question of why it is dark at night - or, to put it another way, why the night sky appears "black" to us and not as bright as the sun.

151. Why is the night sky black?

You may now ask yourself, why should the night sky be bright? This is the seemingly inevitable consequence of an idea that dates back to antiquity and a consideration that arose during the Copernican revolution (point 312), namely the idea of an infinitely extended universe filled with stars...

That a universe with a "limit" is logically impossible was already shown by the famous Roman poet and philosopher Titius Lucretius Carus, known as "Lucretius" for short (probably 97-55 BC), who wrote in his work *"De Rerum Natura"*:

"If space is finite, it has a limit. Imagine that someone penetrates to this last point of the world and hurls an arrow against the boundary. Either the arrow will fly beyond the boundary, or something will stop it; something that must itself lie beyond the boundary. In either case, something is beyond the boundary. This demonstration can be repeated as often as you like and thus the alleged boundary can be pushed back infinitely."

In line with this argument, astronomers since Copernicus have tacitly taken it for granted that cosmic space is infinitely extended in all three dimensions. But the concept of the "infinite" has its own characteristics, as the infinitely many digits of the number pi have already shown (point 4). If there are an infinite number of stars (or, more modernly, galaxies) in an infinitely extended space that are approximately equally distributed, then every possible ray of light would hit a star somewhere in any direction. It is similar to

---

[37] In everything that concerns our weal and woe, we should keep the imagination in check, so first of all we should not build castles in the air, because they are too costly.

being in the middle of a sufficiently large forest (best seen in a spruce forest), where a tree trunk can also be seen somewhere in every (horizontal) direction.

152. Olbers Paradox

And it was precisely such an analogy that prompted the Bremen physician and amateur astronomer (who practiced his "hobby" like a professional astronomer) Wilhelm Olbers (1758-1840) in 1823 to send a paper entitled "Über die Durchsichtigkeit des Weltraums" to the editor of the Astronomisches Jahrbuch, J. E. Bode (1747-1826), who then published it. (The article can easily be found on the Internet via Google Books.) The essential facts, later known as "Olbers' Paradox," read as follows:

"... If there really are suns in the whole of infinite space, whether they are at approximately equal distances from each other or distributed in systems of galaxies, their number becomes infinite, and the whole sky would have to be as bright as the sun. For every line that I can imagine drawn by our eye will necessarily meet some fixed star, and so every point in the sky would have to send us fixed starlight, i.e., sunlight..."

An infinitely extended universe, in which stars are evenly distributed (it doesn't even depend on star density!), should provide a night sky with the brightness of a medium-sized star like the sun. And this is obviously not the case.

Solving this paradox proved to be extremely difficult, and even today it is still used to test various cosmological models, serving as a kind of knock-out criterion. The reason for this is that the conclusion drawn by Olbers is simply compelling, assuming an infinitely extended universe that is uniformly populated with stars and has always existed. The argument that stars become fainter with increasing distance does not help either. Although the brightness of a star decreases with the square of the distance (i.e., $1/r^2$), the number of stars increases with r^3. People then tried "dark clouds," which would ultimately shine just as brightly as the stars for physical reasons (keyword "Kirchhoff's radiation law"); they tried a certain hierarchical distribution of stars in the cosmos (world models according to Carl Charlier (1862-1934)) and came up with a few other explanations, all of which have their own quirks. However, a solution began to emerge around 1854 in the form of a colloquium paper by mathematician Bernhard Riemann (1826-1866) entitled "Ueber die Hypothesen, die der Geometrie zu Grunde liegen" (On the Hypotheses Underlying Geometry) - albeit without making any reference to the aforementioned paradox.

153. Curved spaces and the parallelism problem

There are (mathematically speaking) "spaces" that are finite but unlimited. Olbers and his contemporaries naturally started from the only "type of space" that they could imagine, namely a so-called "Euclidean space," which is known to be "infinite" and without limits. This "space" is characterized, among other things, by the fact that two parallel lines never intersect, not even at "infinity." In Euclid's geometry (the basic geometry taught in school), it is introduced as the fifth postulate or axiom, which is why it is also referred to as the "axiom of parallels." It had to be introduced as an axiom because it could not be proven as a theorem within the framework of the other four axioms, although many mathematicians had tried to prove it since Archimedes. It was Carl Friedrich Gauss (1777-1855) who first recognized that this "parallel problem" could not be solved in principle. Today, we also know why. This is because Euclidean geometry is only one very special geometry among an infinite number of other conceivable ones, which also describe "curved" spaces and in which the axiom of parallels does not apply.

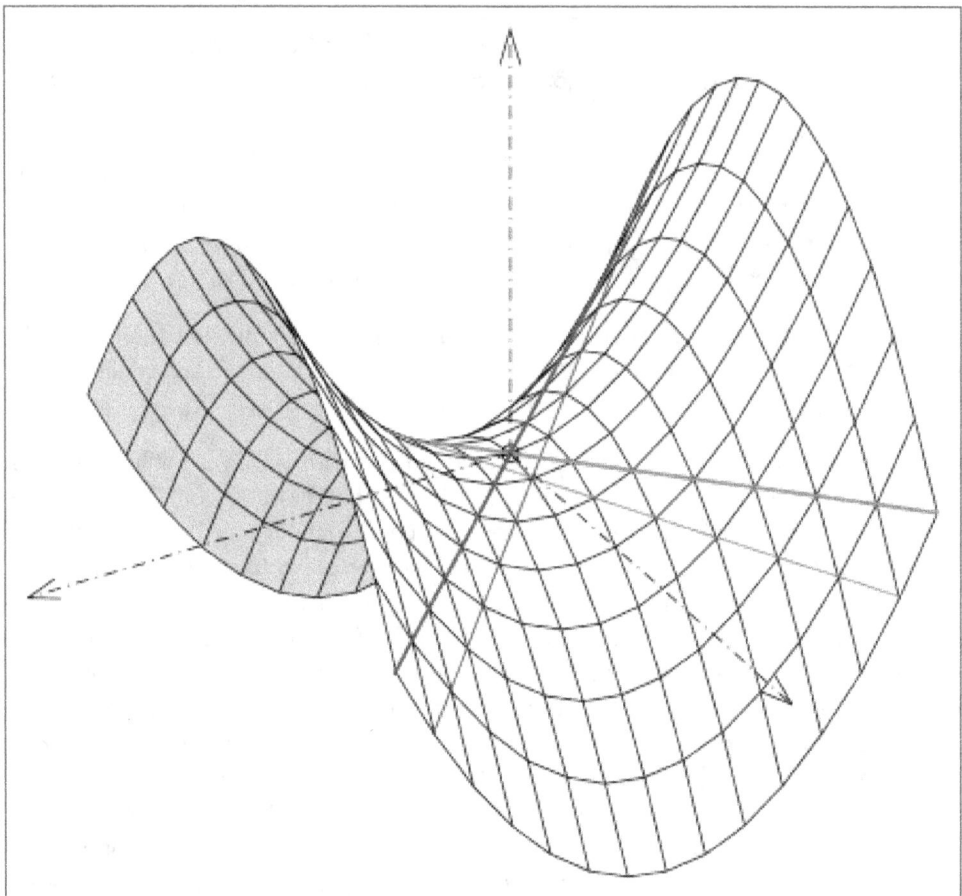

Fig. 58: Hyperbolic paraboloid shell - a negatively curved surface

The discovery and research of such "non-Euclidean geometries" is associated with such famous names as Carl Friedrich Gauss, János Bolyai (1802-1860), Bernhard Riemann (1826-1866), and Nikolai Ivanovich Lobachevsky (1792-1856). Even a seasoned mathematician finds it difficult to imagine a curved three-dimensional space. For this reason, two-dimensional "spaces" are often used for illustration, which can be imagined as surfaces embedded in a three-dimensional space. In such a model, Olbers' paradox is represented as follows: Imagine an infinitely extended "plane" (in Euclidean embedding space with the coordinate axes x, y, and z; here, the relationship z = const. should apply for this plane x, y), which is uniformly occupied by an infinite number of points (they are supposed to represent stars). Now let us imagine that we are a two-dimensional being (naturally without extension in the z-direction and without a digestive tract!) at a point (x, y) of this infinitely extended plane. If you draw a straight line in any viewing direction, this straight line will inevitably meet a point (=star) somewhere. And this applies to every conceivable direction on this "Euclidean plane." Now let's imagine a spherical surface as a "plane" that is also embedded in a three-dimensional space. For the two-dimensional inhabitant of this spherical surface, it does not differ locally from the Euclidean plane. Here too, they can move freely in any direction. However, there are significant differences: 1. the surface is finite (exactly $4\pi r^2$ when r is the radius of the sphere), and 2. the surface is unlimited. If you move on it, you will never find a boundary. The surface of the sphere can therefore be regarded as the two-dimensional analog of a "space" that is finite but unlimited. If you cover this spherical surface with a finite number of points, there will always be a visual ray that does not intersect any of these points. Furthermore, this "visual ray" would be a closed curve (in this case, a circle).

An interesting question in this context is how, as a "two-dimensional person" (again, of course, without a digestive tract!)[38], you can determine quasi-experimentally whether you are living on a spherical surface or a Euclidean plane. The simplest option is to keep walking straight ahead. If you return to your starting point at some point, you have walked along the great circle of a spherical surface. However, if the diameter of the sphere is very large (but not infinite), this attempt can take a very long time. A simpler method is to draw a circle around yourself (e.g., using a taut string representing the radius of the circle), measure its circumference, and calculate the ratio between the circumference and diameter. If the result is exactly the circle number Pi, then you are living on a Euclidean plane; if not, then you are not. Alternatively, the "two-dimensional" person could construct a triangle and then measure the sum of its interior angles as precisely as possible. If it is exactly equal to 180°, then they know that they live on a Euclidean plane; if not, then they do not.

Strictly speaking, there are two "types" of curved spaces. One, where the sum of the interior angles of a triangle is greater than 180°, is called "positively curved." The surface of a sphere is a two-dimensional example of this. The "spaces" where the sum of the interior angles of a triangle is less than 180° are called "negatively curved." The surface of a "saddle" is an example of this, or ideally, that of a so-called pseudo-sphere. A separate type of "geometry" can be constructed for each of these types of spaces. These "geometries" are called "non-Euclidean geometries." The mathematical methods for their detailed description were developed by Bernhard Riemann (1826-1866) in Göttingen in the mid-19th century, which is why they are sometimes called "Riemannian geometry" (as far as positively curved spaces are concerned). Cosmic space can also be "curved," even if it is much more difficult to imagine in this case.

---

[38] It would turn the "two-dimensional" into two "two-dimensional" without a digestive tract...

154. The Spacetime

After Riemann, mankind had to wait about half a century before a scientist provided a justification for the idea of a "curved" cosmic space. That scientist was Albert Einstein (1879-1955). He recognized that the actual world stage is a four-dimensional structure, which he called "spacetime," where a "point" in this space is uniquely determined by four independent coordinates - three spatial coordinates and one time coordinate. Such a point is also called an "event." The shortest path between two events, a "straight line" so to speak in Einstein's geometry of spacetime, is the path that a ray of light or, more generally, a material body moving without forces (think of Galileo's principle of inertia) travels between them. This "path" traces the "curvature" of spacetime. And the curvature itself is caused by masses and mass flows, or, as Albert Einstein himself once put it: *"Matter tells spacetime how to curve, and spacetime tells matter how to move."* This can also be made comprehensible using a simple two-dimensional model. Let's imagine the three dimensions of space are reduced to a two-dimensional rubber sheet stretched in a frame. Now we place an iron ball in the middle of this sheet. The effect is a "dent," which becomes more pronounced the heavier the ball is. If our "two-dimensional person" now constructs a circle with a constant radius at different distances from the sphere and then carefully determines the ratio of circumference to diameter, they will notice an ever-increasing deviation from the circle number π the closer they get to the sphere. This "deviation" can be used directly as a measure of the curvature at the point of measurement. And this is not only the case on the rubber sheet. The sun also "bends" spacetime in the same way, which means that a ray of light from a star that passes close to it is deflected in a precisely calculable way. This effect can be observed during a total solar eclipse by photographing the stars that become visible in the daytime sky around the eclipsed sun and later comparing their positions with those of the same stars photographed six months later in the night sky without the sun. The deviations are in complete agreement with the deviations that result purely mathematically from Einstein's theory.

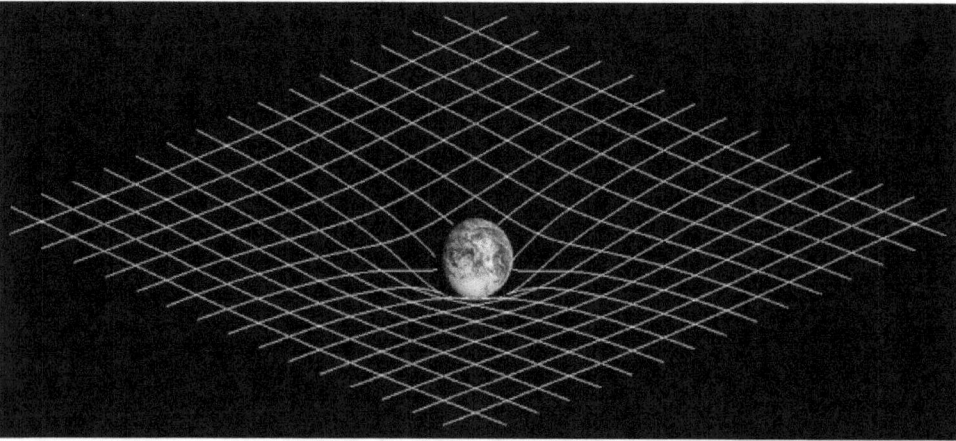

Fig. 59: Our earth also curves "space-time" around itself. If this were not taken into account, our GPS positioning system would be much less accurate (Wikimedia)

What is considered "local" curvature in this context is actually a global curvature on a large scale, the size of which depends solely on the average mass density in the universe. It is therefore possible that the entire cosmic space is "flat" (Euclidean), or positively curved and therefore finite but unbounded, or perhaps even has a negative curvature on a global scale. Which of these three possibilities is true can only be clarified by corresponding observations, which brings us back to "Olbers' paradox".

155. The expansion of cosmic space - galaxy escape

This is because the global curvature of cosmic space is linked to another effect that is accessible to observation and predicted by Einstein's theory: cosmic space is expanding. This phenomenon was first deduced from Einstein's equations in 1921 by the Russian mathematician Alexander Alexandrovich Friedmann (1888–1925) and confirmed shortly thereafter by Edwin Hubble (1889–1953) through observations of distant galaxies. The conclusions drawn from this research led to the theory of the "Big Bang," which, put simply, states that our cosmos must have had something like a "beginning," where space and time, as well as all matter, emerged from a "singularity" (i.e., a "point" of infinitely high energy density).

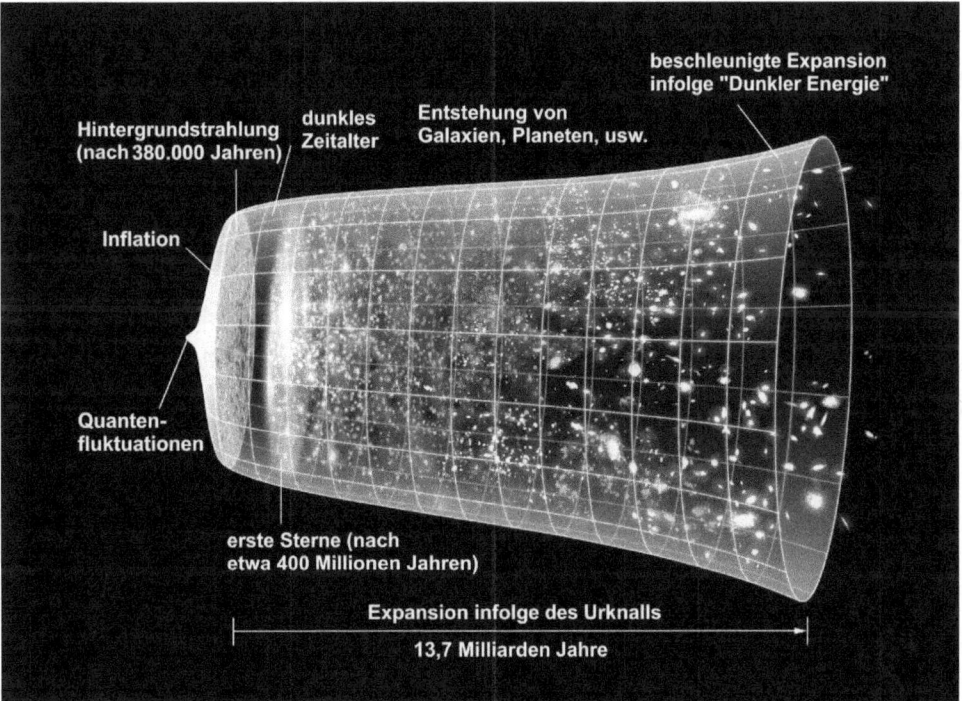

Fig. 60: Illustration of the evolutionary stages of the universe as derived from theoretical considerations and observations (Wikimedia, NASA / WMAP Science Team)

This picture, which was first sketched out by the Belgian theologian and physicist Georges Édouard Lemaître (1894–1966), has since been developed very precisely, but still suffers critically from the fact that no one has yet succeeded in uniting Einstein's general theory of relativity (which predicts a space-time singularity) with quantum theory (which can prevent a singularity) to form a new theory, quantum gravity.

In our context, it is only of interest that this "Big Bang" occurred a finite amount of time ago (almost exactly 13.798 ± 0.037 billion years ago) and, due to the finite speed of light, implies a "horizon" for us as observers that is at a finite distance (approximately 46 billion light-years away - greater than the distance resulting from the light travel time, since space itself has expanded over the last ~13.8 billion years). However, this does not mean that there is nothing beyond this "horizon." Here, too, Udo Lindenberg's motto "Hinterm Horizont geht weiter..." ("Beyond the horizon, it goes on...") applies. But the paradox is that we are looking in the wrong direction, so to speak. Due to the finite nature of the speed of light, we always see objects as they were in the past relative to our present. We see the Moon as it looked 1.28 seconds ago, we see the Sun as it looked 8.32 minutes ago, we see Polaris as it looked around 430 years ago, and we even see the nebula in the sky known as the Andromeda Galaxy as it looked 2.5 million years ago.

This should clarify the issue. The farther away a cosmic object is from us, the less "recent" it is in relation to the Big Bang. At some point, you reach a distance where there were no stars yet, only a hot plasma that had cooled down just enough for the light particles (photons) to separate from the rest of the matter. This occurred around 380 thousand years after the Big Bang. So, when we look into this "distance" with our telescopes, we have a "wall of light" in front of us, behind which the "Big Bang" is virtually hidden. This is exactly what Olbers predicted under different premises (point 152). But why can't we see this "wall of light"? Quite simply because it is not visible to our eyes, and this is due to the aforementioned expansion of space. In the past, when the cosmos became "transparent," the universe was still small and the radiation was extremely short-wavelength. It corresponded to the radiation that an extremely hot plasma of several million degrees would emit. With the expansion of space, the distance between the wave crests (i.e., the wavelength of light) increased progressively (this corresponds "classically" to a redshift but is referred to as "cosmological redshift" here to differentiate it).

156. The cosmic background radiation

Today, the wavelength of the most intense radiation is already approximately 0.2 cm. It corresponds to the radiation maximum of a so-called "black body" with a temperature of around 3 K, which is why this omnipresent cosmic background radiation is often called "3 degrees Kelvin radiation" in German-speaking countries. And since we can't see microwaves with our eyes (otherwise, you could use the microwave oven as a spotlight), the night sky is pretty dark in this respect too, except for the stars. Even if our cosmos is infinite and unlimited (i.e., its geometry is Euclidean), it has not yet had time to expand so far that we can see a "star" in every direction, because the "visible universe" is principally limited for us due to the horizon 46 billion light-years away (this space is called the "Hubble bubble"), although with every second, the horizon moves a little further away and thus new areas of space become visible again and again.

But back to the background radiation. Its cosmological significance lies, among other things, in the fact that it is highly isotropic and has structures on small scales (which, however, if you convert the radiation spectra into temperatures, are only a few hundred thousandths of the average temperature of 2.725 K), which can even be measured and mapped with special observation satellites. These "sky maps" of this "wall of light" make it possible to determine or estimate some important cosmological parameters. These include, for example, the "age of the universe," the ratio of baryons (i.e., protons and neutrons) to photons, and the proportion of matter in the total energy density of the universe. All these parameters are fundamental to our understanding of the universe and its development since the Big Bang. They help us to better understand the structure and composition of the universe and provide important information for cosmological models.

Fig. 61: Map of the temperature fluctuations of the cosmic background wave radiation across the entire sky, recorded by the space probe WMAP (Wilkinson Microwave Anisotropy Probe) between 2001 and 2010. The detected temperature fluctuations are in the micro-Kelvin range and provide deep insights into the earliest structure formation processes in the universe (Wikimedia, NASA / WMAP Science Team)

The detection of cosmic background radiation by Arno Penzias (1933–2024) and Robert W. Wilson in 1964 also provided a solution to another cosmological puzzle, the "Olbers' paradox" discussed in point 152. There is no objection to the assumption of an infinitely extended universe, but since the universe has a finite age and light only propagates at a finite speed, light can only reach us from a finite region since the Big Bang. Thus, the resolution of the paradox lies in the finiteness of the visible universe over time.

157. What is "time"?

But what is "time"? Einstein once defined it quite correctly as follows: "*Time is what a clock shows.*" But this definition is somehow unsatisfactory. Augustine of Hippo (354-430) already recognized this, stating frankly in his "*Confessiones*" with regard to "*What is time?*" "*If no one asks me about it, I know it; but as soon as I want to explain it to a questioner, I don't know it.*" So what is "time"? For Isaac Newton, it was clear that there must be such a thing as "universal" time, for which he proposed the following definition:

"*Absolute, true and mathematical time passes in itself and by virtue of its nature uniformly and without relation to any external object.*"

Physicists and the general public were able to live with this definition for a long time until Albert Einstein came on the scene and developed a completely new, unexpected concept of time based on the fact that the speed of light in a vacuum is a constant for every observer regardless of their own state of motion. This new concept is the content of the special theory of relativity from 1905. Among other things, it introduces the so-called twin paradox, explains the extension of the lifespan of elementary particles when they move very quickly relative to us, and the formula $E=mc^2$ is also a direct consequence of this theory.

Fig. 62: Time is what clocks show (Creative Commons)

The special theory of relativity shows us how "time" (represented by clocks) behaves in our world, but not what time "is". Formally, however, everything is clear. It adds a fourth dimension to the three spatial dimensions, which enables the definition of an event, i.e., something that takes place at a given point in space at a given point in time (see point

154). For an observer, there is always a distinct point in time that separates events in the past from events in the future. This distinct point in time is called the "present" or the "now". It is physically and classically infinitesimal, but psychologically corresponds to a small period of time, as the experience of the present also takes a certain amount of time. This "time dimension of perception" is around 0.1 seconds. We perceive time itself as the present, remember it as the past, and anticipate it as the future. This sensation is often associated with a "flow of time," as Isaac Newton expressed it in his "*Principia Mathematica*". And this "flow" obviously has a direction, namely from the past to the future - a fact that is referred to in physics as the "arrow of time."

158. Motion and the Zeno arrow paradox

The infinitesimal character of the "point in time" present, which separates the past from the future, has something problematic about it upon closer inspection, which Zeno of Elea (490-430 BC) was one of the first to recognize in all its sharpness. I am referring to the "arrow paradox." It is quickly explained: An arrow flies through the air. At any given moment, the arrow remains motionless, i.e., the "momentary" arrow resembles a single photograph of itself. However, "time" consists of an infinite number of moments, and in each moment, the arrow stands still. So where is the movement? Zeno concluded from this: *"That which moves moves neither in the space in which it is, nor in the space in which it is not."*

One could now assume that spacetime (i.e., space and time) has a "granular" structure. But in light of the arrow paradox, this also leads to a contradictory consequence, namely that the arrow is at rest in every single moment but performs movement across many moments. This excludes the possibility of the arrow jumping from point to point, as it were, in such a way that it disappears at point A at time A and reappears instantaneously at point B at time B. The absurdity that arises from such considerations is that space and time can therefore have neither a continuous nor a discontinuous structure, which runs counter to common sense. Nevertheless, *"There is no movement, said the bearded sage. The other was silent, began to walk before him..."* Alexander Pushkin (1799-1837) once wrote.

There are various approaches to solving the arrow paradox. The best known is the concept of limits, which goes back to Augustin Cauchy (1789-1857) and on which modern infinitesimal calculus is based. This quickly resolves the problem because it is true that "motion" cannot be captured by a series of snapshots. The reason for this lies in the fact that it is not sufficient to always consider only one point in time (instant) in isolation. Rather, the concept of instantaneous velocity is essential, which is derived from an infinite sequence of average velocities using a limit value approach. It only makes sense if the spatial and temporal neighborhood is included, because without it, rest cannot be distinguished from motion. The flight of Zeno's arrow can therefore only be understood under the premise that space and time form continua (which is why classical physics also speaks of a "space-time continuum").

The crux of this argument lies in the fact that the concept of limit value initially refers to a special mathematical space consisting of a continuum of points. Who actually tells us that physical space is really equivalent to this abstract mathematical space? And this is where quantum mechanics comes into play to explain the arrow paradox in the form of Heisenberg's uncertainty principle, which states that position and momentum cannot be measured together with arbitrary precision.Namely, whenever the arrow is exactly at location A at a given moment, it has an instantaneous velocity resulting from the limit crossing $t \rightarrow t'$. The more precisely the location A is measured, the more indeterminate the velocity is (and vice versa). In contrast to the Eleatic argument, which claims that the arrow is at rest at a given moment at location A, quantum mechanics states that the arrow has no definable velocity at all at point A.

159. The "flow" of time

Let's leave it at that for now and return to the question, "What is time?", specifically to the aspect of the "flow of time." This describes the constant passing of the present into the past and the dawning of the future as an unstoppable process. We experience time as a constant progression in which one moment merges into the next, seemingly like a river that never comes to a standstill. However, as Kant already noted, the concept of "flowing" only makes sense in this context if it can be contrasted with the alternative of "non-flowing" - an idea that is difficult for us because we only know time as movement.

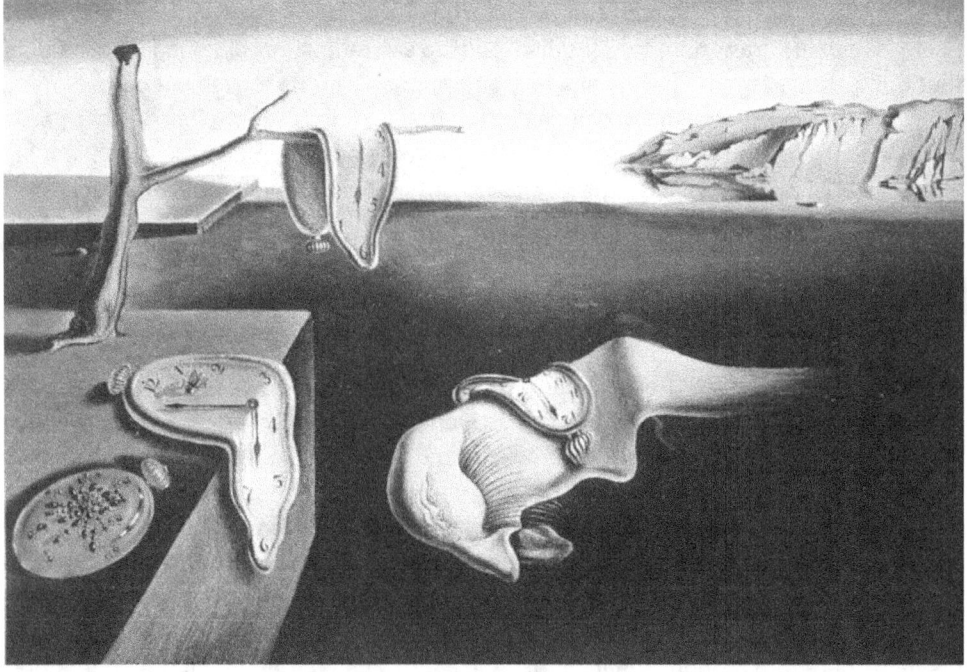

Fig. 63: The famous Spanish artist Salvadore Dali (1904-1989) tried to show the "flow of time" in his 1931 painting "The Persistence of Memory" (source "Museum of Modern Art", New York)

For seasoned physicists, time in this conventional sense is therefore an illusion, albeit a persistent one. Physics views time differently: in its formulas, there is explicitly no "flow" of time. Instead, it is merely a parameter, analogous to the coordinates of a place, with which we can describe changes. In this view, time is therefore not something that "flows," but a tool for quantifying and understanding the change of states. Interestingly, the concept of time becomes meaningless if there is no change because, without change, there is nothing that could be associated with time. Under this condition, it is also no longer possible to define a measurement rule for this parameter. Time itself thus becomes a concept that is closely linked to the dynamics of the world - where dynamics are absent, time also seems to disappear. Wherever there are no changes of any kind, the concept of time becomes meaningless, since under this condition, it is no longer possible to define a measurement rule for this parameter.[39]

160. The thermodynamic and cosmological arrow of time

This is also in line with the so-called thermodynamic arrow of time. Of two temporally separated states of a system, it defines the one that is more in the past as having the lower entropy of the two. A physical system in thermodynamic equilibrium (i.e., in which no more changes take place) thus knows neither past nor future; it is, to a certain extent, timeless.

Another type of arrow of time can be determined by the development of the visible (i.e., visible to us) part of the universe. In this case, time can be measured in terms of the size achieved by cosmic expansion (the "world radius"). This is specifically referred to as the "cosmological arrow of time."

However, the "problem of time" also has its own significance beyond physics, for example, in psychology (why do the days in "old age" seem shorter than the days in our childhood?), in metaphysics (see, for example, Heidegger's Being and Time), or in biology (evolution) and has lost none of its fascination even today, at least for those who are prepared to think about it.

161. Thomas Mann and his "Zauberberg"

We need only recall the words of Hans Castorp in Thomas Mann's novel "Der Zauberberg" (The magic Mountain), in which he wrote the following thoughts "in his head":

"What is time? A mystery - insubstantial and omnipotent. A condition of the world of appearances, a movement, coupled and intermingled with the existence of bodies in space and their movement. But would there be no time if there were no movement? No movement if there were no time? Just ask! Is time a function of space? Or vice versa? Or are

---

[39] This is the case, for example, when the universe has died the "heat death" (point 143)

the two identical? Just asking! Time is active, it has a verbal nature, it "brings about". What does it bring about? Change! Now is not then, here is not there, because there is movement between the two. But since the movement by which time is measured is circular, resolved in itself, it is a movement and change that could almost as well be described as rest and standstill; for the then is constantly repeated in the now, the there in the here."

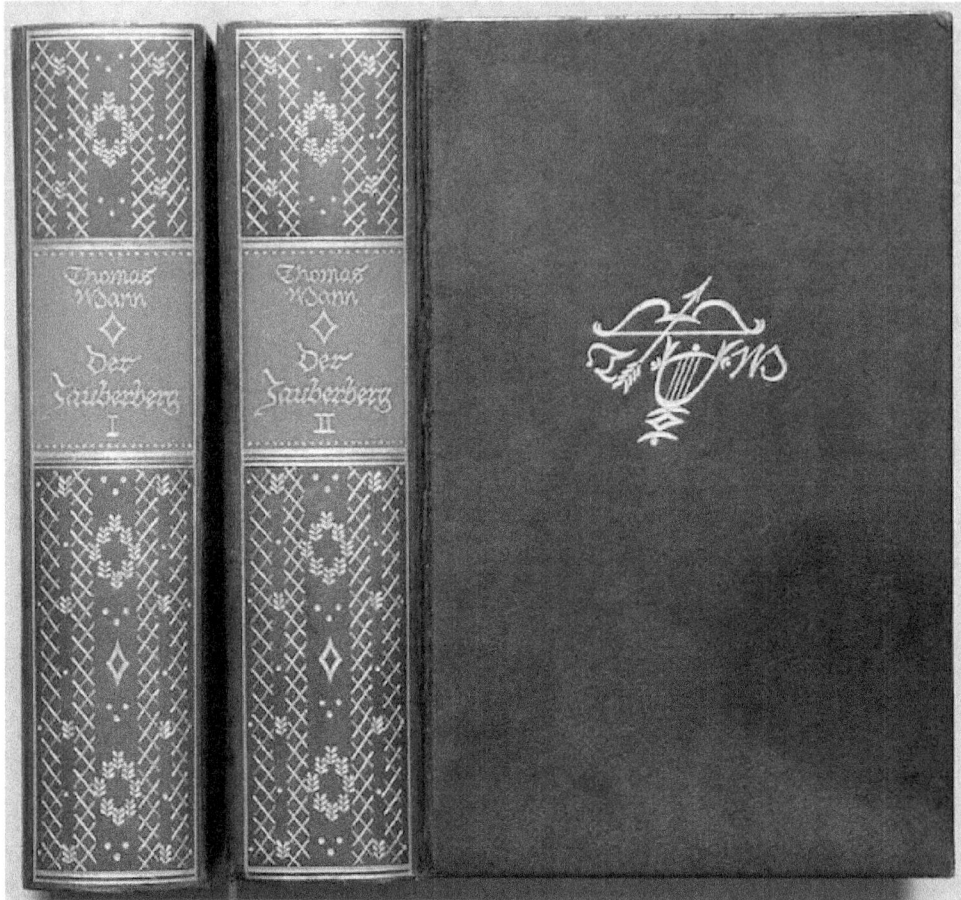

Fig. 64: Belongs in every well-stocked collection of novels (and not only as decoration because of the cover - but also read)

This book, the third major novel by Thomas Mann (1875-1955) after "Buddenbrooks" (Nobel Prize for Literature 1929) and "Königliche Hoheit" (Royal Highness), was one of the bestsellers of the Weimar Republic and beyond. In the tradition of the German Bildungsroman (or rather, a parody of it), an epic narrative is developed in the confines of a lung sanatorium in Switzerland, which revolves around the young Hamburg resident Hans Castorp and, in a finely chiseled manner, is set in relation to a multitude of individual characters. A visit by Castorp to Davos, planned as a "short" one (three weeks), evolves into a seven-year stay during which the protagonist's personality is shaped. The "central theme" of the novel may seem simple, but Thomas Mann's language, the interspersed philosophical and theological debates, some of which are at a very high level (and therefore not always easy to follow), combined with the intense atmosphere of that time, shaped by the First World War, will leave you wanting more. So if you haven't read this book yet, you should put it aside for now and switch to "Der Zauberberg". You won't regret it...

162. „Als wär's ein Stück von mir"

In my student days, when I read a great deal, another book made a strong impression on me, even if it cannot compete in this literary league. It is the autobiography of Carl Zuckmayer (1896-1977), entitled "Als wär's ein Stück von mir" (As if it were a piece of me). Few people will remember this playwright, who made his breakthrough with the comedy "Der fröhliche Weinberg" in 1925. Two years later, his success was consolidated with the play "Schinderhannes," and in 1931 he scored a major coup with "Der Hauptmann von Köpenick," which made him a wealthy man within a very short time. Due to his clear anti-militaristic stance, this play was of course no longer suitable at the time of National Socialism, which prompted him to flee Austria in 1938. He described all of this in detail in his autobiography - the conditions in Vienna after the "Anschluss," his almost unsuccessful escape via Switzerland, which led him first to Rotterdam and then to the United States, where he lived for a time as a screenwriter in Hollywood (to name one well-known location), and his return after the war.

Zuckmayer's autobiography brings the first six decades of the last century to life as vividly as Hans Castorp's life in Davos on the eve of the First World War in Thomas Mann's novel. Step by step, you encounter famous contemporary witnesses such as Max Reinhardt, Bertolt Brecht, and Stefan Zweig. No section is boring. So here's a tip: if you only have a little time to spare for reading, you should postpone it until your vacation. Believe me, Zuckmayer's autobiography is better than any history book about this eventful period. Incidentally, the title "Als wär's ein Stück von mir" is a line from the poem "Der gute Kamerad" by Ludwig Uhland (1787-1862), which he wrote in 1809 under the impression of the deployment of Baden troops against the Tyrolean freedom fighters revolting against Napoleon Bonaparte. This poem, also known as "Ich hatt' einen Kameraden..." ("I had a comrade..."), is still played today in the setting by Friedrich Silcher (1789-1860) at Bundeswehr funeral ceremonies. It was "the" song at the time of the First World War. "The" song of the Second World War was undoubtedly "Lili Marleen," originally sung by Lale Andersen (1905-1972) ("An der Kaserne, vor dem großen Tor..."). Popularized by the army radio station Belgrade, a large number of variants in other languages appeared after 1941, as the song was also extremely popular among Nazi Germany's opponents. An English-language version, for example, is closely associated with the name of Marlene Dietrich (1901-1992), the famous UFA actress who emigrated to the USA in the Heinrich Mann film adaptation of the book "Prof. Unrat" - the "Blue Angel."

Fig. 65: Unfortunately only known to older people - Carl Zuckmayer

As Hitler's empire slowly came to an end, the Red Army pushed the German Wehrmacht further and further back to the west, and finally, the Allies opened a second front in Normandy. In short, the situation within Germany became increasingly unbearable due to the bombing campaign. During this time, another song became a popular hit and provided fatalistic encouragement: "Davon geht die Welt nicht unter" by Bruno Balz (1902-1988), sung incomparably by Zarah Leander (1907-1981) in the film Die große Liebe. And as a final example, "Brothers in Arms" by Mark Knopfler (of Dire Straits, another term for "broke") must be mentioned, which can be described as the song of the Yugoslav wars from 1991 to 2001.

163. The job description of the censor

Throughout history, "songs," as well as poems, pamphlets, and, of course, books, have often been a means of conflict between rulers and the ruled. This led to the emergence of the role of the "censor" in the 16th century, who was often employed by the police authorities. His task was to check the content of certain printed publications to see whether the information or opinions they contained conformed to the prevailing laws, whether they were ethically justifiable, or whether they would simply offend "higher-ranking persons." Of course, these are not all the reasons that can lead to "censorship." Even today, this topic is definitely an issue, although this job title (and the associated activities) is no longer used or carried out, at least in Germany.

Fig. 66: The "censor" has always had to serve as a figure of fun - which is why such caricatures were quickly censored away...

Although the relevant article of our Basic Law states that *"there shall be no censorship"*[40] political correctness and "self-censorship" have found their way into media practice on a massive scale. And lest there be any misunderstanding, "censorship" is not necessarily linked to the profession of "censor". The prohibition of unpopular writings is as old as there are "unpopular" writings. The Catholic Church was particularly prominent in this respect with its *"Index Librorum Prohibitorum"*, the "Index of Forbidden Books", which was first made public in 1559 and only abolished again in 1966. But that is not the point here.

164. The business of "censorship"

This is more about the "small minds" who carried out the business of "censoring" quasi-officially and inevitably came up against "great minds" that they were not in the slightest mentally equal to. They left the evidence for this in the texts, which they disfigured with great meticulousness and imagination.

Let's start with an undoubtedly "great mind", Johann Wolfgang von Goethe (1749-1832) and his tragedy "Faust I". The censor was particularly fond of the "youth-endangering" verses. He sometimes indulged in poetry himself, as the following examples prove: Goethe, for example, says *„Ach, kann ich nie - Ein Stündchen ruhig dir am Busen hängen, - und Brust an Brust und Seel an Seele drängen?"* (Ah, can I never - An hour quietly hang on your bosom, - And breast to breast and soul to soul press?). What, hang on your "bosom" for a whole hour? No, that's definitely not possible - so the verse was deleted and quickly rephrased by the censor: „*Ach, kann ich nie – Ein Stündchen ruhig bei dir sein, - Doch ungestört wir beide nun allein, - Man hat sich doch so manches Wort zu sagen, - das keine Zeugen will!*" (Oh, can I never - Be quiet with you for an hour, - But undisturbed we're both alone now, - We have many a word to say to each other, - that wants no witnesses!").

When it came to ecclesiastical matters, the censor was usually in top form: in the well-known student scene, we read Mephisto's advice „*Doch Euch des Schreibens ja befleißt, - Als diktiert ʼEuch der Heilig ʼGeist!*" (But be diligent in your writing, - As if the Holy Spirit were dictating to you!), which of course could not be allowed to stand: „*Doch Euch des Schreibens ja befleißt, - Weil dies allein studieren heißt.*" (But be diligent in your writing, - Because this alone means studying).

A censor proved that "censoring" is also an intellectual achievement by defusing the mocking song "*Es saß ein Ratt im Kellernest*" (There sat a rat in the cellar's nest), the first verse of which reads in full "*Es saß ein Ratt im Kellernest, - Lebte nur von Fett und Butter, - Hat' sich ein Ränzlein angemäst't - Als wie der Doktor Luther*" (There sat a rat in the cellar's nest, - Lived only on fat and butter, - It fattened itself up quite well - Just like Doctor Luther). "Dr. Luther! - No, that's not possible," it must have flashed through the censor's cerebellum like lightning. But after a couple of hours of brain teasing, he had the solution: „*Es saß ein Ratt im Kellernest, - Lebte nur von Fett und Butter, - Hat' sich ein Ränzlein angemäst't - Wie der gelehrteste Chinese.*" (There was a rat sitting in the cellar nest, - Lived only on fat and butter, - Has fattened up a little skirt - Like the most learned Chinese).

Today, in some parts of the world, the line "*Uns ist ganz kannibalisch wohl - Als wie fünfhundert Säuen!*" (from the "Flohlied") would most likely be classified as questionable and the corresponding literary work would at least be qualified as "harām", possibly even banned in its entirety, although a version censored ~200 years ago already exists in which you can read "*Tralleralla, - Tralleralla!*" in the compromised passage. But since that also somehow heralds a kind of joie de vivre, even that might be worthy of a fatwa...

A censor was normally a respectful official who, like any other official, had to follow instructions. However, they were usually formulated in such general terms that the censor had a great deal of leeway in their interpretation. The German literary scholar Heinrich Hubert Houben (1875-1935), for example, quotes a decree by the Austrian and arch-Catholic Empress Maria Theresa (1717-1780), which contains detailed directives on the treatment of Protestant and anti-Catholic writings and their authors. He writes:

"*The authors of Protestant and anti-Catholic writings await banishment and imprisonment. Even the possession of Lutheran, heretical and generally un-Catholic writings was strictly frowned upon; they were outside all property rights, every clergyman was allowed to confiscate them wherever he found them, every private citizen was obliged under penalty to declare wherever he had seen them. Anyone who bought a book had to present it to his parish priest for inspection within four weeks, otherwise he was fined 3 guilders, which increased considerably in the event of a repeat offense. A third of the fine went to the informer, so the most vile espionage was in full bloom. House searches were the order of the day. Travelers' suitcases were searched at the customs offices, all question-able books were taken away, forbidden ones burned. Disguised officers of the clerical book police visited the bookshops as harmless customers, wormed their way into the confidence of the dealers and forced their way in to procure forbidden books for them; if the booksellers allowed themselves to be persuaded, the informers revealed them-selves as police officers, confiscated the works and punished the sellers."*

As you can see, authors, booksellers, and even readers were in a bad position if they opposed the censorship authorities. This is why censorship remains an effective means of exercising power today. On the other hand, what was considered "censored" in earlier times seems rather embarrassing from today's perspective. The best examples come from theater censorship. In the well-known play "Kabale und Liebe" (by Friedrich Schiller), for example, a censor made the court marshal von Kalb the "head clothes keeper," since in the opinion of the time, a "marshal" could never be an intriguer. Schiller had to battle with censors throughout his life anyway, who found his work not "politically correct" enough to be presented to the general public unscathed. Take "The Maid of Orleans" as an example. The censors had a problem with the heroine of the play, Agnes Sorel, because Schiller introduced her as the mistress of the French King Charles VII (which, of course, she was). But in the opinion of the censorship authorities, this was not allowed, so she had to appear in the censored version of this drama as the king's legitimate wife. And so many more examples could be found.

165. The practice of censorship

But the practice of censorship, which relied to a not insignificant extent on the cooperation of informers, even had to deal with "inns" if their billboards did not comply with the censorship guidelines. This was the case, for example, with the *"Boeuf à la mode"* inn, which opened in Paris in 1816. It is reported that on the day of its opening, it was adorned with a sign showing an ox dressed up in a scarf and a straw hat. What followed was a report by a police informer, which resulted in a whole slew of follow-up activities, investigations, consultations (right up to the police minister!) and even a secret inspection of the sign in question. Because the informer's report was quite something:

"The ox depicted on the sign is known to be the symbol of being fattened. The scarf that adorns him is red, the decoration on his hat consists of white feathers and blue ribbons. A ribbon hangs from his neck together with a decoration similar to the Order of the Golden Fleece, which is worn by princes. The hat is obviously meant to represent the crown and is about to slip off. These allusions undoubtedly serve as proof that this fig-urehead is nothing more than a vile caricature of Her Majesty's person."

- here Louis XVIII. In this case, the fact that the censorship authority refrained from pursuing the matter further after the secret inspection (presumably in connection with a visit to the inn) speaks in its favor. The restaurant, located in Versailles (on the Rue de la Paroisse), still exists today and has an excellent reputation.

166. "Censorship does not take place" - political correctness

Closely related to state censorship is the concept of "political correctness," which - although not codified - encourages self-censorship by attempting to ban certain words deemed inappropriate from the vocabulary. The original aim was to counteract verbal discrimination against minorities by introducing neutral language - a thoroughly laudable endeavor if it had not degenerated (another politically incorrect word!) into an ideological weapon. For example, it is certainly correct and therefore not wrong to use the term "Sinti and Roma" to refer to "Gypsies" (German "Zigeuner") in official publications. However, if a "Zigeunerschnitzel" (which refers to a specific dish and certainly not a "schnitzel made from the meat of a gypsy") can only be called a "Balkanschnitzel" (or, as has been highly official in Hanover since 2013, "Paprikaschnitzel") from now on, then it just becomes ridiculous. It should not be concealed that the pursuit of political correctness can certainly promote imagination and creativity and open up completely new fields of activity - especially for Germanists and political scientists. And there are truly great challenges. Take, for example, the "Klofrau" - we all know her, we have to visit the places she supervises, looks after, and keeps in perfect hygienic condition from time to time for purely biological reasons. It would never occur to us that this term somehow discriminates against the profession or the person practicing it. On the other hand, it is quite true that among some people (mostly professionally "higher-ranking" individuals who believe that their "job" is somehow "more valuable" because it is better paid), this term is used to disparage jobs of lower social prestige. And in order to stand up to this minority opinion, people who have nothing else sensible to do during the day once came up with the term "toilet attendant." But unfortunately, it still contains the unpleasant word "toilet." But here, too, the experts in the German language quickly found a remedy: "toilet attendants" are actually "facility managers" who carry out their useful and socially important work in the "McClean" of a city train station or away from a hotel lobby. So, it's no wonder that normal people increasingly associate "politically correct" language with ridiculous euphemisms and dogmatic, intolerant politics.

Incidentally, the whole debacle started with the word "Negro," which all of us older people still remember from the bakery in the form of the "Negro kiss" (the politically correct term here is "chocolate kiss"). In Grimm's dictionary, we read "NEGER, m. der schwarze, der Mohr, from French *négre* (Latin *niger*)." It is a term that has been accepted since the beginning of the 18th century for people from Africa whose skin is "black" (for reasons that are easy to understand today). With the emergence of various racial theories that focused on the alleged differences between "lesser" lower races (all those who did not have white skin) and a "higher" master race - for example, to justify colonial exploitation or slavery in the American southern states - the semantics of the word "Negro" (or, in the USA, "nigger") increasingly developed into a racially motivated insult. Towards the end of the 20th century, progressive forces attempted to counter this with a new term: "Negro" became "black." But at some point, the term "black" also had negative connotations, so the word "colored" was brought into play. But dark-skinned people don't necessarily think of themselves as "colorful" like Aras, which is why the new suggestion "African-American" found its way into the dictionary of political correctness. But unfortunately, the typical "African" was not an "African-American," which narrowed the meaning of the term too much.

167. The euphemism treadmill

And this is where the dilemma of politically correct language becomes apparent. If words with a negative connotation are replaced by "new" ones, these "new words" will themselves take on a negative connotation over time, as long as the social environment or conditions surrounding the term do not change. They then have to be replaced by an even "newer term" ... ad infinitum, one could jokingly say - "foreigners" - "people with a migration background" - "people with a migration history" - ..., or, another example: "bullies" - "difficult-to-educate children" - "children with behavioral disorders" - "children with behavioral problems"... The experienced linguist refers to this phenomenon as a "euphemism treadmill," which, once set in motion, is not so easy to stop. No one, not even those affected, would get upset about words like "Negro" or "gypsy schnitzel" or feel discriminated against if social reality had no room for racism, sexism, or other types of discrimination. After all, euphemism (i.e., the "whitewashing" of problems) does not make any real contribution to solving them.

168. Biologically speaking, there are no human races

If, for example, it were unanimous in society that there are, in fact, no "human races" and that skin color is merely a phenotypic expression of certain genes adapted to ecological conditions (modern genetics can prove that, genotypically, a "black" and a "white" individual can, under certain circumstances, be more closely related to each other than any two "white" or any two "black" individuals), then "red," "white," "yellow," or "black" would simply be neutral terms for this fact and we could spare ourselves the constant reinterpretation of terms. It would be better to make an effort to change social conditions accordingly (e.g., by investing in education and upbringing). On closer inspection, political correctness is nothing more than a disguise of reality to make it look better than it is and to create illusory problems that can then be argued about to the exclusion of the really important social problems. However, it cannot be maintained in the long term, because the old saying by Abraham Lincoln (1809-1865) still applies: *"You can fool all of the people some of the time and some of the people all of the time, but you can't fool all of the people all of the time."*

The original meaning of the phrase "to make a fool of" actually only means "to keep someone as a jester," i.e., as a "jester" who is allowed to take certain liberties of a parodic nature with his "keeper," which can quickly become fatal for others...

And by the way:

"The downfall of the king announces itself as soon as the king can no longer stand the sayings of his court jester."

And who wants to monitor and censor the "social media" even more?

169. Fools and jokers

In the Middle Ages, such jesters were actually reserved for secular princes, as no less a figure than Charlemagne (747-814) forbade the clergy to employ such people, alongside certain animals used in hunting (dogs and falcons), as early as 789 AD, to put it in modern terms. The "fool" was allowed (and required) to present himself as stupid and clumsy in order to fulfill his role. However, he also had to be clever enough not to exaggerate. The aforementioned Edgar Allan Poe (yes, we always have to come back to this gifted writer of horror stories) beautifully described this in his story "Hop-Frog," where the cleverness and courage of such a "fool" lead to the downfall of a domineering and cruel king.

If a jester was not employed as a "court jester" at court but had to earn his living as a juggler at fairs with jokes and cunning tricks, he was referred to as a rogue since the end of the Middle Ages. Even Goethe knew this (in the prologue to Faust I), stating: „Von allen Geistern, die verneinen, ist mir der Schalk am wenigsten zur Last" (Of all the spirits that deny, the prankster is the least of my burdens).

170. Dyl Ulenspegel

One of these jokers even got his name on the cover of a fun GDR magazine that still exists today: Dyl Ulenspegel (Till Eulenspiegel, 1300-1350).

In GDR times, the "Eulenspiegel" was not always easy to obtain (it was referred to as "Bückware"). After reunification, the paper and print quality improved, but not necessarily the content. In GDR times, it was particularly popular because of the "Funzel". Today, in the age of the general availability of Playboy and the Internet, this is unnecessary for obvious reasons.

But back to Till Eulenspiegel. Whether he ever really existed is still a matter of debate, but it is very likely. There is a very old book about him (first edition!) in the British Museum from 1515 as well as a copy in the Gotha Library from 1519 entitled *"Ein kurtzweilig lesen von Dyl Vlenspiegel geboren vß dem land zů Brunßwick. Wie er sein leben volbracht hatt .xcvi. seiner geschichten."* This is the source of the famous Eulenspiegel stories, which still make not only children laugh (and hopefully also think) today.

It is believed that the author of this exceptionally early form of humorous literature may have been the then bailiff of Brunswick, Hermann Bote (1467-1520). But this is not certain. But if he was, then, without knowing it, he landed one of the greatest literary successes of a Lower Saxon - with repercussions to this day. In the 16th century alone, more than 30 editions of these tales were published in Germany. Even Hans Sachs (1494-1576) drew on them in his famous carnival plays.

Fig. 67: Cover of the first edition of the "entertaining stories" from 1515 and one of the last German-language editions of "Till Eulenspiegel" from 2019

The funny name remains to be clarified. We don't even want to write down the most original interpretation. In Low German - according to the Grimm dictionary - ule means "to clean" and the spegel = mirror is the light-colored area around the anus of the deer in hunting language. However, "Ule" also means "owl" and the "spegel" is a "mirror" in which you are reflected when you look into it. And this is how Till Eulenspiegel is usually depicted - with a two-cornered jester's cap adorned with bells, an owl on his shoulder and a mirror in his hand. The owl is the symbol of wisdom (Minerva) and the saying "*hold a mirror up to someone*" is probably familiar to everyone. What is less well known, however, is that the "mirror" is also linked to some of the great mysteries of science.

171. Is mirror milk poisonous?

I'm not talking about the answer to the trivial question, *"The right and left are reversed in the mirror. Why not top and bottom?"* but the serious question of whether "mirror milk," for example, is digestible or not. By this, I mean the milk that can be seen "behind" the mirror as soon as I place a glass of milk in front of the mirror. Assuming, of course, that the glass of milk behind the mirror is just as real as the glass of milk in front of it. In this case, however, I would be careful about sipping it. At best, it would be completely devoid of nutritional value, and at worst, it would be unpalatable and perhaps even poisonous. Why is that the case? To answer this question, you have to examine the molecular components of this natural product. I am referring to the molecules of the proteins and carbohydrates contained in milk. They are produced biologically - in this case, in the udder of a cow - through biochemical processes and only occur in a left-handed form in the case of proteins and in a right-handed form in the case of carbohydrates.

172. Das Chiralitätsproblem

Why this is the case (and not the other way around) is called the "chirality problem." However, "chirality" is just another way of saying "handedness." Just as there is a right and a left hand or screws with right-handed or left-handed threads, there are both right-handed and left-handed variants of otherwise chemically identical molecules in many organic substances. Such special, mutually mirror-symmetrical molecules are called "enantiomers" in chemistry.

In normal chemical reactions outside of living organisms, both forms usually occur in a ratio of 1:1. Such a mixture is called a racemate or racemic mixture. The fact that life as we know it can only do something with one type of molecule of a certain handedness is completely obvious if you consider how proteins are produced in living cells, for example. Especially in the case of proteins, which are essential as catalysts (enzymes) of biochemical reactions, the structure (often referred to as "folding" of the chain molecule consisting of amino acids) is crucial. A perfect fit is required here, and this is known to depend closely on the symmetry between "lock" and "key." A world in which all living beings are made up of right-handed amino acids and left-handed carbohydrates, instead of left-handed amino acids and right-handed carbohydrates, would function biochemically identically. And so the question arises as to whether the enantiomeric selection of proteinogenic amino acids took place before or after the emergence of life. The answer to this question has yet to be found. In any case, "our" biochemistry cannot do anything with "mirrored" amino acids, such as those contained in the proteins of mirror milk. Perhaps this is why mirror milk is neither nutritious nor poisonous, and the caution of sipping it is rather unfounded. Who knows...

173. The Ozma Problem

But let's stay a little longer with the two directions "right" and "left". As a road user and in everyday life, it is fundamentally important to distinguish between these two directions as an observer - which, experience has shown, is not a problem when you are sober. It only becomes a problem when you have to explain to someone else where right and left are - if the "other" is an "alien" with whom you can only communicate by radio and who is so far away that you cannot identify any common star patterns in the sky. Try it! You'll soon realize that it's about as difficult as explaining the difference between "red" and "green" to a blind person without referring to examples (which they can't see anyway). This problem has even been given a name: the "Ozma problem," after the princess who rules the *Land of Oz* in L. Frank Baum's book "The Wonderful Wizard of Oz." However, "Ozma" is also the name given to the undertaking begun in 1960 by Frank Drake (1930-2022) to search the skies for radio transmissions from intelligent beings with the help of the large radio telescope at Green Bank (point 497). This name is also obvious to those familiar with this famous children's book, as it features the character "Long-Eared Hearer," a being that can listen for hundreds of kilometers.

One could now give the "aliens" the usual explanation that is often given to children: "Left is where the thumb is on the right hand." But as you can find out for yourself by thinking about it, this doesn't really help. After all, the task is to explain - in an absolute sense, so to speak - what is meant by "right" and "left" - and we can certainly assume that the "aliens" know these "directions" and have also named them. It's just that "right" or "left" is what they might call "osiotun." Until 1956, nobody had any idea how to solve this problem. Then, in the field of elementary particle physics, an oddity was discovered that had something to do with spatial reflections. Tsung-Dao Lee (1926-2024) and Chen Ning Yang (Nobel Prize 1957) expressed the assumption in a much-noticed paper that under certain circumstances, a quantity called "parity," for which there is a conservation law, can be violated. In other words, there are apparently physical laws that operate differently in a mirror-inverted world than in our usual world. But can this really be the case? The Chinese-born physicist Chien-Shiung Wu (1912-1997) devised an experiment (the "Wu experiment") and was able to show that this is indeed the case.

Now it is relatively easy to explain to the "extraterrestrials" what we mean by "right" and "left." We simply send them a set of instructions for the Wu experiment by radio, which they can use to reproduce the experiment and thus find out where the direction we call "right" is for them. This effectively solves this important problem of communication theory. But is that really the case? If we assume that a mirror not only reflects directions but also what physicists call "charges" (of which there is more than just the electric charge), then the negative electric charge of an electron "behind the mirror" should become the positive charge of a positron. And then the Wu experiment looks completely different. In order to clarify once and for all where the right and left sides of the aliens are, we must first find out (without flying there) whether the "aliens" consist of "normal" or perhaps "antimatter." And so the dilemma that is the Ozma problem continues...

Fig. 68: The British astrophysicist and radio astronomer Frank Drake first wanted to search for "extraterrestrials" with radio telescopes (SETI), which he then did, albeit unsuccessfully...

174. Buridan's donkey

Symmetries also relate to another dilemma that is nothing less than a matter of life or death - and this in connection with an animal that is often seen as frugal and considered "stupid": the donkey.

The medieval philosopher Johannes Buridan, who lived in the north of what is now France between around 1300 and 1358, posed a profound question (in scholastic fashion): *"Would the will, faced with two completely identical alternatives, be able to choose one alternative over the other?"*

To make this abstract consideration more vivid, the famous fable of "Buridan's Donkey" was invented. In this story, the donkey faces a fatal dilemma: he finds himself exactly in the middle between two equally large and equally tempting haystacks. As both options are completely identical, the donkey cannot decide which of the piles to eat first. Unable to make a choice, and with neither pile holding any particular attraction for him, the poor donkey eventually starves to death - a tragic fate brought about by his indecision and the symmetry of the situation.

The story of Buridan's Donkey illustrates not only the philosophical question of freedom of choice and free will, but also the paradox that can arise when a situation is too perfectly balanced. If there is no clear preference, indecision itself can become a trap, which, in the worst case, as with our donkey, can even end fatally. The fable has inspired numerous discussions over the centuries and is still used today in philosophy and decision research as a symbolic example of the difficulties of choice. And so the poor donkey starved to death (though it was just a stupid donkey).

Fig. 69: This dear guy is not Buridan's donkey...

175. Dilemma and Trilemma

A trilemma, on the other hand, is when there are three alternatives, each of which is unacceptable on its own. So a decision between plague and cholera is a dilemma, and a decision between plague, cholera, and smallpox is a trilemma. The latter takes us back to Hans Albers, who famously played "Münchhausen" in the 1943 film, or rather to Hans Albert (1921-2023), the main representative of the philosophical movement of critical realism. His most important tenet is, "All certainties in knowledge are self-fabricated and therefore worthless for grasping reality." Superficially, this concerns the question of ultimate justifications.

176. Ultimate truths

These are "ultimate truths" that cannot be further substantiated and therefore represent the end of chains of reasoning. "I think, therefore I am" is one of these, on which the philosophy of René Descartes (1596-1650) is famously based. If we examine such chains of reasoning more closely, as Hans Albert has done, we find three variants for justifying a given statement. The first variant justifies the statement with another - and this, in turn, with another - and this, in turn, with another - without ever coming to an end (infinite regress). The second variant is best expressed by the well-known folk song *"Wenn der Topf aber nun' n Loch hat, lieber Heinrich ..."* (But if the pot has a hole, dear Heinrich ...), whose verses (apart from the last!) form an unresolvable circle of argumentation.

That leaves the third variant. It consists of an arbitrary interruption of the reasoning process - and this is achieved in the aforementioned song by the last verse, which is known to cast doubt on the cognitive abilities of the "questioning" "Liese" addressed. These three variants of argumentation chains are known to every epistemologist as the "Münchhausen trilemma". Although it does not pose a major problem a priori, it supports the popular wisdom that "everyone and everything is fallible" as well as the realization that "there can be no absolute certainty," even if there is such a thing as "absolute truth." In this respect, one is reminded of the "brain in the nutrient broth" (point 101), which is led to believe that a simulated world is the real world without being able to recognize this.

177. Cosmological proof of God

The category of "infinite regress" also includes the "cosmological proof of God", which is often put forward by theologians and exists in many variations. This line of argument attempts to prove the existence of God through a logical conclusion and can be summarized as follows:

1. everything that exists has a cause.
2. an infinite sequence of causes in the sense of an infinite regress (i.e. the cause of the cause of the cause ...) is inconceivable.
3. There must therefore be a first cause which itself has no cause.
4. This first cause is God.

At first glance, this argument seems conclusive and convincing, but a closer analysis from a logical point of view reveals considerable weaknesses. This begins with a fundamental contradiction between the first two premises. Point (1) asserts that everything must have a cause, while point (2) immediately relativizes this statement by ruling out an infinite chain of causes as unthinkable. This contradiction is serious because, as is known in logic, almost anything can be proven with two contradicto-ry premises - and thus the argument loses its consistency.

Another problem lies in the fundamental assumption of point (1) that everything must have a cause. This assertion is taken for granted without being really tested or questioned. Who says that there must necessarily be causal relationships for everything in nature or in the universe? Modern physics and quantum mechanics show us that events can occur at a subatomic level without a recognizable cause. These phenomena raise serious doubts about the validity of the first premise of the cosmological argument.

Finally, the conclusion in point (3) leads to another logical dilemma by postulating the assumption of a "first cause" that itself has no cause. This assertion directly contradicts premise (1), which states that everything must have a cause. Why should this "first cause" be exempt from this rule? This is an arbitrary exception that further undermines the logical stringency of the argument.

One final remark should be made: Why should this chain of evidence prove precisely the existence of "God" - and not that of, say, the "devil"?

178. All about the devil

Over the centuries, a great deal has become known about this apparently fictitious person with great influence on humanity through the work of many scholars, most of whom are forgotten today. Let's just take the Christian cultural sphere. Through intensive research, it has been discovered that he

a) a fallen angel who once rebelled against God;
b) is the fallen son of the dawn ("Lucifer" - the bringer of light);
c) appeared as the "witness" of the Nephilim (Book of Enoch);
d) is abysmally evil and tempts people to sin;
e) can change his shape at will, but is not particularly attractive in the original;
f) is plagued by an intense body odor that is somehow strongly reminiscent of the smell of burning sulfur;
g) rules over "flies" ("Mephistopheles") and is the boss of the hell;
h) has a walking disability (special clubfoot);
i) is powerful, but not as omnipotent as the "Almighty" himself;
j) is not unwelcome in heaven from time to time (Prologue to Faust I);
k) likes to mate with witches (as an incubus);
l) has three golden hairs and a grandmother (Brothers Grimm)

and sometimes appears as a sly but not unpleasant contemporary with a high level of education (Goethe's "Faust") and can even put the Stasi in need of an explanation (for example, when, shortly before the fall of the Berlin Wall, he visited East Berlin (disguised as Prof. Jochanaan Leuchtentrager from the Hebrew University of Jerusalem) and told the Prof. Siegfried Beifuss from the "Institute for Scientific Atheism", only to send him straight to hell (which, according to Professor Beifuss, should not even exist) - a very vivid eyewitness account of the first flight phase there - over the "Wall" - can be found in Stefan Heym's novel "Ahasver", which otherwise deals with the "eternal Jew").

Fig. 70: This is how the Dutch painter Jacob de Backer (ca. 1540-1600) imagined the devil (Satan) at work (depicted in his painting "The Last Judgement"). The expert in the field of the study of hell can easily recognize all the essential attributes of the devil, which are: Horns of a goat and a ram, a goat's hide and ears, a pig's nose and canine teeth and a dog's mouth... (Wikimedia, National Museum Warsaw)

The "research on hell" (whose criticism, according to a well-known Protestant bishop, is precisely one of the strengths of modern theology) has discovered even more, but its seriousness has unfortunately been increasingly questioned since the beginning of the Enlightenment (Kant).

179. Hell topography, demonology and torture science

After all, throughout history, a number of luminaries have dealt extremely meticulously with this branch of science, which today is only rarely studied. As a result, completely new fields of research have opened up, such as hell topography (which deals with the physical location of hell and its diverse interior), demonology (which concerns itself with the cataloging and description of the appearance as well as the various responsibilities of the infinitely rich specialist personnel of hell), and torture science (which examines the physical and spiritual punishments (poena positiva and poena privativa) that sinners are to expect in hell). Important philosophical questions were also addressed in this context with great subtlety across many folio pages in extensive scientific works, such as the important question of how long such an "eternity" actually lasts, which is closely related to the "eternal punishment in hell."

One of the most important topographers of hell was undoubtedly the Milan university professor Antonio Rusca, whose work *"De inferno et datu daemonum ante mundi exitum..."*, published in 1621, can be studied in a digital edition on Google Books by the inclined reader who understands Latin.[41] In it, he refutes a large number of incorrect assertions made by his fellow experts in the sense of scientific source criticism, only to then prove with impressive logic that hell is to be found neither at the North nor South Pole, neither on the moon nor on the sun, nor on any comet, but where it has always been suspected anyway - in the womb of the earth. The proof is obvious - the ventilation chimneys of hell are nothing other than volcanoes!

In contrast, the work *"Grausame Beschreibung und Vorstellung der Hölle und der höllischen Tor - oder des andern und ewigen Todes - in teutscher Sprache nachdenklich und so vor die Augen gelegt - daß einem gottlosen Menschen gleichsam die höllischen Spken an noch in dieser Welt ins Gewissen stieben - und Rükk-Gedanken zur Ewigkeit erwekken können"*[42] [2] by Justus Georg Schottelius (1612-1676), who made it as far as the Brunswick-Lüneburg court and chamber councillor and "prince educator" at the Wolfenbüttel court. His literary legacy is enormous and the aforementioned work is thematically rather an outlier. But it has it all. Even the opening copper makes your blood run cold (Fig. 71).:

„Als der Prophet Esajas V.v.14. sagt: Die Hölle hat die Seele weit aufgesperret / und den Rachen aufgethan ohn alle masse / daß hinunter fahren beide die Herzlichen und Pöbel / beide die Reichen und Frölichen: Solches deutet der KupfferTitul an: Auch der in der Hölle ewig herrschender Anderer Tod sitzet oben auff / und hält die Höllen-Schlange / den nichtsterbenden Gewissens-Wurm / unzertrennlich bey sich. Was in dem Grunde und Boden / in denen grossen brennenden Schwefel-Pfulen der Hölle für ewigwehrende Marter / Angst / und Betrübniß verhanden / solches steht geschrieben auff denen unten heraußstehenden drei grossen BakkenZähnen / als die brennende Marterqual / die drukkende Angstqwal / und die beissende Reuquaal: welche die drei andere oben heraußragende grosse HöllenZähne gleichfals zustimmen / und durch den grausamen Zusammenbiß und Zuschluß des erschrecklichen HöllenRachen andeuten / wie es unaußsprechliche Qwaalwesen sein und bleiben müsse / unendlich / unvergleichlich / unabwendlich: O weh / und ewig weh! wegen dieser Unendlichkeit / Unvergleichlichkeit / Unabwendlichkeit: Darin zugleich die allergrausamste bitterste Verzweiffelung mit eingeschlossen bleibet. In den HöllenRachen kann man zwar hinnein schauen / und die in feuriger Angst und Qwaal winselnde Menschen erblikken / aber wie man nicht kann das

---

[41] See https://tinyurl.com/bdz6y8ja

[42] See https://tinyurl.com/mtmeup3v

Ende / also kann man auch keine Enderung ersehen / und wird alles mit der allergrau-samsten Ewigkeit um und eingeschlossen."

("When the prophet Esaias v.v.14. says: Hell has opened wide the soul / and opened the jaws without all measure / that down go both the hearty and the rabble / both the rich and the cheerful: Such is indicated by the KupfferTitul: Even the Other Death, who reigns eternally in hell, sits on top / and keeps the hell-snake / the non-dying worm of con-science / inseparably with him. The eternal torment / fear / and affliction that is present in the bottom and soil / in the great burning sulphur pools of hell / is written on the three great beacon teeth standing out below / as the burning torment / the oppressive anguish / and the biting remorse: to which the three other great hell-teeth protruding above equally agree / and by the cruel biting together and closing of the terrifying hell-mouth indicate / how it must be and remain unspeakable qwaal beings / infinite / incomparable / inevitable: O woe / and eternal woe! because of this infinity / incomparability / inevita-bility: in which at the same time the most cruel and bitter despair remains enclosed. One can indeed look into the maw of hell / and see the people whimpering in fiery fear and anguish / but just as one cannot see the end / so one cannot see an end either / and everything is surrounded and enclosed by the most cruel eternity.")

Fig. 71: Title and "title copper" of the standard work on hell by Justus Georg Schottelius from 1676, illustrated with "Schrekknißvollen Kupferstükken"

After this encouraging introduction, we get down to the details. We learn what punishments await the damned there: hunger and thirst, as well as stench and darkness, illuminated only by pitch flames for 1,000 years (to acclimate, so to speak). This is followed by roasting in sulfur flames for up to 20,000 years. If you are not yet consumed, you are tortured with red-hot iron for the next 100,000 years and so on. And when you have finally completed the whole program, it starts all over again - because the punishments in hell last forever...

An even more detailed description of the torments of hell can be found in the Dominican monk Battista Manni. In his work La Prigione Eterna Dell'Inferno, published in Italian in 1677 (available on Google Books), he accompanies his detailed descriptions of the torments of hell with corresponding illustrations, which even today can still make tender-hearted souls (unless they are die-hard fans of such comics) shudder.

Fig. 72: Here it is shown how the tormented sinner in hell is not only tortured by roasting, skewering and biting, but also how his ears are constantly blasted with a horn. The hell expert Battista Manni has included corresponding illustrations in his book so that believing sinners can really imagine and empathize with this while they are still alive.

Incidentally, he came to the conclusion that even the pure and unadulterated sight of the devil was already an unbearable punishment. He thus confirms the truth of the visions of hell by Catherine of Siena (1347-1380), an extraordinary woman who, in one of her many visions, was allowed a glimpse into hell and saw some very horrible demons.

She said afterwards *"that she would rather walk barefoot on streets covered in glowing coals until the Last Judgement than have to face the sight of them again"*. As she was canonized, her wish not to have to encounter these demons again may have come true.

As far as the detailed descriptions of hell in the early modern period are concerned, it is noticeable that the devil himself usually only plays a subordinate role in them, although he is the catcher of souls, head of administration and supervisor all in one. But the solution to the riddle can be found in Faust I, where he frankly confesses „*Da dank ich Euch; denn mit den Toten - Hab ich mich niemals gern befangen. - Am meisten lieb ich mir die vollen, frischen Wangen. - Für einen Leichnam bin ich nicht zu Haus; - Mir geht es wie der Katze mit der Maus.*" (My thanks! I find the dead no acquisition, And never cared to have them in my keeping. I much prefer the cheeks where ruddy blood is leaping, And when a corpse approaches, close my house: It goes with me, as with the cat the mouse).

I think you could have a good and classy chat with him over a glass of wine about God and the world. Even at the beginning of the 18th century, nobody doubted his high level of education and erudition. In 1715, it was possible to gain a doctorate in theology at the venerable Rostock University by answering the question of whether the devil had the necessary skills to successfully study theology (the answer was "yes", by the way).

Fig. 73: No one else has shaped the image of the devil (Mephistopheles) in the German national consciousness as much as Gustaf Gründgens (1899-1963) (National Film Archive)

As is well known, the devil often appears in animal form, e.g. as a submissive dog (quasi as the poodle's core) or as a purring black cat.

180. Black cat - white spot

As a result, almost all black cats today have more or less large white throat patches. Just look at that! The reason for this is actually sad and has to do with human ignorance and natural selection. If you gradually eradicate the carriers of a phenotypic trait (in this case, coal-black cats), then this phenotypic trait will no longer be able to persist in the population. In the Middle Ages and early modern times, completely black cats were thought to be playmates of witches and the devil, and sometimes even Satan himself - and were killed as soon as they could be caught.

Fig. 74: Humpel's buddy (see item 143) is coal-black, but with a white throat patch and white paw tips. You can find out why here.

It is said that they were locked in wicker baskets and even burned at the stake, sometimes with or without the witch. Only black cats that were not completely black - those with a white throat patch or white paws - escaped this fate. In such cases, they were referred to as white "angel hair," and only these cats remained unharmed and were ultimately able to reproduce. In modern evolutionary theory, this is known as positive selection with regard to the corresponding phenotypic trait. This is why almost every black cat today has at least a rudimentary light spot or sometimes white paws. Completely black cats (i.e., those without any white hairs) are extremely rare.

181. Unsolved problems in cat research

While research into hell, at least in the West, has reached a certain conclusion some time ago and is hardly being pursued any further, the same cannot be said of "cat research," the most important but not sole subject of which is the domestic cat, commonly known as the "house tiger." There is still a lot of research work to be done by cat researchers.

When I think of my cat Humpel, for example, I immediately think of a few important research topics that should at least be the focus of a doctoral student. First of all, I would be very interested to learn why cats always lie down on the newspaper you are reading (or on the rather uncomfortable computer keyboard at precisely the moment you want to work on your computer or notebook). Or why some cats change their favorite sleeping place in the apartment every few weeks. A friend of mine also asked me why cats, when they have to vomit (don't worry, this is quite normal for cats), always do so on the expensive carpet and only rarely on the parquet or stone floor. This is a question of great practical importance for cat lovers, as they are the ones who have to clean up the unappetizing vomit...

182. Dead cats don't purr

The real mystery of the domestic cat, however, lies in its ability to "purr" - and despite almost 200 years of research on this subject, there is still no truly satisfactory answer to the question of how it achieves this. It is precisely this low-frequency vocalization that makes cats so likable, in addition to their sometimes, admittedly rather intrusive nature (especially when it comes to petting, scratching, or food). In any case, this vocalization seems to originate somewhere in the throat region. So far, researchers agree. Thus, it was quite logical that the first step was to take a closer look at this area with scientific curiosity using various cutting instruments. However, the only really certain result of such fine anatomical investigations was the realization that dead cats, unlike living ones, do not purr.

Around 1960, there was finally a first modest breakthrough in this important field of research, thanks to a dog that had bitten through the throat of a poor cat, severely damaging its larynx. The cat in question lived for a few more weeks, as an experienced vet had secured her breathing with a tube. However, she was no longer able to meow. Nonetheless, the tube in no way prevented her from purring, which empirically proved that the larynx could not be involved in this particular sound. What followed were a number of unappetizing experiments on live cats, about which a true cat lover would not really want to know anything more detailed, which is why I have deliberately refrained from describing them here. Even so, they did not lead to an exact localization of the cord apparatus. However, at least some hypotheses could be put forward for future research, such as the "hyoid bone hypothesis" and the "false vocal cord hypothesis."

The "12th International Conference on Low Frequency and Vibration and its Control," which took place in 2006, shows that research into this phenomenon, which in the animal kingdom is only possible in species from the Felidae family, was considered fundamental to biological science. However, the matter is rather complicated and not necessarily comprehensible even to a layperson, which is why it cannot be discussed in more detail here. Anyone who would like to know more about it is referred to the relevant specialist articles.

183. The "seven lives" of the cat

Just this much: it has something to do with the legendary self-healing powers of cats (i.e., their "seven lives"). The expression "seven lives" dates back to the late Middle Ages and early modern times, when the German countryside was overrun by a witch craze that is difficult to explain. At that time, people not only slaughtered and burned black cats but also sometimes threw them from church towers. Because cats have developed a special reflex that causes them to turn in the air during a fall so that they eventually land on their paws (the so-called "stance reflex"), they often survive such a fall from a great height, usually a little dazed but otherwise undamaged or only slightly injured. A human or other animal (unless it is a bird) would certainly break all its bones and would therefore hardly survive a fall from a church tower. The people of the time could not explain this effect and thus assumed that the devil granted cats seven lives.

184. Pisans can't even build straight towers

A certain Galileo Galilei (1564-1641) carried out drop tests of a completely different kind, but at around the same time, on the Leaning Tower of Pisa. Even in his time (around 1590), this tower, whose foundation stone was laid in 1173, was already leaning. It was certainly not the intention of the master builders to erect a "leaning" tower ("The Pisans can't even build straight towers" - as was mockingly said in Genoa, which was at enmity with Pisa at the time).[43] The fact that this nevertheless happened was due to a criminal neglect of the precise investigation of the building ground. This is because the "leaning" of a tower - and a leaning tower does not only exist in Pisa - usually only happens after the building is completed and then only very gradually. The Tower of Pisa (more precisely the one planned as a free-standing bell tower for the cathedral) is rather an exception, as it began to lean during the first construction phase, which caused construction to stop for around 100 years. Only then was a master builder found who had the courage to complete the ruined building - and the city of Pisa, which was dependent on income from tourism, is still grateful to him today...

---

[43] For more details, see e.g. Mosaic by Hannes Hegen, issue 95 ("As a prisoner of the Pisans"), in which Knight Runkel von Rübenstein (Heino) was captured together with Dig and Dag by the Pisan admiral Totalo Flauti...

Fig. 75: Isn't it leaning, the "Leaning Tower" of Pisa? (Wikimedia)

185. Friesland - the land of leaning towers

There are many more leaning towers in the world, even in Germany. The deviation of the Pisa bell tower from the vertical is currently 3.97°, and it will hopefully not change once the foundations have been successfully renovated. By contrast, the steeple of the Midlum church in western East Frisia, which is said to date back to the 14th century, is much more "leaning" (6.74°). As a brick building, it is certainly not as graceful as the Tower of Pisa, but since 2010 it has been officially recognized: the bell tower in this town, which is hardly known compared to Pisa, is the most leaning of all leaning bell towers in the world. If you don't believe it, you can read it for yourself in the Guinness World Records or, even better, travel to Midlum yourself.

Incidentally, East Frisia has a particularly high density of leaning buildings in Germany. The leaning tower of Suurhusen (5.19° inclination) and the free-standing bell tower of the "Johannes des Täufers" church in Engerhafe are particularly noteworthy. The reason for this lies in the mostly spongy building ground near the coast.

While there was not a single case where a tower was conceived as "leaning" in earlier times, today we are correspondingly more advanced. The leanest of all leaning buildings in the world is the Capital Gate Tower in Abu Dhabi, with an inclination of ~18°.

Fig. 76: The brick bell tower of the Midlum church in Friesland is nowhere near as deli-cate as the bell tower in Pisa (and still largely unknown), but it is almost twice as leaning (Wikimedia, F. Vincentz)

186. The Tower of Hanoi

The "trickiest" tower in the world - though not in architectural terms (in this respect, it is more accurately described as "simple") - but in mathematical terms, is the Tower of Hanoi. It typically consists of wooden disks of decreasing radius, each with a central hole designed to fit onto a central post according to size (from "large" to "small"). Initially, there are two additional posts positioned next to it (as shown here), but they remain empty, meaning no disks are placed on them at the start.

The objective is to dismantle the tower from the initial post, following two simple rules: *"You may only move one disk at a time"* and *"You may never place a larger disk on top of a smaller one,"* and then rebuild it in the same configuration on the middle post. The third post serves only as temporary storage.

The problem is relatively straightforward with three disks, requiring exactly seven moves to achieve the shortest solution. I'm confident you can figure out the solution on your own.

Fig. 77: This is it, the Tower of Hanoi. Unfortunately only known to fans of tricky mathematical games - and computer science students who have to deal with the paradigm of recursive programming (Wikimedia)

It gets interesting when you increase the number n of disks. You then need at least 15 steps, 31 steps, 63 steps and 127 steps. If you take a closer look at the series of numbers 7, 15, 31, 63 and 127, you quickly realize that you are dealing with a formation rule of the form (2^{n-1}).

In the original version of the story, it is said that there was a tower in the great temple of Varanasi (in the Indian state of Uttar Pradesh, long regarded by Hindus as the center of the world) composed of 64 golden discs. As you can easily deduce, it would require at least 18,446,744,073,709,551,615 moves to transfer it from the first post to the third post according to the rules. Theoretically speaking, if the monks had begun rebuilding at the start of time (i.e., 13.79 billion years ago) and moved one disc every second in exactly the correct manner, what do you think they would have accomplished today, or would they still be struggling?

187. Mersenne numbers and prime numbers

The number series with $n \geq 0$ shows that the Tower of Hanoi is not just a mathematical gimmick but also hides great secrets of number theory. It provides very special numbers, which are called "Mersenne numbers" after the mathematician Marin Mersenne (1588–1648). But what makes these numbers so interesting for number theorists? The reason lies in the discovery, which in turn goes back to the aforementioned French monk, that prime numbers are particularly common among the Mersenne numbers. As we know, these are natural numbers that can only be divided by 1 and by themselves without a remainder. It has been known since antiquity (through the "sieve of Eratosthenes") that there are an infinite number of prime numbers among the infinite number of natural numbers. It is also known that every natural number can be decomposed into prime factors (a result known as the fundamental theorem of arithmetic) and that this is an extremely difficult task for large numbers (prime factorization). However, it is still not known whether every natural number greater than 2 can be written as the sum of two prime numbers (Goldbach's conjecture) and the connection between Riemann's zeta function.[44] and the so-called "prime number function" (which specifies the number of prime numbers that are smaller than any given number) has still not been proven (Riemann's conjecture).

At first, Mersenne assumed that all numbers of the form 2^{n-1} are prime numbers, but he was quickly able to disprove this by trial and error. Because with you get 15, and that is obviously not a prime number. He also discovered that 2^{n-1} cannot be a prime number if it is not a prime number itself. Checking the assumption that a Mersenne number is always a prime number if it is also a prime number required quite a lot of arithmetic before he realized that $2^{11-1} = 2047$ is and that this number can be broken down into the two factors 23 and 89. A counterexample had thus been found and the conjecture was a thing of the past. There are now 48 such prime numbers known, the largest of which is $2^{57885161-1}$.[45][2] For good reason, I will not list them here in full...

Mathematicians believe (the proof is still pending) that the number of Mersenne primes is just as large as the number of all prime numbers together, namely infinity. It therefore makes perfect sense to check numbers with n>57885161 to see whether they might be prime.

---

[44] It looks like this (harmless): with for: $\zeta(c) = \sum_{n=1}^{\infty} \frac{1}{n^c}$ mit $c \in \mathbb{C}$ für $Re(c) > 1$

[45] There are currently (2024) three more numbers – those with $n = \{74.207.281; 77.232.917; 82.589.933\}$ on the list of "presumed" Mersenne numbers. However, the final proof that these are prime numbers is still pending.

188. Citizen science

And there is a citizen science project on the Internet for this purpose, in which anyone with a computer can participate. You can find it, including instructions and many other things about this type of prime number, at www.mersenne.org. If you succeed in finding a new Mersenne prime number, then you are guaranteed eternal fame among math enthusiasts. However, if you are not so interested in numbers, there is now a large selection of other scientific projects where your help is needed.

For example, if you (like the author) are well versed in the local bird life, you can enter your observations at www.ornitho.de. If, on the other hand, you are someone who likes mosquitoes, then the www.mueckenatlas.de project is the right place for you. In the field of astronomy, there are also many opportunities for laypeople to become scientifically active. The best known of these is the SETI@home project, which was ostensibly set up to sift through the immense flood of data produced by radio telescopes worldwide on a daily basis for suspicious signals. Normally, this would require the computing capacity of supercomputers, but of course these are not available for such projects, which, strictly speaking, are not particularly promising. However, there are huge amounts of computing capacity that lie idle most of the time. I'm referring to the hundreds of millions of personal computers that are now sitting around in offices and home offices everywhere. The Internet has made it possible to pool their capabilities in order to tackle problems that would otherwise be almost impossible to solve. The magic word is "distributed computing". The task of mathematicians (and physicists) is to develop or select algorithms that are designed in such a way that the problem can be broken down into many discrete "chunks", each of which is then passed to a computer on the Internet for processing. Once it has finished, the computer sends the result back to the control center and gets the next piece of work. Ingenious, really. A problem that would otherwise take a computer a year to work on is completed in one day using 365 computers.

It is thanks to the Internet that virtually anyone interested can now take part in the processing of tasks that are otherwise reserved for supercomputers due to the amount of data to be processed (e.g., from high-energy physics or genetics). It has now become so integrated into everyday life that nobody really knows how it all started and how it works at all.

189. Computer networks

Well, a computer network began at some point with two computers that could communicate with each other via a (telephone) cable. One of the two computers was located at the University of California in Los Angeles and the other at Stanford University in San Francisco. The memorable date on which these two computers exchanged data for the first time was November 21, 1969. Shortly after, additional computers were connected in Salt Lake City and Santa Barbara, and as the whole project was funded by the Penta-gon as part of the *Advanced Research Projects Agency* (ARPA), this early version of the Internet was called "ARPAnet."

As this computer network was initially mainly under the aegis of the military, it grew only slowly. By the early 1980s, only a few hundred computers scattered across the USA were online, most of which were so-called "mainframes." More and more civilian research in-stitutions and universities also began to take an interest in this technology, as it allowed for quick and easy data exchange. In the early 1980s, the era of low-cost personal com-puters began with the IBM PC, and Ethernet (1983) was introduced as a connection pro-tocol that enabled simple networking for local computer networks. By the end of the decade, there were already more than 28,000 local networks connected to each other via ARPAnet. This was also when the military began to withdraw from network activities, and the network, now called the "Internet," became a largely civilian institution. The con-tinued success of this network with its world-changing impact is due to the way the com-puters communicate with each other.

190. How does the Internet work? - TCP/IP

And this "way" is completely independent of the networked computers / smartphones, the communication channels and national characteristics and is defined by a so-called "protocol" that only all network participants have to strictly adhere to. This protocol is the "*Transmission Control Protocol / Internet Protocol*", or TCP/IP for short. And the "*Hy-pertext Transfer Protocol*" (HTTP), on which the "user interface" of the Internet, the "World Wide Web" (WWW), is based, is also based on this protocol.

So, how can we imagine data exchange on the World Wide Web (WWW)? The first re-quirement is that each network participant (computer) be uniquely identifiable; in other words, it requires an identification number that exists only once in the entire network. This identification number is the IP address. Its exact structure is not the focus here, but what's important is that it serves as both the sender's address and the recipient's ad-dress for the data packets transported via the network. Essentially, "sending data pack-ets via IP" is analogous to dividing a book into individual pages and placing these pages, or a "pack" of pages, in envelopes marked with the sender's and recipient's IP ad-dresses. Even if some of these letters are sent by airmail, while others travel by ship or car, they will all eventually reach the recipient. At that point, the book can be reassem-bled, regardless of whether page 1 or pages 12-24 arrived first.

Strictly speaking, IP addresses are just unique strings of numbers, which are, of course, difficult to remember. That's why something like an "Internet telephone book" has been created. For example, this system assigns the website "de.wikipedia.org" to the IP ad-dress 185.15.59.224. However, I cautiously assume that the former is easier to remem-ber than the numerical sequence it represents. And, of course, the Internet telephone directory also has a name: DNS (Domain Name System). Furthermore, computers on the Internet that translate "understandable" web addresses into cryptic IP addresses (and vice versa) are called "DNS servers."

There are also different protocols for "envelopes" on the Internet, depending on what data they are transporting. For example, if the "envelope" contains email data, the corresponding protocol is SMTP (Simple Mail Transfer Protocol). On the other hand, if it contains data that opens a website for the recipient, then the HTTP protocol is used. If you are sending individual files, such as PDFs or Excel files, the FTP protocol (File Transfer Protocol) is commonly used. Of course, there are many more protocols beyond those mentioned here. As long as the "envelopes" (data packets) contain recipient and sender addresses in standard format, the design of the content inside the envelopes remains flexible. The devices that transport the data stream ("routers" and "switches") are ultimately only interested in the address details, not the content of the data packets. From this perspective, the Internet acts merely as a service provider that transports data from one computer to another, regardless of the type of device (mainframe, PC, smartphone...), the location of the computer, or the type of data being transported. That's all there is to it. The "miracles" made possible by this technology become evident as soon as you switch on your PC, notebook, or smartphone and go "online."

191. A sufficiently advanced technology...

The Internet in general and the World Wide Web in particular can certainly be regarded as a *"sufficiently advanced technology"* in the sense of Arthur C. Clarke (1917-2008), which is virtually indistinguishable from "magic".

Arthur C. Clarke, together with Isaak Asimov (1920-1992) and Robert A. Heinlein (1907-1988), was the person in the "West" who helped science fiction literature achieve its breakthrough. Just think of the great novel *"2001: A Space Odyssey"* from 1968, which was also made into a movie (not by just anyone, but by Stanley Kubrick (1928-1999) himself!)

192. Solaris

In the "Eastern bloc", his counterpart was undoubtedly Stanislaw Lem (1921-2006), who was just as brilliant a utopian as Clarke. His most famous work is Solaris from 1961, which many connoisseurs are familiar with through the film version by Andrei Arsenyevich Tarkovsky (1932-1986). The fact that this is an extraordinary work is evident from the three film adaptations and numerous stage plays that have been adapted from it. Additionally, the fact that it was not allowed to be printed in the GDR in 1962 further underscores its significance. It was not published by the East German publishing house "Volk und Welt" until 1983, by which time two Soviet film versions were already available. The version that the great Soviet director Andrei Arsenyevich Tarkovsky released in 1972, despite significant resistance, is considered a highlight of film history. It is not necessarily a film intended purely for entertainment. Its significance lies in the fact that it quietly yet powerfully realizes the parable about death, love, and resurrection (which is only a minor theme in Lem's novel). However, the US film adaptation by Steven Soderbergh, starring George Clooney in the leading role, is also a notable interpretation of the complex material. Nevertheless, Stanislaw Lem was not satisfied with either film adaptation, as they deviated too much from his original novel.

193. For friends of Soviet film

In Soviet times, a large number of films were made in the former Soviet Union that definitely deserve the title "masterpiece". I am not just referring to the many fairy tale films that shaped the childhoods of many GDR citizens and which have not lost their unique charm even today. I will only mention *"Abenteuer im Zauberwald"* (Adventure in the Enchanted Forest) from 1964 as one example among many. And if you want to understand Russian sensitivities, you should watch the 1956 fairytale film *"Ilja Muromez"* by Alexander Ptuschko (1900-1973), among others. In this context, I am referring to films that criticize society in a subtle way and had a hard time getting through the censorship of the time. They were usually only shown at various festivals and were hardly ever shown on GDR television.

But Russian history also repeatedly harbored themes that Soviet directors such as Sergei Fyodorovich Bondarchuk (1920-1994, *"War and Peace"*, *"Waterloo"*), Sergei Mikhailovich Eisenstein (1898-1948, *"Ivan the Terrible"*, *"Alexander Nevsky"*), Andrei Arsenyevich Tarkovsky (*"Andrei Rublev"*, *"The Mirror"*, *"Nostalghia"*, *"Stalker"*) and Elem Germanovich Klimov (1933-2003, *"Agonia"*). I don't want to mention the heroic war films, which glorified the "Great Patriotic War" with massive material expenditure, in detail. I am referring more to silent films such as *"Farewell to Matyora"* (Elem Germanovich Klimov, Larissa Shepitko (1938-1979) from 1983, *"Dersu Usala"* (Akira Kurosawa (1910-1998)) from 1975 and *"Letters from a Dead Man"* by Konstantin Lopushanski from 1986, which is more or less the Soviet answer to the Hollywood film *"The Day After"* (1983) with Jason Roberts (1922-2000) in the leading role.

The movie *"Letters from a Dead Man"* is even more disturbing than *"The Day After"*. I saw the film in 1986 at the "Festival of Soviet Film" in Leipzig, where I was visiting a friend who had also managed to buy the coveted tickets. What stuck with me was not so much the content of the movie, but its ending. Normally, the first guests start to leave the theater as soon as the credits roll. Not here. Only when the lights went up did the first guests begin to get up - without saying a word. The theater emptied silently... It was pure madness. However, I doubt that the movie would still evoke such emotions today. An accidental nuclear war (or a nuclear war that has been caused or - see Ukraine - escalated) is no longer a pressing concern for the population today. The generation of those who were involved in the war and those who had to clear away the rubble has naturally thinned out, and for younger generations, wars are more like exciting video games.

When it comes to "dangers to humanity," the topics here tend to be "Castor transports," "nuclear waste disposal," and "genetically modified food," to name just a few of the noteworthy issues alongside the much more horrific. Nobody wants to admit that there is another "elephant in the room" that nobody wants to see. The phrase *"We haven't escaped it yet"* still applies today. The "Doomsday Clock" is currently (2024) at 1.5 minutes to midnight. The last time it reached a similar reading was in 1984 at the height of the nuclear arms race. But nobody seems to care. People seem to believe that advances in information technology will somehow make the accidental triggering of a nuclear war less likely. However, this is a misconception. As decision-making chains become more and more automated, the scope for human influence diminishes correspondingly.

194. And it went "boom"…

A nuclear attack on a major city such as Berlin would be an event of catastrophic proportions, and would, strictly speaking, be beyond anyone's imagination. To fully understand the impact of such an explosion, it is necessary to consider the immediate and long-term consequences from both a physical and human perspective. First of all, the explosion of a nuclear warhead would create a huge pressure wave that would spread at supersonic speed. Buildings within a radius of several kilometers around the detonation point would be completely destroyed. Massive high-rise buildings and infrastructure such as bridges and tunnels would be severely damaged or destroyed by the immense blast wave. People within this radius would have little chance of survival; they would be vaporized by the heat, or burned and killed by the pressure, flying debris, or collapsing buildings. This would happen so quickly that the victims would not even be able to register it.

The resulting heat wave could also trigger fires even further away from the explosion site, which could spread quickly and set fire to large parts of the city. The blinding brightness of the explosion could cause temporary or permanent blindness in people looking directly into the blast.

A danger that persists long after the smoke has cleared is the large amount of ionizing radiation released during the explosion. This radiation can cause cell damage, leading to acute radiation sickness. The symptoms include nausea, vomiting, diarrhea, and bleeding, which occur within hours to days after exposure. In the immediate vicinity of the explosion, the radiation dose is so high that it can be fatal within minutes to hours. After the explosion, a radioactive cloud would rise into the atmosphere and fall back to earth as fallout. The fallout contains radioactive particles that can spread over large areas depending on the weather conditions. People exposed to fallout risk long-term damage such as cancer and genetic mutations. The contamination of soil, water, and food chains could last for decades and make the surrounding area uninhabitable.

Survivors of the initial explosion and fallout would be at high risk of chronic disease. Cancer rates, particularly thyroid, lung, and leukemia cancers, could rise sharply. Mental trauma and long-term health problems such as cardiovascular disease would place an extreme burden on survivors and healthcare systems. The destruction of infrastructure could disrupt the supply of water, electricity, food, and medical aid. Emergency services and hospitals would likely be overloaded or destroyed, making effective aid impossible. The destruction of transport routes could make evacuations and the transportation of relief supplies much more difficult.

In addition to this are the global consequences that are to be expected, particularly in the event of a global nuclear exchange between the nuclear powers, such as the emergence of a "nuclear winter" with unforeseeable consequences for the earth's entire biosphere. In short, the outbreak of a global war using nuclear weapons would be a real threat to the survival of human civilization.

At present (2024), the conflict between Ukraine and Russia in particular has evolved into a complex geopolitical problem that is influenced by numerous factors. Tensions between the two countries began with the annexation of Crimea by Russia in the course of the so-called "Orange Revolution" in 2014 and escalated into an ongoing conflict in eastern Ukraine. Due to the history and development of the conflict, this case can certainly be described as a "proxy war" between the USA (NATO) and Russia, as both sides consider Ukraine to be strategically important territory.

Global political changes, such as the observed transition from a unipolar to a multipolar world, have also contributed to the tensions. The USA, once the undisputed superpower, is now confronted with the rise of other powers such as China and Russia. This loss of hegemony has led the US and its allies to try to maintain and expand their spheres of influence, which could lead to conflicts with other powers.

Russia, for its part, is attempting to maintain and strengthen its own sphere of influence, particularly in the former Soviet republics. Ukraine, which has both cultural and historical ties to Russia but at the same time seeks Western values and rapprochement with the EU and NATO, finds itself in a geopolitical field of tension in which it is now at risk of being caught in the middle.

In the current Ukraine conflict, we can observe in slow motion how a local conflict is being escalated further and further, especially from the outside, how diplomacy is being deliberately avoided, and how we are sleepwalking ever closer to the brink of a nuclear conflict (in which, for example, strategic facilities of the Russian Federation are increasingly coming under fire), all to the encouraging drumbeat of the modern mass media.

For most people, a possible World War III does not seem to be an issue because of the mistaken assumption that such a thing simply cannot happen because a) we have responsible politicians, b) there are international agreements and institutions that ensure control over nuclear weapons, and c) the horror and destructive power of these weapons are so daunting that no rational-thinking person would use them. These assumptions are based on a deep trust in the rationality and stability of the world order. However, it is important to question these assumptions critically. Firstly, history shows that even responsible politicians can make mistakes. Misunderstandings, miscommunication or unforeseen escalations in times of crisis can quickly spiral out of control. The Cuban Missile Crisis of 1962 is a striking example of how close the world came to a nuclear conflict[46] and this was not the only incident in which miscalculations or even accidents involving nuclear weapons ("broken arrow events") occurred. As far as the latter is concerned, an unofficial list of over 1000 such accidents in the USA alone was published in 2001 (the "official" list contains exactly 32 entries). And there have also been a number of miscalculations which, under unfavorable conditions, could well have led to nuclear war.

195. We haven't gotten away with it yet

The textbook example is linked to the name Stanislav Yevgrafovich Petrov (1839-2017), who the world only learned about in 1998. It could be that we all owe him our lives.

Fig. 78: Could it be that we all owe our lives to this humble gentleman? (Wikimedia)

1983 was the year when a Soviet interceptor shot down a Korean passenger plane that had entered Soviet airspace for reasons that are still not fully understood. All 269 passengers aboard the Boeing 747 were killed. At the time, Ronald Reagan (1911-2004) was the President of the United States, and Yuri Vladimirovich Andropov (1914-1984), who was already seriously ill, was in charge of the Soviet Union. The political atmosphere between the superpowers was more heated than it had been for a long time. During this tense period, on September 26, 1983 (26 days after Korean Air 007 was shot down over Sakhalin), the sirens began blaring at the command post of Serpukhov-15, the computer center of the satellite-based missile warning system "Oko." Colonel Stanislav Petrov was the officer on duty that day. The huge screen showed the launch of an intercontinental ballistic missile from an American military base, registered by the military spy satellite Kosmos-1382. In around 25 minutes, it was projected to reach a city in the Soviet Union.

---

[46] Today, there is no longer a secret direct line between the leaders of the nuclear powers and no confidential negotiators. There is only mistrust, speechlessness, almost personal-ized hatred, especially towards "Putin", as well as a lot of speculation about mutual inten-tions and "what ifs". That should scare us all.

Every military officer on duty at the complex knew what had to be done according to regulations. The countdown to the apocalypse began. But Colonel Petrov, who had an engineering background and had switched careers, wondered how likely a first strike with just one missile was, considering it would trigger a massive counter-strike. He concluded that it must be a false alarm. He was not dissuaded when the satellite reported a second, third, fourth, and fifth missile launch in quick succession.

According to American military analysts, the world was never again as close to nuclear self-extinction as it was on that night of September 25-26, 1983. Although Petrov did not have the final decision-making authority (this responsibility lay with the seriously ill Andropov), he was able to interrupt a potentially fatal chain of events at a critical moment that could have inadvertently led to nuclear war. The following morning showed that he had made the right decision, contrary to protocol. This was because the "Oko" system had misinterpreted reflections of the rising sun on clouds near Malmstrom Air Force Base in Montana, where US intercontinental ballistic missiles were also stationed, as missile launches.

The question that concerns us all, but is ignored by the majority of the population, is whether such a situation, which could unintentionally escalate into nuclear war, could happen today. The answer to this question is clearly "yes." As long as nuclear weapons with overkill capacity exist in the world, the danger that they will be used - whether intentionally or unintentionally - remains present. There should be no illusions about this.

196. Self-made disasters

The 21st century is ushering in troubled times. Global financial crises are shaking the social cohesion of entire states, and population growth is accompanied by increasing scarcity of resources. The balance of power between economically influential states is shifting, while enemy stereotypes, previously thought to be overcome, are being revived. Anachronistic religious conflicts, in which the deepest aspects of the Middle Ages collide with modernity, are becoming increasingly uncontrollable. Wars that stay below the nuclear threshold are becoming more viable options, while the art of diplomacy continues to deteriorate. In addition, the ever-advancing networking of all areas of life through information technologies not only brings obvious advantages, but also opens new gateways for terror and destruction with consequences that are difficult to predict.

Just as worrying is the fact that mere technical mishaps can lead to catastrophes that develop a momentum of their own and ultimately become almost impossible to control. Just think of the accidental release of a lethal pathogen designed as a biological weapon[47] or a software error that might cause a chemical factory to blow up like Bhopal. And many more examples can be found here.

---

[47] Think of COVID-19, where the origin of the pathogen is still not 100% clear and where there are serious indications that molecular biologists had a hand in its "creation"...

One term that is often heard in this context is "security vulnerability." Some software companies can no longer keep up with patching the kind of security vulnerabilities found by ambitious hackers. When they only affect home PCs, they may be bearable. However, if they allow access to important, life-sustaining systems such as those in the energy industry or military, then all it takes is one lunatic with the same mindset as those who spread computer viruses to make disaster scenarios come true - scenarios we don't even want to imagine. Of all the global disasters that could befall us (meteorite strikes, supervolcano eruptions), the man-made ones are undoubtedly by far the most likely.

197. Disturbed risk perception

This reveals an interesting phenomenon related to risk perception. Risks are always perceived in an inherently subjective way. Objectively, however, a risk can be quantified, for example, by the probability of occurrence and the potential damage balance. For instance, an individual may perceive an objectively improbable event, such as a plane crash, as a greater risk than the significantly more likely event of becoming the victim of a traffic accident on the way to the airport. This discrepancy between risk perception and objective, precisely quantifiable risk has now also become a tried and tested means of exerting political influence.

A notable German term, which the Brothers Grimm would certainly have enjoyed and which describes this situation perfectly, is "German Angst." It is intended to describe a specific form of anxiety and apprehension that Germans are said to experience. It is used whenever there is a need to express a particular tendency of the German population towards deep, often exaggerated concern and pessimism regarding social, political, or economic developments. It became particularly popular from the second third of the 20th century when Germany became known for its intense debates on environmental and security policies. Since then, the Germans' biggest enemy still seems to be the "atom" (albeit with a downward trend) alongside the "gene." The strong opposition to nuclear power and genetic engineering vividly illustrates the deep-rooted skepticism and the need for security and control in German society - and the suppression of rational considerations regarding a safe and environmentally friendly energy supply. Green indoctrination has done a great job here, far removed from any objectivity. This is always disastrous when pure ideology comes up against deficits in the education system.

198. Pisa sends its regards

Pisa[48] sends its regards, because those who know nothing must believe everything, including the fears that the media spread every day in unison with, according to my observations, not only green politicians. The gap between the findings of experts in the field of nuclear energy and radiation protection (based on decades of research work) on the one

---

[48] PISA is a school performance test initiated by the OECD, in which Germany just reaches the average of all OECD countries

hand and "popular opinion" on the other has become so wide, fueled by politics, that it no longer seems possible for both parties to have a sensible exchange of arguments that goes beyond a hearty "then move to Fukushima!". Incidentally, such "defiant responses" are typical of people whose half-knowledge no longer allows for reason-able argumentation due to a lack of intellectual mass. They can often be found in the comment columns of online articles and blogs.

Incidentally, the underlying reason for the particularly high-risk perception in relation to radioactive radiation is easy to understand. Radioactive radiation cannot be seen or tasted - it can only be "heard" if you have a Geiger counter to hand. Many people would not only be surprised but might even be frightened if they were allowed to walk around their home with such a device in their hand. *"Fear of cancer eats away at the soul"*, was recently written in the German magazine "Der Spiegel" . There real-ly is something to it. According to the WHO, the main health problems in both Chernobyl and Fukushima are psychological problems that have to do with this "cancer fear". Not much can be done about this, even with education. It is easy to understand that people who have lost their homes and their lives suffer from stress symptoms, depression and other psychological problems. Treating these ailments became a far greater challenge for Japanese doctors than treating the virtually non-existent radiation patients.

199. Chernobyl, Fukushima and their nuclear deaths

The assessment of the objective health consequences in relation to the Fukushima reactor accident (it is always astonishing to note how some politicians and media representatives suggest in a difficult way and with remarkable chutzpah that the ~18. 500 victims reported dead or missing from the tsunami caused by the Tohoku earthquake have anything to do with the Fukushima reactor accident) and even more from that of Chernobyl diverge completely, depending on whether one refers to official figures from studies commissioned by the WHO and the UN, to the information provided by non-governmental groups with a negative attitude towards nuclear energy (Green Peace, Doctors for Peace) or self-proclaimed nuclear critics. It is precisely in this order that an exponential increase in the number of victims can be observed. Unfortunately, as an unbiased observer, it is impossible to assess which figures come closest to reality. What does emerge, however, is the realization that the acute radiation sickness caused by particularly high radiation exposure has claimed surprisingly few lives (Chernobyl according to official figures around 30 as a direct result of explosion trauma and acute radiation syndrome and an estimated approx. 4000, which can be seen as a long-term consequence of the increased radiation exposure;[49] Fukushima one case of lung cancer four years later, 2202 deaths as a result of the evacuation measures), while in the case of Chernobyl the cancer rate for certain types of cancer in the radioactively exposed population is

---

[49] "The impact of Chernobyl's nuclear disaster 33 years later". PBS NewsHour Weekend. Public Broadcasting Service.

statistically well proven to increase over time (this applies in particular to thyroid cancer in children). The same is expected in the case of Fukushima. However, due to the radiologically justifiable only slight increase in the probability of illness there, only the future will show whether the resulting objective increase in cancer cases can be proven to be statistically significant at all. Here too, the objectively verifiable risk and the "perceived" risk are far apart.

It was not uninteresting to follow the media treatment and "processing" of the Fukushima reactor disaster over a longer period of time in the quality press with a background in physics studies. In short, it was frightening and amusing at the same time - if the occasion hadn't been so sad. I would like to illustrate this with an article from "Welt online" with the worrying headline "*7400-fold dose of caesium in fish from Fukushima*" (March 16, 2013).

200. The dead fish from Fukushima

The reason for this was the following press release, which is quoted here in the original:

"Tokyo Electric Power Co. said Friday it detected a record 740,000 becquerels per kilogram of radioactive cesium in a fish caught in waters near the crippled Fukushima Daiichi Nuclear Power Station"

Without question - for someone who is not familiar with the unit of measurement "Becquerel"[50] (and that will have been most of the readers of the above article), a figure of 740,000 in relation to radioactivity will most likely seem somehow threatening. The reference to the fact that this figure corresponds to 7400 times the dose[51] of "caesium" (whatever that means) is obviously intended to emphasize how dangerous it is to have to live near Fukushima. Someone who knows nothing about the unit of measurement Becquerel (Bq) and therefore has no comparative values in mind will understandably be shocked by this headline and perhaps even wish every "nuclear power advocate" a "vacation in Fukushima" - just read the comments sections on such articles...

So what is this report (I always refer to the Welt Online version) all about? First of all, the author's emphatically unfortunate wording needs to be corrected and the fact described as follows: "Near the Fukushima nuclear power plant (where fishing is not allowed anyway), the power plant operator (!) reports that a fish was caught with an activity of 740,000 Bq/kg, which can be attributed to (radioactive) caesium." The picture provided in the article is probably intended to show such a fish. In my opinion, however, it can only be a golden trout (Oncorhynchus aguabonita) - which immediately raises the question of how it may have ended up in the Fukushima Daiichi harbor basin from the rivers of California. But the editors probably just wanted to add a suitable picture to their report... However, if you look at the original sources, the ominous fish is probably a specimen

---

[50] The "becquerel, abbreviated as Bq, is the SI unit for the activity of a radioactive sub-stance and indicates the number of atomic nuclei that decay radioactively per second.

[51] Correct is "7400 times the amount of"...

from the family of "greenlings" (i.e. grop-like, bottom-dwelling fish that are very common around Japan). These fish take up a lot of radioactive caesium-137 with the bottom sludge and accumulate it in their muscle flesh for a while. All the extremely heavily contaminated fish from the area were obviously greenlings, with the original record being around 510,000 Bq/kg. These fish all lived near a water outlet from the NPP into the harbor basin there.

Fig. 79: As you can already see from the color, the "dead fish from Fukushima" pictured in the article cannot be a "greenling" (of which several species live in the sea near Fukushima)...

740,000 Bq only means that 740,000 beta decays of the caesium isotope Cs-137 take place every second. This figure in itself says nothing at all. You have to relate it to something - in this case to one kilo of "fish". Then it is easy to calculate how much Cs-137 is in this one kilogram of "fish". All you need to know is the specific activity of one gram of caesium-137, which "Wolfram Alpha" immediately tells us: 3,214 TBq/g. One kilo of muscle meat from the aforementioned fish therefore contains 230 ng (nanograms) of caesium-137. The question that now arises is: is this still healthy or will it inevitably kill you? The only way to answer this question is to calculate the radiation dose that you will be exposed to in addition to the natural radioactivity if you incorporate 1 kg of this fish (e.g. broken down into a large number of sushi rolls, although I don't know whether "greenlings" are even suitable for this). In this context, "(energy) dose" refers to the energy that is released into the surrounding volume in the form of ionizing beta radiation (electrons) during the decay of caesium-137. It is measured in joules per kilogram. The unit of measurement for this is the gray (Gy).

What is much more interesting, however, is the biological effect of the radiation, expressed by the so-called equivalent dose, which is given in sieverts (Sv). To determine it, a conversion factor is used, which is known as the dose conversion factor and is $1.3 \cdot 10^{-8}$ Sv/Bq for caesium-137. This can now be used to determine the dose of the fish irradiated with 740,000 Bq:

Dose=dose conversion factor*activity*mass of the fish

This means that if I eat the entire kilo of fish, I will receive an additional dose of 9.62 mSv. In order to interpret this value correctly, comparative values are necessary. And for this you need to know that the average (natural+artificial) radiation exposure of the population in Germany per person and year is ~3.1 mSv. The fish therefore increases my annual exposure by about twice as much. So you shouldn't eat such heavily contaminated fish too often - and preferably not at all. However, it certainly won't cause death, unless you choke on one of its bones (according to hearsay, at least). In any case, I am not aware of any documented cases - so please take my statement polemically).

What is interesting in this context is that the (specific) activity limits (i.e. the Becquerel values per kg of food) are set completely differently from state to state. In Germany and the USA, the exposure limit for adults in relation to food is generally 1000 Bq/kg, while in Japan it was lowered to 100 Bq/kg after the tsunami disaster and power plant accident to reassure the population. This explains the headline above. The question now is, what is a reasonable limit? I think the European limit value of 1000 Bq/kg is a value that we can live with.[52] A lower limit value, such as the "fear limit value" in Japan, would of course appear to be better at first glance. However, it must be borne in mind that even an activity of 1000 Bq/kg has not been proven to be a health hazard in any way. Radon spas (e.g. in Bad Kreuznach) even advertise an activity of 300,000 Bq per m^3 of breathing air for inhalation cures. If the limit value is set too low, this can lead to certain foods - fish with low levels of contamination in Fukushima - simply being thrown away, even though eating them only marginally increases the annual dose to the consumer (see Science, vol. 338, p. 480, 2012).

In this context, it is important to point out the widely unknown fact that a large part of a person's natural radiation exposure (around 10,000 Bq) comes from themselves, primarily from radioactive potassium-40 in their bones.[53] This means - and is certainly worth a headline of the kind above - that a well-filled soccer stadium emits around 400 million Bq...

---

[52] A limit value of 370 Bq/kg applies to milk and dairy products in the EU

[53] Again, please do not be fooled - the activity of potassium-40 in an adult human is about 10,000 Bq, but the resulting radiation dose (about 0.17 mSv/year) is significantly lower than the total natural radiation exposure of about 2.4 mSv/year.

201. Earth radiation and earth radiation detection

Incidentally, the pathological "fear of radiation" dates back to times before the concept of radioactivity even existed (the term was coined by Marie and Pierre Curie in 1898). It first appeared in the 19th century when it was possible to systematically detect so-called "earth rays" using a special measuring device, a particularly sensitive "antenna" for earth rays - the divining rod. Today we know that this is a non-existent but highly problematic form of radiation for health, which was believed to emanate from geological faults, water and ore veins in the ground, and special "earth radiation networks". It is considered dangerous if a person's sleeping area is located directly above such a source of earth radiation. Permanent indisposition, a shattered psyche, heart and circulatory problems, etc., are thought to be the inevitable consequences, from which you can only protect yourself to a limited extent with an appropriate protective amulet (available in abundance on the Internet). Fortunately, however, there are plenty of experts in the esoteric sector who can either use their expertise to help you find the places in your home where the effect of earth radiation is at its lowest or, even better, sell you appropriate shielding devices and install them in the critical areas. These are usually very expensive, technically sophisticated (sometimes with an LED intensity display!) but otherwise completely non-functional tin or plastic boxes, which usually contain a coil and a capacitor as well as some wire. These act as "receivers" of the earth's rays, absorbing them, so to speak, and then returning them to the ground via a protective contact, now rendered harmless. You can buy such a box for as little as €129, which you can make yourself for far less than €10 with a clearly identical effect...

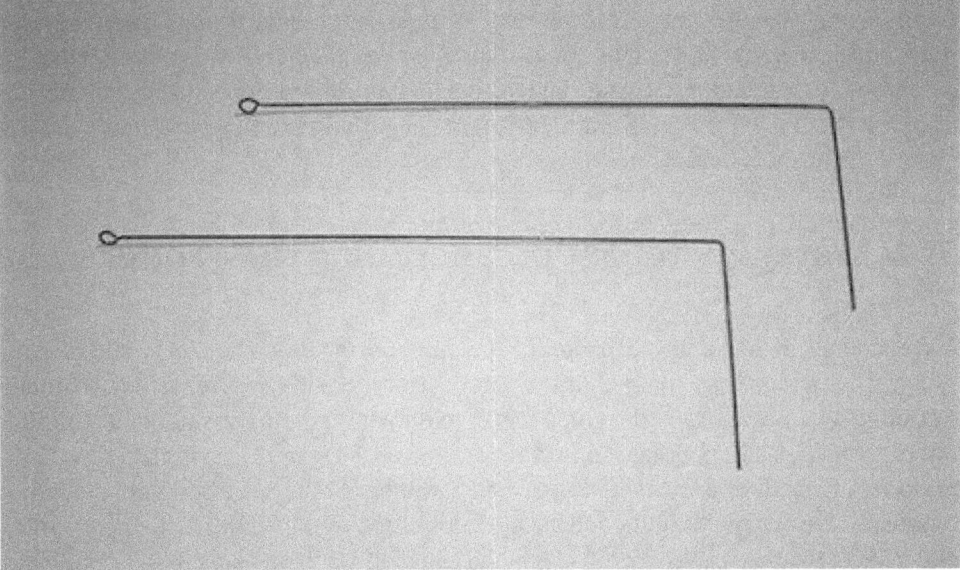

Fig. 80: The relevant specialist shops stock various types of dowsing rods for detecting earth radiation for very little money... (Free advertising image - the two bent wires cost just € 27.50 + shipping)

202. The eco power filter

In this context, we should also mention the eco power filter ("green electricity filter"), which, when plugged into a socket, is supposed to completely block dirty nuclear and coal-fired electricity so that only ecologically safe electricity can pass through to the consumer (e.g., a table lamp). Unfortunately, this invention did not catch on. This was due to competition from energy suppliers who offer 100% green electricity (making the filter obviously useless) or to the (eco) power fluctuations, which are said to have led to a strong flickering of the aforementioned table lamp and sometimes even to it going out completely. This was too much for true eco-enthusiasts, because even they can't find much use for fidget spinners.

It is quite astonishing that in a society that often proudly calls itself a "knowledge society," it is possible to establish business models successfully that are based on dubious claims from otherwise completely normal citizens...

203. Decadence indicator esotericism

Just think of the esoteric industry. What you can find there. There are pendulums with which you can test which foods or cosmetics, or which detergents are compatible or not. There are "pendulum courses" where you can learn how to use these high-tech devices correctly. Alternatively, "one-handed rods" (derivatives of the classic dowsing rod) are also available for this purpose. One of their great advantages is that they understand German (or a other language). All you have to do is hold them carefully balanced over your soup bowl and then ask loudly and clearly "Ist das genießbar?" (*Is this edible?)* According to the instructions for use, only questions that allow a yes or no answer may be asked). If the rod then wiggles "correctly", there is nothing to stop you eating it. Otherwise, vomiting must be expected...

204. Water revitalization according to Grander

And of course, in the context of esoteric hocus-pocus, the broad field of "water revitalization" must not go unmentioned. This is a special process that was revealed personally by God to a Mr. Johann Grander (1930-2012) to transform ordinary water into "Grander water" with special properties. For example, "revitalized water" is said to kill bacteria, "detoxify" the body after ingestion, and when "revitalized water" is discussed, the "word vibrations" are claimed to influence the life force of the water according to the meaning of the word. Moreover, even a drop of it in the washing machine helps to save detergent because it reduces the surface tension of ordinary water without the addition of surfactants.

Attempts have been made to objectify the differences between ordinary and revitalized "Grander water" using scientific methods in so-called double-blind tests. However, these attempts have met with only modest success. The only difference that could be determined between the two "water types" was the price per liter. After all, the Swiss company Grander made an annual turnover of almost 13 million euros with this "water" in 2010. A really nice figure for the mental health of esoteric followers who have money to spare for such things. Irrationality is all the rage - and it comes at a cost.

Fig. 81: The thirsty esoteric layman might assume from the seal that "water revitalization" indicates a particularly high content of microorganisms in the fruit spritzer and wonder whether it is really healthy?

The disciples of esotericism seek the unexpected energies of Mother Nature in ordinary stones and try to channel them through meditation. With dowsing rods from the specialized trade firmly in their hands, they stumble around trying to locate water veins that emit negative earth rays, and some of them seriously try to make contact with the hyper-intelligent pilots of so-called "unidentified flying objects." Forums and self-help groups are then used to discuss this at length; the latest "gemstone waters" are sampled, tarot cards are used to predict the future, intensive chats are held with long-dead people, and conspiracy theories are developed that are bursting with half-baked knowledge. And as an astonished observer, you might ask yourself: can they still be normal people? The answer is simply, "yes."

205. Stupidity is not a disease

Stupidity is not a pathological mental defect that can somehow be treated with psychotropic drugs. It is simply stupidity and an expression of the fact that people are doing too well. Late Roman decadence perhaps sums it up best. And it's best to leave people alone as long as their actions are not life-threatening. Because esotericism is a completely humorless (and extremely profitable) business. An esoteric who is a staunch opponent of vaccination, for example, had to learn this...

206. The opponents of vaccination - bankrupt

He offered a prize of €100,000 on the Internet for anyone who could prove the existence and size of measles viruses on the basis of scientific studies. A young doctor saw this as easy money. The relevant specialist articles were easy to find on the Internet and so he put together a corresponding "portfolio of evidence" without much effort. The esoteric "anti-vaccinationist" (surprisingly a biologist himself) had (as expected) only one answer ready: "He doesn't believe these papers - and therefore won't pay the prize money offered." Well, the doctor went to court.[54] The criteria of the competition had been met in terms of form and content, the Ravensburg Regional Court ruled. Based on the statements of an expert witness, there is no doubt about the existence of measles viruses and that the plaintiff had succeeded in providing the required proof. The opponent of the vaccination therefore has to pay the 100,000 euros to the doctor - a classic own goal. Instead of taking the loss lightly, the alternative medicine doctor now wants to appeal to the next instance without any sense of humor to continue his campaign.

In addition to its entertainment value, however, this ruling has another significance. Under the guise of esoteric secret knowledge, opponents of vaccination have been trying for some time to gain more and more influence on public opinion by claiming, without evidence, that vaccinations against infectious diseases such as measles, rubella, or mumps have no demonstrable protective effects or that the vaccinations would lead to long-term effects such as asthma, diabetes, cancer, HIV, and multiple sclerosis - a preposterous argument for anyone who has enjoyed a comprehensive education, has not lost their critical thinking, and is able to inform themselves. The measles epidemic that broke out in Berlin in spring 2015 shows that a lack of vaccination protection for children can have fatal consequences.

---

[54] The case number at Ravensburg Regional Court is 4 O 346/13 - 100,000 euros prize mon-ey for the detection of the measles virus

The question of the pros and cons of vaccinations has been widely discussed in countless scientific studies worldwide with the clear result in favor of vaccination recommendations. For constitutional reasons, the former compulsory vaccination for certain diseases (in this case smallpox) was abolished in Germany in 1983. Instead, only vaccination recommendations are issued, which may or may not be adhered to. Experts and legal scholars are still arguing about whether this was a good idea overall, after several measles outbreaks have occurred in Germany since 2005, where the only partial immunization among children and adolescents at that time proved to be a serious problem.

Let there be no misunderstanding. So-called "anti-vaccinationists" are not always found among the purely esoteric. Their intentions are usually entirely honorable. The problem lies more in a distorted perception of risk, which, as is often the case, is related to risk awareness and education. Our current high life expectancy in industrialized countries is largely due to the introduction of vaccination against dangerous infectious diseases that tend to spread epidemically. With its help, for example, smallpox (vaccination had been compulsory in the German Reich since 1873 and was abolished in 1983) could be completely eradicated and diseases such as diphtheria, mumps, measles, and poliomyelitis (to name a few) could be eradicated to a large extent. However, the increasing vaccination fatigue among the population could well bring this positive development to a halt again, as the example of measles shows.

The situation is different when it comes to "experimental vaccines," such as those used during the coronavirus pandemic. It was highly problematic to bring RNA vaccines onto the market outside of the approval rules and to use them in a politically driven manner. In this case, it was certainly wise not to get vaccinated - which subsequently proved to be the right decision (see "RKI files").

But back to esotericism and its disciples. As soon as it is given a scientific veneer, this otherwise harmless nonsense can quickly degenerate into a game between life and death - I am referring here to certain "alternative" healing methods, the inefficacy of which should be immediately obvious to anyone with a basic understanding of science.

207. Miracle Mineral Supplements

Take, for example, the new "miracle cure" MMS that has arrived from America. "MMS" stands for 'Miracle Mineral Supplement' - and chemically, it is nothing more than a highly corrosive 28% sodium chlorite solution (caution: do not confuse it with sodium chloride, i.e., ordinary table salt!). Sodium chlorite is usually used as a bleaching agent and disinfectant and is sometimes used medically to treat external wounds. This miracle cure was "discovered" by a certain Jim Humble, a former Scientology activist and self-proclaimed "bishop" of the "Genesis II Church of Health and Healing," who claims to have cured dozens of natives of malaria in the South American jungle. The secret of his miracle cure, as he has allegedly discovered through constant prayer or meditation, lies in the "stabilizing oxygen," which only needs to be activated - e.g., with citric acid. The resulting chlorine dioxide is then the actual agent that is effective against malaria, dementia, tuberculosis, autism, AIDS, Lyme disease, cancer, ADHD, warts, flu, coronavirus, impotence, boils, hepatitis, etc. It is therefore understandable that Mr. Humble's presentation at the "Spirit of Health 2014" congress in Hanover led to storms of enthusiasm among the followers of alternative "medicine" gathered there. After all, there is now a remedy for virtually every conceivable ailment. And on the homepage of the "healer" from America you can then read in highly scientific terms

"MMS is too weak an oxidizing agent to oxidize healthy cells in the body. It therefore cannot damage the beneficial bacteria in the body, they are too strong for that. It can only oxidize weaker things in the body and these are the pathogens that are made up of weaker constructs."

Fig. 82: MMS - the packaging looks respectable, but you shouldn't swallow the stuff under any circumstances (Creative Commons)

Well then, off to the nearest esoteric pharmacy, where you can buy 100 ml of bleach and 100 ml of 50% citric acid (to activate!) for a mere €20 (100 ml of sodium chlorite solution costs around 40 cents in specialist chemical stores). Business appears to be thriving, and the risk of coming into conflict with the law seems to be manageable for the vendors. The German health authorities are rather helpless in the face of "MMS." Although attempts are made here and there to shut down online stores, the lack of a coordinated approach means that the successes achieved are rather modest. MMS is banned in several European countries, while health authorities in Canada and the USA also advise against its use. Since February 26, 2015, the sale of MMS has been illegal in Germany, as it has been classified by the Federal Institute for Drugs and Medical Devices as requiring approval. As is well known, medicinal products requiring approval may only be marketed if their efficacy, safety, and quality have been proven in an approval procedure. And this is clearly not the case with MMS.

208. Quackery

MMS is modern quackery. Originally, the term "quackery" referred to a professional group who, in the Middle Ages and early modern times, offered a wide variety of ointments, tinctures, and miracle cures at fairs in the style of a market barker, thus competing with doctors. These self-proclaimed "healers" traveled from place to place and exploited people's limited medical knowledge to sell their often ineffective or even harmful products. Later, all people who claimed to treat illnesses without any training or official license were referred to as "quacks." The term comes from the Dutch "kwakzalver" and means "boastful ointment seller."

Sebastian Brant, who lived in Strasbourg between 1457 and 1521 and was immortalized by his work Das Narrenschiff (The Ship of Fools), already had the following to say about the "quack":

„Des Quacksalbers Praktik sei so gut, - daß sie allen Siechtum heilen tut... - Solch Narr kann dich in'n Abgrund stürzen, - eh du's gemerkt, dein Leben kürzen!".[55]

Or to put it in more modern terms: the primary concern for the doctor is the well-being of his patient. The primary concern for the quack doctor, on the other hand, is the highest possible fee, which is why an important rule for quacks is *"Demand your money while the patient is writhing in pain!"*. Any improvement in the patient's state of health can therefore even be counterproductive for their wallet.

Today, in addition to common sense, which is often sufficient, there are a few simple criteria for reliably diagnosing pseudomedicine or quackery. At a time when the internet facilitates access to information and even disinformation, it is more important than ever to critically question who you entrust your health to.

---

[55] "Let the quack's practice be so good, - that it cures all infirmities... - Such a fool can plunge you into the abyss, - before you know it, shorten your life!".

209. Diagnostics of ineffective alternative remedies

When it comes to "alternative remedies", for example, completely ineffective remedies are almost certainly betrayed by some of the following points:

a) by an explicit reference to their exotic origin ("Himalayas", "rainforest").

b) by claiming that they have "proven themselves thousands of times" - without any clinical evidence or studies to back this up.

c) that they bring healing when "conventional medicine" fails (this is especially true for alternative cancer remedies).

d) that they help universally against a multitude of diseases that have nothing to do with each other (as with MMS).

e) that they are linked to certain people or organizations that have developed the therapy with these "remedies" (e.g. "Granderwasser").

f) that they are advertised with case histories in which "cured" patients - people like you and me - describe in flowery terms how the product has allegedly achieved the most astonishing results for them.

210. The boom in pseudoscience

At first glance, it is astonishing how many people are attracted to esotericism and pseudoscience. On closer inspection, however, this is merely a kind of counter-reaction to the alienation processes in the so-called knowledge society. According to Robert E. Lane, this is a social formation in which conscious individual and collective knowledge and its acquisition and organization form the basis of social and economic coexistence. It produces products that are "indistinguishable from magic" (Arthur C. Clarke), but also increasingly leads to a mindset that is no longer interested in the "how" of functioning. It polarizes society into producers of knowledge and mere consumers of their products. The accompanying commercialization, politicization, and medialization of knowledge lead to problem areas, the failure to address which can ultimately lead to the destabilization of social structures. One need only think of the "energy transition," which is understood and organized by German politics less as a technical problem and more as a purely monetary and administrative issue. The fatal consequence is that the laws of nature cannot be overridden by parliamentary resolutions - no matter how large the majorities may be. The repercussions are already evident across the country and in our own wallets. All of this goes hand in hand with a difficult form of scientific and educational ignorance, which is cultivated in the media, particularly in the lower echelons of private TV providers. In some circles, it is now considered chic to "have no idea about hoots and hollers." This could be seen as a kind of harmless cult of ignorance if it did not conceal a long-term danger for the knowledge society. And this long-term danger lies in the increased emergence of "para-knowledge," i.e., "knowledge" that lacks scientific justification. It becomes dangerous when it leads, for example, to a child suffering from cancer being denied conventional medicine by the parents and instead seeking help in a kind of "Bach flower therapy." At this point, harmless nonsense quickly becomes deadly serious.

211. Placebo effect and spiritual healing methods

Interestingly, in Germany in particular, there is a strong tendency toward spiritual healing methods which, if they work, can at best be attributed to the placebo effect. A placebo is usually understood to be a substance that consists only of potato starch or sugar, plus flavoring and coloring additives, and otherwise contains absolutely no therapeutic chemical substances. Nevertheless, it is effective for certain complaints in about one-third of patients, provided they are unaware that they are taking a placebo. This observation is part of the placebo effect, the mechanisms of action of which have still not been fully elucidated. In principle, we only know that psychosomatic correlations play an important role here.

212. Why do glues stick?

Incidentally, there are a large number of things for which it is hard to believe that there is not yet a completely satisfactory answer as to how they work. Take, for example, the question of why adhesive tape (or "glue" in general) actually "sticks." The key physical and chemical terms here are adhesion and cohesion, each of which makes its own contribution to the bonding process. Cohesion refers to the binding forces between the atoms and molecules of a substance, which lead to the internal cohesion of a liquid or a solid. In liquids, this is noticeable in the form of surface tension and also determines their viscosity and rheology (i.e., flow behavior). Adhesion, on the other hand, is a force that acts between two condensed phases and leads to their mechanical cohesion. Both phenomena have been well studied experimentally and in engineering terms and are of great practical importance as basic principles of adhesives. However, this is diametrically opposed by a lack of theoretical understanding of the microphysical functioning of the bonding process itself, which makes the theoretical prediction of the bonding effect on given surfaces and materials almost impossible except through experiments. It is therefore not surprising that a number of adhesion theories exist, but they do not form a unity and each of them can only describe partial aspects of this amazing phenomenon. What they all have in common is that so-called secondary valence forces, or more precisely Van der Waals forces (which in themselves are extremely weak compared to chemical bonding forces), are held responsible for the bonding process. To achieve this, the parts to be joined must come as close as possible to each other to less than one micrometer. This is usually only possible with very flat (smooth) surfaces, which can be observed very clearly on a broken sheet of slide glass. If you press the broken surfaces against each other again, you will not find any adhesion, because the surface is too rough for that. If, on the other hand, you place two clean panes of glass exactly on top of each other, you will notice an adhesion - if they are really flat - that can sometimes be so strong that it is difficult to separate the two panes of glass again. The adhesive now occupies an intermediate position between the two surfaces to be joined. Since it is usually liquid, it can penetrate the irregularities and interlock with them during the curing process. The adhesive forces bind the surface to the adhesive, and the cohesive forces of the adhesive itself hold it together. In "ideal" adhesives, the adhesive and cohesive forces are therefore approximately equal. Furthermore, the adhesive must be suitable for the materials that are to be bonded together. This is because (even if advertising claims otherwise) there is no such thing as an all-purpose adhesive. The adhesive must always be suitable for the respective material.

Of course, this also applies to the adhesive that the "climate stickers" use to stick themselves to roads, buildings, airports, or other public places in order to draw attention to their environmental concerns.

213. Climate glue protesters

In a world where "scientists" are developing new and revolutionary methods to save our planet, they stand out as shining examples of dedication and creativity. Their secret weapon? An inconspicuous but highly effective glue - called superglue. These activists have decided that the most effective way to raise awareness of so-called "man-made climate change" is to literally glue themselves to roads and other public places. The logic of using modern adhesives as a powerful tool in the fight against climate change is easy to understand: by sticking themselves to busy streets and places, they force the public and the media to focus their attention on the pressing issues of climate change. However, it hardly occurs to them that they are only getting on the nerves of their fellow human beings, endangering lives, and causing scorn and ridicule. The astonished outsider fails to realize that the preparation of such actions requires careful planning and a good dose of courage. After all, one first has to choose the ideal place and time to attract maximum attention and ensure that the adhesive is really strong enough to prevent immediate removal, but of course not so strong that it could cause serious injury. It sounds like a difficult balance - and it is.

Some might argue that sticking to asphalt is not exactly the most effective way to influence policy. But the climate stickers seem to act according to the motto "Any attention is good attention." And indeed, there is arguably no better way to ensure that the evening news covers climate change than through the gridlock that these actions trigger.

Ironically, blocking roads with climate stickers often actually leads to an increase in CO_2 emissions, as cars are stuck in traffic jams and emit exhaust fumes.

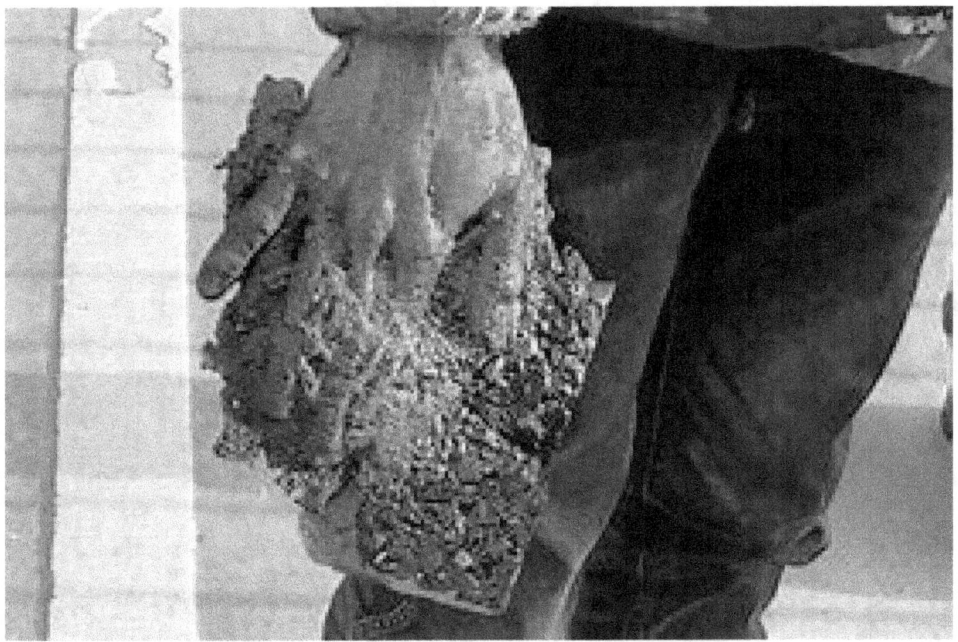

Fig. 83: Climate-glued hand of a species of the self-proclaimed "last generation" after the adhesive bond with the road asphalt was dissolved by a police service provider (© daa)

214. Miracle glue „gecko feet"

One of the most interesting - and for a long time, most puzzling - adhesives is not an adhesive at all; it is the foot of a small lizard, the gecko. These lively little animals are known to be able to run up a vertical pane of glass without any problems and even cling to the ceiling, defying gravity as it were.

This astonishing phenomenon is related to the weak Van der Waals forces, which, while individually weak, can develop considerable adhesive strength when acting collectively. Electron microscope studies have shown that the ability to adhere even to polished glass is due to the presence of millions of fine hairs on the gecko's toes. Each hair tip splits into thousands of even smaller ends, known as spatulae. These form small pads that interact with surfaces through Van der Waals forces. Due to their sheer quantity per unit area, an enormous adhesive force is created.

Such investigations may enable the development of a "dry adhesive tape" in the near future, which could have countless applications. It could be used both underwater and in the vacuum of space and would probably have a self-cleaning effect, much like gecko feet.

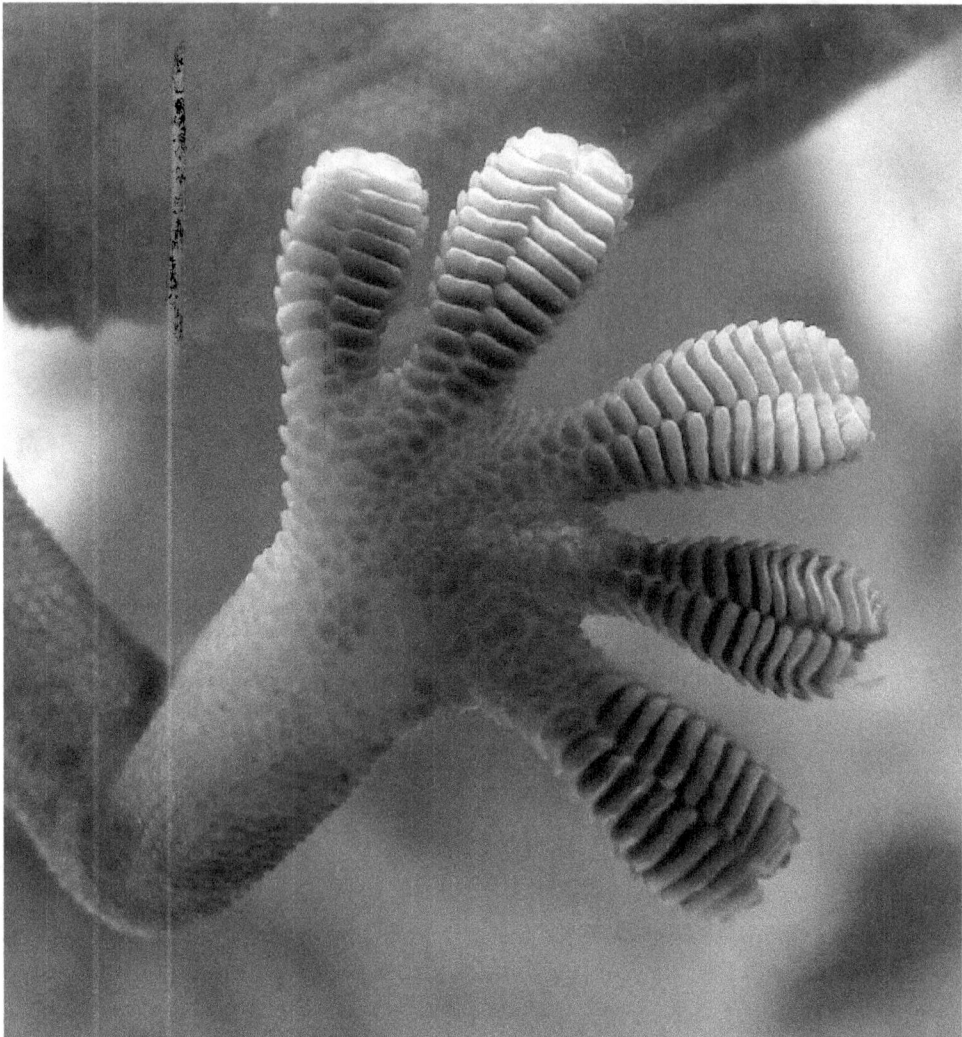

Fig. 84: Gecko foot underside (Wikimedia)

As interesting as the explanation of the adhesive force of gecko feet and tape using Van der Waals forces may be, a closer look reveals certain contradictions, especially with standard adhesive tape, that defy theoretical explanation. This begins with the observation that when removing a strip of adhesive tape, you first have to give it a strong tug before it can then be pulled off relatively easily with a steady pull. If you compare the adhesive forces measured with the molecular forces predicted by molecular adhesion theory, you quickly realize that they are far too low to explain the observed adhesive behavior. Therefore, there must be at least one other effect at play that causes Post-It notes to stick to a computer monitor, among other things. According to cavitation theory, one possible explanation is that numerous microscopically small bubbles are hidden in the adhesive, acting like suction cups and thus providing additional resistance to the removal of an adhesive strip. At the beginning ("jerk phase"), these bubbles would offer greater resistance to tearing off due to the resulting negative pressure than later, when they merge during peeling, enlarge, and then burst. After this initial phase, only pure adhesion needs to be overcome. If this theory is correct, an adhesive strip should be much easier to peel from its backing at low air pressure. Despite researching on the Internet, I have not yet found a description of an experiment testing this hypothesis.

215. Bird glue and bird murder

A very nasty glue that was once made from mistletoe berries is known as "bird glue." Spread on thin branches, it was used in the form of "glue rods" to catch small birds, which were then either sold as caged birds or plucked and roasted side by side on shashlik skewers. Incidentally, this trapping method dates back to the Romans and is now, of course, banned in the EU (Italy was the last country to enact a corresponding hunting ban in 2014), as is trapping songbirds for culinary purposes (with national exceptions).

However, some traditional bird-trapping areas such as Malta, Cyprus, southern Spain, southern France, southern Italy, and the Adriatic countries do not adhere to this ban. There, hundreds of thousands of robins, finches, starlings, thrushes, warblers, and lapwings (to name just a few species) are still caught during bird migration seasons using horsehair snares, snap traps, nets, shotguns, and even glue rods (in Catalonia), only to be eaten or sold to animal traders. It is estimated that around 120 million birds are killed legally in Europe every year and another 100 million illegally.

Despite the now severe fines, it seems that this annual bird murder can hardly be curbed. Every spring and autumn in the Mediterranean countries, a real petty war develops between bird lovers (most of whom have traveled here)[56] and illegal bird trappers, in which one side tries to find and disable the illegal traps and the other side sometimes even shoots at the bird lovers - or, more often, punctures the tires of their vehicles or smashes their windshields.

216. Wind turbines and bats

What poachers are in southern Europe, wind turbines are in Germany, to which countless birds (especially birds of prey such as the red kite) and bats fall victim every year.[57] And let's not forget the many millions of insects that fall within the range of the rotors. Do you remember the discussion about the construction of the Waldschlösschen Bridge in Dresden, the building of which cost the Elbe Valley its UNESCO cultural heritage title? In 2007, the start of construction was halted by the Dresden Administrative Court because the Lesser horseshoe bat (*Rhinolophus hipposideros*) - a strictly protected bat species - was said to be living in the area of the construction site. Bats living near wind turbines, on the other hand, rarely receive the attention of the highest court. According

---

[56] https://www.komitee.de/de/ueber-uns/

[57] The wingtips of fifty-metre rotors reach speeds of up to 340 kilometers per hour!

to conservative estimates, eco-windmills "shred" around 250,000 bats a year (though the number could exceed 400,000, as the Journal of Wildlife Research recently found).

For a bat to die, it doesn't even have to be hit by one of the giant blades. It is sufficient for it to come close to them. In this case, the large changes in air pressure rupture its internal organs - experts refer to this as barotrauma. However, bats that suffer only "mild barotrauma" may not die immediately but could continue to fly for a few minutes or even hours. Their remains are, of course, not found when a wind farm area is searched.

Apparently, there are two types of bats in the eyes of eco-ideologists. If one type lives in a farmer's barn, the farmer must not seal the roof, because the bat will then no longer be able to find its daytime roost. If, on the other hand, the bat lives in a landscape or nature reserve where wind turbines are erected or in the immediate vicinity due to the (usually alleged) favorable "wind harvest factor," then it dies due to its own fault (why does it fly into the rotors?) or because of a "higher purpose," namely "saving the climate," as you can read everywhere in the gazettes. After all, as a well-known Green from Tübingen said: "Without climate, no bats and birds...!" Perhaps one or two people might argue that the few bats and birds lost are acceptable, given that far more supposedly die in road traffic.

But what certainly cannot be tolerated in the long term is the disfigurement of the landscape associated with the "energy transition" by technical monsters which, despite their huge numbers, fail to make a sensible contribution to the necessary energy base load.

217. Zapping electricity from wind power

Just one example that should give food for thought: In 2023, the actual feed-in capacity of all wind turbines, integrated over the entire year, amounted to a whopping 18% of the installed nominal capacity.[58] If this is supposed to be a success for the wind energy sector, then good night!

On one day of the year (granted, it was quite windless), the power fed into the grid was only 24.0 MW, which is a mere 0.06% of the 39,612 MW installed at that time (2014). On the day of maximum feed-in capacity (when it was quite stormy in Germany), 74.9% of the installed nominal capacity flowed into the electricity grid. The problem that grid operators face is that the amount of electrical energy fed into the grid at any given time must match the amount being used by consumers. Otherwise, the voltage in the grid can drop or rise, which in extreme cases can lead to a power grid collapse ("blackout"). The bottom line is that behind every MW of "wind power" and every MW of "solar power," there needs to be a traditional thermal power plant based on coal, gas, or uranium to compensate for their volatility. To afford alternative energies, one must also afford appropriately dimensioned backup capacity. After all, it is dark at night and sometimes there is no wind. Furthermore, to be clear, there are still no practical storage technologies that can address the problem of volatile energy sources beyond the level of pipe

---

[58] Currently, the installed capacity of wind turbines in Germany is around 70 GW

dreams, despite what some may claim. Sometimes, it helps to understand the problem by breaking down complicated issues to a level that everyone can understand.

218. How the German Renewable Energy Sources Act (EEG) works...

As far as the problems with wind and solar power are concerned, I found the following analogy on the net, which I can fully endorse:

"Imagine a consulting agency advising an internationally successful company to implement an "employment turnaround". The concept is to gradually replace the core workforce with so-called fair workers. Fair workers are said to be socially advantageous. The Recruitment Facilitation Act (EEG) stipulates that fair workers - once hired - always receive full pay, regardless of whether they work or not. However, their work ethic is characterized by fickleness and laziness. Sometimes they arrive almost on time and "really get stuck in", sometimes they don't show up at all for days on end. A colleague from the core workforce must therefore always be on call to replace the absences of the well-paid slacker. Ten years later, the company management, together with the consultants and the Fair Workers' Union, is celebrating the fact that more Fair Workers have been contracted in 2024 than ever before - with the conditions described being fixed for 20 years. An independent analysis by a labor market researcher now shows that Fair Workers were on duty for 14.8% of the collectively agreed working hours over the record year. The "workhorses of the employment turnaround" therefore worked significantly less than a one-day week. If this company is called Germany, then the fair workers are called Ökostromanlagen, the consulting agency AGORA and the trade union "Bundesverband Windenergie" - whereby the founders of the consulting agency have now secured an influential position on the company's management board."

While most people don't care whether birds and bats die at wind turbines, it's a completely different story when their own health is threatened - e.g. by infrasound.

219. Wind turbines, crazy minks and infrasound

As is well known, infrasound is a type of sound with a frequency below 100 Hz that is usually inaudible to humans. Only when its level (i.e., its volume) is very high can fluctuations in air pressure be felt and sometimes even heard. The official view is that infrasound is only harmful to people when they can hear it. This justification is often used to dismiss complaints from those affected, and the concerned individuals are usually denigrated as hypochondriacs. However, the fact that people in the vicinity of large wind turbines are likely not merely imagining their persistent headaches, sleep disorders, and depression is supported by a number of clinical studies as well as a notable incident in Denmark. In this case, wind power operators had installed turbines near a mink farm. When these turbines were activated for trial operation, the animals began to scream wildly and bite each other. When the vet in charge called the police the next morning to have the turbines turned off, half a dozen animals were already dead in their cages. More than 100 minks had inflicted such severe wounds on each other that they had to be euthanized. It soon became clear that the vibrations from the wind turbines, which have a frequency below 20 Hz and are therefore imperceptible to the human ear, had driven the animals nearly insane. Since there are also videos of the minks running amok, this phenomenon garnered significant attention beyond Denmark. The fate of the Jutland mink farmer not only made headlines online but even occupied the Danish Parliament. Since then, the energy transition has faced a new problem. It is no longer credible to tell people that infrasound is merely a harmless side effect of wind turbines. Even if some people have grown accustomed to the new landscape aesthetics, they can no longer ignore concerns about their and their children's health. For this reason alone, the acceptance of wind power is slowly but steadily declining.

But sound waves are also associated with another, sometimes even frightening phenomenon.

220. Thunderstorm

I am referring to thunderstorms that occur particularly frequently in our latitudes during the summer months. One particular type is the frontal thunderstorm that approaches my property in Upper Lusatia, usually towards the evening and always from the west. This is always a good reason to sit comfortably in a garden chair with a bottle of beer and follow its approach with interest. It begins with a sultriness that builds up over the course of the day and - when dusk falls - with more or less intense weather glow in the west. This glow is the reflection of lightning behind the horizon or lightning that cannot be seen due to the low-hanging clouds. At some point, the first roll of thunder can be heard. The closer the front approaches my observation point, the more often I see lightning flashes between the clouds and between the clouds and the ground. It then takes a while before the accompanying thunder can be heard. At first, it does not sound like a bang but is still muffled and rolling due to many refractions and reflections on the ground.

In the meantime, the zenith and the eastern sky have also become covered with dark clouds. Then, all of a sudden, the calm that was just present is interrupted by an increasingly stormy downdraft. The spruce trees begin to bend in the wind, and the branches of the mighty beech tree above me sway ever more menacingly. When the grain is in bloom during the storm, milky clouds of pollen rise over the fields. Lightning flashes intermittently, and as you count from three to six until the thunder rumbles, you realize that the storm front is only one or two kilometers away. As the wind picks up, the first large drops of rain hit you, indicating that it's time to clear the field. Just as the wind intensifies into a brief storm, it disappears as quickly as the heavy rain begins. The atmospheric pressure has moved on, and the ambient temperatures have noticeably dropped. This is followed for just under half an hour by heavy rain (sometimes accompanied by hail), illuminated by lightning flashes in quick succession, followed immediately by loud thunder.

After that, the storm is over, and it usually just rains for a little while until it finally clears up (or showers continue). The storm front has moved on, and sparkling stars can now be seen between the first gaps in the clouds in the exceptionally clear air.

Fig. 85: A fast approaching storm front (Author)

221. Heinz Erhardt and the thunderstorm

If you would like to experience the full drama of such a thunderstorm in a lyrical way, I highly recommend the poem "Das Gewitter" by Heinz Erhardt (1909-1979) - not in its written form, but in the dramatic interpretation by Heinz Erhardt himself (available on YouTube, for example). His impressive description of a thunderstorm almost demands an explanation, which I will now attempt.

Frontal thunderstorms always occur when two air masses with different temperatures meet at a "front". Most frequently, a cold front meets warm air, but the reverse process is also possible. Cold air is known to be denser and heavier, which is why it pushes underneath the warm air at the "front", forcing it to rise. Meteorologists refer to this as an "occluded front." As a result, the unstable temperature stratification of the lower atmosphere causes intensive, convective cloud formation, and precipitation occurs behind the cold front. The cold air flowing in at ground level lifts the water vapor-saturated warm air, causing the water vapor to condense out, releasing latent heat (which increases the sultriness on the ground in front of the approaching front) and forming rapidly rising clouds (in this case, thunderstorm towers). It should be noted that not every cold front is associated with thunderstorms. The cold front must flow into an area prone to thunderstorms; in other words, it is only the trigger for the formation of a thunderstorm front. Additionally, the sometimes very large differences in air pressure between the areas separated by the front create intense winds due to the pressure gradient force, which herald the immediate passage of the thunderstorm front.

222. Lightning and thunder

The most important characteristics of a thunderstorm are thunder and lightning. Both are interrelated, with the thunder always following the lightning for an observer. In fact, the time difference between the lightning and the accompanying thunder can be used to estimate the distance from the observer. The speed of sound in air is approximately 340 m/s. Therefore, if the thunder follows the lightning after 10 seconds, the lightning strike was about 3.4 km away from the observer.

Lightning flashes are electrical discharges between clouds or between a cloud and the Earth's surface, and a charge separation must have previously occurred between these objects. The dipole nature of the water molecule plays an important role in this process. As warm, moisture-saturated air rises along the storm front, friction and the atomization of water droplets cause charge separation within the cloud. At high altitudes, the water droplets freeze into ice crystals, which become positively charged due to vertical, upward winds, while the remaining water droplets carry a negative charge. Consequently, a positively charged area forms in the cold upper part of the cloud, while a negative charge predominates near the ground. An electric field develops between these regions due to the potential difference, growing until the voltage reaches several hundred million volts. Once a critical voltage, around 170,000 V/m, is exceeded, a massive short circuit occurs, resulting in a lightning bolt. The current can reach up to 100,000 A and heat the air in the lightning channel to up to 30,000 °C within fractions of a second - more than five times the temperature of the sun's surface (photosphere). This tube of hot air expands rapidly, creating a sound wave that is heard as thunder.

Nature lovers and hikers should avoid being struck by lightning at all costs, primarily for personal safety. The most important rule is to avoid elevated locations and isolated trees. The safest option is to stay inside a car and watch the storm from there (a Faraday cage). Otherwise, if caught in an open field during a thunderstorm, it is best to lie flat on the ground or at least crouch down. Returning home with soaked clothes is always preferable to ending up in the hospital with trauma, severe tinnitus, or extensive skin burns (or not waking up at all).

On average, Germany experiences only three to four lightning-related deaths per year. A century ago, this number was much higher (~300), when many people still worked in the fields during summer. Viewed in this context, death by lightning is now a negligible risk compared to other dangers, though it remains higher than dying in a "nuclear accident." More commonly, a lightning strike can cause a fire in a building.

223. The lightning rod

But one of the founding fathers of the United States of America, Benjamin Franklin (1706-1790), also invented the lightning rod. Incidentally, the first lightning rod in Germany was installed on Hamburg's main church, St. Jacobi, back in 1769. Since then, most churches (and later also residential buildings and barns) have gradually been equipped with lightning rods, with the result that the number of church fires caused by thunderstorms has decreased significantly.

Fig. 86: Did you know that a Bohemian Premonstratensian canon and scholar, who was in correspondence with such intellectual greats as Leonhard Euler, invented the "lightning rod" independently of Benjamin Franklin?

Less well-known than Benjamin Franklin in this context is the Bohemian Prokop Diwisch (1698-1765), who also invented a lightning rod around 1754. Unfortunately, the farmers in southern Moravia, where he worked in a Premonstratensian monastery, were not impressed by this invention, as they attributed a severe drought with corresponding crop failures to its use on a rectory near Znojmo (Znojmo). Today, we can say with a degree of certainty that a "drought" has nothing to do with the presence of lightning rods.

224. What is Ball lightning?

But there is still a question related to lightning rods that remains unanswered: Does it also help against ball lightning? We do not know; in fact, we do not even know with any certainty what "ball lightning" actually is and how it occurs. There are now some theories that can be seriously considered. However, they are so complex that they are only accepted as explanations by a minority of experts.

So what is "ball lightning"? In any case, it is a rare phenomenon that is usually, but not always, associated with thunderstorms. Observers most often describe it as a glowing sphere with a diameter ranging between a decimeter and a meter that moves relatively slowly through space, sometimes suddenly changing direction. Photographs of this phenomenon are now also available, so there is no longer any doubt about its authenticity. However, little is known about the damage it causes, with some exceptions. One of these exceptions is St. Pancras Church in Widecombe-in-the-Moor in the county of Devon in England. On October 21, 1638, during a violent thunderstorm, a bolt of lightning struck the church, which was well attended by 300 worshippers, killing four and injuring around 60. According to contemporary sources, the interior furnishings were badly damaged. Others, however, believe that it was not a bolt of lightning, but the Incarnate One himself, who was attending a service in the church...

GLOBE OF FIRE DESCENDING INTO A ROOM.

Fig. 87: Depiction of a ball lightning floating through a room...

Most explanations of the formation and physics of ball lightning assume that it is a type of "plasma ball." However, this contradicts physics (Archimedes' principle), as a heated gas bubble should both rise and dissipate fairly quickly. Nevertheless, ball lightning has been observed for up to half a minute. Of all the theories about ball lightning, the one published in 2000 by John Abrahamson and James Dinniss remains the most plausible, especially since it is supported by experiments. According to this theory, ball lightning occurs whenever an ordinary bolt of lightning strikes the ground with a high silicon content ("sand"). The silicon oxide is reduced to its metallic form and vaporized. The resulting silicon gas is then oxidized in the air, creating the observed luminous phenomenon. Additionally, various extremely complex processes are thought to stabilize the oxidation process in a spherical area through self-organization, allowing the phenomenon to last for several dozen seconds. Shortly afterward, researchers from Brazil tested this theory experimentally by vaporizing silicon wafers (such as those used in semiconductor production) under the influence of an electric current of 140 A and then igniting the vapor with a spark. They observed a glow the size of a table tennis ball and color gradations ranging from orange-white to blue-white, similar to what eyewitnesses have reported for "real" ball lightning. The "balls" lasted up to eight seconds and could both float and burn holes in a lab coat. At least the hypothesis that ball lightning was "UFOs" from extraterrestrials, as some esotericists still claim today, seems to be a thing of the past. Nonetheless, this theory cannot be considered the final explanation. Reports of ball lightning and its unusual properties are too varied, and only some aspects of the phenomenon can be theoretically explained in a reasonably plausible manner.

225. Sprites, blue jets and elves

So there is still a lot of research to be done in this area. This also applies to another thunderstorm phenomenon, sprites (i.e., "gremlins"), which have been known since the 1960s. Sprites are a mostly red luminous phenomenon that emanates from the upper edge of a thunderstorm cloud and shoots into the high atmosphere, so to speak. They can only be effectively observed from orbit or from high-flying aircraft, which is why they remained undetected for so long. Here, too, there are various theories that attempt to explain this and similar phenomena (such as the so-called "blue jets" and "elves"). One theory suggests that during particularly intense thunderstorms, a correspondingly intense voltage field builds up above the cloud cover. This field may discharge in the ionosphere itself due to the oversaturation of electrons, triggered by incoming cosmic gamma radiation. The ionosphere is the part of a planetary atmosphere where large numbers of free charge carriers, such as ions and electrons, are permanently or temporarily present.

226. Ionosphere and shortwave broadcasting

The ionization of air molecules is caused by short-wave solar radiation (UV and X-rays) as well as particle radiation and cosmic radiation. Since the solar component is the largest contributor to ionization, the degree of ionization in the lower and middle layers of the Earth's ionosphere (i.e., in the altitude range between ~70 km and ~200 km) exhibits a typical diurnal cycle: during the day, ionization dominates due to solar radiation, while recombination dominates at night. This "recombination glow" leads to a slight brightening of the night sky and thus limits, for example, the maximum exposure time of astronomers' sky photographs.

However, what makes the ionosphere particularly interesting is that it acts like a mirror for radio waves in the short-wave band (frequency between 3 MHz and 30 MHz). This phenomenon enables planet-wide radio reception; with a simple shortwave receiver, you can receive radio programs from over 100 countries. It is important to note that there are two types of radio waves relevant to propagation: first, ground waves, which are limited by the horizon, and second, space waves, which are reflected by the ionosphere and can thus reach locations "beyond the horizon." Additionally, the water surfaces of oceans and the groundwater on land also reflect these waves, allowing shortwaves to reach virtually any point on the Earth's surface through multiple reflections. This is why shortwave receivers are often referred to as "world receivers."

The fact that shortwave reception remains effective at night, despite the absence of solar radiation as an ionization source, is due to the so-called F-region, located between 250 km and 400 km in altitude, which maintains a sufficiently high concentration of free electrons at night.

227. Meteor scatter

But there is also another interesting source of ionization in the Earth's high atmosphere, which is exploited by amateur radio operators in particular. I am referring to the short-lived traces of ionized air left behind by meteors (shooting stars) entering the earth's atmosphere. They also reflect short-wave radio radiation (especially ultrashort waves in the frequency range between 30 MHz and 300 MHz) very well, which is exploited in the meteor scatter radio method to achieve long ranges. Distances of up to 2500 km can be bridged for a few seconds to one or two minutes. This method, which was invented in 1929 by a Japanese man (Hantaro Nagaoka, 1865-1950), is of course no longer relevant for today's radio communications. Instead, the reflection of ultrashort waves is used today in some places on earth for daytime and night-time observation of meteor streams.

228. Laurentian tears - or the invention of barbecuing

A meteor stream is understood to be a certain accumulation of "shooting stars" whose orbital extensions in the sky meet at a certain point, which is called a "radiant". These streams of particles crossing the Earth's orbit are usually named after the constellation in which these radiants lie.

Fig. 88: The martyrdom of St Laurentius in a painting from the St. Laurentius Basilica in Deutsch Gabel (Jablonné v Podještědí) ... (Author)

It has long been known that conspicuous clusters of shooting stars occur on certain days every year. For instance, Chinese astronomers have recorded a shooting star event in 36 AD, which can today be clearly attributed to the Perseids (whose radiant lies in the constellation Perseus). Exactly 222 years later, in 258 AD, another event occurred that is indirectly (but not causally) related to the Perseids. This was the death of St. Lawrence in August of that year. Since then, it is said, the heavens have been weeping "tears of St. Lawrence." And not without reason, because St. Lawrence had been wronged by the Roman Emperor Valerian, and - how could it be otherwise? - over a financial matter. Pope Sixtus II (195-258) had entrusted the church treasury to his archdeacon Lawrence (who was not yet a saint at the time), and Valerian also needed money, which is not something only emperors require. When the emperor discovered this, he demanded that Lawrence hand over the treasure. However, Lawrence distributed it to the poor, which ultimately earned him great sympathy from posterity but infuriated the emperor so much that he had Lawrence roasted alive on a large iron grill (as we would say today). This is where the name "Lawrence's tears" for the meteors of the Perseid stream comes from - and you have something interesting to share whenever a shooting star flits across the sky. Of course, you can also make a wish, and if you wish for "good weather" for the next day, this wish usually comes true.

229. Shooting stars, bolides and meteorites

Most meteors that we see at night as shooting stars are rock particles just a few millimeters in size that burn up completely in the air. However, when they reach "cherry size" or larger, they become so bright that they are referred to as bolides or fireballs. These are quite rare but extremely impressive. The larger ones often disintegrate or fragment during their flight through the atmosphere, and in the case of particularly large ones - i.e., those whose original mass exceeds a few kilograms (depending strongly on the type of meteoroid, its entry speed, and angle) - fragments can even reach the Earth's surface. In this scenario, we speak of meteorites, and the people who search for and possibly collect them are called meteorite collectors.

Reports of anyone being hit by a meteorite are very rare and usually cannot be substantiated, except for one notable case: Elizabeth Hodges. The then 31-year-old was hit quite severely on the hip by an orange-sized meteorite, which had smashed through the roof of her house with a loud bang on November 30, 1954, in Alabama. All other reports of meteorite strikes in populated areas either caused property damage, indirect bodily harm (as with the Krasnodar meteorite), or led to an increase in the value of a car that was otherwise destined for the scrap heap. For instance, on October 9, 1992, a Chevy Malibu (built in 1980) parked north of New York was hit by a meteorite, leaving a noticeable hole in the trunk. Since then, the car has been used only as a "showpiece," which saved it from being scrapped.

Incidentally, you don't need to worry about a meteorite hitting your car, as long as you're not in it when such an event occurs. The damage is fully covered by comprehensive insurance. Buildings, however, are a different matter. Only if a fire occurs, which can certainly happen, will homeowners insurance help, provided it includes coverage for fire damage. Otherwise, you might have to pay for the damage yourself due to "force majeure." But there's nothing stopping you from taking out additional meteorite damage insurance with a trusted insurance agent.

230. Meteorite impacts

However, there are also conceivable meteorite impacts where even insurance companies would be powerless - I am referring to those that could devastate large areas of land (like the impact that once formed the Nördlinger Ries in the Swabian Jura) or even lead to mass extinctions of animals and plants (such as the extinction of the dinosaurs 65 million years ago). Such events are thankfully quite rare and have not yet occurred in the brief history of humanity. The Tunguska event of 1908 is an exception, but it only affected an extremely sparsely populated area in the Russian taiga. If, however, the impact site of the bolide had been somewhere in the middle of Europe, the consequences would have been catastrophic in the truest sense of the word.

Fig. 89: When the Soviet geologist Leonid Alexeyevich Kulik (1883-1942) reached the "Stony Tunguska" region in 1927, he still saw the effects of the 1908 meteorite impact in the form of huge areas of fallen trees (Wikimedia)

Major meteorite impacts still hang like a sword of Damocles over the Earth today, although the probability of catastrophic events in the medium term is not too great. For some decades now, observational technology has also begun to record the population of planetoids crossing the Earth's orbit as completely as possible, provided they are larger than one kilometer, and to take a particularly close look at the more dangerous of them, such as (99942) Apophis (it has a diameter of around 340 m). This celestial body, discovered in 2004, caused some excitement not only among astronomers, because the exact calculation of its orbit revealed a small probability that it could possibly collide with Earth at one of its closest approaches[59]. And that wouldn't be pretty, because such a body would release an energy of ~800 megatons of TNT, which is about 16 times the explosive power of the most powerful hydrogen bomb ever detonated on Earth. Such a celestial body would not wipe out humanity. However, the damage would be immense, regardless of whether it fell on land or into one of the oceans. In the latter case, tsunamis with an onshore wave height of up to 100 meters would have to be expected. However, compared to the "dinosaur killer" that hit the Gulf of Mexico 65 million years ago (Chicxulub), Apophis still appears to be relatively harmless. Even the impact that formed the Nördlinger Ries basin around 15 million years ago was caused by a celestial body 5 times the size.

---

[59] In the meantime, an impact in the 21st century can be completely ruled out.

231. The Yucatan dinosaur killer

The Chicxulub impact, which created a crater with a diameter of around 180 km in what is now the Gulf of Mexico, was of an entirely different magnitude. It was caused by an asteroid 10 to 15 km in diameter that struck the Earth unimpeded on a fine day 65 million years ago at a speed of several dozen kilometers per second. The resulting climatic changes wiped out all land creatures on Earth down to the size of a well-fed cat over a relatively short period of time. For the sake of completeness, however, it should be noted that at the same time, there was extremely strong volcanic activity in present-day India (known as the "Deccan Traps"), and it is still not completely clear which of the two events - or perhaps both together, the meteorite impact and the environmental pollution from the volcanism - ultimately led to the extinction of these creatures. Of the reptiles, only a few crocodiles and tortoises survived, as well as some small feathered dinosaurs, whose descendants now populate my property as house sparrows, blackbirds, green-finches, collared doves, jays, magpies, and rooks. Thus, small feathered dinosaurs became the ancestors of today's diverse bird world.

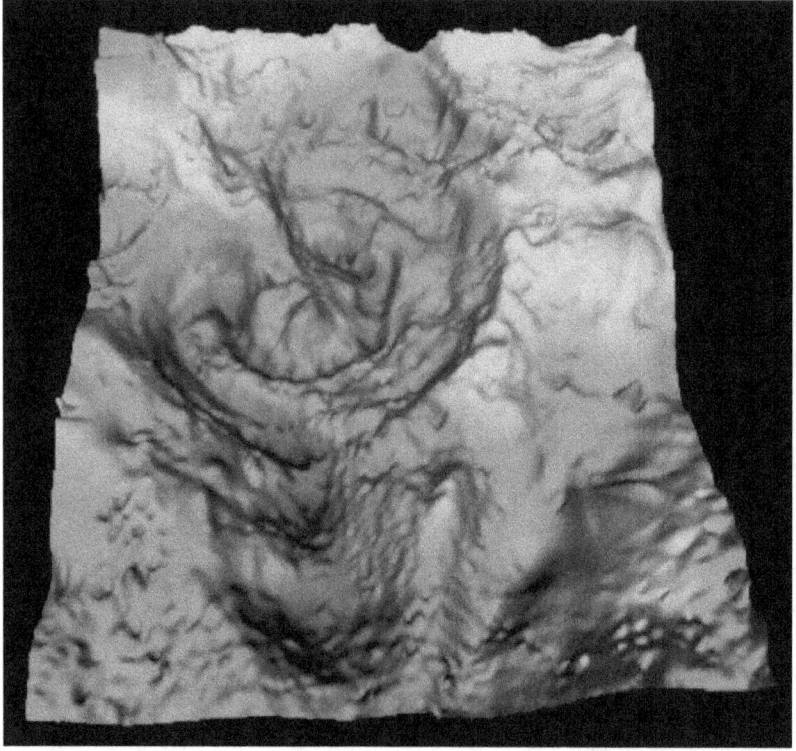

Fig. 90: The location and shape of the "remains" of the Chicxulub crater in the area of today's Gulf of Mexico can only be recognized by the associated gravity anomalies. The gravity measurements on which this graphic is based have shown that it is a basin with several rings (up to 180 km in diameter) and a fourth, outer ring with a diameter of around 300 km. As the crater is buried under thick sediments and has not eroded, it is one of the best-preserved large impact craters on Earth. (Wikimedia, NASA)

In hindsight, the only beneficiaries of this catastrophe were the mammals, which were then inconspicuous and about the size of a rat. And that was fortunate for all of us. The famous paleontologist Stephen J. Gould (1941-2002) once described this "luck" as follows - and there is really nothing more to add:

"Had not the celestial body destroyed their (the dinosaurs') flourishing diversity, they might still be alive today (Why not? They had been doing well for 100 million years, and only 65 million years have passed since then in the history of the earth). If the dinosaurs were still around, mammals would almost certainly be small and insignificant (as they were during the hundred-million-year reign of the dinosaurs). And if mammals are small, limited in their capabilities and not endowed with consciousness, they certainly don't evolve into people who can express their indifference. Or who name their sons Peter. Or who marvel at heaven and earth. Or who ponder the nature of science and the proper relationship between fact and theory. They would be too stupid to try; too busy getting the next meal and hiding from the evil velociraptor."

232. Mass extinctions

Incidentally, we are currently experiencing another phase of mass extinction of animals and plants. The only difference is that the cause in this case is not a meteorite impact but humans, whose emergence was ironically facilitated by a meteorite impact (possibly in conjunction with the volcanism of that time in the east of present-day India, which may support the following thesis). Most mass extinctions (faunal turnovers) in Earth's history are self-made and have no cosmic causes. Instead, they are linked to increased volcanic activity and the associated climate changes. Life itself can also contribute to the reversal of evolutionary achievements, as may have happened 251 million years ago at the boundary between the Permian and Triassic periods, according to recent research. There are several examples of such mass extinctions in Earth's history.

233. The oxygen catastrophe

One of the most far-reaching consequences was undoubtedly the so-called "oxygen catastrophe," which resulted in the extinction of most anaerobic microorganisms living approximately 2.4 billion years ago (at that time, there were no other forms of life). The cause was the emergence of photosynthesis, which is known to produce oxygen. However, oxygen is an extremely reactive gas, which the creatures of that era tolerated no better than we tolerate prussic acid today. This means that a certain concentration of oxygen was absolutely lethal for these early life forms. Only a very few of them managed to adapt, which was extremely complex from a biochemical point of view. On the other hand, the availability of free oxygen led to the development of a new and more effective method of cell metabolism known as "respiration." This, in turn, was a prerequisite for the evolution of highly complex eukaryotic cells from primitive prokaryotic cells, and subsequently, multicellular organisms. Here, too, a true catastrophe facilitated new evolutionary progress, namely the emergence of multicellular animals and plants.

234. Supervolcanoes

As we have only known for a few decades, there is another phenomenon besides catastrophic meteorite impacts that can seriously endanger the continued existence of our civilization. And this type of catastrophe is much more likely than a Chicxulub meteorite impact. I am referring to the eruption of a so-called supervolcano. The relatively new term "supervolcano" found its way into the geological sciences when, after an in-depth analysis of the eruption of Tambora in 1815 on Sumbawa in Indonesia and Toba on Sumatra around 74,000 years ago, it became clear that the conventional definition of a "volcano" could no longer do justice to this phenomenon.

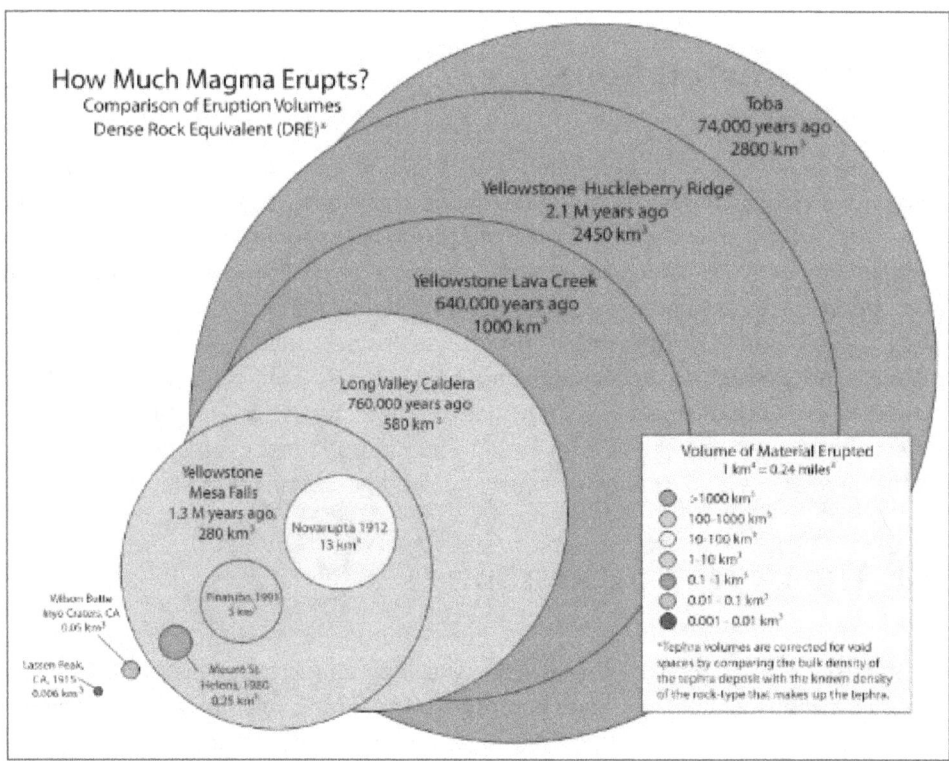

Fig. 91: Einige Supervulkan-Ausbrüche im Vergleich (Creative Common)

Its characteristic feature is a huge stationary magma chamber above a so-called mantle plume or "hot spot", i.e. a region where hot material rises in the earth's mantle and liquefies under the earth's crust due to pressure relief. If this magma chamber breaks through the Earth's surface in the form of a supervolcanic eruption, it empties quite quickly, leaving behind a so-called caldera (a collapse crater) rather than a classic volcanic structure. This caldera is very large and quickly fills with material so that it is barely noticeable on the Earth's surface. A smaller supervolcano dozes away near Naples, for example (I mean the Phlegraean Fields). But once it erupts (and it will!), it will be too late. It can become much stronger than the eruption of Krakatoa in 1883, for example, and there are significantly stronger and therefore more dangerous supervolcanoes on Earth. If their output of magma and tephra exceeds the 1000 km^3 mark, a corresponding eruption can very quickly develop into a global catastrophe. Today, geologists are aware of a total of 20 of these on Earth, but there may well be more, as they are difficult to recognize in a dormant state.

235. Yellowstone

The best-known example of an unpredictable supervolcano is the "hot spot" under Yellowstone National Park in the U.S. state of Wyoming. Its magma chamber, which has been precisely measured in recent years, has a volume of approximately 24,000 km^3 with dimensions of about 60 x 40 x 10 km (length, width, height).

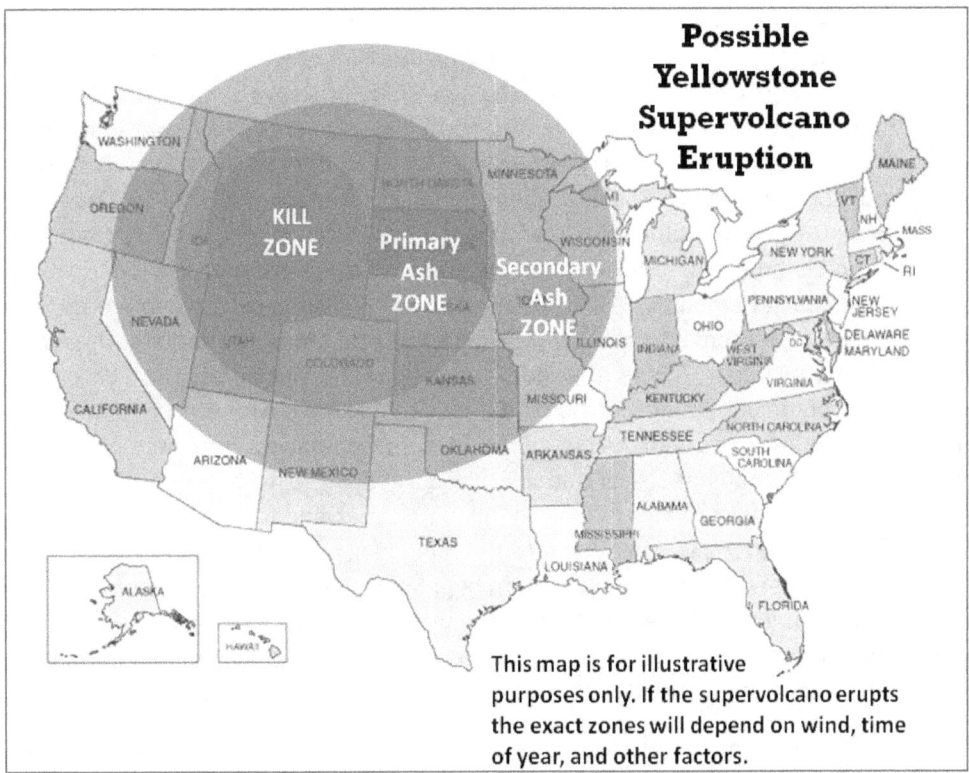

Fig. 92: Possible impact zones in the event of a hypothetical eruption of the Yellowstone super volcano

The concerning aspect is that geologists believe an eruption is already overdue. As the American plate slowly slides over the "hot spot" of the mantle plume responsible for the magma chamber due to continental drift, the periodicity of previous eruptions can be roughly estimated from the calderas lined up like a string. According to this, the Yellowstone supervolcano has erupted nine times in the last 17 million years, most recently

640,000 years ago. Based on the latest caldera formations, it is estimated that an eruption with nearly complete emptying of the magma chamber occurs approximately every 600,000 to 900,000 years. The implications of this can be seen, in a more limited form, in the fictional disaster movie "Supervolcano" from 2005.

236. The genetic bottleneck of mankind

The effects that such an eruption has on the environment are difficult to estimate in detail - but in any case, they are so catastrophic that , with the exception of the Toba eruption 74,000 years ago (in which, as geneticists have established on the basis of the "genetic bottleneck," all but a few thousand individuals alive at the time perished), there has been no comparable natural disaster in the history of mankind.

You might be wondering how such a conclusion can be reached. To understand this, it is important to know that a gene (which codes for the amino acid sequence and therefore the primary structure of a protein) can occur in different forms within a living organism, known as alleles. For example, one allele results in blue eyes, while another results in brown eyes. Both eye colors provide equal visual acuity, so eye color is usually inconsequential for vision. If a sufficiently large population exists, alleles are distributed through sexual reproduction and their number increases through mutations. Their relative frequency in a population is described by the term allele frequency. From a genetic standpoint, Darwinian evolution is essentially the constant change in allele frequency in a population through natural selection or genetic drift. For instance, if a founder population of a species emerges on an island from a few individuals, these "founders" and their immediate descendants only represent a small proportion of the alleles present in the original population. This is referred to as genetic impoverishment, which can be measured concretely. As individuals on the island only mate with each other, with no effective gene exchange with the rest of the species' population, the island population and the remaining population will gradually diverge, leading to the formation of new species (the technical term for this process is "allopatric speciation"). The island population is genetically impoverished compared to the remaining population and is said to have gone through a genetic bottleneck. Modern population genetics methods can precisely quantify this bottleneck (for example, estimating the size of the founder population) and determine the temporal position of the "bottleneck" (i.e., when the founder population established itself on the island).

This process can also occur without an island, such as when a natural disaster destroys the majority (possibly 99%) of a population and the alleles contained within it. The few surviving individuals then represent only a small subset of the alleles that were once present in the population. This also creates a quantifiable genetic bottleneck. When the mitochondrial DNA of humans was examined some time ago, it was surprisingly discovered that there are very few differences between people from different parts of the world. Humanity is genetically exceptionally homogeneous, which suggests a genetic bottleneck (the concept of "race" based on phenotypic differences is not supported from a biological-genetic perspective) that must have occurred around 70,000 years ago.

Within the margins of error, this date coincides quite closely with the Toba eruption approximately 73,900 years ago. Some anthropologists believe this is not a coincidence. It is plausible that this supervolcanic eruption, of which only the massive caldera now housing Lake Toba on Sumatra remains, may have led to the extinction of the majority of modern humans in Africa and Asia at that time. In this case, humanity was evidently fortunate in its path to the present day. But as I mentioned, this is just one possibility.

237. We haven't escaped yet

We haven't escaped yet. At least six supervolcanoes of the highest class (8) are waiting to erupt. This is many times more likely than an impact from a celestial body with a diameter of more than one kilometer. Insurance companies have even calculated that the probability of dying in a supervolcano eruption is several times greater than dying in a commercial airliner crash.

238. It rumbles ominously under the Campi Flegrei

The Phlegraean Fields, Europe's supervolcano, are a fascinating yet disturbing natural phenomenon that (along with neighboring Mount Vesuvius) attracts many tourists to Italy. The volcanic area west of Naples is known for its impressive geothermal activity with hot springs, fumaroles, and solfataras. In recent months (2023/24), a series of noticeable earthquakes have occurred in the region around Naples. This seismic activity has brought the experts at the National Institute of Geophysics and Volcanology (INGV) to the scene. They have published a computer simulation of a possible eruption scenario at the Campi Flegrei, which underscores the significance of this activity. It shows that there is apparently an ascent of magma from the central magma chamber, which triggers increased earthquake swarms.

At the same time, the uplift rate of the supervolcano has quadrupled. The entire bay is rising faster than it has in decades. Scientists speak of a "magma lens" underground that is causing this rapid uplift. In recent months, the ground around Pozzuoli has been rising by an average of one centimeter per month. Between April 9 and 10, 2024, the bay rose by one centimeter in just two days. This rapid uplift has caused concern among residents and visitors to the region, as it is reminiscent of the dramatic events of previous volcanic eruptions.

Despite this worrying activity, scientists emphasize that an eruption is currently very unlikely. They continue to monitor the situation closely and are working tirelessly to minimize the risks and inform the public. Nevertheless, they have now declared a "yellow" risk level, which means that increased vigilance is required. The authorities have also updated contingency plans and launched information campaigns to prepare the population for possible scenarios. The Phlegraean Fields therefore remain a place of great scientific interest and tourist attraction, but also of potential dangers that need to be monitored.

239. The mini super volcano in the Eifel

Anyone who believes that we here in Germany are on the safe side when it comes to volcanoes is mistaken. And it doesn't even have to be a supervolcano erupting somewhere in the world. We have our own volcano in Germany. In the Eifel region. It last erupted approximately 12,930 years ago, reaching an "explosive force" that exceeded that of Vesuvius in 79 AD (the so-called Plinian eruption) by up to 5 times. The "bang" must have been heard throughout much of Europe and likely startled many people of the Neolithic period (Upper Palaeolithic). The "hole" left by this eruption now forms Lake Laach. It is slightly oval, averaging 3.3 km in diameter and just over 50 m deep. Although Lake Laach is referred to as a "maar," from a geological point of view, it is not a true maar, but represents the collapse caldera that formed above the magma chamber which was completely emptied during the aforementioned eruption.

Fig. 93: Exposed tuff layers at Lake Laach in the volcanic Eifel. They date from the last eruption just 12,930 years ago (Wikimedia)

The fact that the last eruption was truly remarkable can still be experienced today along the volcano nature trail. The geological evidence from that time describes a truly apocalyptic event which, interestingly enough, could happen again at any time, as the Eifel volcanism is only in a dormant phase. This is because there is a "hot spot" deep underground and a magma chamber at its tip. A gas-rich phonolitic magma collects there under high pressure. When it rises through fissures and cracks in the rock - as happened approximately 13,000 years ago - it is depressurized and the gases are released explosively, which in this particular case is amplified many times over by contact with water. This is known as a "phreatomagmatic eruption," in which pyroclastic clouds (such as those formed during the eruption of Mt. Pelé in 1902, point 240) are created, which can devastate large areas of land.

Fig. 94: Glowing cloud of the Montagne Peleè volcano on the Antilles island of Martinique, which completely destroyed the city of Saint Pierre (~40,000 inhabitants, 3 survivors) in 1902... (see point 240) (Creative Common)

How should we imagine such an eruption in the Eifel? An eruption occurs very suddenly;
i.e., there are probably only a few visible indications that an eruption is imminent. This
could be increased mofette activity or - as the geologist Ullrich C. Schreiber describes
very nicely in his Vulcano thriller Die Flucht der Ameisen - that these creepy-crawlies
start to leave in large flocks before things really get going (but it's no use)...

In any case, approximately 12,930 years ago, the earth's interior suddenly began to rum-
ble because the rising magma encountered water, which immediately evaporated and
blew open the existing cracks and fissures in the rock, allowing the thin, gas-rich lava to
rise ever higher. This ascent then suddenly erupted in a phreatomagmatic explosion,
which created a crater in the ground and from which rock ground into powder was hurled
at high speed in the form of a huge hot cloud of ash to a height of over 30 kilometers. This
phase is called the "Plinian main phase" because it was first observed and described by
Pliny the Younger (61/62 - 113/115) during his escape from the catastrophic eruption of
Vesuvius in 79 AD (keyword "Pompeii," "Herculaneum"). It is characterized by a rapidly
developing mushroom-shaped eruption column that rises into the stratosphere and
then collapses under its own weight. This creates so-called "pyroclastic flows" from a
hot (up to 800 °C) and dense emulsion of volcanic dust, pumice, and gas, which is aptly
called "nuée ardente" (pyroclastic flow) in French. They burn and suffocate all life in their
path, leaving behind thick layers of solidified volcanic material (so-called ignimbrites).
As they reach a flow speed of up to 700 km/h, you cannot run away from them. In the
case of Lake Laach, these deposits reach a layer thickness of up to 60 meters, and the
wind-blown dust could still be detected in Sweden and northern Italy. It is estimated that
this eruption produced approximately 16 km^3 of loose material (so-called tephras). This
puts it in a league above the well-known eruption of Mt. St. Helens in 1980 and even
Pinatubo in 1991. So such an Eifel volcano really isn't harmless. Incidentally, after about

a week, the whole phenomenon was over and what remained of the Eifel was nothing more than scorched earth, which was to develop into a tourist magnet at the end of the next 12,000 years. It is worth going there once if you are not afraid of an eruption. After all, many volcanologists believe it is very likely that the Laacher See volcano will erupt again sometime in the next 1,000 years. And this could theoretically happen as early as next week. To give you a better idea of the scenario, a two-part movie with the modest title Volcano was filmed in 2009 under the direction of Uwe Jansen. Although it does not come close to the American thriller Supervolcano from 2005, which deals with an eruption of the supervolcano under Yellowstone National Park, it is otherwise quite realistic - at least as far as the geological facts are concerned...

240. Louis-Auguste Cyparis and the Montagne Pelée

In connection with "Glow Clouds", another name comes to mind: Louis-Auguste Cyparis (1875-1929), who once traveled through the United States with a Barnum's traveling circus under the stage name "Ludger Sylbaris". We have already met Phineas Barnum (1810-1891) in connection with "Barnum's American Museum" and the Siamese Twins (point 59).

Mr. Cyparis' only task was to sit on stage in a replica of a prison cell and dramatically recount a part of his life story to the audience. It all began on the Antillean island of Martinique in the town of Saint-Pierre, where he participated in a bar brawl on the night of May 7 to May 8, 1902, which reportedly resulted in a death. He was subsequently detained and placed in the "Maison d'Arrêt" in a half-timbered cell with only a small barred window. This proved to be his fortune. Just a few hours later, all 40,000 inhabitants of Saint-Pierre were killed, and only he, a shoemaker named Léon Compère-Léandre, and a girl who had the presence of mind to save herself in a coastal grotto survived the disaster. On May 8, 1902, Ascension Day, shortly before 8 o'clock in the morning, the nearby volcano Montagne Pelée erupted, sending a cloud of embers with a temperature exceeding 700 °C racing down its slopes at speeds of up to 800 km/h, killing all but three of the town's inhabitants. Even in his prison cell, red-hot vapors entered through the small window and burned parts of his skin, leaving him with large scars on his arms, legs, and back. These scars would later serve him well in his brief circus career. Unfortunately - or perhaps fortunately - the cell door also withstood the torrent of fire, so Cyparis was unable to escape and was only discovered and rescued three days later. As there were no surviving witnesses to the bar brawl in which he was allegedly involved, he was eventually pardoned by the governor. He emigrated to the USA, where Phineas Barnum was waiting for him to appear in his popular freak show.

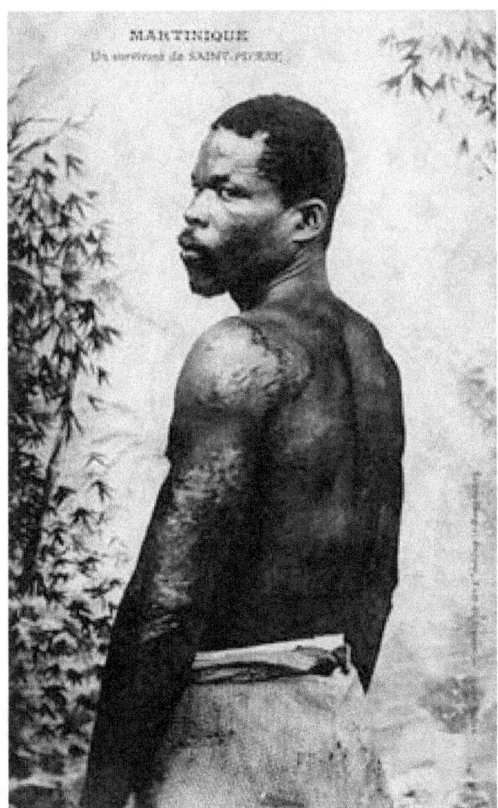

Fig. 95: Louis-Auguste Cyparis (1875-1929), who survived the pyroclastic cloud of the volcano Mt. Pele in a prison cell in 1902...

There, a new life began for the giant-like man of color from Martinique, disfigured by scarred burns. As *"The Most Marvelous Man on Earth"*, he was one of the attractions of Barnum's traveling circus alongside the *"Cardiff Giant"* George Auger, Rob Roy, the "albino and contortionist" and Charles Tripp, "the man without arms". Incidentally, he was the first person of color to have a lucrative circus career in this way, at least for a short time, but it came to an abrupt end on June 5, 1903. On this day, he stabbed a circus guard while drunk, which earned him another, but now more permanent, prison sentence. After his pardon, he is said to have gone to Panama to excavate the Panama Canal with many others.

241. Fever flies and building the Panama Canal

He was certainly also bitten by malaria and yellow fever mosquitoes. At that time, it was already known that yellow fever and malaria were transmitted by certain mosquitoes. The American doctor William Crawford Gorgas (1854-1920) then developed a strategy that, within just 18 months, contained the rampant yellow fever and malaria epidemic in Panama to such an extent that the canal could be built, and deaths from these diseases were kept within limits thanks to improved hygienic conditions in the infirmaries.

Fig. 96: One of the rare photos showing the building of the Panama Canal (1907). (Wiki-media)

The first attempt to dig a canal through the isthmus, which was started in 1881, ended in a fiasco, which was not only the bankruptcy of the canal construction company at the time. Over 22,000 workers also died of the aforementioned diseases within just eight years. At some point, the number of grave crosses reached such an extent that it was almost impossible to recruit new workers.

242. Corpses preservation

This is why the corpses were soaked in vinegar and shipped to Europe in barrels to be buried there. Vinegar was already used as a preservative in ancient times. It is thought to have been discovered when wine or beer was accidentally soured. In any case, vinegar must have been known as early as 6000 BC, as indicated by archaeological findings from Mesopotamia. It was soon discovered that it could be used to preserve more or less cooked crops over a longer period of time, which is known as pickling. An entire region in Germany still benefits from this today (Spreewald). Later, vinegar evolved into a frequently used condiment and, in Rome, diluted with water, into a refreshing lemon-ade - a drink known as "posca". The body of Alexander the Great (356-323 BC) was also preserved when he died in Babylon in the summer of 323 BC. But not in vinegar, instead in honey. This allowed his body to be transported to Alexandria without spoiling, where this great king was finally buried in a place that is still unknown today.

Incidentally, preserving a corpse in honey is an early form of embalming, which was used occasionally in antiquity. The technical term for this is "mellification". As far as I know, honey is no longer used in this manner anywhere today. It tastes too good for that, and the quantity required is probably far too expensive for most people. Cryopreservation with liquid nitrogen, as is occasionally used by wealthy clients in the USA, for example, is the method of choice today.

The method that Vladimir Ilyich Ulyanov (better known by his pseudonym "Lenin", 1870-1924) and later Stalin himself (less known by his real name Iosif Vissarionovich Dzhugashvili, 1878-1953) had to undergo without being asked at Stalin's behest, however, has not stood the test of time. It requires constant care of the corpse in order to protect it from natural decay. Today, for example, a private foundation has to spend 1.5 million dollars a year to keep Lenin looking reasonably fresh in his mausoleum on Red Square in Moscow (he even has to change his wardrobe every three years!). In fact, it's time to give him a proper burial, just as was done for Stalin...

Fig. 97: Although long gone - still a much-visited tourist attraction on Red Square in Moscow - Lenin in his glass tomb (Wikimedia)

243. The Mausoleum of Maussolos II.

The term "mausoleum," as a center of a death cult akin to that of Lenin, actually originates from a lesser-known figure than Vladimir Ilyich. It traces back to the petty king Mausolus II, who resided in the port city of Halicarnassus (modern-day Bodrum, Turkey) for 24 years from 377 BC. Because he was afraid, like many insignificant rulers, of being quickly forgotten, he had a tomb, or Mausoleum, built for himself. It was only completed three years after his death. But as one of the "Seven Wonders of the World," it was meant to serve its purpose throughout the ages. The Roman historian Pliny the Elder (circa 23-

79 AD), who witnessed the eruption of Vesuvius in 79 AD but did not survive it, described what it once looked like. We know this from Pliny the Younger (circa 61-115 AD).

The mausoleum itself, or rather the remains of it, could be visited until 1523 (i.e., for 1873 years). However, in that year, the knights of the Order of St. John, who occupied the site, needed building material for a new fortress, and the ruins of the mausoleum were an extremely convenient source of stone. Since then, only one of the "Seven Wonders of the World" remains that can still be visited on site today. And it's not going to disappear anytime soon, even though a few brainwashed fighters from the "Islamic State" recently threatened to demolish it...

I doubt whether Lenin's mausoleum will last as long as that of Mausolus of Halicarnassus. Firstly, it is architecturally far too simple to even come close to being considered a modern wonder of the world.

Fig. 98: This is roughly what the "Maussolleion of Maussolos" in Halicarnossus might have looked like when the ancient writer Antipater of Sidon included it in Herodotus' list of "Wonders of the World" (Creative Commons)

Lenin himself is also being increasingly forgotten, as his obscure teachings are no longer part of state school and university education. Moreover, the social system he helped establish has spectacularly failed worldwide. Yet, there was once a brief and largely forgotten episode in human history when "Marxism-Leninism" was considered a "science" (much like "gender studies" is today). In reality, however, it was a political ideology upon which a social system with a dictatorial flavor was established - first in the former Soviet Union and then, after the Second World War, in some countries of Eastern Europe (as well as various other countries like Cuba). It survived for several decades until it inevitably failed, primarily for economic reasons. The fact that the corresponding upheavals were largely peaceful still seems remarkable, even from today's perspective.

244. The "genius" of the Carpathians

Only one of the communist "leaders", who called himself the "genius of the Carpathians" but was otherwise just a petty, colorless civil servant, did not survive the peaceful revolutions at the end of the 1980s - Romania's "Great Conducator" Nicolae Ceausescu (1918-1989). Shortly before that, his courtiers had still been lavishing praise on him as the "Titan of Titans", the "sweetest kiss of home soil", the "glorious oak from Scornicesti" and the "light that blinds even the sun". In the end, he stood in front of the firing squad with his wife Elena, the "loving mother of the nation" and "Romania's greatest scientist", and could not believe that he, who had previously had his own people shot, was now being shot himself - after a summary execution.

The irony of the story lies in the fact that Ceausescu himself signed the document declaring a national state of emergency immediately beforehand - and in this way created the legal basis for a military summary court, which promptly indicted him. "*Nicule, we are being murdered? In our Romania?*" were Elena's last words to her husband. "Our Romania" - it had been true for decades, but now it was over. And very few people in Romania will want to remember this unsavory "leader" on the day of his death, Christmas Day 1989.

Fig. 99: Nicolae Ceausescu (left; does anyone else know the gentleman next to him?)

245. Dracula

Incidentally, Ceausescu is popularly known as "Draculescu" for a reason, as his name includes Bram Stoker's (1847-1912) famous novel character "Dracula." The communist dictator of Romania even played a significant role in this himself, as he saw himself as a successor to Vlad Tepes (Vlad II Draculea, 1431-1477), whom he had stylized as a national hero. This Vlad Tepes, whose real name was Vlad Basarab, was in turn the historical model for Bram Stoker's novel "Dracula."

Fig. 100: Thanks to Bram Stoker and "Bram Stoker's Dracula" from 1992, a character from the Ottoman Wars of the 15th century who is known to everyone today: Vlad II Tepes Draculea

Here's what you need to know about this probably most famous voivode of Wallachia: "Tepes" means "the Impaler" and "Draculea" means "son of the dragon." He was given the first name because it was his habit to "impale" the Ottomans who invaded his country - if they fell into his hands alive. This particular method of execution, which dates back to the time of Hammurabi (around 1760 BC), was intended to undermine the fighting morale of the Ottoman army and stop their further advance into the Christian lands north of the Carpathians. It is reported that Mehmet II (1432-1481, the conqueror of Constantinople) had to withdraw in 1462 after the unsuccessful siege of Targoviste along a line of 20,000 impaled Turks.

As far as killing his enemies was concerned, Vlad the Impaler developed a particular passion for this practice. He preferred to dine among the stakes, surrounded by the smell of corpses and sometimes making small talk with someone who was still alive (the death throes could last up to two days!). For example, the Romanian author Marin Sorescu (1936-1996) included Ceausescu quotes in his play "The Third Stake," which only came to the attention of the censors relatively late (after Sorescu's death). In 1477, Vlad II Dracula himself was finally killed - probably during fighting near Bucharest. In any case, his head was cut off and, preserved in honey, taken to the Sultan in Constantinople, where it was to become a major attraction at the victory celebrations.

Due to his cruelty, but also because he had decisively resisted the expansionist drive of the Ottoman Empire towards Central Europe, he was quickly glorified by the people and later - as he was said to have drunk the blood of his enemies from time to time during his lifetime - regarded as a "vampire." However, the belief in vampires itself only emerged around two centuries after Vlad Tepes' death and persisted in the Balkans until the 20th century, when it slowly became a general part of popular culture.

246. The Fearless Vampire Killers

Incidentally, Roman Polanski's 1967 film "The Fearless Vampire Killers" offers a very good introduction to the genre of vampirism and is well worth watching again. In particular, the expert opinions of the famous vampire researcher at the University of Königsberg, Prof. Abronsius, have lost none of their topicality. In this film, you can also admire Sharon Tate (1943-1969) bathing in a wooden tub, who was famously gruesomely murdered by the Manson Family two years after filming was completed.

Fig. 101: The unforgettable, enchanting Sharon Tate - at that time Roman Polanski's wife and seen here bathing as Sarah Shagal in Roman Polanski's film "The Fearless Vampire Killers"... Two years ago she was murdered in her house together with four guests in the act of a bloodlust by the so-called "Manson Family".

Enlightened people know, of course, that there have never been any undead "vampires", nor will there ever be.

247. Vampire bats

However, educated people also know that there are real "vampires" that feed entirely on blood - namely the American vampire bats (*Desmodontinae*). They are not popular with farmers, as they settle on grazing animals at night and inflict small wounds from which blood flows, which they then lick up or suck in. An anticoagulant in their saliva prevents the blood from clotting quickly and an "anesthetic" prevents the animal (usually cattle) from noticing.

Of the three known vampire bat species, only the "common vampire" occasionally attacks a human. However, it is of course a myth that it sucks them dry, as the undead vampires a la Dracula are said to do. Such a bat only consumes around 20 milliliters of blood per meal (roughly equivalent to the contents of a small chicken egg), which it then has to digest in peace. The blood loss suffered by humans and farm animals as a result of bat visits is therefore relatively harmless.

Fig. 102: A cute vampire bat licking blood) on the head of a cow. They would actually be quite harmless if they didn't transmit rabies (Creative Commons)

248. The rabies

What is not without concern is the unpleasant fact that vampire bats are ideal carriers of rabies, which is almost always fatal. It is estimated that well over 50,000 cattle and around 20 people die each year from this extremely dangerous viral infection in Brazil alone. In Germany, this disease (as far as it is not the specific bat rabies) seems to have died out since 2008, which in turn is rather good news. This is because there is still no cure for rabies once the disease has broken out. However, it is possible to protect yourself against it with a preventive vaccination.

249. Homeopathy

Or through homeopathic remedies, as the famous homeopath Constantin Hering (1800–1880) claimed to have discovered over 170 years ago and as the advertising of various "naturopaths" promises. They still refer to his "Hering's Rule" today, alongside the so-called "law of similars," to give their teachings a scientific veneer. However, Hering's Rule is just one of many hypotheses that have been developed over time in homeopathy to substantiate the alleged effectiveness of this method.

Based on the observation that rabid cats or dogs (and humans) develop a fear of water during the course of the disease, Constantin Hering came up with the idea in 1833 of diluting dog saliva (which is also liquid) accordingly and producing a "rabies nosode" from it. This was to be used not only in veterinary medicine but also in "naturopathy" as a universal remedy. Hering even claimed that this "rabies nosode," like any homeopathic remedy, could act universally against a variety of ailments.

The preparation was not only recommended for rabies itself but also for other, seemingly random diseases and complaints: epilepsy, hydrophobia, headaches, arthritis, urinary tract infections, and general phobias were all on the list, as were vaginal cramps in women and, believe it or not, even fits of rage and raving à la Klaus Kinski. The belief in the effectiveness of these "remedies" was reinforced by the postulate of homeopathy: the more a remedy is diluted, the greater its healing power is supposed to be - an assumption that is completely untenable by the standards of modern science.

The paradox here is that the less active ingredient (in this case dog saliva) is contained in the preparation, the greater the effect, and the more expensive the preparation is supposed to be.

250. Potentiate

As is well known, the experienced homeopath achieves this increase in effect through the act of "potentization". This is the process of diluting a "primary substance" in a certain way, either with distilled water, lactose or ethyl alcohol, until the desired degree of dilution is achieved. A common unit of measurement for this is the D-potency (decimal potency). It indicates that each step is always diluted to 1/10. The potency D21 then corresponds (i.e. after 21 corresponding dilution steps) to a dilution of 1 to one trillion - which, incidentally, is a very common dilution of "homeopathically" effective substances.

251. Homeopathic potency experiment

But what does that actually mean? To find out, I recommend the following simple experiment, perhaps accompanied by a few elementary calculations with a calculator. According to pure doctrine, the effect of a homeopathic substance should increase the more it is diluted. As a unit, we use so-called C-potencies here, where the "C" stands for "centesimal potencies", because in this case one step dilutes the solution to exactly 1/100. The ingredients are some Prima Spirit (or a high-proof schnapps such as the infamous 80% straw rum) and 1.2 liters of tap water (if you want to convert the tap water into healthy and vital mountain spring water beforehand, then I recommend the VitaJuwel gemstone stick for a mere €49.90 for stirring and simultaneously "vitalizing" the water in the à la Grander method). For equipment, we also need 12 normal drinking glasses (we only want to venture as far as a C12 potency of the alcohol), each filled with 1/10 liter of water. We also need a pipette that allows us to take up 1 milliliter (=1/1000 liter) of liquid at a time.

And now the experiment can begin. First, fill the pipette with one milliliter of schnapps and transfer its contents to the first glass of water, which should then be shaken in a certain way rather than stirred (remember, according to Johann Grander, water has a "memory" that can be refreshed by shaking it in the right way). Then use the pipette to take a sample of one milliliter from the first water glass and empty it into the second water glass (don't forget to shake!). Then take a sample from the second water glass and empty it into the third. Repeat this process until you have finally reached the desired C12 potency of the schnapps in the 12th water glass. And then to your health! But be careful, because according to homeopathic principles, you should now feel much more "tipsy" than when drinking the same amount of "original substance". Just try it out, because empiricism usually wins out over theory.

But what do these dilution levels mean in concrete terms, when asked "scientifically"? This is where the calculator comes in handy. Here are the following results, made somewhat clearer by comparison: C1-potency - a bottle of schnapps diluted with the contents of a bathtub; C2-potency - a bottle of schnapps diluted with the contents of a tanker full of water; C3-potency - a bottle of schnapps diluted with the contents of a swimming pool; C4-potency - a bottle of schnapps diluted with the contents of a large tanker full of water; C5-potency - a bottle of schnapps tipped into Lake Erie; ... C9-potency - a bottle of schnapps poured into the ocean; ... C12-potency - a bottle of schnapps diluted in water with the volume of the planet Jupiter. Will it really get you drunk? After all, there is still a good chance of finding a fairly large amount of ethyl alcohol molecules from the "original substance" in the last glass of water. But of course, no real homeopath stops at C12. Only from C24 onwards can you really be sure that the water glass no longer contains even one ethyl alcohol molecule from the first pipette. But it should be C75 - a bottle of schnapps diluted in water with the volume of our manageable universe!

Incidentally, common C-potencies in homeopathy are C6, C12, C30, C200, and C1000. For some ailments, however, it is advisable to prescribe even higher potencies of a suitable homeopathic remedy. C10,000 and C100,000 are often used here. It is said that such high potencies work on a particularly deep spiritual level. And after what has just been said, one can only agree with this...

252. Effect without active ingredients

We also understand now why homeopathic remedies do not fall under the Medicines Act but only under the Food Act. This is because, according to all human experience, something that contains no or virtually no active ingredients cannot "work" against or on anything. Of course, everyone is free to believe in such an "effect" (which obviously cannot work for pets, although veterinarians often sell a few "globules" for dogs and cats for a lot of money). So if you happen to be bitten by a dog while on vacation in southern climes (especially in the Balkans, Turkey, or outside Europe, such as in Morocco), you should consider getting a quick vaccination against rabies rather than swallowing a "rabies nosode." If you survive in the latter case, you can be sure that it was only because the dog was not rabid. This is because a successful infection, starting with the first appearance of rabies symptoms, is usually almost 100 percent certain to be fatal within seven days.

In this context, it is interesting to note that, in addition to the highly infectious rabies virus that causes rabies, there is also a non-pathogenic form of the virus. It can also travel along the nerve tracts into the brain, but it does not cause fatal meningitis there. Incidentally, some brain researchers are using this to elucidate the neuronal wiring of certain brain sections, and there is still much research to be done in this area.

253. Island talents - Savants

We now know quite well how this 1.3 kg piece of matter called the "brain" works. It consists of around 100 billion neurons, which are connected with each other via around 100 trillion connections. Interspersed among them are a large number of so-called glial cells, which provide energy to the neurons. The interconnection of the neurons is not random. However, if malformations occur, this can result in certain cognitive disabilities, but sometimes (and often in connection with this) also truly amazing insular talents.

Let's start with memory. Can you imagine being able to reproduce around 12,000 books of normal size at any time and virtually word for word from memory (with an error rate of around 1%)?

254. Kim Peek - the "Rain man"

Kim Peek (1951-2009) could do it. His only shortcoming was that he was unable to grasp the meaning. Kim Peek was a barely viable contemporary, but one with a phenomenal memory. He could reproduce content verbatim, recount the corresponding events for any given date in a matter of seconds, and also mastered the art of calendar calculation to perfection (what day of the week was July 13, 1746?). Incidentally, one of his favorite pastimes was memorizing zip code lists and telephone directories. It only took him a few seconds to read a page, after which he knew every telephone number and address by heart.

Fig. 103: Kim Peek (1951-2009) - one of the best-known savants with an extraordinary memory thanks to the movie "Rain man" (Wikimedia)

But despite these impressive skills, his life was anything but easy. His extraordinary memory was offset by massive limitations in other areas. He was often unable to cope with simple everyday tasks on his own and was dependent on the help of others. The fact that his abilities were a double-edged sword is shown by the fact that he did not really understand the content of the information he stored. What may seem like a gift to many was often a burden for him, which severely limited his independence.

We mere mortals would be happy if we only had a fraction of Kim Peek's memory. But would that really be so great? Doesn't nature have a good reason for not giving our memory such prominence, so to speak?

255. Photographic memory

Take, for example, the Russian memory artist Solomon Shereshevsky (1886-1958) or the American Jill Price, who is still alive today. Born in 1965, she can recall all (and really all!) of the events she has witnessed since then, starting with the date February 5, 1980, without any gaps and with all the details. It's as if someone asks you what you saw on TV on April 23, 1996 at around 9:15 p.m. - and you answer, demonstrably correctly, "*I was watching the movie 'Rain Man' with my dog Bobby and I remember that Charlie Babbitt, played by Dustin Hoffman, was sitting in the Las Vegas casino playing blackjack with his brother Charly, who was played by Tom Cruise. By the way, on the table to the left of the TV was my blue-dotted flower vase with two yellow daffodils and a red tulip...*". What is interesting about this case is that this insular talent developed very gradually, and the woman (unlike Solomon Shereshevsky, who could not even read) otherwise shows no cognitive anomalies - she works quite normally as the managing director of a private educational institution. Apart from her phenomenal photographic memory, she is as normal as you and me in terms of other mental abilities. This is rather unusual for savants (the term given to people with exceptional insular talents).

The question that science is asking here is what happens in the brain of such exceptional people and why these abilities are not accessible to "normal" people. Incidentally, this has led to the myth that we "normal people" only use 10% of our brain's potential capacity, but this myth has since been refuted. Strictly speaking, such a claim is pretty pointless, as it is not even known what this 10% is supposed to refer to.

Insular talents are most frequently expressed in exceptional memory and arithmetic abilities as well as in artistically relevant talents such as a photographic memory or absolute pitch. Most of these talents are innate and sometimes go hand in hand with considerable deficits in other areas. Savants are often autistic. However, there are also cases where these talents suddenly appear, for example, after a sports accident. American Orlando Serrell is a notable example of this phenomenon; he was hit in the head by a baseball at the age of 10 and has had a "calendar memory" similar to Jill Price's ever since. There are currently around 100 people living on Earth who can be described as savants. Thus, they are truly few and far between. In most cases, however, savants have not advanced very far in science (at most as study subjects).

256. Geniuses

There, "normal" highly gifted people - and their enhancement, geniuses - are more in demand (the word is derived from "ingenium," which is also found in "engineer"). Classification in the category of "genius" is and has always been extremely subjective, as it is not purely intellectual ability that plays the key role here, but rather creative power: the ability to discover or create something new - technically or artistically. Today, attempts have occasionally been made to attach the term "genius" to a relatively objective measure, the intelligence quotient. According to this interpretation, all people whose IQ exceeds 130, for example, are considered "geniuses." It is therefore not surprising that there are genius criminals, genius military leaders, genius actors, genius writers and, very occasionally, even genius statesmen. But a high IQ alone is anything but an indicator of genius - at best, it can be helpful on the path to true genius. In New York, for example, there is the elite Hunter College, which only admits children who score at least 130 points on a standardized intelligence test. Almost all of the graduates achieved extremely lucrative positions in their lives, but none of them stood out due to such an extraordinary life achievement that they would deserve the title of "genius." So there must be other character traits that make a gifted or talented person become a "genius." And these include "divine curiosity" (Albert Einstein), a deep interest in the subject matter, iron perseverance in pursuing one's goals (Johannes Kepler), "practice, practice, practice" (Wolfgang Amadeus Mozart), and the good fortune to be able to pursue one's interests away from everyday worries. However, being a genius often also means going against prevailing views in society and having to accept the consequences.

There is now a social consensus about the genius status of some important people. I am referring to people such as Leonardo da Vinci and Michelangelo Buonarroti, who have significantly shaped our image of the Italian Renaissance. Among natural scientists, Johannes Kepler, Isaac Newton, Gottfried Wilhelm Leibniz, Charles Darwin, and Albert Einstein come to mind. Among artists, Johann Sebastian Bach, Joseph Haydn, and, of course, Ludwig van Beethoven were undisputed geniuses. Among philosophers, my personal favorite in this respect is Ludwig Wittgenstein, alongside Plato and Aristotle. An extraordinary density of geniuses can also be observed among mathematicians, as the names Joseph-Louis Lagrange, Leonhard Euler, Carl Friedrich Gauss, David Hilbert, and, more recently, Gregory Perelman (a truly exceptional example of this genre!) prove. They all certainly had some kind of talent, whether in the artistic or mathematical field. But they had to work hard for their "genius" themselves. Talent alone doesn't get you very far, even in sports. And even talent is first and foremost just potential inherent in the genes, which has to be brought to light through hard work and perseverance, as well as benevolent encouragement from parents and teachers.

257. Deficits in the education system

Unfortunately, today's (German) education system is no longer up to this task, as it has been increasingly geared towards mediocrity by politicians in recent de-cades. In such an education system, uniquely gifted pupils have a hard time, as they are encouraged to put a lot of time and effort into the subjects they are not good at in order to at least reach the average, instead of supporting them in the subjects they are particularly good at in order to become exceptional there. It is preci-sely the de-individualization of school, where it is all about class and school avera-ges and where everyone has to be "taken along," that is causing the scope for creativity and original thinking - in other words, what makes a unique individual - to shrink more and more. Yet it is precisely the indi-vidual promotion of special talents that is necessary in order to have enough people to drive cultural and scientific-technical progress in the future.

Another problem of our time seems to me to be that education is increasingly being equated with academic performance. For example, it is interesting to note that the num-ber of 1.0 school-leavers (average grade!) in the German capital Berlin quadrupled be-tween 2006 and 2012. This is countered by the fact that universities are increa-singly complaining about the lack of many basic skills - starting with proper written expres-sion, including error-free writing - and are now having to offer special cours-es for some of their students in order to compensate for these deficits as quickly as possible. A more detailed analysis of this situation has revealed that it is not the growing intelligence of the students and the special pedagogical skills of the tea-chers that have contributed to this shift in grades towards an ever-improving avera-ge, but first and foremost the grad-ual reduction in the level of the subject matter pursued by the ministries of education together with "experimental didacticians." And there is obviously a method to this, be-cause the political demands for "social justice" superficially mean that the Abitur is be-coming easier and easier, as it should no longer be a privilege of those who learn better and know better, because ulti-mately all pupils should get it.

258. Abitur oder Matura

Incidentally, the idea of creating a uniform entry requirement for admission to secondary school that would be accepted by universities originated in Prussia. The Abitur was in-troduced as the highest Prussian school-leaving qualification as early as 1788 as the so-called "Abitur Regiment" in the face of considerable opposition from the nobility and the church. Its later humanistic form can be traced back to Wilhelm von Humboldt (1767-1835) and Johann Wilhelm Süvern (1775-1829), who gave the Abitur a humanities-lin-guistic and mathematics-science orientation. From 1834, passing the Abitur examina-tion became compulsory in Prussia for admission to university. It was only in 1896 that women were allowed to take the Abitur. Individual doctorates by women (although rare) were already awarded at some German universities (especially Göttingen) in the 19th century. However, it was only at the beginning of the 20th century that it gradually be-came possible for women to pursue a regular degree. Women gradually began to gain a foothold in medicine in particular, while they remained the exception in mathematics and the natural sciences.

One of these exceptions was Emmy Noether (1882-1935), who studied mathematics in Erlangen from 1903 and was the second woman ever in Germany to gain a doctorate in this subject. Her first habilitation attempt at the University of Göttingen, supported by David Hilbert (1862-1943) and Felix Klein (1849-1925), failed due to the conservative legislation of the German Empire, but this did not prevent her from lecturing in Hilbert's name. It was only after the First World War that she was finally able to habilitate in 1919 and thus obtain at least a non-tenured professorship. Her mathematical specialty was the so-called invariant theory and modern (abstract) algebra.

Fig. 104: Emmy Noether, around 1910 (Wikimedia)

259. Noether's theorem

She is known to physicists in particular through Noether's theorem, which is a fundamental theorem with far-reaching applications in theoretical physics. It is worth dwelling on it for a while without being able to fully appreciate it here.

$$I_j(q_1, q_2, \ldots, q_N, \dot{q}_1, \dot{q}_2, \ldots, \dot{q}_N) \equiv \sum_{k=1}^{N} p_k \frac{dQ_k}{ds_j}\bigg|_{all\ s=0} = a\ constant$$

Fig. 105: For those who don't know it yet - the "Noether theorem"

A key concept in this theorem is the term "invariant." This refers to a quantity that does not change even if a mathematical structure associated with it or a physical system described by such a mathematical structure changes. Energy is one such example.

Let's take a frictionless pendulum. The pendulum bob swings back and forth in a gravitational field, whereby it gains potential energy during the upward movement, but the kinetic energy decreases by the same extent. During the downward movement, it accelerates. The pendulum bob becomes faster and faster until its speed and thus its kinetic energy reach a maximum at the lowest point. However, the sum of both types of energy is always constant throughout the entire pendulum motion. It is also said that a law of conservation applies to the invariant quantity of "energy." Emmy Noether discovered why this is the case. She investigated symmetries in a more abstract sense and discovered, when applied to physical systems, that there is always an associated conservation variable for every continuous symmetry. "Continuous" here means, among other things, that the laws of nature do not change if I move my experimental setup from one location in the universe to another. This symmetry, which exists due to the homogeneity of space, implies, for example, the law of conservation of momentum according to Noether's theorem. If, on top of this, the laws of nature are independent of the direction of space, then this leads to the conservation of angular momentum. It therefore only applies in so-called isotropic spaces. Suppose the laws of nature were dependent on time. Then it might take less energy to boil a liter of cold water today than it did yesterday or the day before. A capable engineer could then come up with the idea of capitalizing on this difference by creating energy out of nothing, so to speak. But nothing like this has ever been discovered. Even if you observe galaxies many hundreds of millions of light years away, whose light was also emitted many hundreds of millions of years ago due to the finite speed of light, you will find that the same laws of nature prevailed there at that time as they do here today. Physically, this is a sign of the homogeneity of time (no point in time is distinguished from another) and the famous law of conservation of energy, first formulated by Robert Mayer (1814-1878) in 1842, follows from Noether's theorem. The conservation laws of classical mechanics thus necessarily follow from fundamental properties of physical space and time. And we have Emmy Noether to thank for this insight.

Symmetries play an even greater role in modern elementary particle physics, a domain of quantum physics, than in classical mechanics. It can even be said that the theory can only be formulated on the basis of specific symmetries, which in turn predict the conservation variables observed in elementary particle interactions (quantum numbers, different charges, parity (mirror invariance), spin). It becomes particularly interesting when certain symmetries no longer apply under certain conditions, i.e., they are "broken." This is the case, for example, when the "mirror image" of a physical process is not realized anywhere in nature. And it was precisely such a violation of mirror symmetry that was discovered by Tsung-Dao Lee and Chen Ning Yang in 1956. You may remember that I have already reported on this in connection with the Ozma problem (point 173, how to explain to an alien by radio where right and left are).

260. Symmetry and beauty

The term "symmetry" is rarely associated with physics but rather with "beauty." Something - a thing, a living being, or a pattern - that exhibits symmetry is perceived as pleasant and balanced, in short, as beautiful. This sense of symmetry as an aesthetic ideal is deeply rooted in us and goes beyond mere preference. Even in ancient times, symmetry in architecture and art was seen as an expression of harmony and perfection. Greek temples such as the Parthenon are well-known examples of the pursuit of perfect symmetry, which was seen as the embodiment of the divine and the order of the universe.

Fig. 106: The "Judgment of Paris" led to the Trojan War - because Aphrodite promised him the most beautiful woman in the world (and, unfortunately, she was already married, namely to the king of Sparta, Menelaus) (Pinturicchio - "Judgment of Paris", fresco, Paris, France)

The preference for symmetry is completely independent of culture (at most of the form), as the preference for symmetrical patterns can be found everywhere, whether in the decoration of everyday objects or in the construction of buildings and temples. In nature, the preference for symmetry can be seen in the evolutionary development of living beings, where symmetrical features are often seen as a sign of health and vitality. This also explains why people find symmetrical faces more attractive - they may reflect genetic fitness.

The equation of symmetry and beauty therefore seems to arise less from reason or a learned cultural tradition than from feeling. It is an intuitive reaction based on deeply rooted biological and psychological principles. The opposite of symmetry is asymmetry. And it too has its very own charms.

Fig. 107: Symmetry has played a major role in architecture since time immemorial (Wikimedia)

261. The golden ratio

A line can only be divided symmetrically in one way, but an infinite number of times in such a way that the right-hand part cannot be aligned with the left-hand part. But one division stands out among these countless asymmetrical divisions because it seems somehow unique to us. To do this, you have to divide the distance into two parts in such a way that the smaller part relates to the larger part in the same way as the larger part relates to the whole. This special division is known as *Proportio divina*. Since the 18th century, the term "golden ratio" has become established. It can be found in buildings, paintings and in many different ways in living nature. If you examine it more closely, you realize that it is represented by an irrational number

$(1+\sqrt{5})/2 \sim 1.61803398874989484482045868343656...$

can be represented. It is not a transcendental number like pi, but an algebraic number, as it can be constructed on the number line using a compass and ruler. And it is also the most "irrational" number there is, because it can be calculated using the simplest conceivable continued fraction (it contains only ones as digits).

Fig. 108: Do you recognize it, the "golden ratio"?

What man has unconsciously incorporated into his works of art and architecture for thousands of years has been demonstrated in many ways by living nature. Take our arm, for example. The length of the hand and the length of the forearm are divided according to the golden ratio. Consider an ivy leaf. Here, the longest vein of the leaf relates to the shortest vein in the same way as the longer section relates to the shorter section of the golden ratio. Thus, you can find countless examples where the golden ratio somehow makes an appearance: the arrangement of leaves on a plant, cones of conifers, shells of mussels and snails, proportions in animals, etc. Even in astronomy, or more precisely, in celestial mechanics, the golden ratio plays a certain role. As you may know, there is a zone between the orbit of Mars and the orbit of Jupiter in which there are a particularly large number of small planets (planetoids). This area is therefore also known as the planetoid belt. Their orbits cannot be accurately predicted over long periods of time, as they are "disturbed" by the other planets in the solar system - especially by the massive Jupiter. As a result, the distribution of planetoid orbits in this zone is not uniform but instead clusters at certain distances from the sun or even forms conspicuous gaps in which virtually no planetoids can be found.

262. Kirkwood gaps and KAM theorem

Incidentally, these "gaps" are called "Kirkwood gaps" after the American mathematician and astronomer Daniel Kirkwood (1814-1895). If you calculate the ratio between the orbital period of a planetoid and the orbital period of the interfering body (in this case the planet Jupiter), you will find that the positions of the gaps coincide with the ratios of two small integers such as 2:1, 3:1, 5:2 ... . In this case, we speak of "orbital resonances" or "commensurabilities".

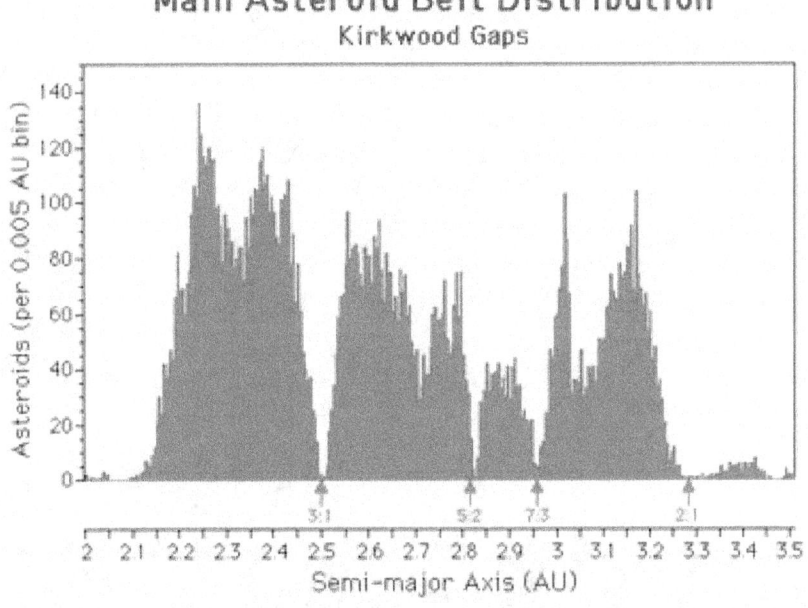

Fig. 109: Kirkwood gaps in the planetoid belt between Mars and Jupiter

In a 3:1 resonance, for example, a planetoid moves around the sun exactly 3 times when Jupiter moves around the sun once. In this case, both celestial bodies always approach each other at the same point in their orbit and gravitationally attract each other most strongly there. It's as if you were on a swing whenever it reaches its highest point. In this

227

way, the disturbances caused by Jupiter systematically build up - a resonance. The planetoid will therefore change its orbital parameters over time until it adopts a new, non-resonant orbit or passes close to a larger planet and is possibly thrown out of the solar system. In this way, the resonance gaps tend to be emptied. This means that the radial positions of the planetoid orbits are most stable in which the orbital ratio is a ratio of an irrational number to an integer, e.g. $\sqrt{2} : 1$ or :2.

A precise analysis of this fact was carried out in the last century by the mathematicians Andrei N. Kolmogorov (1903-1987), Vladimir I. Arnold (1937-2010) and Jürgen Moser (1928-1999). According to their results, which are laid down in the so-called KAM theorem, the most stable orbits in the planetoid belt are precisely those in which the resonance ratio corresponds exactly to that of the golden ratio, i.e. ~1.618:1.

263. The golden ratio in photography and painting

The 3x3 grid that can be displayed in live view mode on better digital cameras also has something to do indirectly with the golden ratio. It is approximated here by 1/3 to 2/3 (which is why it is also called "the rule of thirds") and helps to position objects in the photo in such a way that they create as harmonious a view as possible for the viewer. It is therefore always a good idea to position the object that is to dominate the photo (e.g., a face or a flower) in one of the two intersections of the upper lines. For landscape shots, on the other hand, the lower line should coincide with the horizon (or the upper edge of the water level of a lake).

Fig. 110: Pictures and photos composed according to the "golden ratio" have a special aesthetic appeal

The rule is also well known: two-thirds earth, one-third sky. I think that makes the principle clear. This rule is actually just an adaptation of the same rule that has been used in painting for centuries. Many famous painters such as Raphael (1483-1520) and Titian (ca. 1488-1576) studied the golden ratio intensively, but often accepted deviations from the ideal proportions for artistic reasons. These deliberate deviations enabled them to create unique compositions which, despite or precisely because of their imperfect proportions, have a special effect on the viewer. This enabled them to give their works an individual touch that went far beyond mathematical perfection.

264. Albrecht Dürer

Albrecht Dürer (1471-1528) wrote a major treatise in which, among other things, the Proportio divina is treated in detail. His posthumously published work from 1528, entitled „Vier Bücher von menschlicher Proportion" (Four Books on Human Proportion) and known as "Dürer's Theory of Proportionality", is particularly noteworthy here.

Albrecht Dürer, certainly one of Nuremberg's most famous sons, was born on May 21, 1471. At the age of 16, he became an apprentice to the painter Michael Wohlgemut (1434-1519), and in 1490, he embarked on his obligatory travels, visiting places such as Basel, Strasbourg, and Venice. In 1495, he settled in his hometown as an independent master and became the most versatile German artist of his time as a draughtsman, painter, and engraver. The woodcut of the Apocalypse, published in 1498, already displays his very personal style. Around the same time, he was commissioned by Elector Frederick the Wise of Saxony (1463-1525, Luther's protégé) to create the Epiphany Altarpiece for the castle church in Wittenberg.

In 1506, Dürer undertook his second study trip to Italy, which brought him together with Gentile Bellini (1429-1507) and the young Titian (ca. 1488-1576) in Venice. The painting "Feast of the Rosary," which is now in the National Gallery in Prague, was also created there.

From 1511, he produced a large number of engravings, hand drawings, and wood engravings, among which "Knights, Death and the Devil," "Melancholy," and "Saint Jerome in the Casing," as well as the "Great" and "Little Passion," are probably the best known. This not only reveals Dürer's mastery but also his art of composing pictorial content. Just consider the engraving "Melancholy." Here, we see an angel in the form of a stately woman sitting amidst the instruments of earthly craftsmanship and research with a downcast expression. An hourglass on the wall and a bell remind us of the transience of all earthly things. A magical square of numbers may also point to the secret relationships that are valid in the infinite cosmos and yet elude the human mind. It is guarded by a tired dog beneath a truncated parallelepiped as a symbol of loyalty. A definitive interpretation of this 24x18.8 cm copperplate engraving with its mysterious content seems hardly possible - every era has its own, and Albrecht Dürer himself can no longer be asked...

Fig. 111: Albrecht Dürers „Melancholia" from 1514 (Wikimedia)

The last period of his work began in 1520 when he and his wife traveled to Antwerp to attend the accession of Charles V. Here, he painted portraits of himself and others. He painted a portrait of the painter Bernard van Orley (also known as "Raphael of the Netherlands," 1491-1542). This memorable portrait, which has adorned many a stamp, can be seen today in the "Old Masters" Picture Gallery in Dresden. In this context, it is interesting to note that for a long time, there was a difference of opinion among experts regarding the authorship of a painting, with complete disagreement as to whether it was by Dürer or van Orley.

265. The Shroud of Turin

This painting is a copy of the Shroud of Turin, which today is almost certainly attributed to Bernard van Orley, who is said to have painted it in 1516. Much more interesting - and certainly more mysterious - is the original itself, which has been kept in the Basilica of San Giovanni in Turin since 1578. It is a 4.36-meter-long and 1.10-meter-wide linen cloth on which - barely visible - a negative image of an obviously crucified man is depicted. Some consider this cloth to be the one in which Jesus Christ was wrapped for burial after he was taken down from the cross, while others believe it to be an artifact from the late Middle Ages. To cut a long story short, the issue remains unresolved. Experts cite reasons, evidence, and plausibility for both interpretations, but there is a tendency towards the "artifact" hypothesis.

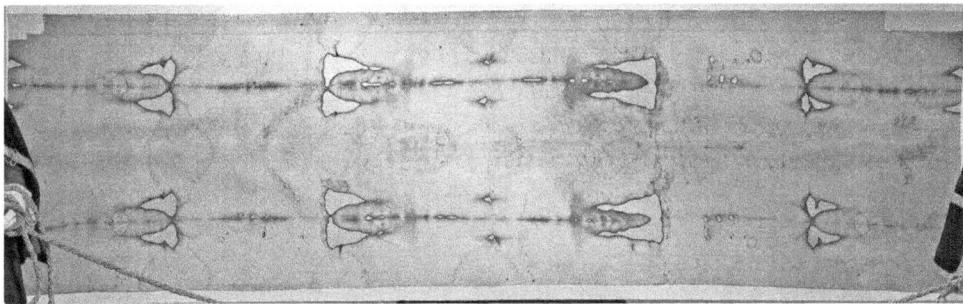

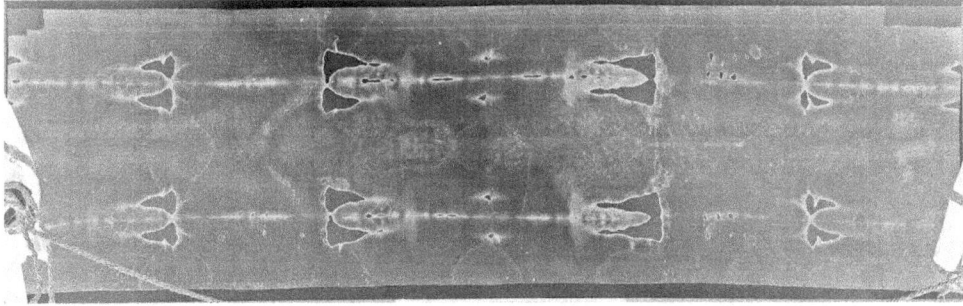

Fig. 112: Positive and negative of the Shroud of Turin (Creative Commons)

The first interpretation is highly theological, while the second relies more on objective scientific studies of the Shroud. However, the Catholic Church, which is otherwise not particularly concerned with certificates of authenticity for relics, takes a somewhat indifferent stance in this case. It refuses to classify the Shroud as a relic, as many believers request, but allows it to be venerated as an icon. Despite this, the Shroud remains one of mankind's great scientific mysteries.

266. Sindonology

In order to unravel the mystery of the Shroud of Turin, "Sindonology" was founded, an interdisciplinary science that deals with the Shroud (from the ancient Greek "sindon"). But what makes the Shroud of Turin so mysterious? At first glance, hardly anything is recognizable on it, except for the faint outline of a human being, which becomes clearer the further you move away from the cloth. But the real sensation came in 1898, when the photographer Secondo Pia (1855-1941) took the first photographs of the shroud. A positive suddenly appeared on the negatives of his photographic plates: the face and body of a tortured man with clear signs of crucifixion became visible. This discovery elevated the shroud from a religious cult object to a scientific enigma.

The first major study of the Shroud took place in 1900, but the pinnacle of scientific research occurred in 1978, when samples of the Shroud were taken to determine its age using the radiocarbon method ((item 267)). The result was shocking for many devout Christians: the linen of the shroud was dated between 1260 and 1390 AD. This dating seemed to disprove the thesis that the Shroud dated back to the time of Jesus. However, these results were later called into question, as possible contamination or later repairs could have altered the radiocarbon dates. However, the owner of the "Sacra Sindone" did not allow any further samples to be taken, meaning that the question of age can probably never be definitively clarified.

The mystery surrounding the creation of the image became even greater when modern image analysis methods discovered 3D information in the photographs. This discovery reinforced doubts that the image was created by an artist, as it has a depth not found in normal paintings. Furthermore, this hypothetical artist would not only have had to possess exceptional talent but also an understanding of negative images, which was simply unthinkable in the late Middle Ages and early Renaissance.

Another fascinating aspect is that the face on the Shroud is strikingly similar to the "classical" image of Jesus Christ as it has been depicted in church painting and iconography for centuries. Interestingly, this particular image of Jesus can first be traced back to the 14th century, suggesting that it may have been influenced by the Shroud of Turin, which is known to have been first publicly exhibited in Lirey near Troyes in 1357.

The last public exhibition of the Shroud took place in Turin in 2015. The next opportunity to see the Shroud will probably not be until 2025, provided there are no further changes. Until then, there is time to wait for further discoveries by sindonologists, which could perhaps finally shed light on the centuries-old mystery surrounding the Shroud of Turin.

The linchpin of scientific research into the Shroud of Turin has always been its exact dating. There are a number of approaches that can be used to check the plausibility of the assumed dates (year of Christ's death) and also to "absolutely" narrow down the production of the shroud in terms of time.

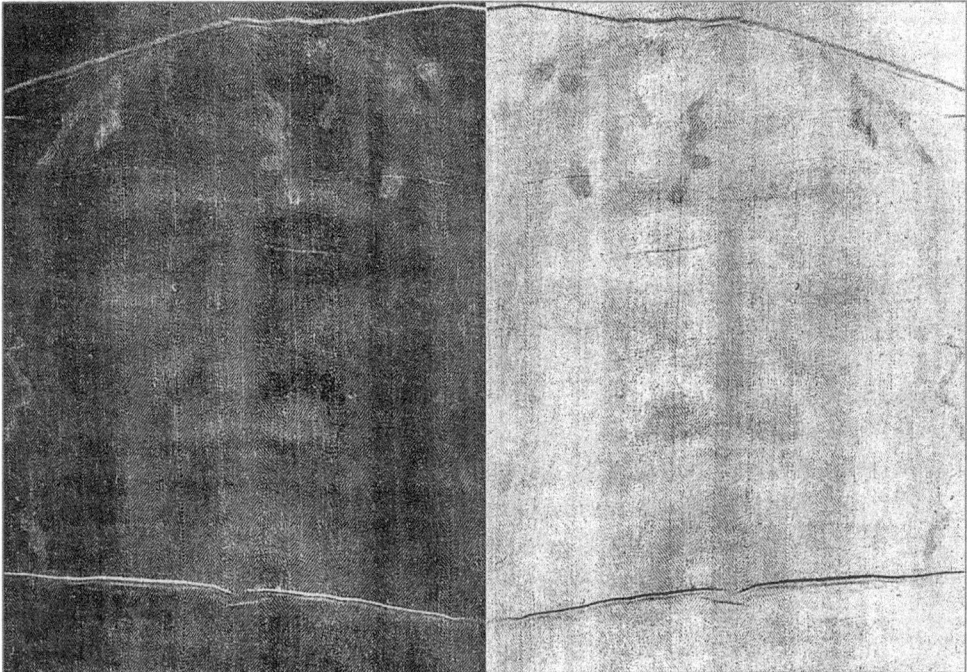

Fig. 113: Original image (negative) and the image developed from it (positive) of the face on the Shroud of Turin in a photograph by Giuseppe Enrie from 1931. It is difficult to understand how a Renaissance artist could have developed the idea of a "negative" in his time. (Wikimedia)

267. Radiocarbon dating

One of these is radiocarbon dating, which is now used almost as a standard in archaeological and historical research due to its high reliability and time resolution. Its principle is not difficult to understand. The key terms are isotopes, radioactive decay, and half-life.

The first thing to know is that the element carbon - a fundamental basis of all life - occurs in nature in three isotopes, which differ in the number of neutrons they contain, and one of which, C14, is unstable with a half-life of 5730 years. The number "14" here refers to the mass number, i.e., the sum of the number of protons (always 6 for carbon) and the number of neutrons in an atomic nucleus. The isotopes C12 and C13, on the other hand, are stable; however, we only need to be concerned with the isotope C12 here. Because the isotope C14 is constantly being formed in the Earth's upper atmosphere as a result of cosmic radiation, an equilibrium has been established between new formation and decay, so the concentration of this isotope in the Earth's atmosphere essentially remains constant over long periods of time. This concentration is approximately one C14 atom for around one trillion C12 atoms. These atoms are primarily bound in molecules, which are known to be a starting material for photosynthesis in green plants. As long as a plant grows and thrives, it absorbs both C12 and C14 in a ratio identical to the isotopic composition of the air, fixes the carbon it contains, and uses it to build up its own substances, such as various carbohydrates. The same C12/C14 ratio will also occur in animals that eat plants (e.g., antelopes or cows) and animals that eat animals (e.g., lions and humans). A continuous supply of food also keeps the C12/C14 ratio constant during their lifetime. This changes when a plant or animal dies. Then the radioactive clock starts ticking because the C14 atoms decay slowly, and the C12/C14 ratio changes in proportion to the time that has elapsed since then. By measuring the ratio between C14 and C12 in a sample, the age can be determined - the less C14, the older the sample.

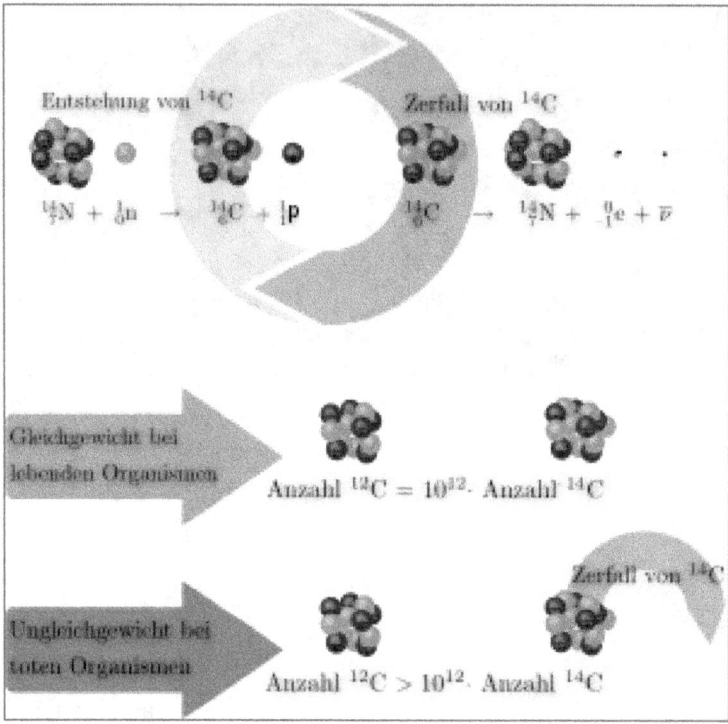

Fig. 114: The carbon cycle of C14 (Wikimedia)

The simplest method to determine the carbon isotope ratio is to hold a Geiger counter to it and measure the radioactivity of the sample. However, this requires large quantities of corresponding organic material. For this reason, this method usually only plays a role in physics practicals in natural science courses. Modern methods (of which there are several) are mainly based on mass spectrometry, such as the AMS method (AMS stands for *Accelerator Mass Spectrometry*). In this method, individual atoms of a sample that has been converted to a gaseous state are ionized and accelerated by electric and magnetic fields, thereby separating them into their isotopes (which have different masses) and determining their proportions. Tiny amounts of the carbon to be analyzed are usually sufficient to determine the important C12/C14 ratio and thus ultimately the age of the sample. In the case of the Shroud of Turin, a total of 12 samples were obtained from a 7 cm x 10 cm strip of fabric and an age of origin (when the linen plants for the shroud were harvested) of 691 ± 31 years was calculated from all the dating carried out in various research institutes. The range of this method, developed by Williard Frank Libby

(1908-1980, Nobel Prize 1960), is a maximum of ~10 C14 half-lives and thus only extends to the middle Pleistocene. The radiocarbon method is therefore not suitable for dating dinosaur bones.

268. Dendrochronology

The special methods of dendrochronology do not go quite as far back in time, but at least as far back as ~10,000 BC. The term is derived from the Greek word "dendron," which means "tree" - and it can be used to date wooden buildings, but also ships whose wooden planks and keel timbers are still well preserved, in some cases back to the year in which the corresponding trunks were felled. The so-called annual rings, which anyone who has seen a smoothly sawn tree trunk is familiar with, are used for this purpose. In temperate regions, a tree produces a more or less wide ring every year, the thickness of which reflects the respective growing conditions. Since the living conditions are roughly the same for all trees of a species in a certain area, all these trees in the corresponding region have roughly the same characteristic sequence of narrow and wide annual rings.

Fig. 115: Annual rings (Author)

By placing trunk slices of trees with overlapping lifespans next to each other, the annual rings of these trees can be synchronized, so to speak. The idea and first practical implementation of this method were conceived by an astronomer named Andrew Ellicott

Douglass (1867–1962), who wanted to research solar activity in the past rather than archaeologically valuable artifacts. Tree rings synchronously show the same eleven-year period as the sun shows its sunspots. His method was later adopted by archaeologists and developed into a tried-and-tested means of dating. Today, they can draw on a complete series of tree rings recorded in various tree ring tables, which - as in the case of the Hohenheim ring calendar - now go back more than 12,000 years into the past. So if you find the remains of an old Slavic rampart somewhere in Lusatia, you will try to dig up a tree trunk from the former fortification that is still reasonably intact in the marshy ground. If the tree rings can be identified from its cross-section, then the archaeologists have struck gold. This is because by comparing it with a suitable "ring calendar," it is now possible to determine the exact date, down to the year, when the tree was felled for the post.

By the way, the plum tree that my brother sawed down without being asked is 67 years old. I know that. I counted its rings...

269. Regional climate history

The idea of using tree rings to reconstruct solar activity and the climate history of the last 1000 years is of great importance today. At a time when "man-made climate change" is being propagated ever more loudly by politicians of all stripes and exploited by the media, dendrochronology offers a valuable, objective method for analyzing past climate conditions. This method makes it possible to precisely document both natural climate fluctuations and regional changes over long periods of time.

While "natural climate change" is well documented scientifically, "anthropogenic climate change" is often used as a political tool to justify far-reaching changes - from the energy transition to the disfigurement of old cultural landscapes in Europe. The analysis of historical climate data thus becomes not only a scientific but also a socially relevant task[60].

270. Milankovic cycles

From this perspective, it may be useful to understand the natural cycles of ice ages and interglacial periods, known in technical jargon as "glacials" and "interglacials." This is because "astronomical" cycles play a role here, which humans cannot change, even if a party executive were to push through a corresponding resolution at a more or less green party conference. The Earth's global climate is primarily determined by the energy input from solar radiation into the atmosphere. It is the primary source of energy that drives the atmospheric dynamics we experience daily as "weather." Therefore, even small changes in the solar constant (the energy that reaches us per square meter per

---

[60] See, for example, Reichholf, Eine kurze Naturgeschichte des letzten Jahrtausends. Fi-scher, Frankfurt am Main 2007

second from the sun at the top of the atmosphere) in the per-mille range can lead to serious changes in the climate system. The paleoclimate of the last million years has been very well researched in this respect, with the Earth undergoing several more or less pronounced glaciation cycles separated by interglacial periods.

The questions that arose regarding the causes of these cycles formed the so-called "ice age problem," which was the subject of intensive research at the beginning of the 20th century. In this context, Alfred Wegener's (1880-1930) study of the Greenland ice sheet and Vladimir Köppen's (1846-1940) reconstruction of the climate from the end of the Pliocene (around two million years ago) should be mentioned. An explanation for these climate cycles and their observed lengths was proposed in the 1930s by the Serbian scientist Milutin Milanković (1879-1958). His theory, which has since been referred to as the "Milanković theory," is widely recognized today. It attributes cyclical climate changes to variations in the Earth's orbit around the sun, which are linked to changes in the solar constant.

As is known from celestial mechanics, the undisturbed orbital shape (ellipse) and the position of the orbital plane (given by the vector of orbital angular momentum) each represent invariants of motion in a 1/r potential. This means that the orbital parameters (major semi-axis), (numerical eccentricity), and (orbital inclination to the ecliptic, by definition 0° for the Earth's orbit) are stable variables - provided that there are no gravitational disturbances from other planets. However, this is not the case in the solar system. The large planets of the outer solar system, in particular, have a strong influence on the Earth's orbit, which can be quantified in the short term but cannot be predicted in the long term (keyword: "deterministic chaos"). Using perturbation theory, these changes in the orbital parameters can be investigated with sufficient accuracy to establish a correlation between changes in the Earth's orbit and certain climate parameters (such as the mean annual temperature). Such calculations were first carried out and published by Milutin Milanković.

There are essentially three variables that are relevant in this context: the eccentricity of the Earth's orbit, the inclination of the Earth's axis relative to the ecliptic, and the lunisolar precession. The eccentricity of the Earth's orbit (i.e., the degree of deviation of the orbital ellipse from an ideal circle with a radius of 1 AU) can fluctuate due to the noncentral gravitational perturbations of the other planets. These changes are non-harmonic, representing the superposition of different "frequencies," which lie in a range between 90,000 and 105,000 years, with a clear center of gravity at approximately 93,000 years. Additionally, there is another period with a duration of approximately 400,000 years. Large eccentricities cause a smaller perihelion distance and a larger aphelion distance, which leads to strong seasonal fluctuations in energy irradiation. Changes in the eccentricity itself (i.e., the change in the shape of the orbit from an (almost) ideal circle to an ellipse and back again) lead to fluctuations in the solar constant of around 1 per mille (approximately 0.7 W/m^2), which corresponds to fluctuations in the Earth's equilibrium temperature of a few degrees. Currently, the solar constant is 1325 W/m^2 at aphelion and 1420 W/m^2 at perihelion. A change in the shape of the orbit therefore causes a change in the temperature contrast over the course of a year as well as cyclical changes in the average annual temperatures in the cycle of eccentricity fluctuations with periods of approximately 100,000 or 400,000 years. If, during an epoch of increased orbital eccentricity, winter in the northern hemisphere coincides with perihelion (as it does now), then the "winter" season is shorter than in the opposite case (winter in the northern hemisphere, Earth at aphelion) due to Kepler's second law. This is why northern winters today are shorter (currently seven days) and warmer than southern winters. During times of low orbital eccentricity, however, the differences in the lengths of the seasons disappear.

A second variable that can change is the inclination of the Earth's axis relative to the Earth's orbital plane. This variable, known as the "obliquity of the ecliptic" (epsilon), is currently 23.5° and can fluctuate between 21.9° and 24.2°. The period of this nutation movement is approximately 41,000 years. The reason for this is the gravitational interaction of the sun and the moon with the asymmetrical mass distribution of the Earth. Nutation primarily influences the annual cycle of solar radiation entering the uppermost layers of the atmosphere. At high geographical latitudes, this can amount to an annual average of up to 16 W/m^2. Although the obliquity of the ecliptic has no influence on the total energy input over a year, the latitude-dependent temperature contrast (difference between maximum and minimum temperature) increases in both hemispheres as the inclination of the axis increases (remember, with an epsilon value close to 0°, there would be no seasons). Among other things, this process impacts the formation and melting of polar ice sheets. In periods of low temperature contrast, the polar ice sheets tend to be reduced, only to grow again in times of high temperature contrast. A correlation is therefore expected between the ice cover of the poles and the epsilon value of the Earth's axis.

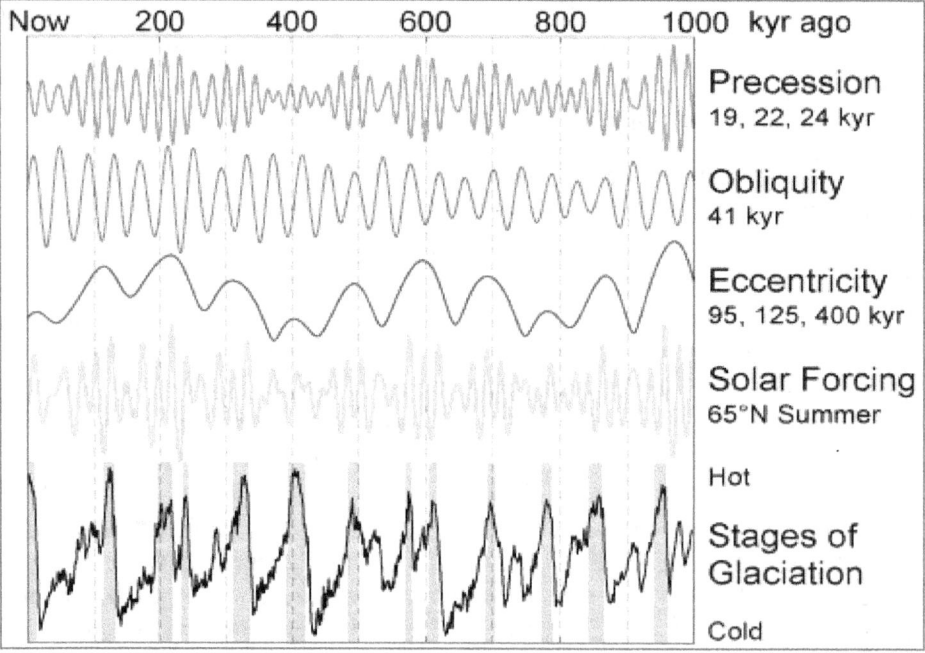

Fig. 116: The superimposition of different astronomical cycles correlates with the occurrence of ice ages and interglacials on Earth

The effect of lunisolar precession on the Earth's climate is related to changes in the eccentricity of the Earth's orbit. It determines the time of year when the Earth is closest to the sun, i.e., at perihelion. As previously mentioned, today the Earth passes through its closest point in orbit to the sun during winter in the northern hemisphere. 11,000 years ago, it was exactly the opposite. At that time, winter in the northern hemisphere was significantly longer than it is today. Due to the apsidal rotation of the Earth's orbit, the precession period is slightly shorter than the Platonic year of 25,700 years in terms of the change of seasons, at around 21,000 years. It has no influence on the Earth's net energy input from the sun but can change the strength of the annual temperature cycle.

The astronomical cycles for orbital eccentricity (approximately 100,000 and 400,000 years), the tilt of the Earth's axis (approximately 41,000 years), and the apsidal rotation (approximately 21,000 years) are therefore responsible for the total amount of radiation that the Earth receives from the Sun, as well as its spatial and seasonal distribution. Even relatively small changes in the energy input can trigger processes in the atmosphere that can lead to a relatively strong restructuring of the climate system. According to recent research, the ice ages at the end of the Pleistocene cannot be explained by the Milankovitch cycles alone. However, they are very strongly correlated with them. This means that astronomical causes, in combination with complex feedback mechanisms, can strongly influence the Earth's global climate.

The exact effects of solar constant variability on the Earth's climate system are still insufficiently researched, although solar energy is the primary energy source for all atmospheric processes. This is particularly true for the microvariability of solar luminosity over different solar activity cycles (of which the 11-year cycle is one).

However, more recent studies, somewhat apart from politically ambitious climate research, seem to indicate that the simplification of a constant solar luminosity (expressed by the solar constant S at a distance of 1 AU from the Sun) is inadmissible. The possible range of variation of S can certainly cause climate effects that are of the same order of magnitude as the postulated anthropogenic influences.

Milankovitch cycles also play a major role in the climate development and climate history of the planet Mars. Since this planet lacks the stabilizing effect of a large moon on the axial tilt, the climatic effects are even more extreme compared to Earth.

271. Martians

Of course, this was not yet known in the middle of the 19th century and Mars was imagined as a true paradise for its inhabitants, the "Martians" (who were not yet thought of as "small and green" at the time). By consistently applying the principle of similarity, an editor of a gazette from 1859 ("Die Fackel - Literaturblatt zur Förderung geistiger Freiheit") was able to deliver the following description of the "Martian inhabitants" to his astonished readers as a matter of course:

"As already reminded, it may be as warm on average on Mars as on Earth. But since the summer half-year lasts almost as long as a whole year on Earth, the heat in summer and the cold in winter must be much higher than here. Some types of fruit will produce two or three harvests every summer. Noble wines and tropical fruits ripen in the long summer, their wines acquire a fire, a spicy smell and flavor that we cannot imagine. (...)

Since every body is ten times lighter than on Earth, it must be very easy for the inhabitants of Mars to move themselves and large masses of bodies from one place to another. Perhaps they do not even need highways for this, their main highway is the air. Their balloons carry 10 times more than ours, they do not need such a fine, expensive type of air to fill them as we do. As we fall through 15 feet in the first second, they only fall through 6 feet; a fall from a considerable height therefore brings danger to few. The art of sailing in the air was closer to them, was invented earlier, developed more eagerly and with more rewarding success. Trade and commerce are in full bloom; people are busy exchanging the products of the various warm and cold countries. Journeys to foreign parts of Mars, journeys around the whole of Mars, which is five times smaller than ours, are something very common. Every inhabitant of Mars knows all the cities and curiosities of his planet from his own experience. On Mars, on the other hand, because oral communication is so easy, because all business can be transacted through personal meetings, there is perhaps no need for the art of writing or bookmaking. (...)

Since Mars is 5 times smaller than the Earth, its people are presumably also 5 times smaller, dwarfs by our standards, but well-built, even winged, so that they can easily take to the air".

Almost half a century later, some astronomers believed that they had discovered large irrigation systems on this planet, which appeared to be predominantly dry in terms of color, the ominous "Martian canals." Thus, it is certainly not surprising that at the beginning of the twentieth century, a large part of the educated middle classes in Europe and North America came to agree that there was a highly developed civilization on Mars that was struggling with an acute shortage of water and therefore had to build huge canals to channel the meltwater from the poles to the warm equatorial regions. In some circles, the question of how to establish contact and communicate with them was seriously debated. The proposals - often put forward by reputable scientists - seem rather quaint to us today. For example, there was a serious suggestion that large clear-cuts should be made in Siberia in the form of letters and words that could be recognized by Martian astronomers with their telescopes.

Fig. 117: You don't seem to be really welcome on Mars, as this image from the US space probe Curiosity clearly shows...

272. War of the worlds

It is therefore not surprising that a special media event (as we would say today) took an unplanned course on Halloween day in 1938 (even if the drama was later greatly exaggerated). On this day, the radio play "Invasion from Mars", based on Herbert George Wells' (1866-1946) novel "War of the Worlds", was broadcast on the American radio station CBS. It was arranged so realistically by the famous actor and director Orson Welles (1915-1985, "Citizen Kane") that a mass panic broke out among listeners in the eastern states. What had happened? All the people who were listening to the comedy program *"The Chase and Sanborn Hour"* were surprised by an interruption with an important announcement. First an astronomer reports huge gas explosions on Mars, then a little later a radio announcer says that a huge flaming object has fallen near Grovers Mill near New Jersey. Then music follows again until a reporter on the scene reports: *"Good God - something is crawling out of the shadows like a gray snake. It looks like tentacles. The body is as big as a bear and shines like wet leather. But the face - it - it's indescribable!"* As a result, the phones started ringing in most police stations on the American East Coast and worried citizens wanted to know from the equally surprised officers what was going on and what they should do. In some places, panic attacks are even said to have occurred, but this is only confirmed in a few cases. Because hardly anyone knew that these were well-planned interspersions in the ongoing program - only those who had consciously caught the brief hint at the very beginning...

Fig. 118: The book "The War of the Worlds" from 1897, which is still worth reading today, was published in a first German edition in 1901. Here is the 1906 edition.

The really ingenious thing, however, was that Orson Welles placed the fictitious news clips in the CBS program just as the rival channel NBC went into a commercial break. All those who "flipped" (as we would say today) to CBS while bypassing this commercial break suddenly found themselves in a completely different movie, in which many could not easily distinguish truth from fiction. The radio play found fertile ground, because war was in the air. The press and radio had been preparing the population for months with worrying reports from the Far East (Japan and China) and Europe (Germany). Naturally, quite a number of citizens thought it was quite possible that aliens from Mars would visit America with flashes of light and poison gas attacks and that they would somehow have to get themselves and their families to safety.

What H.G. Wells imagined in his fantasy can be seen today in the cinema or on the television or computer monitor, because Hollywood has of course made several films of this piece of "world literature". Only the 2005 version, which was directed by Steven Spielberg and set in the present day, is worth mentioning here due to its impressive array of special effects.

Astronomers in the 1930s were well aware that there could be no "Martians" on Mars, as it was now known that Mars only had a thin, non-breathable atmosphere consisting mainly of carbon dioxide and that it was also quite cold there. However, the existence of microbial life is still considered quite possible today and the targeted search for traces of "life" on Mars is still the primary mission objective of unmanned Mars landers.

273. Life on Mars?

When, in the 1970s, the technology was finally ready for landing probes on the red planet (Viking 1 and Viking 2), one of their tasks was to search for any traces of life on the surface of Mars. A number of extremely sophisticated experiments were carried out for this purpose, but the results were difficult to interpret. Almost 50 years have passed since then, and our knowledge of Mars has exploded. However, evidence of "life" has still not been found. In 1996, there was a brief glimmer of hope that we might have made a discovery after all. The then US President Bill Clinton took the opportunity to inform the world at a special press conference that NASA had found microscopic structures that appeared to resemble fossilized bacterial life forms when analyzing a meteorite found in Antarctica that was clearly from Mars. In summary, this interpretation could not subsequently be upheld, although the elongated structures, which can be clearly seen on an electron micrograph of a fracture surface, really do look like mineralized bacterial cells that have shrunken. However, scientists now believe that they can explain these structures in purely inorganic terms.

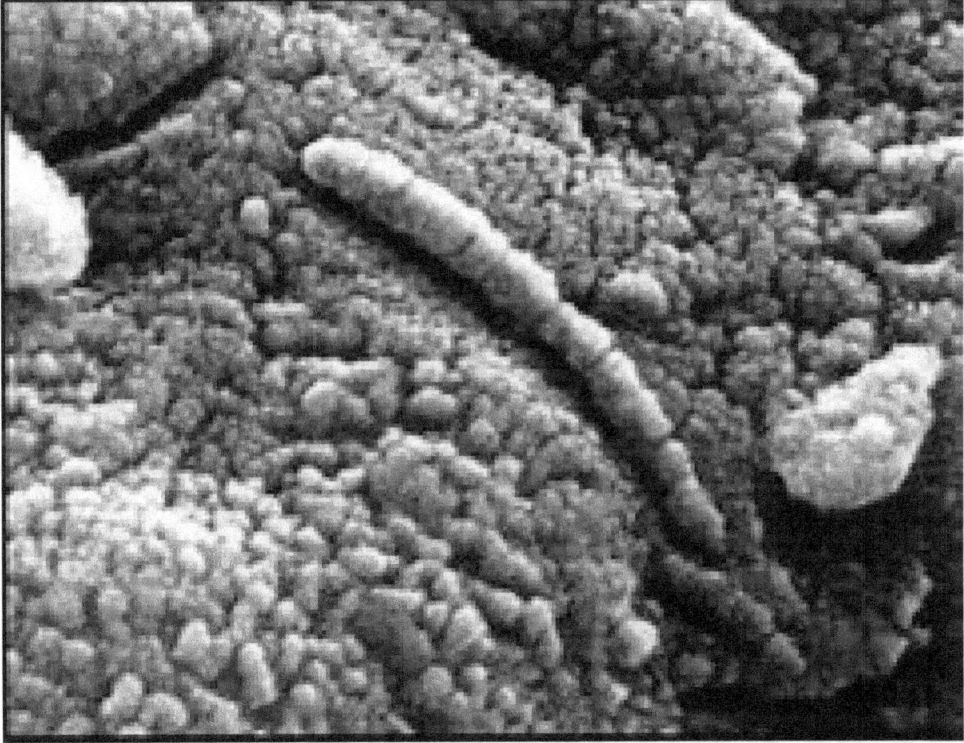

Fig. 119: Bacteria-like structure (here colored green) on an electron micrograph of a sample of the Martian meteorite ALH84001. It is now known that this is an inorganic structure.

A major difficulty with such investigations is always the possible contamination of a sample with terrestrial bacteria.

274. Earth - the planet of bacteria

Biologically speaking, our Earth is a planet of bacteria. The first form of life to emerge on Earth over 3.6 billion years ago was a primitive bacterium. The last living being that will likely end the era of life on Earth in the distant future (because the sun is getting bigger and hotter) will also be a bacterium. Archaea and bacteria are the organisms on Earth that are best able to adapt to extreme habitats. Some of them would even have a good chance of surviving for many generations on today's Mars.

If you think more closely about the role of bacteria on our planet Earth, you will come to the conclusion that, according to all reasonable criteria, bacteria (more precisely pro-karyotes) are the predominant form of life. This applies both to the number of individuals as well as the biomass contained in them, which is at least as large as the biomass of all "non-bacteria," i.e., all animals and plants together that currently live on Earth. With-out bacteria, humans would not be viable (and would be a lot lighter). A few hard-to-believe figures are perhaps of interest here. A human body consists of around 10 trillion body cells and around 100 trillion bacterial cells. Although the latter are much smaller than a typical body cell, they still weigh a total of two to three kilograms in an adult. Most bacteria reside in the gastrointestinal tract and are essential for our lives. However, as everyone knows, our skin and even our teeth are also densely colonized by bacteria. Ef-fective digestion of our food would be completely impossible without the assistance of bacteria. The well-known intestinal bacterium Escherichia coli is just one example.

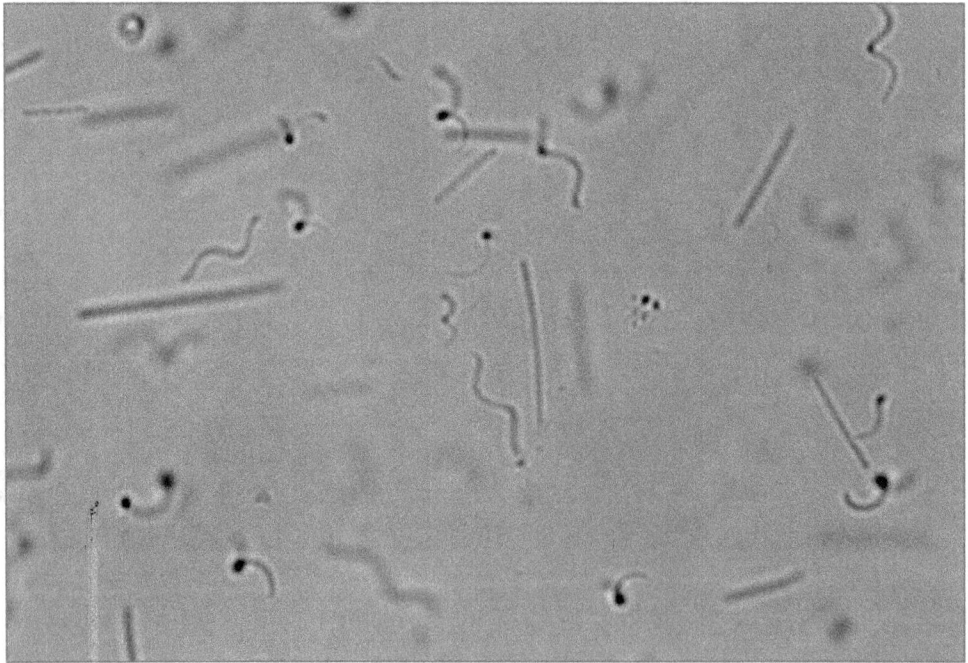

Fig. 120: Even a light microscope can be used to visualize bacteria, which occur in a wide variety of forms (e.g. as bacilli, cocci, spirochetes, etc.). (Author)

Genetic researchers have also discovered that the microbes that colonize humans contribute more than three million genes that add vital functions to the human genome. All these microorganisms - likely several thousand species (difficult to say because the term species is unclear due to the possibility of horizontal gene transfer in prokaryotes, which include archaea and bacteria, and is therefore ambiguous in contrast to eukaryotes) - form the so-called "microbiome" of humans. Such a microbiome differs from person to person, and each age also has its own bacterial colonization. Today, scientists are even able to group microbiomes according to certain criteria so that "enterotypes" can be defined in this way. Different people can then have different enterotypes of microbiomes, which can even be consciously changed, for example, by switching their diet from primarily steak to primarily spinach and salad.

275. Deadly bacteria

Unfortunately, among the many bacteria, there are also a few that are after our lives. These include (since they were very prominent in the past) the bacterium *Yersinia pestis*, the plague pathogen transmitted by fleas, the bacterium *Corynebacterium diphtheriae*, which caused the incurable, child-killing diphtheria over 150 years ago, and, as a final example, the anthrax pathogen *Bacillus anthracis*, which has had an inglorious career as a bioweapon. The fact that these pathogens have nevertheless largely lost their terror is due to the work of many microbiologists and researching physicians, of whom Robert Koch (1843-1910), Emil Adolf von Behring (1854-1917) and Alexander Fleming (1881-1955) are particularly worthy of mention.

276. Penicillins

The latter found "the" weapon against bacterial infections of all kinds more by chance and thus indirectly saved the lives of countless people. I am referring to the discovery of the penicillin class of substances. In 1928, he observed that bacteria-free fringes formed around certain molds that had colonized bacterial cultures in Petri dishes. This was a major breakthrough in medicine, as countless soldiers lost their lives during the First World War not directly as a result of combat, but due to bacterial infections. Fleming was awarded the Nobel Prize in Physiology or Medicine in 1945 alongside Ernst Boris Chain (1906-1979) and Howard Walter Florey (1898-1968) for their work on penicillin.

The actual significance of these antibacterial substances (antibiotics) lies in the fact that they inhibit the growth of the bacterial cell walls after cell division, causing the bacteria to burst due to their increase in volume. The industrial production of these active substances, which are obtained from certain molds, has given doctors an extremely effective remedy that has alleviated the fear of many bacterial infectious diseases. Unfortunately, however, this sword is beginning to dull. Just as with many other antibiotic substances, there are now more and more pathogenic strains of bacteria that have become resistant to penicillins in particular (e.g., penicillin G) or to various antibiotics in general. Sometimes one treatment helps when another no longer works.

277. Multi-resistant germs

If, on the other hand, no antibiotics are effective at all, we speak of a multi-resistant bacterial strain. These strains, to be honest, are virtually home-made. The development of such resistance results from Darwinian evolutionary mechanisms and can be easily explained within the framework of evolutionary biology. It teaches us that organisms best able to adapt to external conditions are also the most capable of surviving and, therefore, reproduce the most. So if antibacterial substances are used too often or not strictly as prescribed (e.g., in animal husbandry), there will always be a few bacterial mutations that survive their effects. With this advantage, they form a sub-population in which this resistance is inherited, which then slowly displaces all other sub-populations lacking this characteristic. And this is precisely what will become an extremely serious problem in the future. In Germany alone, around 30,000 to 40,000 people already die every year from multi-resistant "hospital germs" - and the number will rise if we do not succeed in developing new classes of antibacterial substances. Research in this area is running at full speed - with some successes. Around seven large and more than 20 small and medium-sized companies worldwide are currently working on new antibiotics (particularly broad-spectrum antibiotics) and other antibacterial drugs. But whether it will be possible to win the race between the hare (us) and the hedgehog (the multi-resistant germs) in the long term is still questionable. Strictly speaking, only the extremely responsible use of existing antibiotics and results-oriented research can help here. However, pharmaceutical research is extremely complex and costly. As a result, even today, there are still no drugs for many very rare diseases - simply because there is no profit to be made.

278. Orphan drugs and Ebola

But this is slowly changing due to legislative measures that encourage or even force pharmaceutical companies and taxpayer-funded research institutions to look for such orphan drugs. An instructive example in this case is the development of a vaccine against a hemorrhagic fever that usually occurs only rarely in Africa - specifically, Ebola, which is named after a river in the Congo. Anyone who falls ill with this viral infection (no antibiotics are primarily helpful here - at most against infections that may develop as a result of the disease) suffers from extremely high fever, severe fluid loss, internal bleeding, and even organ failure, which increases the death rate of people infected with it to up to 90%. As a viral disease, it is also highly contagious, which makes it even more difficult to treat. On the other hand, there have only been relatively small local outbreaks in Africa in the past, affecting only a few dozen to a few hundred people, mostly poor individuals - in other words, not exactly a clientele to encourage pharmaceutical giants to invest a lot of money in the development of an effective Ebola vaccine. However, this quickly changed with the explosive outbreak of the disease in some West African countries in 2014 and with it the danger that it could also be introduced into Europe or North America. Although a number of other events have since pushed Ebola out of the headlines, the development of at least an experimental vaccine against it has been fairly quick. And in the case of filoviruses,[61], this was not easy, and the researchers had to dig deep into their molecular biological and biochemical bag of tricks.

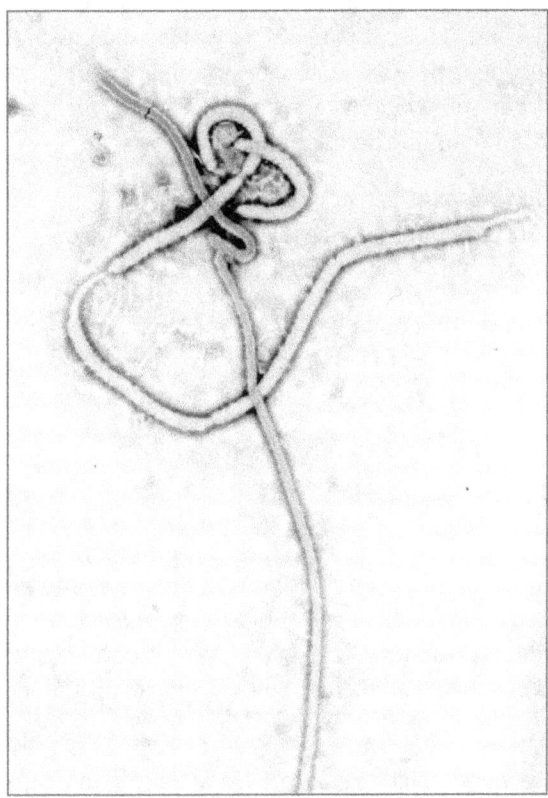

Abb- 121: If you look at this virus under an electron microscope, you are looking at the Ebola virus... (Wikimedia, Public Health Image Library (PHIL) of the Centers for Disease Control and Prevention)

The aim of a vaccine against filoviruses is to induce the production of so-called antibodies[62] against them in the human body. The basic principle is known from smallpox, a disease that has since been eradicated by vaccination. It is called "deception": the immune system is tricked into believing that it has a specific viral infection and induces it to produce corresponding antibodies. In the case of the Ebola virus, people are injected with viruses whose genetic material has been altered so that they merely look like a real Ebola virus but do not have the same effect. The pathogen that causes chimpanzee rhinitis has proven to be a suitable vehicle (it is not combated by human immune cells),

---

[61] Filoviruses are a family of viruses that form long, thread-like virus particles and can lead to severe hemorrhagic fevers in humans, such as Ebola and Marburg.

[62] Antibodies are proteins that are produced by our immune system to recognize and neu-tralize specific antigens of pathogens such as bacteria or viruses.

whose viral envelope has been equipped with special proteins using modern molecular genetic methods, which the Ebola viruses also carry in their envelopes. The antibodies that form in humans in response to this virus are then just as effective against real Ebola viruses. Initial clinical trials have been extremely satisfactory, as you can read, and the first effective Ebola vaccine called "Ervebo" was approved in 2019. Nevertheless, we must not lose sight of the fact that Ebola, with its approximately 22,000 deaths (2014 to April 2015), has relatively low case numbers compared to the poverty-related diseases malaria, tuberculosis, and AIDS, with their more than six million deaths worldwide. Only the latter, probably because they are no longer a problem in the highly developed industrialized nations, are not or only rarely reported on. Yet they are the real "killers" (alongside civil wars, etc.) of the African population.

Incidentally, infectious diseases can have immense political consequences. This can be observed in Guinea, Liberia, and Sierra Leone, which were recently hit by Ebola, in different ways depending on how efficient the respective healthcare systems were.

In the past, one disease in particular has decisively changed the political landscape in the West: the plague. In contrast to smallpox, it still exists today but is no longer a major problem in countries with a developed healthcare system and higher hygiene standards.

279. Plague pandemics

This was particularly different in the Middle Ages and early modern times in Europe, when veritable "waves of plague" swept through the countries and wiped-out large parts of the population, especially in the cities. Even today, ossuaries containing the remains of tens of thousands of plague victims can still be found in the catacombs of many cities (e.g., Vienna, Paris).

As the connection between plagues and rats (especially in the cities - Constantinople, for example, was long regarded as a kingdom of rats) was not recognized at the time, the "Black Death" was mostly regarded and accepted as a "punishment from God." This only changed in the 15th and 16th centuries, when more and more doctors recognized that it was a "contagious" disease transmitted by rats, from which one could not protect oneself by fleeing to the countryside. Anyone interested in this subject is familiar with the "plague masks" with their bird's beak-like extensions, which were filled with fragrant herbs to at least somewhat reduce the smell of the pestilence. The quite successful quarantine procedure was also developed to at least contain local outbreaks of plague. In 1377, the council of the city of Ragusa (now Dubrovnik) decided that all people and goods coming from places where the plague was raging were to be quarantined on a small island off the city for a month. There are similar reports about Venice and its "plague island."

Fig. 122: In 1585, 648 people died of the plague in Schlocknau (Schluckenau, Sluknov, North Bohemia) and 465 in the surrounding (parish) villages. This plaque is located on the west side of St. Wenceslas Church from 1711 and tells of one of the great catastrophes of that time. (Author)

280. Isaac Newton and his axiomatic mechanics

Otherwise, if you could afford it, you fled the city for the countryside, as Isaac Newton (1643–1727) did in 1665 when England was struck by the Great Plague. At home with his parents, as a student about to graduate from Trinity College, Cambridge (which was temporarily closed due to the plague), he used the time to think about mathematics, physics (especially optics), and philosophy, laying the foundations for his fame as one of the most important scientists of modern times. In a sense, Newton can be described as the founder of mathematical physics, just as Galileo Galilei (1564–1642) can be described as the founder of experimental physics. Both loved experiments, and both recognized that mathematics is the true language of physics. It is therefore not surprising that Newton made ample use of it in his major scientific work, *Philosophiae Naturalis Principia Mathematica*, published in 1687, albeit in a way that takes some getting used to for us today. In his book *Newton's Principia for the Common Reader*, the famous astrophysicist Subrahmanyan Chandrasekhar (1910–1995) explained Newton's work in his inimitable way for physicists like us, who have mastered differential and integral calculus (also invented by Newton!) but are no longer accustomed to thinking in terms of proportions.

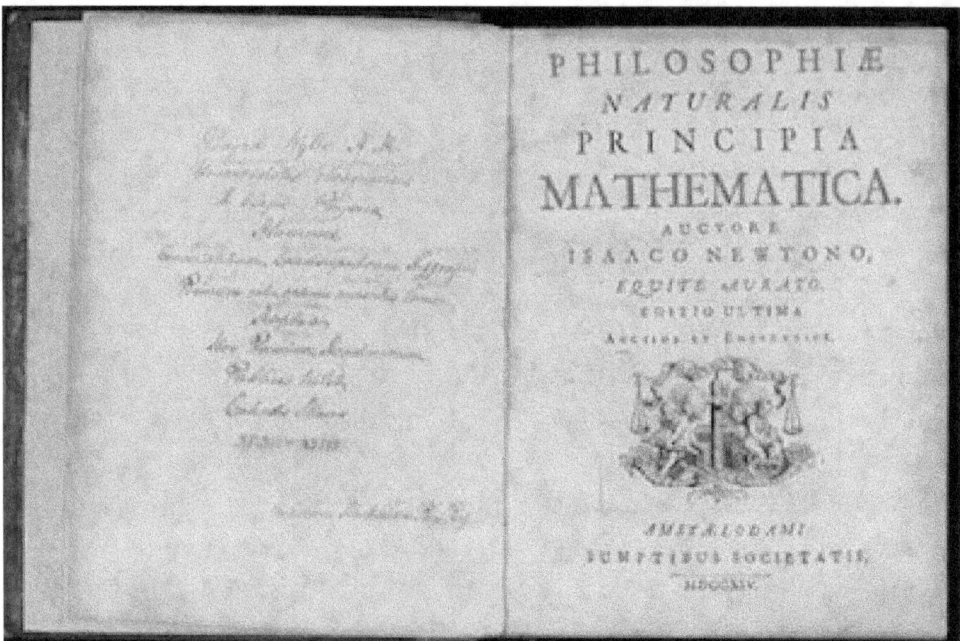

Fig. 123: One of the truly important books in the history of mankind

To construct his "mechanics," Newton used a method that goes back to the Greek mathematician Euclid (3rd century BC) (keyword "Euclidean geometry," whose parallel axiom we have already briefly touched on). Like Euclid, he started with a few principles that were recognized as true and independent of each other. They include a mathematical equation for the movement of a body (equation of motion), the Galilean principle of inertia, and an assumption about the force that describes the influence of two bodies on each other (action = reaction), and with the necessary mathematical knowledge, he was able to deduce exactly what the movement of a body must look like depending on its initial conditions. This type of mathematical description of nature, which is based on a carefully selected set of basic assumptions supported by experiments, is known as axiomatic reasoning. However, it must be borne in mind that, just as in mathematics, axioms in physics can never be formally proven, but (in contrast to mathematics) can be verified or falsified through observation and experimentation. Incidentally, this is what happened to Newton when Albert Einstein was able to show at the beginning of the 20th century that his axioms are no longer valid at speeds approaching the speed of light. For all "everyday movements," however, Newton's theory is completely sufficient, as demonstrated by the countless applications of technical mechanics alone.

But let's return to the role of mathematics in physics. As in mathematics, it is also possible in physics to produce new statements that are true within this axiomatic system by correctly applying mathematical rules, starting from an axiom system that is free of contradictions. For example, after introducing the law of gravity, Newton quickly succeeded in deriving Kepler's three laws. A good student today might need just one A4 page to do this, whereas Johannes Kepler had to spend years of intensive calculation and thinking before he was able to derive them from Tycho Brahe's observational data and finally write them down. It is well known that one of the great strengths of the mathematical method is the ability to derive new statements from already proven ones, as this makes it possible to predict new outcomes within a theory. In contrast to pure mathematics, where proven derived statements ("theorems") are inherently true, this is not necessarily the case in physics. Therefore, the rule applies here that theoretical predictions must always be verified experimentally.

281. The Higgs boson

A prominent example of this is the Higgs boson, postulated purely theoretically in 1964 by Peter Higgs (1929-2024), which follows from the standard model of elementary particles. In his case, around 50 years had to pass before the experimental foundations in the form of a huge particle accelerator (LHC, Large Hadron Collider) and the corresponding detector technology (ATLAS) were available to test this theoretical prediction experimentally. And the result was positive, not only to Peter Higgs' delight...[63]

282. Reality and imagination

Mathematics is therefore not only the language in which the basic laws of physics are formulated but also the language from which other laws are derived. It is also the language used to arrive at new hypotheses and insights about nature that are, at least in principle, accessible to experimental testing.

Another aspect of the mathematical description of nature must also be pointed out. We usually describe the world with terms that evoke "images" in our minds. These "images" come from our experience of a "middle" world, whose typical scales can be measured in millimeters to kilometers. However, we also use these "images" when, for example, we mentally venture into the microworld by imagining atoms, and even elementary particles, as small spheres. Typically, a non-physicist may even consider this "image" to be a real, adequate conception of micro-objects. However, this is not the case, as quantum mechanics teaches us. Ideas from the world of medium scales will not get you very far when it comes to superposition, entanglement, or wave-particle duality. Here, we need completely new structures and concepts that are far removed from our usual imagination. These, in turn, come from abstract mathematics (think of the abstraction of Hilbert space or the wave function), and they help us navigate the world of quanta with complete logical clarity. In this context, mathematics proves to be the most suitable language for describing the realm of the unimaginable to humans.

Now, let's return to the "axioms" that the famous Greek mathematician Euclid introduced around 2300 years ago as the basis for a mathematical theory (in this case, the geometry of the plane). His "Elements" are still regarded today as an exemplary mathematical theory, and it is particularly astonishing how, with just a few basic assumptions,

---

[63] He was awarded the Nobel Prize in Physics together with François Englert in 2013

an excellently functioning mathematical model can be developed that describes the geometry of our world with great precision.

The essence of the axiomatic method is perhaps best understood by translating "axioms" as "rules of the game." Chess, for example, is a game based on rules that have been established once and for all and whose origin is neither questioned nor scrutinized. Instead, players go to great lengths to determine whether they can win the game in a specific situation by adhering exactly to these rules. In principle, this is the same method used by a mathematician or theoretical physicist who ultimately wants to know whether or not the axiom system (or physical theory) being used can solve a problem of interest.

283. The (failed) Hilbert program

Incidentally, the idea that every mathematical theory should be based on axioms goes back to David Hilbert (1862-1943). He believed that every mathematical statement that can be formulated within a contradiction-free axiom system can be assigned the predicate "true" or "false." But interestingly, this is not the case, as the brilliant logician Kurt Gödel (1906-1978) proved in 1931 (remember "Gödel's Theorem, swamp, horse, and crested head..." point 100). In his famous incompleteness theorems, he proved that it is always possible to find statements in an axiomatically constructed theory whose truth value cannot be determined within the system. There are also mathematical theorems - for example, in arithmetic - that are true but whose truth cannot be proven within the axiom system. In the meantime, mathematicians have discovered some statements that cannot be proven within their corresponding axiom systems.

If you want to make the effort to understand Gödel's theorems and their implications, we recommend the cult book by Douglas R. Hofstadter, "Gödel, Escher, Bach: An Eternal Golden Braid." In it, you learn, among other things, that Kurt Gödel's ingenious discovery is not only important in mathematics but is of a general nature. A key example here is self-referential statements, which can lead to semantic paradoxes of the type *"Everything I say is a lie."*

284. Maurits Cornelis Escher

The text, which is not always straightforward but truly well-composed, is broken up by the graphics of the Dutch graphic artist Maurits Cornelis Escher (1898-1972), which are also repeatedly referred to in the text. They are also highly paradoxical, as they represent either impossibilities (think of the graphics "Treppauf, Treppab" (Stairway Up, Stairway Down) and "Waterfall") or astonishing surface parquetry (e.g. "Metamorphosis I" or "Day and Night", so-called "Escher tiles").

Fig. 124: M. C. Escher in the work "Sphere Surface with Fishes" 1958 (Wikimedia)

Escher's attempt to depict "infinity" is also remarkable. Based on his tessellations, he found two solutions for this. One was to move from the plane to a boundless surface, and the other was to use non-Euclidean geometries, such as those developed by the Russian mathematician Nikolai Ivanovich Lobachevsky (1792-1856). Incidentally, his famous motifs of snakes and fish were created in this way.

Escher is also a good example of how mathematics and art can complement each other. Even formulas that seem dry and meaningless to a casual observer can develop their own aesthetic appeal for an expert in the subject.

285. Euler's identity

The prime example of this is the relationship discovered by Leonhard Euler in 1748

$$0 = 1 + e^{i\pi}$$

It links two transcendental numbers, namely Euler's number e (the base of the natural logarithms) with the number Pi, the ratio of the circumference of a circle to its diameter with the root of (-1) (= the imaginary unit) to a value of -1. And this is really the case, as can be easily proven. Nevertheless, it is somehow astonishing and mysterious how two numbers that cannot even be represented as fractions and in which one is responsible for "exponential growth" and the other for everything to do with circles, together with the "impossible number" ($\sqrt{-1}$) result in exactly -1 (mathematicians call this "Euler's identity").

286. Exponential growth and the Club of Rome

Speaking of "exponential growth," do you remember the Club of Rome? This rather elitist association, founded in 1968, became famous in 1972 when it published the study *"The Limits to Growth."* In it, the authors attempted to show that unchecked population growth, industrialization, and ecological overexploitation would eventually reach their natural limits (i.e., resource depletion) and lead to a global humanitarian catastrophe. The justification for this was clearly based on the dynamics of exponential growth processes. However, the time predictions made in this first study all proved to be wrong, while the methodology used - simulation of future developments based on systems theory - was increasingly refined over the following decades. Interestingly, the latest report dates from 2012 and is titled *"2052: A Global Forecast for the Next Forty Years."* It can certainly be regarded as serious in terms of both content and methodology and may be considered a political guide for the near future. Of course, it also suffers from the problem that, according to tradition, no one has formulated as aptly and wittily as Karl Valentin (1882-1948):

"Forecasts are difficult, especially when they concern the future"[64]

But let's get back to the core issue: exponential growth. Most people can hardly imagine what it actually means; otherwise, they would hardly participate in such nonsense as chain letters and other Ponzi schemes.

---

[64] This saying is sometimes also attributed to Niels Bohr

287. Chess and rice

Here is an insightful legend with a surprising result: the inventor of chess was able to inspire his king with his game to such an extent that he granted him a wish. And his wish initially sounded quite harmless. The inventor of the game of chess wished for a lot of grains of rice, the number of which was determined by a simple rule - as follows: One grain should be placed on the first square, two on the second, four on the third, and so on (i.e. there should always be twice as many grains of rice on one square as on the previous one) until all 64 squares have been played. This request seemed rather modest to the king, so he promised to comply without giving it much thought. When the court mathematician began to calculate the number of grains of rice, the monstrosity of the "modest" request quickly became clear. So we calculate: $1+2+4+8+16+32+...+2^{64} =$ 18446744073709551615. That's around $1.8 \cdot 10^{19}$ grains of rice. So if a grain of rice weighs 0.03 grams, then $1.8 \cdot 10^{19}$ grains of rice weigh $5.5 \cdot 10^{11}$ tons, or 550 billion tons. This corresponds to around 1150 global rice harvests (as of 2014). Roughly speaking, if all these grains of rice were distributed evenly over the surface area of Germany, this would result in a layer thickness of around a quarter of a meter. And yet we are only dealing with a simple geometric series here.

The category of "exponential growth" also includes bacterial reproduction (a bacterium divides every half hour), the population explosion (see Thomas Robert Maltus (1766-1834) "*The Principle of Population*") and the extraction of raw materials (just think of the unregulated exploitation of oil reserves).

288. Limited growth

All these "types of growth" have the characteristic that they are limited, as they consume resources - and these resources under consideration here are all finite. As the Club of Rome correctly points out, the earth's natural resources limit the maximum number of people who can be fed at a certain level of prosperity. This number is likely to be somewhere around eight billion - a figure that was already reached on November 15, 2022. For ecological reasons, however, even this figure is too high. The current use of natural resources by the world's population exceeds the earth's annual regeneration capacity. If all natural resources - such as wood, food, or plants - were distributed evenly among all countries in the world, Germany would have used up its share by May 2, 2024. This day is known as "Earth Overshoot Day." However, there are major differences in resource consumption between various countries. If the entire world population lived like the Germans, we would mathematically need three Earths, while the lifestyle of the United States would "consume" around 4.9 Earths if applied to the global population. Demographers assume that the world population will rise to a maximum of 9.4 or 9.7 billion during this century, while UN experts believe a maximum of 10.4 billion is possible. In any case, population growth will lead to problems in the medium and long term, although no one knows whether, and if so, how they can be overcome.

289. The ecological footprint

This insight comes from studies that show the so-called "planetary boundaries" in relation to nine selected parameters - and, in a methodologically similar way, by calculating the "area" that a global citizen needs in hectares to sustain their life. This is done, again, in relation to a selection of parameters and is shown as the individual's "ecological footprint". But all these studies have only academic value if they do not lead to politically implementable measures. And experience shows that skepticism is warranted. "Spaceship Earth" has many captains but no helmsman, and the crew is bursting at the seams. In most of the cabins, there is abject poverty, and in some, violence reigns. The fact that we still feel so comfortable is only due to the fortunate circumstance that we don't live in the slums of a Third World city, as the majority of people on Earth do. But the thunder of the guns is approaching, as you can see if you pay attention to the daily news.

When you look at the population growth figures outside the industrialized nations, it becomes clear that the order of the day is to slow down the population explosion in a humane way. The solutions to this dilemma have long been known: more education, better training, and equal opportunities - including for women - targeted family planning with the provision of necessary resources, improved healthcare, and a little more comfort and security in life. These factors (as the industrialized nations have abundantly demonstrated) will cause the birth rate to tumble. But all this is easier said than done, and the conditions for implementing such a program seem to be deteriorating, especially considering the many conflict regions around the world.

Yet, with a bit of luck and understanding, those of us still living in paradise may only pass through purgatory - not hell - on our way to the stars. After all, nature will find a way to address the problem of overpopulation. The mathematical formula that nature will follow in this respect, describing the development of a population of living beings with limited resources over time, has been known since 1845.

290. The logistic equation

It was discovered by the mathematician Pierre François Verhulst (1804-1849) during an analysis of Malthus' theory of population growth. This "Verhulst equation" is better known in German-speaking countries as the "logistic equation." It was studied in detail in the second half of the 20th century, as it is closely related to dynamic systems that tend to be chaotic. The following example shows how it works. This is about the nun. No, not the one in the convent, but the butterfly *Lymantria monacha*, whose caterpillars used to devour entire coniferous forests (the resource on which they feed) within a very short time. Today, this butterfly from the moth family is controlled with poison if the foresters deem it necessary. This is why it is no longer seen as often as it used to be.

Fig. 125: The "nun moth" used to be prone to catastrophic mass reproduction ("calamities") in the coniferous forests of Central Europe. Today, this moth has become quite rare. The formerly destructive value of its caterpillars has now been taken over by the bark beetle. (Author)

The life cycle of this moth is as follows: Female moths lay their eggs in late summer, from which the caterpillars hatch in spring. The caterpillars then immediately feed on spruce and pine needles until they have consumed around 1000 spruce or 300 pine needles. After they are fully grown, they look for a cozy spot where they spin a cocoon and pupate. Two or three weeks later, after completing their metamorphosis, the moths emerge. They mate, lay eggs on the trunks of conifers, and die. However, the eggs overwinter, and the annual cycle begins anew.

You can imagine that an initial population of this insect doubles every year - just like the grains of rice on a chessboard. However, there are also processes that counteract this population growth, such as predators, weather conditions, parasites, and the finite number of needles in a coniferous forest. Mathematically, such a scenario is best described by population ratios, i.e., the quotient of the population size in the current year to the population size in the previous year, with normalization so that the largest possible population is set equal to 1 (or 100%). Simple exponential growth can then be formally described using a recursion formula, where a constant is chosen and can be interpreted as the "birth rate." Verhulst's idea was to supplement this simple relationship with an additional factor on the right-hand side, which formalizes the processes opposing growth. Its effect is easy to recognize: For a very small value, the factor is close to 1, so the equation is not too different from the original growth equation. However, if the value becomes very large, the factor approaches zero, ensuring that the right-hand side disappears in the limiting case. Clearly, both terms work against each other. The first term tries to increase the population more and more, while the second term suppresses this unbridled growth. We are dealing with a non-linear feedback loop, which makes the Verhulst equation extremely interesting. It is now straightforward to use a computer to investigate how an initial population develops over the years with different birth rates. If, for example, the birth rate is below 1 (as in Germany, where 1 signifies a simple reproduction rate), then the population will eventually die out. With a birth rate of 1, both directions, regardless of whether you start with a population of 100% or with a small population of 1% of the possible 100%, lead to a stable limit value of approximately 66%. Here, the population development stabilizes and does not change in the following years. If the value is increased further, e.g., to 2.5, then a certain "oscillation behavior" is observed, but ultimately the stable value of approximately 66% is also reached after a certain time. Things become particularly interesting when you select a value of approximately 3.449. From this point, the stability of the 66% population size becomes unstable and splits into two branches, one above 66% and the other below 66%. This splitting is referred to as "bifurcation" in technical jargon. It can also be easily explained with the nun moths: In the first year, the female nun moths lay so many eggs that, by the summer of the following year, after the caterpillars have eaten excessively and the forest has become increasingly sparse, a particularly large number of moths hatch from their pupae. These moths lay even more eggs, leading to a correspondingly high number of barnacle caterpillars the following year. However, by the midpoint of their lives, the caterpillars have nothing left to eat, as their food - spruce and pine needles - has already been consumed. The population collapses due to overpopulation and resulting famine, and only a few surviving caterpillars eventually develop into butterflies. The following year, there will be only a few nun moths, and the cycle starts all over again. The population obviously fluctuates between years of mass reproduction and years when the moth is relatively rare. If you increase the birth rate in the Verhulst equation to approximately 3.449, the two previously stable values become unstable again and create a population that tends to experience exceptional mass reproduction every four years. Such a cycle can actually be observed in natural barnacle moth populations. If allowed to continue growing, further bifurcations occur at certain values until, eventually, the butterfly population changes its size completely unpredictably from year to year. This state is usually referred to as "chaos," and the mathematical theory that deals with it is called "chaos theory." If you would like to see the path to "chaos" graphically, you should search for the "bifurcation diagram" on the internet. Speaking of "googling"...

291. Big numbers - googol and googolplex

Did you know that the term "Google" is derived from the word "googol", a special neologism for a number with 100 zeros (i.e. 10^{100})? Pronounced correctly in German, it is "zehn Sexdezilliarden", and written out

10,000,000,000,000,000,000,000,000,000,000,000,000,000,000,000,000,000,000,000 ,000,000,000,000,000,000,000,000,000,000,000,000,000,000.

Incidentally, this makes it easy to

further potencies, e.g. $10^{10^{100}}$ - the Gogolplex, or $10^{10^{10^{100}}}$ the Gogolplexplex.

At this point I would like to hazard a guess, which of course I cannot prove: Among the first googolplexplexplexplexplexplex digits of the number pi will be found in coded form, as described at the beginning of this book (point 5), some essential paragraphs of this book. But there is no point in looking for them. There is not a single computer in the world that would be able to store any number in the Gogolplex range digit by digit. And another thing, in the manageable cosmos there are only about 10^{80} protons, i.e. a number that is tiny compared to a Gogol or even a Gogolplex...

292. The Observable Universe

But what does "observable universe" mean? Are there parts of the universe that are not "observable" by us? This is where it gets complicated, as cosmic space extends over three spatial and one temporal dimension. Spacetime can be "curved" in much the same way as a two-dimensional surface can be "curved." I will not go into this in detail here (see point 153). Just this much: As Albert Einstein demonstrated in 1915 with his General Theory of Relativity, the "straightest" lines that can connect two "events" (defined by three spatial coordinates and a point in time) are "geodesics," along which one can move without any force acting upon them. Rays of light, for example, follow such a path in a vacuum. This makes it conceivable to have a static space that is curved similarly to the surface of a sphere. In such a space of positive curvature, a beam of light sent straight ahead would return to its starting point (backwards, so to speak!) after a finite amount of time. Such a space would be finite (like the finite surface of a sphere) but without boundaries (like the surface of a sphere). However, there are also spaces with negative curvature. They are analogous to the surface of a pseudosphere (a cross-section resembles the surface of a saddle) and, like the flat "Euclidean space" of our experience, are infinitely extended.

But what "bends" this cosmic space (more precisely, spacetime)? It is the energy distributed in space in the form of mass (stars, galaxies) and momentum. The question of the geometry of cosmic space is therefore essentially a question of the quantity and distribution of the galaxies it contains and must therefore be determined by observations.

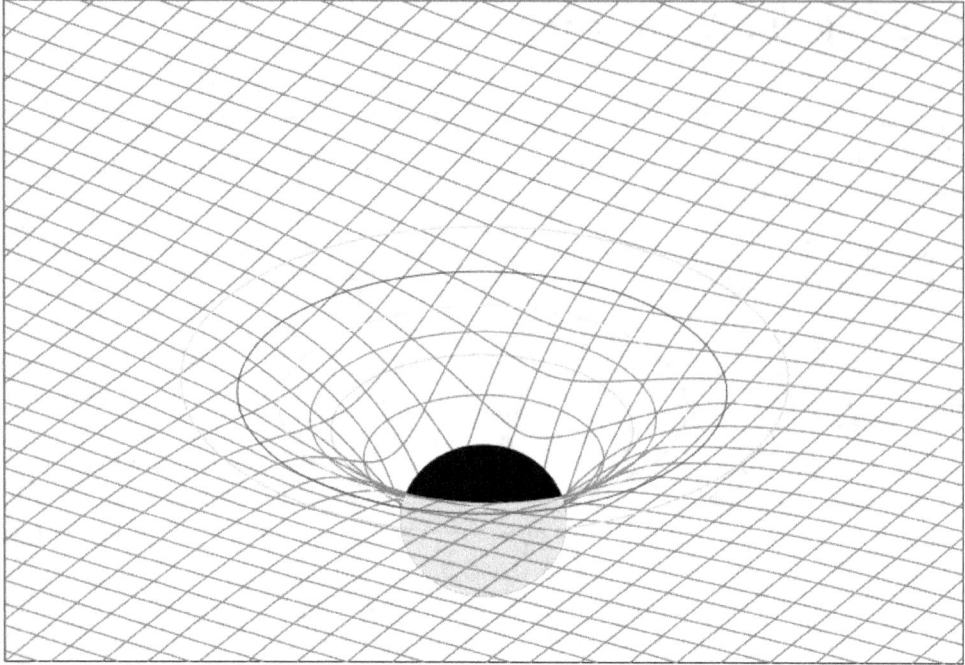

Fig. 126: According to the general theory of relativity, masses are able to bend space-time in a specific way, which can be observed, for example, in the deflection of light at the edge of the sun or in so-called "gravitational lenses" (Wikimedia)

However, cosmic space has another unexpected property: it expands. This can be observed by the fact that virtually all galaxies are moving away from us, and the further away they are from us, the faster they recede. This is similar to evenly distributed raisins in the yeast dough of a raisin cake, which slowly rises. Or, as a two-dimensional analogy: draw evenly distributed dots on a balloon to represent galaxies. If you now blow it up and imagine yourself as an "observer" on one of these points, you can also observe how all the others move away - and the further away they are from the observer's point, the faster they move away. This effect, transferred back to the cosmos, is described by a simple law known as the "Hubble Law", which establishes a connection between the escape velocity of galaxies (which can be measured using the redshift of their spectral lines) and their distance.

293. Big Bang and Hubble bubble

It implies that, in the past, all distances between galaxies were once smaller than they are today and that this "expansion" - viewed backwards - can be traced back to a singular point, which is usually referred to as the "Big Bang." We now even know that this "Big Bang" occurred almost exactly 13.787 billion years ago. It is precisely this fact that limits our observable cosmos to a quasi-finite region, which is bounded by a horizon and moves further and further away from us over time. The reason for this is the finite speed of light, which means that a beam of light can only travel a distance of 13.798 billion light-years in 13.798 billion years. In reality, however, the "horizon" is much larger due to the expansion of the universe (it has continued to expand during the lifetime of light). It is estimated that the diameter of this region of space, known as the "Hubble bubble," is approximately 93 billion light-years. This is the "observable universe" for us. And since there is no "center" of the universe, each galaxy has its own Hubble bubble. In this way, the universe really can be infinitely large, and it is a good working hypothesis to assume that it looks similar everywhere in it (given the same age of the universe, all galaxies are the same distance in time from the Big Bang) - even outside our own Hubble bubble.

In our "finite" but unlimited balloon model, this means that its surface is relatively evenly covered with "galaxy points" and that there is a circular "light front" around each galaxy, which indicates the respective horizon and expands further and further into cosmic space at the speed of light plus the speed of expansion, thus bringing more and more galaxy points into its field of view. If we now blow up the balloon, we can simulate the cosmic expansion and recognize that the distance to the horizon increases faster than the finite speed of light. For further argumentation, however, it is only important to understand that it continues beyond the horizon - perhaps even to infinity...

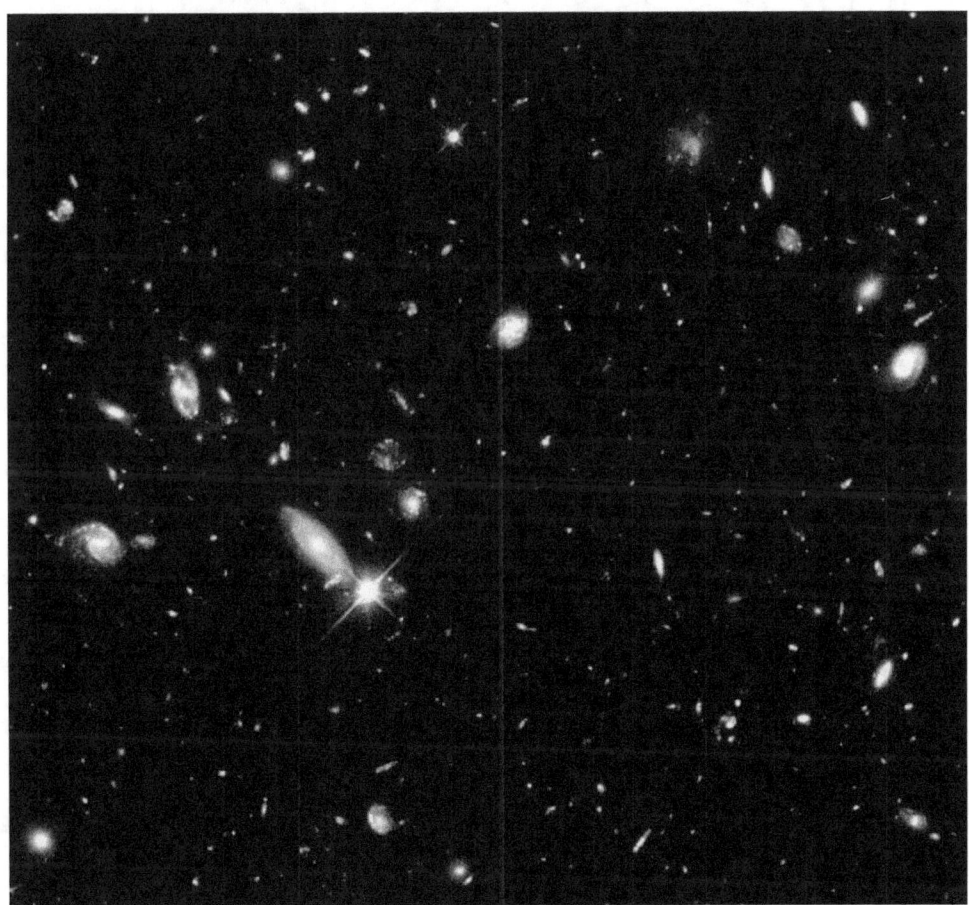

Abb: 127: The Hubble Deep Field contains several thousand galaxies, the furthest of which are over 12 billion light years away from us (NASA, Wikimedia)

294. The Multiverse of Doppelgangers

And this leads to some interesting consequences, which are based on the fact that every part of space (e.g. the contents of our Hubble bubble) is itself finite and every finite number is known to be small compared to "infinity". Physically speaking, every finite volume can be described by the number of quantum states that can be realized in it. This number is huge, as all possible combinations have to be taken into account. But it is finite and therefore still tiny compared to infinity. To make this a little clearer, imagine a huge chessboard consisting of 8x8 squares lined up without gaps. On each of these squares, 0 to 32 pieces are arranged completely randomly (diced, so to speak). Now you can ask yourself the question, what is the maximum distance you have to move away from a randomly selected 64-square board until you are 100% sure to find another 64-square board with the identical arrangement of pieces (exercise). And what applies to the chessboard with its pieces also applies to the quantum states in a volume the size of the Hubble bubble, which represents "our" universe. So where in the infinite (or sufficiently large, should it be finite) cosmos can we find another area of space the size of a Hubble bubble whose quantum state is indistinguishable from ours? A rough calculation tells us: a little more than a Gogolplex meter ($\sim 10^{10^{120}}$)). And what does that mean? It means that at this distance we can expect a world that is indistinguishable from ours at any given moment. There sits an alter ego of yours, completely identical to you, dear reader, with the same life story and the same memories, reading these lines from a book completely identical to this one. Crazy, isn't it? And in an infinitely large universe with the aforementioned characteristics, there are again an infinite number of doppelgangers of all of us - and not only completely identical ones, but also with every conceivable variation.

The repetition of our Hubble bubble is of course a maximum requirement. A "double" should be found a lot closer. It should be possible to find an area of space with a radius of 100 light years with an identical design in about $10^{10^{92}}$ meters. And since a human being is a lot smaller, you can expect to find a perfect doppelganger of him at a distance of just a few $10^{10^{28}}$ meters. But that is still far beyond the horizon...

According to these considerations, the universe consists of an infinite number of "Hubble bubbles" that overlap and together form an infinite (or very large, if finite) multiverse. The physicist Max Tegmark, who has studied this subject in detail, speaks in this case of a level 1 multiverse.

295. Big Bang theories

Science still has a lot more to offer in this respect. For example, it is still a mystery as to how something like a "big bang" could have happened in the first place. An important discovery in this context was that shortly after the ominous point in time zero (roughly between 10^{-35} and 10^{-30} seconds after the actual Big Bang), there was an enormous "inflation" of space, which is called "inflation". It answers the question of why the universe is so large, so uniform, and so flat. A rapid expansion of space can explain these and other otherwise puzzling properties in one fell swoop. As it is also supported by observable implications, the cosmic inflation phase is now an integral part of any Big Bang theory. However, if you think further, this comes at the cost of other peculiarities. For example, the theory of "eternal chaotic inflation" postulates the existence and continuous formation of new, parallel, coexisting Level 1 universes. In each of these universes - some exist forever, many others collapse again immediately - different values apply to the fundamental constants of nature, and only in those where they happen to assume values that allow the creation of living matter are there also observers. And here, too, there are universes (this time "universes" that form Level 2 multiverses) that are similar or largely similar to ours, where there are also doppelgangers of us. These universes, which the theory predicts, are in principle no longer even detectable from within our universe.

Even crazier, isn't it? Wherever empirical data is lacking or difficult to obtain, or where you don't even know what to look for, bizarre ideas and theories flourish. The only question is - and Niels Bohr once asked this in connection with quantum theory - whether the idea or theory is really crazy enough to be true. Ultimately, however, it is observation or experiment that decides whether a description of nature is right or wrong and not the opinion of a scientist...

296. Theory and empiricism

As the famous philosopher of science Karl Popper (1902-1994) stated, theories about any subject can never be regarded as "true" in the final analysis - their "truth" results from an as-yet unsuccessful falsification. Or to put it another way: scientific theories can never be proven in the final analysis but can be refuted at any time by empirical evidence. Experiments and observation are the instances against which a theory must be measured. It is therefore perfectly legitimate to invent theories. However, they must then also be measured against reality. If there is only one discrepancy, you either have to limit their validity accordingly (such as classical mechanics to speeds that are small compared to the speed of light), extend them accordingly (such as special relativity), or you can throw them straight into the garbage (such as the once very popular Phlogiston theory).[65]

At this point, it is perhaps interesting to point out a theory that saw the light of day in the book "*Physics for the Inquiring Mind*", written by Eric M. Rogers and published in 1960 (easy to find on the "web").

---

[65] The phlogiston theory is an outdated scientific theory that assumes that combustible materials contain a fire-like element called "phlogiston", which is released during com-bustion.

297. The demon theory of friction

It claims to deal conclusively with a very difficult phenomenon that is of fundamental importance for technology - friction. This theory is best presented in the form of a dialog between a supporter of the theory (A) and a skeptic (S). It is the "Demon theory of friction". So let's begin:

S: I don't believe in demons.

A: I do.

S: Anyway, I can't imagine how demons create friction.

A: They go against things and bring them to a standstill.

S: Even with the best will in the world, I can't see any demons on the roughest table surface.

A: They are too small and completely transparent.

S: Friction is stronger on rough surfaces.

A: There are also more demons there.

S: But if I add oil, the friction is reduced.

A: Oil also drives the demons away.

S: But if I polish the table, the friction is reduced and the ball keeps rolling.

A: You are also cleaning away the demons. There are then fewer of them pushing against the ball and the ball only comes to a halt after a longer distance.

S: But a heavier ball creates more friction and it crunches when rolling.

A: More demons push against it and their bones shatter, which is clearly audible.

S: If I put a brick on the table and try to push it against the friction with increasing force, it won't move up to a certain point because the friction will cancel out the force I exert.

A: Of course, the demons push against it just hard enough so that the target does not move despite the force they exert. However, their powers are not unlimited and at some point they collapse.

S: But if I push hard enough and set the brick in motion, friction occurs that slows down the brick's movement.

A: It's logical, because when the demons have collapsed, they are crushed by the brick. It is their crushed bones that oppose the sliding movement of the brick.

S: But I can't feel them.

A: Rub your finger on the table.

S: But friction follows certain laws. It has been experimentally proven that friction brakes the sliding tile with a speed-dependent force.

A: Of course, the same number of demons will always be crushed, no matter how fast you drive over them.

S: If I drive a brick over the table again and again, the friction is the same every time. The demons would have been crushed the first time.

A: Remember, demons multiply incredibly quickly.

S: And then there are also other laws of friction. Braking, for example, is proportional to the pressure with which the two surfaces are pressed together.

A: Demons live in the pores of the surface. A higher pressure challenges a larger number of them. They then push against it and are crushed. Demons work exactly like the forces you found in your experiments.

You can, of course, invent many other "theories" to explain specific phenomena using this scheme (have you seen through it?). These theories are artificial, offer virtually no reference to other related physical processes, and can only be mathematically defined to a limited extent. In short, such a theory is superfluous but not useless - it is at least entertaining, which is why I have presented it here. Nevertheless, I would like to conclude by drawing your attention to the presentation of the demo theory of friction.

298. Quaestio disputata

This is a method of disputation known as "Quaestio", which was used by scholastics in the Middle Ages to clarify questions that are still not completely irrelevant today, such as "*Which is greater: the joy of the wolf when it sees the lamb or the fear that the lamb feels when it sees the wolf*" or the more practically important "*Is it healthy to get drunk once a month*". Such questions are also known as "scholastic questions". At the early European universities, they were discussed with great seriousness and zeal in speech and counter-speech in strict compliance with logical laws - and were often recorded in letters, treatises and dissertations. This gave rise to the literary genre of the "*Quaestio*", which even Galileo Galilei used in his "*Dialogo sopra i due massimi sistemi*" (Dialogue on the two main world systems, 1632).

It is not possible to say with certainty exactly when the Quaestio method originated. It developed gradually from the 6th century (A. M. S. Boethius, around 480–526 AD) until it was finally canonized and became an integral part of university education. The earliest mention of it by name can be found in the statutes of the University of Paris, which date back to 1215. The Quaestio was conducted after a lecture (lectio) as a disputation between teacher and student or between two students (i.e., as opponent and answerer, above "S" and "A" (point 297)), primarily to consolidate the subject matter and to train logical thinking as a quaestio disputata. The oral examination for obtaining an academic degree also took the form of a mostly public academic debate. Incidentally, Martin Luther attempted to force such a public debate with his 95 theses, which, as is well known, did not succeed due to the explosive nature of the topic, but ultimately led to the Reformation.

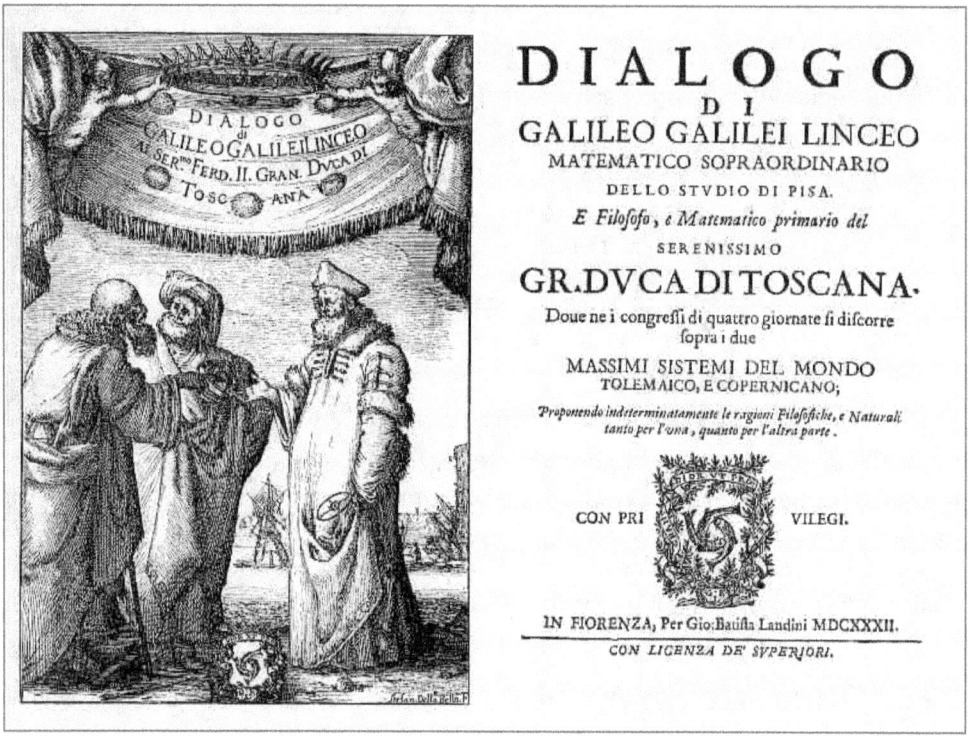

Fig. 128: Cover picture of the "Dialogo" by Galileo Galilei from 1632

But back to the Quaestio. Its main subject matter was the *Holy Scriptures*, i.e., the writings of the Bible, the Church Fathers, and the ancient philosophers. Questions were derived from these texts and submitted to a disputation. The Quaestio always consisted of three parts, each involving an "*opponens*" and a "*respondens*" alongside the professor. One party would formulate a statement, which would then be debated in an alternating discussion between thesis and counter-thesis. Today, a classic panel discussion, where people discuss politely and reasonably with each other, comes closest to this methodology. From today's perspective, the only disadvantage was that in this thoroughly scientific method, the premise - e.g., a quote from a Church Father or a line from the Bible - was never questioned, leading to the most outlandish conclusions from today's standpoint (at least for non-theologians).

Questiones were, of course, not only reserved for the Faculty of Theology. Particularly beautiful "flowers" (i.e. "questions" or even dissertation topics) have been preserved from the Faculty of Medicine. They contributed to the fact that this field of knowledge took a step backwards rather than developing more progressive ideas that surpassed those of Galenos of Pergamon (around 130-215 AD), for example.

Fig. 129: Scholastics during a lecture... (Wikimedia)

Here is a small selection of topics that the academic physicians of the time dealt with and subsequently attracted the ridicule of the humanists: "*Did God create the human skull as the seat of the brain or rather as a window for the eyes?*" - "*Is the embryo more like the mother or more like the father?*" - "*Does debauchery cause baldness?*" - "*Is it good for health to live only on bread and water?*" - "*Is the woman a more imperfect creature than the man?*" etc. pp. Incidentally, these topics (with the exception of the former) were all discussed with the necessary seriousness at the University of Paris in the 14th and 15th centuries, as is documented in writing.

299. The scholastic problem of Almighty

However, there were certainly topics in scholasticism that are still of interest today. Let's take the question of whether the "Almighty" really is omnipotent, which is not entirely unimportant in theological circles. This problem can be broken down into a simple question that anyone can understand and which - in the scholastic tradition - reads as follows:

"Can God, as an omnipotent being, create a stone so heavy that he cannot lift it himself?".

If you try to examine this question logically in terms of its alternative answers, you quickly realize that it presents a paradox. If God can create such a stone, He is not omnipotent (since He cannot lift it). However, if God cannot create such a stone, He cannot be omnipotent according to the question. You can twist and turn it however you like; the paradox cannot be resolved without scholastic mental acrobatics - and God cannot be omnipotent for logical reasons alone. Therefore, the term "Almighty" can certainly be seen as presumptuous. But joking aside, the paradox can be avoided if it is reformulated slightly. All you need to do is specify a concrete weight for the stone (however large it may be). In this case, the statement would be: Yes, He can create a stone with this weight, and yes, He can lift a stone of exactly this weight. Of course, no one would pose such a question today, because weight is known to be a gravitational force that only makes sense if a mass is subject to a gravitational field.

Today, we tend to ridicule scholastics for their sometimes bizarre questions, but this does not do justice to them or their time. The scholastic method originally traced back to Aristotle involves attempting to solve scientific or theological questions by constructing purely theoretical and logical chains of argument that can either be traced back logically and stringently to a statement by ancient luminaries (or church teachers) or show that such tracing is not possible due to conceptual ambiguities. Consider the popular story of how Galileo Galilei once wanted to show his university colleagues and high-ranking church representatives the moons of Jupiter that he had recently discovered through his self-made telescope, only for most of them to refuse to look through it, arguing that there was nothing about it in Aristotle. This may seem strange to us today, but in scholasticism, empiricism did not count for much.

300. Unity of teaching and research

However, what has developed in the shadow of this method and still persists today is the unity of research and teaching, the lecture as a tried-and-tested method of imparting knowledge, the examination and disputation as prerequisites for obtaining academic degrees, and - not to be underestimated - academic and teaching freedom (oh, I forgot Bachelor's and Master's degrees). What has been overcome, however, is the unconditional belief in authority, the contempt for empiricism, and - most importantly - higher education institutions are no longer pure, and in the truest sense of the word, (Christian) men's clubs.

301. The vagabond literature

The medieval university milieu was of course not only full of inquisitive young people but also of restless contemporaries who felt much more at home in the city dives than in the professors' colleges. They were mostly students who had dropped out, members of the lower clergy who had not found a flock, as well as a multitude of poetic and musical vagabonds. Some of their literary and musical outpourings have survived to the present day. In this case, we speak of so-called "vagabond literature." The best known (but not the oldest) of these works is known to many today, after this treasure was artistically lifted and made known by Carl Orff (1895-1982).

302. Carmina Burana

I am referring to the song collection *"Carmina Burana"*, which was discovered in 1803 in the library of the Benediktbeuern monastery in Bavaria. This collection of song manuscripts contains texts by almost exclusively anonymous writers who lived between the end of the 11th century and the 13th century and can be classified as vagabonds. It contains a large number of hearty mockery and love songs, a number of pub songs, but also two longer texts conceived as plays. Some of the language used is so hearty that some of the songs would probably have ended up on the Index of Publications Harmful to Young Persons if young people understood Latin or Middle High German. Otherwise, the Car-mina Burana would only be known to a few interested literary scholars today, had it not been for the composer Carl Orff (1895-1982). He took a selection of the songs, organized them thematically into three categories and set them to music in a completely new way, because neither he nor anyone else knew what the songs once sounded like. And with the first performance in 1937, he achieved a great success. The opening song "Fortuna Im-peratrix Mundi" (Fortuna, ruler of the world) is a catchy tune that everyone knows today, even if they may not be able to directly associate it with Carmina Burana. In any case, "O Fortuna" has been adapted in many ways, including by rock and pop groups (Linkin Park), gothic metal (Therion, Nota Profana) and Middle Ages (Gregorian).

The flash mob at Vien-na Westbahnhof on April 23, 2012 is also unforgettable and is definitely worth watching on YouTube (watch the faces of the passers-by).

Fig. 130: "Forest landscape" scene (left) in the Carmina Burana song manuscript (Wikimedia)

303. Codex Manesse

In addition to the song manuscript from Benediktbeuern, Germany has an even more famous song manuscript known as the *"Codex Manesse,"* which is rarely seen in the original (like the Turin Shroud). Its insurance value is currently 50 million euros. It is kept in the Heidelberg University Library but can be viewed as a facsimile on the Internet at any time and takes us back to the time of the minstrels. Its outstanding importance lies in its size, the large number of "singers" presented (starting with Emperor Henry VI), the wonderful miniatures preceding each of the "singers," and its excellent state of preservation, as the manuscript, which is written in Middle High German, is now almost 700 years old. It was probably written between 1300 and 1340, at a time when the heyday of minnesong was already drawing to a close. The manuscript is arranged strictly according to the social status of the people whose texts make up the song manuscript. It begins with the Hohenstaufen Emperor Henry VI (1165-1197), followed by kings (beginning with the "last" Hohenstaufen Konradin (1252-1268)), dukes, margraves, counts, barons, representatives of the service nobility, and the "masters" - for the connoisseur, a "who's who" of the 13th century in the German lands. These include well-known names such as King Wenceslas of Bohemia (1271-1305), Henry of Mohrungen (died around 1220), Walther von der Vogelweide (1170-1230), Wolfram von Eschenbach ("Parzival," around

1160-1220), Hartmann von der Aue ("Poor Henry," died around 1220), Gottfried von Straßburg (died 1250), and, of course, Tannhäuser (died around 1265), to name but a few. Their poems form the literary basis that will one day develop into German national literature.

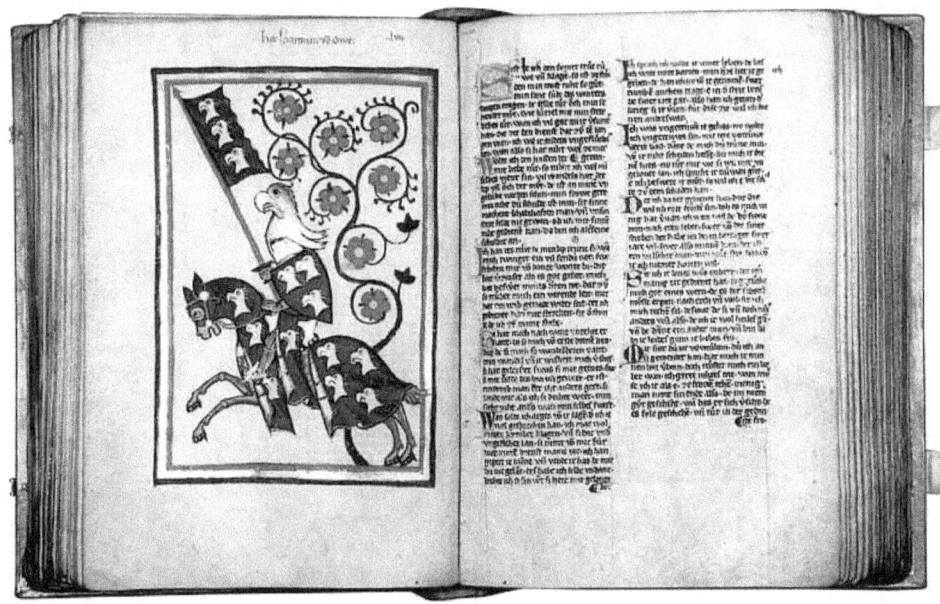

Fig. 131: Page from the Codex Manesse, one of the most valuable books in Germany (Wikimedia)

As far as the epitome of Minnesang (love poetry between a knight and a high-ranking and often married lady) is concerned, Walther von der Vogelweide (around 1170-1230) has become its quintessential representative. Although we hardly know anything about him, his lyrical love songs and singsongs (including political ones) have survived the ages thanks to the "Codex Manesse." Two of them have become better known and are often performed at medieval markets today. I am referring to "Unter der Linde" and the "Palästinalied," which deals with the reception of a crusade and has become particularly well known thanks to the German band "In Extremo" (who have dedicated themselves to "medieval rock").

In keeping with the times, Walther wrote his poetry in Middle High German, the language of the people (more precisely, the literary language) in Germany at the time of the Staufer emperors. After the merging of the North and South German dialects by Martin Luther, our modern High German finally developed from this language via Early New High German and Modern High German. Despite a time difference of around 700 years, the verses are still easy to understand, which shows that languages and dialects are quite conservative systems after all. So just let the original text of "Unter der Linde" sink in, even if some words don't immediately reveal their meaning:

Under der linden an der heide,
dâ unser zweier bette was,
dâ mugt ir vinden
schône beide gebrochen bluomen unde gras.
vor dem walde in einem tal -
tandaradei!
schône sanc die nachtigal.

Ich kam gegangen zuo der ouwe,
dô was mîn friedel komen ê.
da wart ich enpfangen hêre frouwe,
daz ich bin sælic iemer mê.
kuster mich? wol tûsenstunt!
tandaradei!
seht, wie rôt mir ist der munt.

Dô het er gemachet also riche
von bluomen eine bettestat.
des wird noch gelachet innecliche,
kumt iemen an daz selbe pfat.
bî den rôsen er wol mac -
tandaradei!
merken, wâ mirz houbet lac.

Daz er bî mir læge, wessez iemen,
- nu enwelle got - sô schamt ich mich.
wes er mit mir pflæge, niemer niemen
bevinde daz wan er unt ich
und ein kleinez vogellîn!
tandaradei!
daz mag wol getriuwe sîn.

The original can be found in the digitized edition of the "Codex Manesse" (Internet) on page 130 in the second column and a large number of sung versions (also in High German) on YouTube (I recommend the interpretation by "Qntal").

As far as the actual Minnedienst is concerned, there is a better source than Walther, namely the almost completely preserved work "Frauendienst" by Ulrich von Lichtenstein (around 1200 to 1275), which has strong autobiographical features and introduces us to a knight who is in no way inferior to the well-known hero of the Mancha in the Sierra Morena, whom Miguel de Cervantes Saavedra (1547-1616) described so excellently. You can also read his poems in the "Codex Manesse."

The reception of the poetic works of the German High Middle Ages only took place in detail in the age of Romanticism. Sources were sought (such as the Grimm brothers, who discovered the "Codex Manesse" in the Royal Library in Paris in 1815), translations and retellings were produced and published (for example, Ludwig Tieck (1773-1853)), and the time of their creation and the characters were illuminated (such as Ludwig Uhland (1787-1862)).

304. The Nibelungenlied

Another poem, the "Nibelungenlied" ("Siegfried, the Dragon Slayer"), which has survived in three manuscripts and whose creation can be dated to the 13th century, even attained the status of a quasi-national epic of the Germans in the 19th century. Unfortunately, today - like many other "classics" - it is no longer part of compulsory school education in all cases. This explains why a not insignificant proportion of today's generation of schoolchildren cannot relate to this early German epic (the plot is dated to the 10th to 11th century). This is a shame because the stories told in it are certainly no less exciting than those of Harry Potter or The Lord of the Rings. But perhaps the opening words, which are quoted here again in Middle High German, will make those interested in history and literature curious to find out more about this epic and its time:

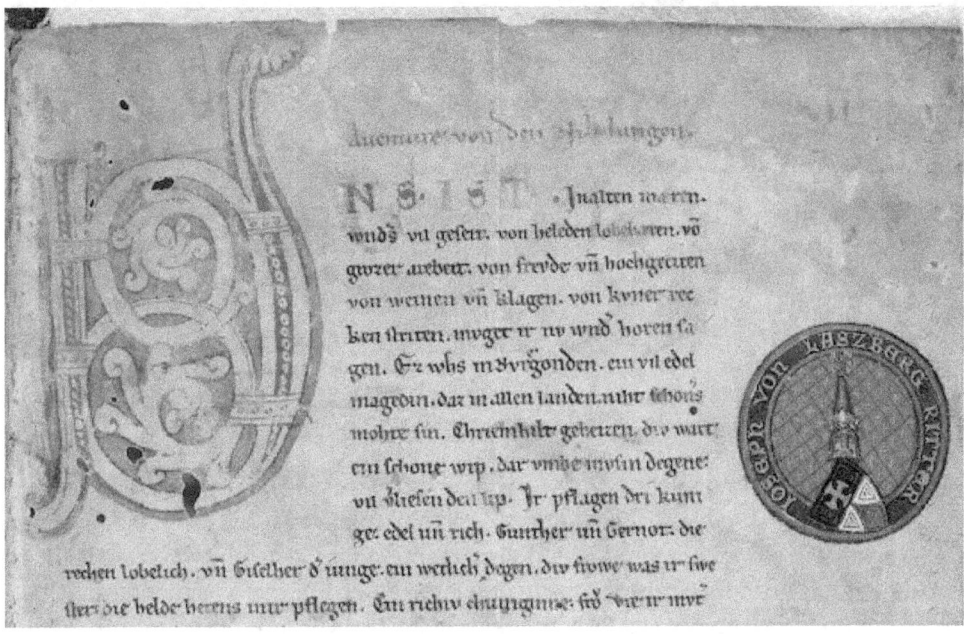

Fig. 132: Entrance page of a copy of the Song of the Nibelungs (Wikimedia)

Uns ist in alten mæren
wunders vil geseit
von helden lobebæren,
von grôzer arebeit,
von freuden, hôchgezîten,
von weinen und von klagen,
von küener recken strîten
muget ir nû wunder hœren sagen. [66]

---

[66] We have been told many wonderful things in old stories of heroic glory, of great work, of joys, weddings, of weeping and lamentation, of bold battles of heroes, you can now hear wonderful things said.

The Nibelungenlied brings together various sagas. The whole is divided into two large parts, the first of which extends to the murder of Siegfried by Hagen von Tronje and the second from Kriemhild's marriage to Etzel (who is identified by historians with the Hun king Attila from the time of the migration of peoples) to the fulfillment of her terrible revenge. Of the film adaptations, I would only like to mention the one from 1924, which was directed by Fritz Lang (1890-1976) and is without doubt one of the classics of film history.

Fig. 133: Kriemhild (played by Margaret Schön (1895-1985)) in Fritz Lang's "Die Nibelungen" from 1924 (German Film Archive)

We have already gotten to know Fritz Lang as the director of "Dr. Mabuse, der Spieler" ("Dr. Mabuse, the Gambler") (item 105). His films from the 1920s were made precisely at the time of the transition from silent film to sound film, whereby the two Nibelung films "Siegfried" and "Kriemhild's Revenge" were still conceived as silent films, albeit with superimposed text panels. The set design of the films was stunning, the characters finely drawn (also in terms of their clothing), the atmosphere medieval and gloomy - and many special effects that were breathtaking for the time (such as Siegfried's fight with the dragon) as well as the proximity to the literary original (which of course most filmgoers knew at the time) made the work a great cinema success. If you want, you can watch both parts on YouTube.

It remains to be seen whether, as a German-speaking educated person in this day and age, it is absolutely necessary to have read the "Nibelungenlied." But there is certainly something like a canon of national and world literature that conveys an indispensable core of education and that one should not only consider reading. It may even be a good thing that as a pupil you are virtually forced to deal with some of the works listed in it at least once in your life.

305. Humboldt's ideal of education

In this context, reference should be made to the so-called "Humboldtian ideal of education," which had a decisive influence on higher education in Prussia after the end of the Wars of Liberation and still persists today - not only in Germany but also in a particularly pure form in American elite universities - in the unity of research and teaching (although it has been significantly challenged by the "Bologna Process"). At this point, it is perhaps interesting to recall what Wilhelm von Humboldt (1767-1835) understood as "education":

"There is worse certain knowledge which must be general, and still more a certain education of disposition and character which no one may lack. Everyone is obviously only a good craftsman, merchant, soldier and businessman if he is a good, decent person and citizen, enlightened according to his status and without regard to his particular profession. If school lessons give him what is necessary for this, he will very easily acquire the special skills of his profession afterwards and will always retain the freedom, as so often happens in life, to move from one thing to another."

And most importantly, according to Wilhelm von Humboldt, every person should be given the opportunity to acquire a basic level of education, which they can then gradually expand according to their talent and abilities (and if they cannot afford it, supported by scholarships) up to a "higher" university education. Thus he writes:

"The highest ideal of the co-existence of human beings would be to me that in which each developed only of itself and for its own sake."

In this sense, education does not only mean participating in the knowledge of the world and reflecting on it for oneself as an individual. It also means orientation, the development of a historical consciousness, the ability and inner desire to educate oneself (the "educated" is a reader), it is also a source of self-knowledge and enables self-determination and should also go hand in hand with moral and ethical integrity.

The literary scholar Dietrich Schwanitz (1940-2004) described the part of "Humboldtian education" that concerns the "humanities" in his highly regarded and sometimes rightly criticized book *Bildung. Alles, was man wissen muss* (also available as an audio book!). It can therefore certainly be recommended as a guide to a comprehensive education. What is missing, however, is a compilation of the mathematical and scientific content of education, which is at least as important. These were added somewhat later by the science historian Ernst Peter Fischer, among others, in his book *Die andere Bildung*.

Fig. 134: Wilhelm von Humboldt (1767-1835) - the founder of the modern university

Today, when the phrase and demand for "more education" has become something of a catchphrase among politicians of all stripes, careful observation reveals a creeping departure from the Humboldtian ideal of education, which has given Germany a leading position in culture and science for almost two centuries (with some interruptions). This departure begins with the decreasing importance attached to the temporal priority of general education over vocational training. Of course, it is still necessary to discuss - **especially in light of current technological possibilities - **what content should be considered general education. It goes without saying that today, as in Humboldt's time, this can no longer be limited to philosophy, philology, and history alone. However, that should not be a reason to neglect these subjects. The increasing economization of educational content, based on the notion that only knowledge benefiting the job market is useful, can, upon closer examination, lead to a dangerous development. It manifests itself in the transformation of universities into mere teaching institutions, in state influence on educational content (consider the stalling of courses related to nuclear power plant technology and the establishment of largely irrelevant, more ideologically driven subjects such as gender studies), in the erosion of university self-administration by the state through financial incentives, and in the calculation of "returns on education," which ultimately reflects the view that the value of educational qualifications is measured solely by their economic benefits.

But that's easy to say. One problem is the political aspiration to provide as many people as possible with university entrance qualifications, which is currently making it almost impossible for high schools to manage student numbers and hindering the promotion of individual talent. This problem will hopefully resolve itself over the next few decades due to demographic trends. Therefore, it would be an opportune time to revisit Wilhelm Humboldt's ideas and to move away from the politically driven tendency to lower the level of educational content and to weaken protective measures in the education sector. Additionally - and I feel this is particularly important - it is crucial to strengthen the social acceptance of non-academic educational qualifications.

306. Knowledge-based society

It is often said that we all live in a "knowledge society" today, but we forget that knowledge and education are actually two different things. The content of these two terms is not contradictory, but knowledge can exist without education (the reverse is not true - education presupposes knowledge). Thanks to technological aids such as the internet and the now ubiquitous smartphone, knowledge can now be accessed at any time without the need to have intellectually acquired, scrutinized, and thought it through. As is well known, Google Glass is currently in vogue, allowing users to display knowledge content from the immeasurable wealth of the internet as required. There are even initial thoughts about transferring the functionality of these glasses to contact lenses. Without appropriate eye control, this could potentially eliminate quiz shows such as "Who Wants to Be a Millionaire," and the examination culture of the education system could also be jeopardized by such a "contact lens." With such technical aids, even the "uneducated" can become "knowledgeable." This person is more comparable to Kim Peek, who could recite an entire library by heart due to his savant talent but was unable or barely able to understand its contents and contexts. However, as with everything in life, there is another perspective. Such an external store of knowledge can be an extremely useful tool for an educated person who has learned to use their mind. This individual differs from the "uneducated scholar" (a term used by philosopher Peter Bieri (1944-2023)) in that they can critically evaluate information, place it in larger contexts, and derive motivations for further research, ultimately aiming to integrate this knowledge.

307. Internet

That's why it's really important to learn to use the Internet critically, as it now enables faster and less complicated access to accumulated knowledge than traditional libraries can. What I particularly appreciate is that with archive.org, Google Scholar, Google Books, the German Digital Library, and Europeana (to name just a few of the "gems"), the Internet brings virtually entire libraries directly to your computer workstation. I am particularly interested in digitized older books, such as those on local history, which you can now research, view, and often download (and, of course, most importantly, read) without leaving your home or workplace thanks to the digitization of entire libraries.

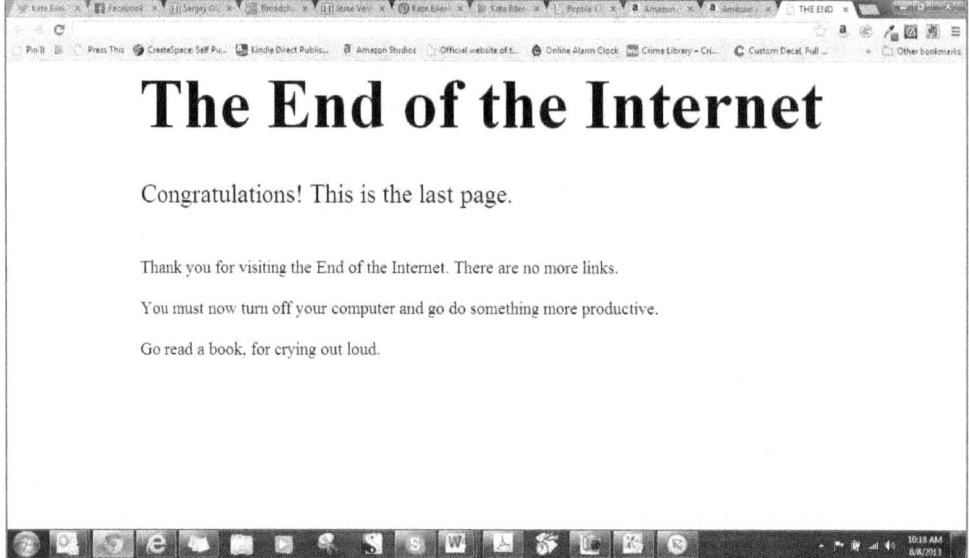

Fig. 135: If this page appears in the browser, the only thing that helps is to switch off the computer...

308. People and books

People's relationship with books has always been somewhat ambivalent. Some people have not touched a book since their school days (apart from perhaps the Bible or a cookbook). For others, it is enough to read one book over and over again (such as the Koran) because they believe it contains everything they need to know. And for still others, books are at the center of their lives. Sociologically, their relationship to books can be divided into three categories: bibliophiles, bibliophages, and bibliomaniacs. For those who are unfamiliar with these Latin terms, here is the explanation: a bibliophile loves books (which doesn't necessarily mean that they read them), a bibliophage devours books; they are a "bookworm," and a bibliomaniac is obsessed with books, often accumulating them in such large quantities in their private library that they would need hundreds of lifetimes to read them all. In some cases, it can even be difficult to categorize someone this way, as the boundaries between book lovers, "bookworms," and book addicts are notoriously blurred.

Fig. 136: This is how Carl Spitzweg imagined a "bookworm"... (Wikimedia)

Let's take the bookworm, someone who finds pure pleasure in reading books. Faust's famulus named "Wagner" was certainly one of them, as Goethe had him say:

> Man sieht sich leicht an Wald und Feldern satt,
> Des Vogels Fittig werd' ich nie beneiden.
> Wie anders tragen uns die Geistesfreuden,
> Von Buch zu Buch, von Blatt zu Blatt!
> Da werden Winternächte hold und schön,
> Ein selig Leben wärmet alle Glieder,
> Und ach! entrollst du gar ein würdig Pergamen;
> So steigt der ganze Himmel zu dir nieder.[67]

The famous Munich painter Carl Spitzweg (1808-1885) depicted a more whimsical version of a "bookworm" in several of his paintings. These works show an eccentric, somewhat distant-looking person on a stepladder in the middle of a library, stubbornly reading a book held close to his nose, while his other hand holds another book and a third, clutched under his left arm, threatens to slip out.

Charles Nodier (1780-1844), a representative of French Romanticism, described another, more pathological case in his novella "Le Bibliomane". In this story, his hero, Théodore, was only able to look at women's shod feet - not with the manner of a pathological shoe fetishist, but with the thought: *"What a pity about this excellent leather. What a beautiful book cover it could have made!"*

Many "bookworms" whose lives are documented lived very modestly to indulge their compulsion to read. In contrast, many bibliomaniacs tended to belong to wealthier classes. They invested all their money in building their libraries and were always on the lookout for new, preferably bibliophile treasures.

309. Bibliomaniac murderer

Some bizarre stories from the past have also been handed down, such as that of the magister Johann Georg Tinius (1764-1846), who was born in Staakow in Lower Lusatia. His excessive addiction to book collecting turned him into a murderer, which resulted in many years in prison after a highly publicized circumstantial evidence trial. Of course, I cannot fully address this judicial case here. Those interested can consult the relevant report in the "Zeitschrift für die Criminal-Rechts-Pflege in den Preußischen Staaten mit Ausschluss der Rheinprovinzen," volume 1830 (29th issue), available on Google Books. The murder of the "Wittwe Kunhardt" described in detail there was one of the first major

---

[67] One soon fatigues, on woods and fields to look,
Nor would I beg the bird his wing to spare us:
How otherwise the mental raptures bear us
From page to page, from book to book!
Then winter nights take loveliness untold,
As warmer life in every limb had crowned you;
And when your hands unroll some parchment rare and old,
All Heaven descends, and opens bright around you!

circumstantial cases to end with a conviction of the perpetrator, a highly respected pastor and manic book lover.

The presumed motive was that the magister needed money to buy more books. He attempted something similar to what we now call a "grandchild trick" in criminal jargon. On February 8, 1812, he visited the obviously wealthy widow Kunhardt under the pretense of delivering a letter. The letter, written by a man unknown to the widow, requested 1000 thalers, which she was to hand over to the bearer of the letter. However, this did not occur. In the end, the 75-year-old lady was found unconscious and covered in blood in her armchair by her housekeeper. She regained consciousness briefly to make a few vague statements before dying from the injuries inflicted on her.

The criminal investigation department in Leipzig (where the crime took place) quickly identified the pastor and Mr. Magister Tinius as possible perpetrators and arrested him without making a fuss. During the investigation, another murder case, that of the Leipzig merchant Schmidt a week earlier (January 18, 1812), came into the investigators' spotlight. As the magistrate stubbornly denied the charges (he never admitted to them throughout his life), an extremely lengthy circumstantial trial developed, which coincided with the division of Saxony into two parts, causing a variety of bureaucratic difficulties and delays. In 1820, Tinius was finally found guilty at first instance and in 1823 at second instance in the "Widow Kunhardt" murder case based on the circumstantial evidence. However, the murder of merchant Schmidt could not be proven with legal certainty. Thus, he was imprisoned for the next twelve years (the time spent on remand was not taken into account), where, despite being far from his library, which had already been sold off, he wrote a bombastic work on the "Revelation of John." He was finally released at the age of seventy-one and died at eighty-two in 1846, not without leaving behind two more works of theological content.

History knows of several other "book lovers" who also became murderers. For example, the Catalan monk Don Vincente, who achieved a certain degree of fame through the short work by the then 16-year-old Gustave Flaubert (1821-1880) entitled "Bibliomania." We will briefly touch on his story. Don Vincente, who ended up in Barcelona after his monastery was plundered, opened a bookshop there. He was hardly interested in the content of the books; only their bibliophile value mattered to him. The rarer a book was, the greater his desire to acquire it. In this respect, he was anything but squeamish. When a series of gruesome murders occurred in Barcelona in 1836, nobody suspected that an obsessed bookseller was wielding a dagger. As the victims were all respected and educated, it was initially assumed that the murders were commissioned by the secret inquisition, which had been stripped of its power. Thus, people suspected of being members of this secret organization were inspected more closely. One of them was Don Vincente. During an unexpected search of his bookshop, a particularly valuable unique copy of a book was found, which was proven to have belonged to the bookseller Patxot, who had been burned in his house along with his books some time previously. Further investigations and interrogations finally managed to convict Don Vincente of ten counts of murder, which he admitted under the weight of the evidence. The "monster of Barcelona" was thus unmasked and brought to justice. He justified his atrocities with the following words:

"People are mortal. They will be called before the Lord anyway, some sooner, others later. But good books are immortal; they must be protected."

Don Vincente met his end at the garrotte shortly after his conviction.

310. The Pitaval

Such and similar stories - i.e., criminal cases with special features - were already being collected and prepared for interested readers in the 18th century, as actual crime literature ("detective stories") did not yet exist at that time. It is only a product of the 19th and early 20th centuries. This genre of literature is called "Pitaval literature" after its founder, François Gayot de Pitaval (1673-1743). In Germany, the jurist Paul Johann Anselm von Feuerbach (1775-1833), who published a collection entitled "Merkwürdige Rechtsfälle" at the beginning of the 19th century, is regarded as its founder. Anyone interested can easily find it in digitized form on the Internet. Incidentally, he was the father of "Feuerbach," whom Karl Marx (1818-1883) once honored with his theses (you know, *"The philosophers have only interpreted the world differently; the point, however, is to change it"*). The "Prague Pitaval" by Egon Erwin Kisch (1885-1948, the "roving reporter") is also well-known, and as a former citizen of the GDR, I am of course also familiar with the Pitaval stories by Friedrich Karl Kaul (1906-1981). Today, film has largely usurped this genre in the form of television documentaries. Think, for example, of the series "The Great Criminal Cases" or "Criminal Cases that Moved Switzerland." But back to books. Because books can change the world...

311. Books can change the world

One of these books was written by Nicolaus Copernicus (1473-1543) from Warmia, whose book *"De revolutionibus orbium coelestium"* I had the pleasure of holding in my hands in a first edition. It is part of the old collection of the Christian Weise Library in Zittau. It is more accurate to say "back in possession" because it was stolen during the GDR era (1988) in a manner that has still not been fully explained. This theft was carried out not by a successor of Don Vincente, but by the Stasi on behalf of the so-called "Commercial Coordination," a department of the Ministry of Foreign Trade of the GDR aimed at procuring foreign currency for the ailing state via the Western art market. Incidentally, Copernicus' work was not the only book to go missing in this way. Also worth mentioning is the extremely valuable copy of a manuscript of the *"Ostfriesisches Landrecht des Grafen Edzard I."* (circa 1518), which later (1992) reappeared and was acquired by the library in Emden (after a legal dispute that ended in a settlement, the copy now belongs 50/50 to the Christian Weise Library in Zittau and the Johannes A. Lasco Library in Emden). Fortunately for the Christian Weise Library, Copernicus' book was also eventually recovered during an attempted sale (in 2008, a similar edition in New York fetched 2.2 million dollars at auction).

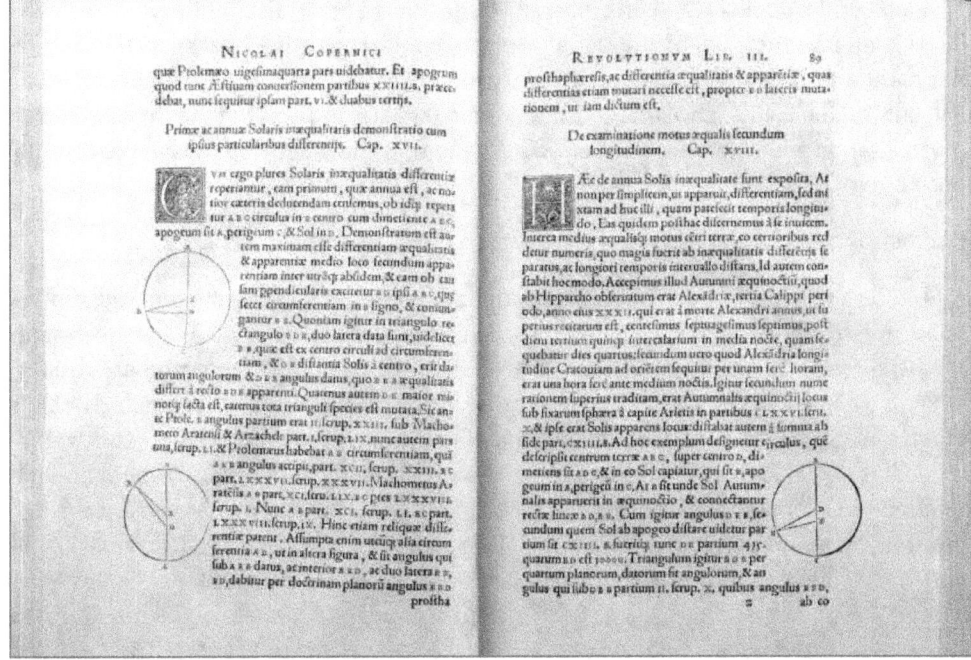

Fig. 137: Copernicus' main work was only accessible to a handful of "specialists" at the time. It was only when his heliocentric world system, based on mathematics, became dangerous to the Church that it was banned.

But beyond its bibliophile value, what makes this book so valuable to the world is its content and impact. Immanuel Kant (1724-1804), the great philosopher from Königsberg, coined the term "Copernican Revolution" to describe its significance. The positions of the celestial bodies, as derived from the *"Prutenian Tables,"* which were partly calculated using the Copernican system, were undoubtedly somewhat more accurate than predictions based on the 300-year-older *"Alfonsine Tables,"* which were grounded in Claudius Ptolemy's geocentric view of the world. However, this increased accuracy was not due to a superior calculation model. Instead, it was because the initial data was more up-to-date and, in some cases, more precise. The real revolutionary aspect of Copernicus' work became fully apparent only a few decades after his death: the idea that the Earth is not the center of the universe and that, by adopting this perspective, it is possible to develop a view of the world that not only accounts for observable phenomena but also explains their physical causes.

312. Copernican revolution as a paradigm shift

The great scientific theorist Thomas S. Kuhn (1922-1996) discusses a paradigm shift that became decisive for the development of natural science from the end of the 17th century, and not without reason. The real work, however, was done by Johannes Kepler (1571-1630), who was able to draw on the excellent observational data of Tycho Brahe (1546-1601) and realized that most of the problems of the heliocentric system could be avoided by introducing elliptical orbits and by assuming that these orbits were traversed by the planets at non-uniform speeds. To achieve this, it was first necessary to free oneself mentally from "obvious" geocentrism, which Tycho Brahe did not yet succeed in doing, but Galileo Galilei and Johannes Kepler did, against the spirit of the times and with much internal and external struggle. Or, as the famous novelist Victor Hugo (1802-1885) once put it,

"Nothing is more powerful than an idea whose time has come."

After that, things went from strength to strength. More and more scholars took up the telescope to observe the sky. In 1675, the Greenwich Observatory was founded under King Charles II, and John Flamsteed (1646-1719) became its first "Astronomer Royal." In 1686, Isaac Newton (1643-1727) presented his work "Philosophiæ Naturalis Principia Mathematica" to the Royal Society, in which he developed a strictly axiomatic mathematical theory based on Euclid's geometry, as we already know, which would later be referred to as "classical mechanics" (point 267). By analyzing Kepler's third law, he discovered the law of universal gravitation and thus created the basis for a physical justification of the heliocentric system. And just 100 years later, "celestial mechanics" was virtually complete, and the positions of the sun, moon, and planets could be predicted with almost arbitrary accuracy from a few initial observations and, as was believed, "for all time."

The establishment of a new worldview, which was not limited to astronomy, in the age of the Renaissance and the beginning of the modern era represented a fundamental turning point in the history of the West. Art and science began to develop. The rediscovery and reception of ancient works, their dissemination and teaching in the art faculties of the flourishing universities gave rise to a learned and inquisitive middle class. Geographical discoveries, the flourishing of maritime trade, and a new interpretation of Christianity contributed further. Humanism became the most important intellectual movement of the time, and the elites attempted to escape the intellectual climate of the late Middle Ages, which was characterized by scholasticism and vulgar theology. In the midst of this period, a work by a canon from Warmia on the *"Revolutions of the Celestial Spheres"* was published, which quickly attracted the attention of scholars of the time. In his "Dialectics of Nature," Friedrich Engels (1820-1895) succinctly articulated the social impact of the publication of a "technical book" that was primarily comprehensible only to the initiated:

"The revolutionary act by which natural science declared its independence and repeated, as it were, the bull-burning of Luther, was the publication of the immortal work with which Copernicus, timidly indeed, and only on his deathbed, so to speak, threw down the gauntlet to ecclesiastical authority in natural matters. The emancipation of natural science from theology dates from then on."

The reason for this was that ultimately the decision between geocentrism and heliocentrism was tantamount to a decision between a religious-idealistic worldview and a scientific worldview. This was also recognized in all its clarity by the official church of the time at the beginning of the 17th century and led to the public burning of Giordano Bruno (1548-1600), the condemnation of Galileo Galilei before the Inquisition Court (1632), and in 1616 (!) to the entry of *"De Revolutionibus ..."* in the list of forbidden books (*Index Librorum Prohibitorum*), where it remained until 1758. And, as we know, it led to a development at the end of which stands today's modern science with all its achievements. This is why we rightly speak of a "Copernican Revolution" with Kant.

313. Industrial revolution - digital revolution

It was still a long way from the "Copernican Revolution" to the "Digital Revolution," the results and technical achievements of which we can enjoy today. In contrast to the "Copernican Revolution," it cannot be traced back to one person but only to a number of individual developments and is roughly comparable in its significance to the Industrial Revolution - which began at the end of the 18th century - when the steam engine replaced muscle power in many areas of industry. What the steam engine was then, the computer is today, and in between lie around 200 years of technical innovation and creativity, based on the scientific findings of the 19th and 20th centuries. In particular, the quantum theory founded by Max Planck and its applications in solid-state physics/solid-state electronics, which gave us lasers, microcircuits, computers, the flat displays of our televisions, PCs, tablets, and smartphones, as well as "robots" in all shapes and sizes. Nobody is surprised anymore that such things exist. And no one - unless they have studied the subject - can plausibly explain how it all works. Terms such as microprocessor, graphics chip, and clock frequency are quickly thrown around, but what they actually "do," how they are manufactured, and what certain technical terms or acronyms such as LCD or TFT actually mean is largely beyond general knowledge. This is certainly not a bad thing as long as there are people who simply want to know and who therefore make the effort to learn the not-so-easy profession. After all, it is up to them to ensure that development does not come to a standstill and that we "users" can enjoy a new smartphone with even cooler features and apps every year.

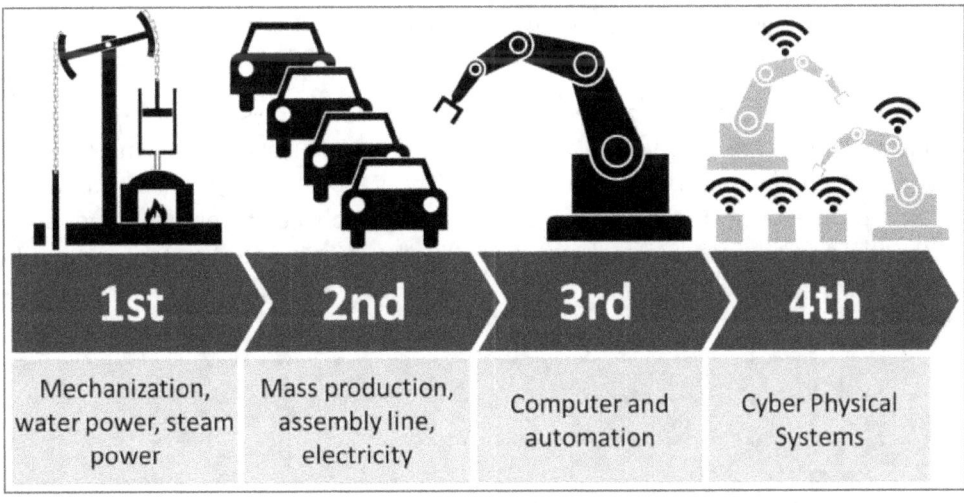

Abb 138: From the first to the fourth industrial revolution (Wikimedia)

The digital revolution has changed our lives in many ways and shapes our everyday lives more than any other technological development before it. The advent of the internet and the networking of computers ushered in a new era of communication and information processing. E-mails, social networks, online shopping, and streaming services are now an integral part of our lives. The ability to exchange information worldwide in real time and access an infinite amount of data has fundamentally transformed both our private lives and the world of work.

One particular innovation at the beginning of the digital revolution is the development and spread of LCD displays (*Liquid Crystal Displays*), which have become an integral part of modern life. This technology has revolutionized the way we visualize and consume information. Let's take a closer look at displays beyond the electron beam tube and explore how they work

314. LCD - Liquid crystal displays

Its original form, familiar to most people from digital wristwatches, is the LCD display, with the acronym "LCD" standing for *"Liquid Crystal Display."* But can there even be such a thing as "liquid crystals"? After all, crystals are the epitome of symmetrical solids, whose variety of colors and shapes can be admired in mineralogical collections. We encounter them as beautifully shaped quartz crystals, monoclinic gypsum crystals, or, quite ordinarily, as grains of salt and sugar. Some people even call a beautifully cut diamond crystal their own. But can crystals really be liquid? They would then lose their most important characteristic: their "crystalline" form. However, that is not the decisive factor here. In our context, we are concerned with another property of crystals, known in technical jargon as "anisotropy." This refers to the directional dependence of a crystal's physical properties. This concerns, among other things, its elastic properties, key figures associated with electromagnetic propagation processes, such as the dielectric constant, magnetic permeability, or the associated refractive index for electromagnetic waves.

The best known of these is probably the birefringent calcite crystal (calcite). Different refractive indices are observed here in the direction of its crystal axes, some of which also depend on the angle of incidence of the light. If a beam of light is sent through a birefringent crystal, it is split into two partial beams, which are appropriately referred to as the "ordinary" beam (which adheres to Snellius' law of refraction) and the "extraordinary" beam. They are each polarized differently, as can be easily determined with a second calcite crystal, which is used like a polarization filter.

Fig. 139: Birefringent calcite crystal (Wikimedia)

315. Liquid crystals

And it is precisely these properties that we are talking about here, as there are also liquids that exhibit anisotropic properties and are therefore referred to as "liquid crystals." In 1888, Friedrich Reinitzer (1857-1927), an Austrian from Prague, discovered the first "liquid crystal" in the form of the ester derivative of the fatty acid cholesterol (cholesteryl benzoate). In extensive investigations, he determined the extraordinary properties of this substance (he also recognized its birefringent properties, for example) but was not quite able to make sense of it. Later, other chemists were able to find other "anisotropic birefringent liquids." The results of this early research were finally reflected in the monograph Liquid Crystals published in 1904 by Otto Lehmann (1855-1922), who succeeded Heinrich Hertz (1857-1894) in Karlsruhe. His groundbreaking results, which in retrospect were certainly worthy of a Nobel Prize, initially met with little response until the end of the 1960s, when they were revisited - and more from a technical background.

316. The Schadt-Helfrich cell

As a result of this research, the Swiss physicist Martin Schadt and the German physicist Wolfgang Helfrich from Munich applied for a patent, which was to become the basis for all subsequent liquid crystal-based display technologies. However, it was rejected by the Munich Patent Office due to a lack of "inventive step." The officials probably considered the functionality of the prototype to be too primitive. In other countries, however, including Switzerland, the potential behind this invention was recognized.

Fig. 140: Digital display, constructed from Schadt-Helfrich cells (Wikimedia)

Do you remember the display elements of an LCD wristwatch or a pocket calculator made up of segments? In principle, each segment is a so-called Schadt-Helfrich cell consisting of two glass plates with a liquid crystal layer only a few 10 μm thick between them. Transparent conductive layers of indium tin oxide are vapor-deposited on the opposing glass surfaces, which form segments that can be structured as required and from which the desired characters (e.g., letters or numbers) can later be assembled. Their surface is also prepared in such a way that the molecules forming the liquid are aligned in their longitudinal axis with an axis of this glass plate. If both glass plates are initially aligned in the same way and the upper glass plate is then rotated by 90°, a correspondingly twisted stack of molecules is created between the lower glass plate, similar to a so-called Reusch mica column. A beam of polarized light whose plane of polarization is parallel or perpendicular to the orientation of the bottom glass plate will then undergo a 90° rotation as it passes through the liquid crystal. The light beam therefore exits through the upper glass plate in a different polarization state. Now imagine that there are polarization filters on the lower plate and the upper plate, which are also rotated by 90° relative to each other. In this case, the light can pass through the arrangement without significant absorption, i.e., it is transparent. If a voltage is now applied to the vapor-deposited electrodes, the molecules of the liquid crystal will slowly align themselves along the electric field as the voltage increases, which destroys their spiral staircase-like arrangement. As a result, the direction of oscillation of the light passing through is less and less influenced by them and finally hits the polarization filter, which then completely blocks it from a certain voltage value (often 5 V). At this moment, the correspondingly switched segments become visible as dark areas - just as in a digital wristwatch or calculator display. The structure of such a display can therefore be imagined as a sandwich consisting of the following layers (from bottom to top): Reflective layer - polarization filter - glass plate - counter electrode in the form of the segments to be displayed - liquid crystal embedded in a corresponding spacer - electrode in the form of the segments to be displayed with contacts - glass plate - polarization filter rotated by 90° relative to the lower polarization filter.

This basic principle has not changed much in the displays of our flat screens, except that today we are able to use the production methods of modern microelectronics to produce large-area displays made up of "pixels" composed of such cells (take a look at your flat screen or smartphone display with a magnifying glass). "Color filters" for red, green, and blue complete the sandwich, so that a pixel can be formed from such a triplet of subpixels, with which any color can be generated according to the RGB model. Additionally, the reflective layer must be replaced by a backlight. Furthermore, an app that takes the voltage from all the pixels of a smartphone display simulates a flashlight...

317. Lippmann's color photography

Incidentally, the idea - two glass plates, the lower one mirrored on the back, and something in between - had already been conceived around 100 years earlier. It was the Frenchman Gabriel Lippmann (1845-1921) who was awarded the Nobel Prize in Physics for this idea in 1908 (and quite rightly, in my opinion - even if the process, which I would like to describe briefly below, has long since been forgotten). I would also like to mention that Lippmann was the doctoral supervisor of Marie Curie (1867-1934) at the Sorbonne, the discoverer of radium, who is one of the very few people to have been awarded the Nobel Prize twice.

This process is a special form of color photography that uses the interference properties of light, not just frequency and amplitude as in conventional analog and digital color photography. The idea was to make incident and reflected waves interfere in such a way that standing light waves form in the gap between the cover plate and the rear glass plate (the back of which was made reflective by a layer of mercury). Darkness prevails at their wave nodes and "light" at their wave crests. Wave crests follow wave crests or wave nodes follow wave nodes at intervals of half a wavelength. If the gap is now filled with a photographic emulsion, as is also used in black and white photography, the silver halide crystals are preferentially reduced to metallic silver at the points where the "wave crests" are located. After developing such a "Lippmann plate," you get something like an extremely complex diffraction grating. If you hold it up to a diffuse white light source (sunlight reflected on a white surface), you will see a color negative image of the previously photographed subject. If, on the other hand, the photographic plate is mirrored again from behind with mercury (or silver), you will see an extremely high-quality color image of the subject. However, as this process also has some significant disadvantages (I would just like to mention the low sensitivity of the emulsions, which are extremely fine-grained for physical reasons - you also want to see some blue sky in landscape photos), it has not been widely used.

Fig. 141: Bouquet of flowers, photographed around 1895 by Gabriel Lippmann (Wikimedia)

The development of orthochromatic and panchromatic sensitized emulsions and the invention of three-layer color film by Agfa in Germany and Kodak in America (from around 1935) made color photography suitable for the masses and left no room for the Lippmann process. Today, "Lippmann plates" are among the most valuable exhibits in museums that collect photographs. The Musée de l'Élysée in Lausanne, for example, owns Lippmann's photographic estate in the form of 130 color photographs. Unfortunately, visitors to the museum cannot admire them in the original, as they are hidden behind thick walls and thus safely in the museum's vault.

Photography is certainly one of the truly great inventions of mankind. Without it, for example, I wouldn't know what my great-grandparents looked like when they had their picture taken in a "photography studio" in Görlitz around 1900.

318. Even technical devices can die out

Today, virtually everyone is a photographer since the original medium of photography - first the photographic plate and then the film - has been replaced by digital image sensors, and such an image sensor is now in every cell phone. And hardly anyone knows anymore (if they didn't even do it themselves in their youth, but then again, you have to be a bit older to have done that...) how to develop film and create paper images from it, whether in black and white or in color. This teaches us that not only animals and plants can become rare or even extinct, but also once-widespread technical devices. This certainly includes analog photography. Or who can still do anything with an 8-inch, 5¼-inch, or even 3½-inch floppy disk today? I still have quite a few of them archived with programs that I wrote back in the 1980s in Turbo Pascal, for example. But I no longer have any devices with which I could read their contents. The vinyl record is also (apart from its "enthusiasts," of course) such a discontinued model. Here, too, you can see that many people still own parts of their record collection for nostalgic reasons, but the playback device, the record player, has long since gone the way of all earthly things. And the record (also an ingenious invention!) that I bought with my "welcome money" on Kurfürstendamm in West Berlin still exists in one of my cupboards today (for those who are interested - Mike Oldfield, "Incantations") - but I listen to it on YouTube when I feel like it. And so I quickly find other "products" that are no longer readily available today - like the humble typewriter, which was a must-have in every office in the pre-PC era (yes, it really existed!). A "delete function" was even invented for them. It was called "Tipp-Ex" and did what the "Del" or backspace key does for a letter today...

319. The ancestry of the automobile

Interestingly, with a little imagination, you can recognize something like evolutionary lines of development in individual products that adhere to the laws discovered by Charles Darwin (1809-1882) and Alfred Russell Wallace (1823-1913). Take the automobile, for example. Today, this vehicle exists in many "species" (car types), which can be divided into different genera, families (e.g. cars, trucks, buses, ..., tanks), etc. Some of them have long been extinct. Some of them have long been extinct (like the phenomobile) and can only be admired in museums, similar to the skeletons of various dinosaurs (tip: Transport Museum in Dresden). Others continue to evolve by engineers changing their shape slightly, improving the drag coefficient (air resistance) or optimizing the engine, etc. What in nature is the struggle for existence is in the automotive industry the customer with his ideas, demands and wallet - and of course the competition.[68] And car types that have serious defects - in which, for example, the brakes regularly fail - are doomed to extinction if the defects cannot be fundamentally eliminated.

---

[68] Unfortunately, this no longer applies in absolute terms since politicians (for example in the form of the EU) have been intervening more and more in the market in a regulatory and controlling manner - just think of the idiotic ban on combustion engines

Fig. 142: Darwinian evolutionary path of the passenger car (Creative Commons)

As is well known, every continuous line of development begins with an original form from which all other lines of development that arise through splitting ("speciation") can be derived. It is difficult to determine this for the automobile. One could, for example, start from the "sledge", from which the "carriage" with wheels was derived and whose primary source of propulsion was the muscle power of humans and animals. Here too, as in biology, there is obviously the problem of the "Last Universal Common Ancestor" (LUCA). Let us simply take propulsion as the primary characteristic. Then we might have the following series: human and animal muscle power, wind, steam engine, internal combustion engine, propeller drive, jet drive, electric motor...[69]

Wind as a source of propulsion, modeled on sailing ships for wheeled carriages, has been around since the beginning of the 17th century. The first steam traction engine, which reached around 5 km/h, was built in 1769 by the Frenchman Nicholas Cugnot (1725-1804). In the meantime, there was also a short development line in which the drive was realized by a clockwork, but this died out as soon as it appeared.

---

[69] Strictly speaking, the electric motor belongs in the list after the steam engine and before the internal combustion engine

Fig. 143: Louis Tuscherer with his automobile

The invention of the internal combustion engine and its use as a replacement for the horse in a carriage gave the automobile a significant boost in development. From around 1880, the German inventor Louis Tuchscherer (1847-1922) worked on the development of a lightweight carriage powered by a simple two-stroke engine running on kerosene. The mechanical engineering entrepreneur Carl Benz (1844-1929) heard about this and had Tuchscherer explain the "carriage without horses" to him in detail. He then developed his own concept from this, which he called the "tricycle," as it ran on three wheels and was powered by a compressionless two-stroke engine that he designed himself. The patent application for this "Benz Patent Motor Car No. 1" ushered in the real age of the automobile. Tuchscherer is (wrongly, in my opinion) only a side note. However, it was not until a year later that Gottlieb Daimler (1834-1900), who was able to cover shorter distances with his motorized carriage, proved its usefulness as a means of transport. As his "carriage," like any other carriage, had four wheels, it is now regarded as the "prototype of the automobile," just as the "riding carriage" he developed together with Wilhelm Maybach (1846-1929) shortly beforehand is regarded as the prototype of the motorcycle. In addition to the engine, it also contained a number of other pioneering innovations, such as front-wheel steering in the form of Ackermann steering. This broke the ice, and other inventors and engineers began to get involved in the lucrative business of automobile construction. In 1892, Rudolf Diesel (1858-1913) registered his engine, which is named after him today. Even electric motors found their way into vehicle construction, and by 1900, this drive, powered by batteries, achieved a considerable market share. And so, as a compromise between different requirements, something like a "basic model" was created, from which the diversification of the automobile into "functional series" began: Vehicles for private passenger transportation with the objective of ever-increasing speed and growing driving comfort; vehicles for transporting a large number of people (buses); trucks for transporting goods; and special military vehicles (tanks). Each of these "functional series" then split again into lines of development that were either continued through innovations or proved to be dead ends and thus virtually "died out."

Fig. 144: The "Phänomobil" of Zittau entrepreneur Gustav Hiller (1863-1913) did not yet have a steering wheel

So when we look at today's mid-range car, it combines the innovations of a long line of ancestors that can easily be traced back to the horse-drawn carriage. Another observation is interesting because it also occurs in the living world. Under similar environmental conditions, natural optimization through mutation and selection leads to similar design solutions, as the shapes of sharks, ichthyosaurs, penguins, and dolphins demonstrate. Even today's mid-range cars are becoming increasingly similar in terms of their shape and equipment, and you sometimes have to look closely for the manufacturer's logo to tell them apart. However, the similarities between the evolution of technical objects and the evolution of life are only superficial - the differences are far too great. The blueprints of automobiles are stored as drawings and documents in archives or on data carriers such as hard disks, and temporarily in the brains of engineers. The "blueprint" of a living being is contained in every basic building block of life, in a cell, in the form of genes. Just as the development of a car can be traced by analyzing archived documents and blueprints, the history of a living being's development can also be learned by analyzing its genes.

320. The mitochondrial Eve

Surprisingly, and in a sense, all humans alive today can be traced back to a single "original mother" - the so-called "Mitochondrial Eve". But how is this possible? To understand this, we need to know that there are two types of genetic material in human cells. First, there is the human DNA in the cell nucleus, which exists in two sets (one set from the mother and one set from the father). Second, there is mitochondrial DNA (mtDNA), which is localized in the mitochondria, the "power plants" of the cells.

This mtDNA is inherited exclusively from the mother, a principle known in genetics as "maternal inheritance". This means that a child always receives its mtDNA from the mother, while the paternal mtDNA is not passed on during fertilization. In this way, the mtDNA is passed on from generation to generation almost unchanged.

Like any other genetic material, mtDNA is also subject to random changes - so-called mutations. Interestingly, the mutation rate of mtDNA is relatively constant. This consistency makes mtDNA suitable as a "genetic clock." With its help, genetic changes can be tracked over very long periods of time, which in turn makes it possible to reconstruct the evolution and spread of humans based on objective criteria.

For example, if a mutation occurs in a less functional section of the mtDNA, this is passed on to all female offspring. Over many generations, this mutation accumulates in the population. By analyzing the degree of enrichment, a method called "coalescent analysis" can be used to determine the approximate time at which this particular allele first appeared.

However, it is also possible to reconstruct an original mtDNA sequence from which the entire current range of mtDNA variation originates. The carrier of this original DNA sequence is referred to as "Mitochondrial Eve." She lived around 150,000 to 200,000 years ago in Africa. Naturally, she was not the only woman of her time, but the mtDNA lineages of all her contemporaries died out over time, and so she became our common ancestor in the female line.

In addition to determining the age of "Mitochondrial Eve," mtDNA also provides insights into the migration history of modern humans. By analyzing the mtDNA, it can be estimated that the first modern humans left Africa around 50,000 to 70,000 years ago. These humans then gradually spread across the entire globe. The diversification of Homo sapiens therefore began around 40,000 years ago, which corresponds to around 2,000 generations.

The findings from the study of mtDNA not only offer astonishing insights into our biological past, but also highlight the common origin of all humans on Earth and the migrations that have led to the current distribution of humanity on Earth.

Fig. 145: Using the methods of genetic analysis, it is possible to show that all people living today are virtually the maternal descendants of an "original mother" who lived somewhere in Africa more than 150,000 years ago

321. Colonization of America

Then, around 15,000 years ago, the colonization of the previously uninhabited continent of America by Asians began via the Bering Strait, which had been exposed due to the Ice Age. Geographers refer to this land bridge as "Beringia," and at that time, it was a tundra-like grassland populated by large mammals (mammoths, musk oxen, reindeer, etc.). Humans probably followed these herds and made their way from Siberia to Alaska, from where they then moved southward. Their numbers must not have been too large, as there are only a few archaeological artifacts clearly linked to these people, known as the "Paleo-Indians." One of the few sites is located in Texas at Buttermilk Creek, where a large number of flint stone tools were found, suggesting an age of between 14,000 and 15,000 years.

Once they arrived in America, they quickly colonized both the northern and southern continents. A similarly ancient site, dated by the radiocarbon method to an age of around 12,000 to 14,000 years, is located in Chile, at Monte Verde, and was discovered in 1977. It is a dwelling site with a tent-like structure that served as a residence for a small group of hunter-gatherers. Finally, another site in Tierra del Fuego is around 2,000 years younger, where waste in the form of animal bones, plant remains, and stone tools from people who once lived there was unearthed in a cave around 1930.

In the following millennia, these first immigrants - particularly in Central and South America - gave rise to great cultures (at least in terms of their architectural and artistic legacies), the remains of which today generate considerable income for these countries through tourism. I am referring here specifically to the Aztecs, the Maya, the Olmecs, and the Incas.

As for the Aztecs and Incas, both cultures were at their peak when a second wave of immigration to America began - caused by the travel plans of a seafarer from Genoa named Christopher Columbus (1451-1506), who wanted to reach the coveted Spice Islands of South Asia by traveling in the "wrong" direction on behalf of the Spanish crown. Columbus was determined to find a sea route to the east, but what he did not know at the time was that an entire continent would block his way. This landmass, later called America, was already home to highly developed civilizations that were just as surprised by the arrival of the Europeans as Columbus himself. He never realized that he had discovered a new continent and therefore believed that he had found islands off the coast of "India." This misconception still survives today in the term "West Indies," despite the fact that there is no trace of "India" anywhere near the Lesser and Greater Antilles or the Bahamas archipelago.

322. Amerigo Vespucci

Only Amerigo Vespucci (1451-1512) realized that he was dealing with a new continent - America, named after his first name. Vespucci was a Florentine explorer and cartographer in the service of the Portuguese and later the Spanish crown. During his travels to South America between 1499 and 1502, Vespucci concluded that the areas he explored were not part of Asia, but rather a separate, previously unknown continent.

The German cartographer Martin Waldseemüller (ca. 1472-1520), who published a world map in 1507, paid particular attention to this insight. On his map, Waldseemüller used the name "America" for the first time to refer to the newly discovered continent. He did this after reading Vespucci's reports and letters, in which Vespucci detailed his findings and the discovery of a new continent. Waldseemüller named the southern part of the new continent after Vespucci's first name, "Amerigo," because he considered him to be the first to recognize the true nature of the New World.

The decision to name the continent after Vespucci was also based on the desire to honor Vespucci's contribution to the discovery and mapping of the New World. In his work "Cosmographiae Introductio", Waldseemüller wrote: "I see no reason why anyone should object to naming it (the continent) after its discoverer and sage Americus, who was the first to establish the nature of these regions". This led to the name "America" being adopted and spread by geography and cartography.

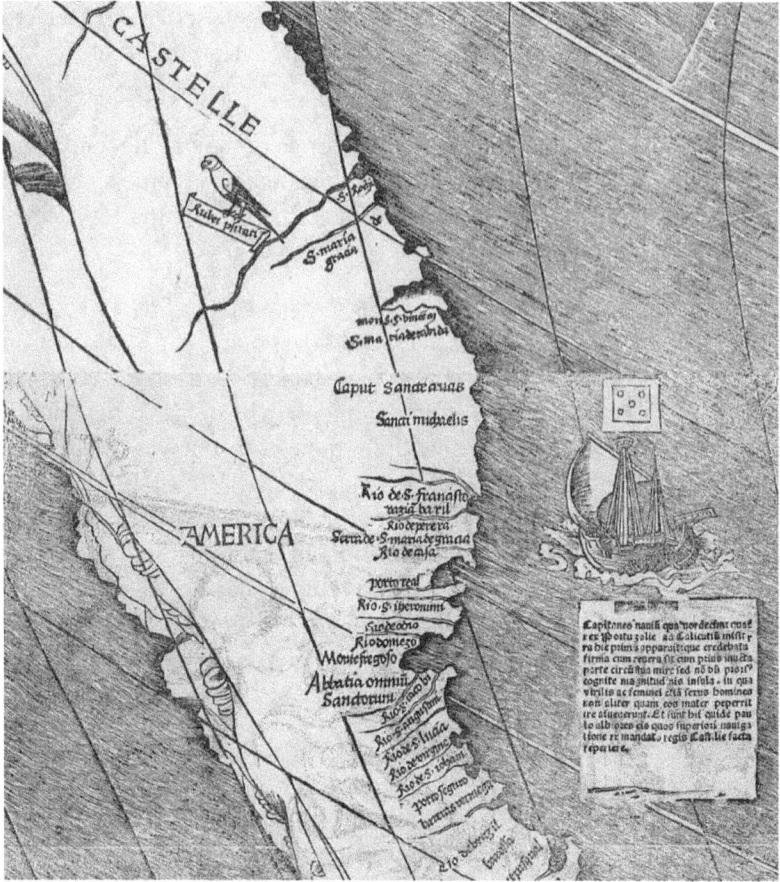

Fig. 146: Waldseemüller's map from 1507, on which the name "America" appears for the first time (Wikimedia)

Interestingly, Vespucci himself was not exactly a well-known figure during his lifetime, but it was only through Waldseemüller's map that his name became permanently associated with the new continent. Over time, his contributions gained recognition, and the name "America" became established for both North and South America. Thus, Vespucci was posthumously honored, and his name became forever linked with "America."

The discovery of the first islands off the coast of the Americas in 1492 set in motion a development that brought unprecedented wealth to several seafaring European states and led to the rapid decline of the American civilizations of the Aztecs (Mexico) and the Incas (Peru) within just a few decades.

323. Cortés and Pizarro

This decline is generally associated with two names: Hernán Cortés (1485-1547), who conquered Mexico, and Francisco Pizarro González (1476-1541), the conqueror of the Inca Empire. The American historian William Hickling Prescott (1796-1859) wrote two remarkable books about the events that led to the astonishing and extremely rapid decline of the Aztecs and Incas, which rival even the best adventure novels in terms of both style and suspense. These are *The History of the Conquest of Mexico* and *The History of the Conquest of Peru,* both of which are still worth reading today if you are interested (e.g., as a free e-book on Google Books).

It is particularly impressive that Prescott, who was completely blind in one eye and almost blind in the other, wrote these works through immense effort despite his severe visual impairment. Using a "noctograph," a special writing frame, he produced these works under great hardship, which still have a lasting impact today.

Although not all the details may correspond to historical reality, Prescott's works still convey an extremely vivid picture of the events. They illustrate in a gripping way how a small but determined group of adventurers, driven by their greed for gold and power, managed to permanently destroy two powerful empires and change the course of history on the American continent forever.

324. Aguirre, the wrath of God

A prototype of such adventurers (and a particularly unpleasant one at that) was certainly the Basque Lope de Aguirre (1511-1561), who arrived in Peru in the wake of Pizarro and from there attempted to search for the "gold country" (Eldorado) that was rumored to exist in the still completely unknown Amazon region. His Amazon expedition, which began in Lima in 1559 and led him, along with around 300 Spanish conquistadors and local Indian helpers, to the middle reaches of the Amazon and from there to its mouth in the South Atlantic, became the model for a remarkable film by Werner Herzog. The film excellently captures the atmosphere and drama of such an expedition. I am referring to Aguirre, the Wrath of God, starring the unforgettable Klaus Kinski (1926-1991) in the leading role.

Although the film does not closely follow the historical record, it vividly illustrates the type of people who might have engaged in such life-threatening adventures under the command of a charismatic, but - as in this case - also maniacal leader. Klaus Kinski was undoubtedly the ideal actor for this role. One might even say he was almost playing himself.

In addition to the perilous descent from the heights of the Andes into the humid, steamy Amazon lowlands (filmed on location!), and the raft trip on the raging upper reaches of the jungle river, Aguirre's climactic monologue (delivered by Klaus Kinski) is likely to remain etched in memory:

„Ich bin der Zorn Gottes. Die Erde, über die ich gehe, sieht mich und bebt. Wer aber mir und dem Fluss folgt, wird unerhörten Reichtum erlangen."[70]

es, the driving force behind the conquistadors was nothing other than greed for wealth, greed for gold. A mythical land located in South America played an important role in this: Eldorado. It was said to be a place of unimaginable wealth, with an abundance of gold. This myth "originated" as a rumor around 1541 and then developed a remarkable life of its own, until it was finally relegated to the realm of fables around 1800 by Alexander von Humboldt (1769-1859). The land does not exist, it never did, but the name has endured - whether as a catchphrase, a place or pub name, a car brand, or even a movie title (just think of John Wayne) - it remains part of our cultural memory.

The "real" Aguirre managed to make his way to the mouth of the Amazon - of course, without finding any trace of this "land of gold" - and fled to what is now Venezuela after occupying Isla de Margarita (Lesser Antilles), where he acted as a tyrant for a brief period. However, when he attempted to return to Peru on foot with his remaining followers, he was confronted by troops loyal to the king and was ultimately shot by his own men after being captured - likely with an arquebus (a type of early firearm).

325. Arquebus

This was the firearm used exclusively by the conquistadors, capable of firing bullets around 20 mm in diameter. This "device" caused more fear and terror because of the loud bang it made when fired than because of its accuracy. For this reason, it was usually only effective in warfare when an entire battery fired simultaneously. Afterward, the shooters often had to immediately switch to close combat, as there was hardly any time to reload (they were muzzle-loading rifles powered by black powder). Interestingly, the name "arquebus" (meaning "hook rifle") comes from the fact that the rifle, which was sometimes over two meters long, originally had a "hook" attached to the barrel. This hook allowed the rifle to be "hooked" (e.g., onto a wall when firing from a castle wall) to absorb the sometimes-enormous recoil.

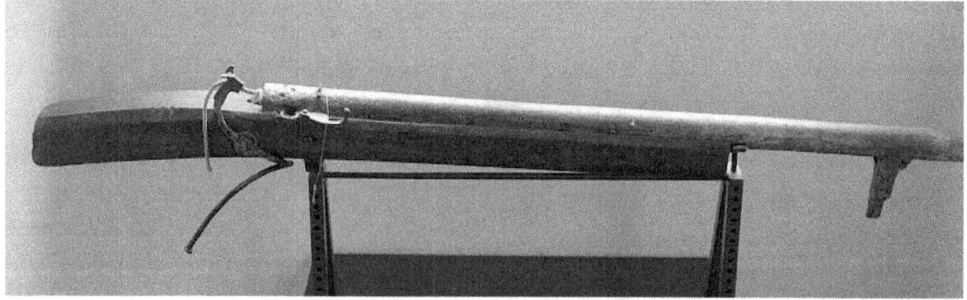

Fig. 147: Typical Arquebuse with "hook" on the foreleg (Wikimedia)

---

[70] I am the wrath of God. The earth I pass will see me and tremble. But whoever follows me and the river, will win untold riches.

326. Gunpowder

The first systematic military use of black powder as "gunpowder" in an open field battle took place during the so-called "Hundred Years' War" (between 1337 and 1453). Various types of cannonballs (often made of stone or metal) were fired using "fire pots," as the first cannons were called. Their advantage was that they had a much greater range than longbows (the most feared weapon of war at the time). At the same time, the so-called "blunderbusses" also appeared, but they were difficult to handle and extremely inaccurate. Nevertheless, they could penetrate a knight's armor at a distance of around 100 meters. A special mixture of potassium nitrate (more precisely potassium nitrate), charcoal, and sulfur was used as a propellant or explosive. It was later claimed that this "invention" came from a monk from Freiburg named Bertold Schwarz (around 1359), but this is not proven and does not coincide with his first use. What is certain, however, is that Roger Bacon (1218-1292), a learned English Franciscan monk, already knew the recipe, as he described the production of "black powder" in his work "Opus Maius" in 1266. However, where he got the recipe from is still unknown to historians today. In any case, black powder already existed in China in the 11th century. As you can easily imagine, the demand for gunpowder increased massively from the 14th century onwards. Its production thus became a lucrative (but also dangerous) business

327. Powder mills

It was the responsibility of so-called "powder mills", which sprang up like mushrooms in suitable locations at the time. Their task was to finely grind the "ingredients" of saltpetre, sulphur and charcoal and mix them in the right proportions. The result (the black powder or gunpowder) was then filled into barrels and sacks and sold via middlemen.

Incidentally, almost all powder mills were water mills built outside of villages in forest valleys. They can be recognized by a horseshoe-shaped protective wall around the mill, which protected adjacent farm and residential buildings in the event of an "explosion" - something that had to be expected once every few decades. When such a mill "burst," it was understandably very conspicuous and was therefore often noted in chronicles. Today, with a few exceptions, the only reminders of their former locations are the names of places and paths.

It is perhaps interesting to note that only charcoal made from the wood of the decaying tree (Frangula alnus) was used for particularly high-quality black powder, as it has a very low ash content. This "tree," which is more of a shrub, has therefore been given the name "powder wood" in some areas.

In connection with gunpowder, a number of idioms have developed in German that are still used today. For example, if you find yourself in a dangerous situation, you are "sitting" on a powder keg, so to speak. If you're a cyclist in a bike race and you're always chasing the person in front of you without taking the lead yourself, you're doing nothing more than "keeping your powder dry." And anyone who couldn't smell powder was simply a coward. If, on the other hand, you want to characterize someone as stupid and simple-minded, it is usually enough to imply that they "didn't invent the powder." Behind these well-known idioms, the connection with gunpowder is still easily recognizable. However, there are also idioms that everyone is familiar with and that everyone uses without knowing where they actually come from or what context they are derived from.

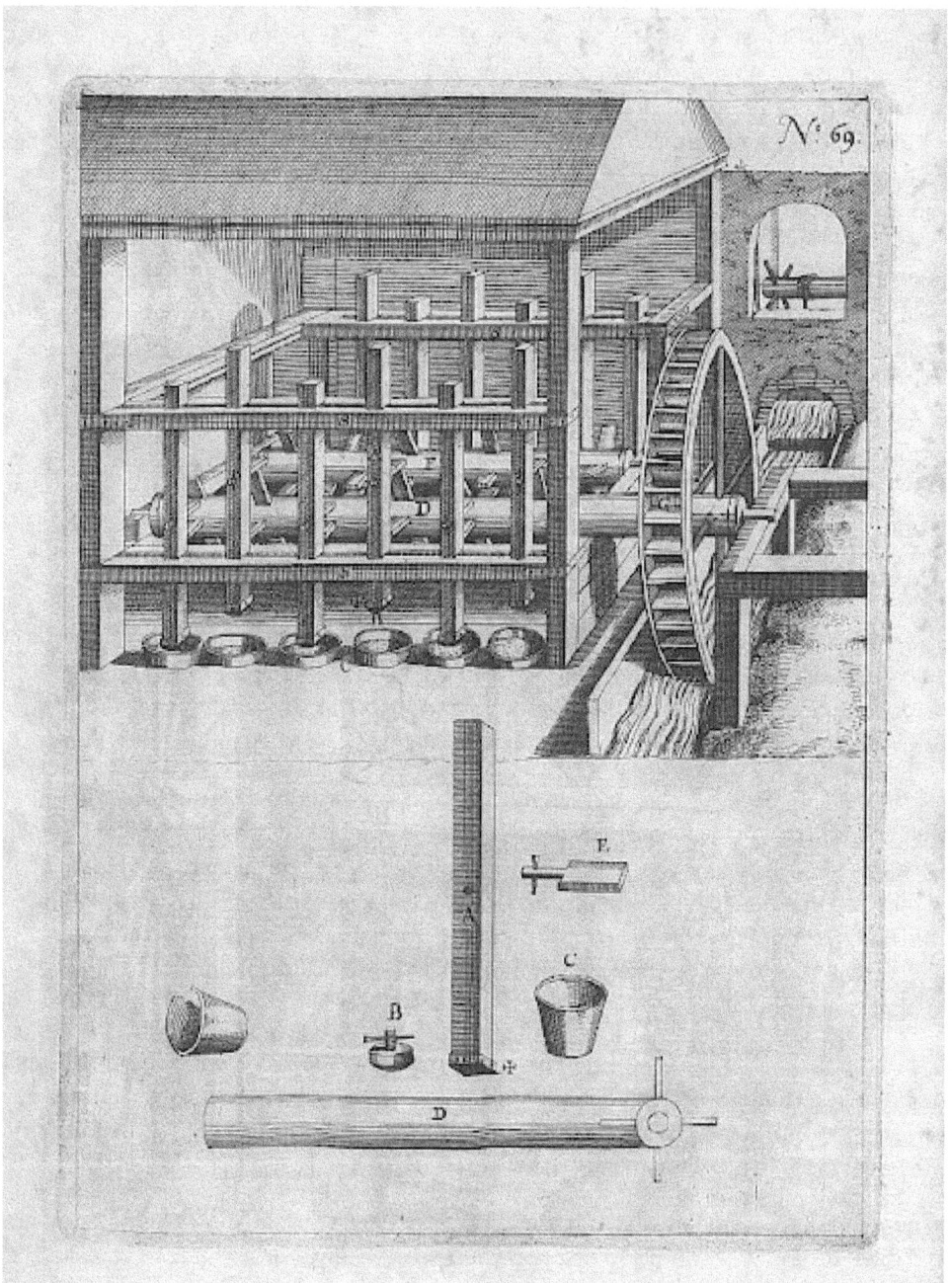

Fig. 148: Structure of a typical powder mill (Deutsche Fotothek)

Let's take the phrase "making blue," which is popular among some employees. Everyone knows what it means: staying away from work today. And if it's a "Monday," it's called "blue Monday." The most likely interpretation refers to the non-working Monday before Shrove Tuesday, when, according to the medieval dress code, people were allowed to wear colorful garments - especially indigo blue - which were otherwise reserved for Sundays. Lent Monday thus became a "blue Monday" and the resulting idiom a term for a day off work. But other interpretations are also in circulation.

328. Sitting in the ink

Or let's consider another saying whose meaning is immediately obvious to us, but whose origin is hardly known: "sitting in ink." The phrase first appears in the sermons of a certain Johann Geiler von Kaysersberg (1445-1510) from Schaffhausen, whose intent and livelihood consisted of delivering powerful sermons, which he had recorded and printed. He is also regarded as a spiritual pioneer of the witch craze in Germany, which does not seem to fit with his otherwise humanist views. In his sermons concerning the "Ship of Fools" by Sebastian Brant (1457-1521), the line *"Du bist voller Sünd, du steckst mitten in der Tinte"* (You are full of sin; you are in the middle of the ink) can be found. This then became the saying that expresses the misery that befalls you as soon as your problems threaten to get out of hand. The fact that good ink sticks firmly and is difficult to wash off probably plays a role here.

Ever since humans stopped writing with a wedge-shaped stylus on clay tablets, they have used "colored water" (because that's what "tincta aqua" means) to write down their thoughts on paper-like materials such as papyrus or parchment. The Latin word "tincta" then evolved into the Old High German "tincte" (which still lives on in "tincture"), until someone finally dropped the "c" - and the word "Tinte" (ink) was born.

Over the centuries, a variety of different inks were developed, but the most important and eventually most commonly used was based on the round balls that adorn some oak leaves in late summer. These are the galls of the oak gall wasp (Cynips quercusfolii). They used to be collected in large quantities to make a decoction consisting largely of gallic acid, which was mixed with iron(II) sulfate and a little gum arabic. This produced a dark liquid that was very good for writing with quills (e.g., sharpened quills). When the ink dried, the bivalent iron was oxidized to produce deep dark trivalent iron, which, together with the gallic acid, formed a deep black complex compound. Documents written with this type of iron gall ink show a particularly beautiful contrast to the white paper. Of course, other colors of ink were also used in the scribes' rooms of medieval monasteries or in the royal and imperial chancelleries. For example, there were red inks made from the secretions of purple snails. Monks, on the other hand, developed expensive gold and silver inks, which were usually only used to artistically design the initial letters of new paragraphs as well as illustrations.

Fig. 149: Dr. Johannes Geiler von Kaysersberg (1445-1510), preacher (lithograph by P. and F. Simon, Wikimedia)

329. Codex Argenteus

An exception here is the "*Codex Argenteus*," a copy of the so-called "Wulfila Bible" from the 6th century, written in Gothic. Only silver and gold inks were used in this manuscript, and these inks were naturally very expensive.

Gold and silver inks were often used in manuscripts to emphasize their special significance. They were made from finely ground metal powder mixed with a solution to create a writable ink. Gold ink often consisted of real gold leaf ground into a fine powder, while silver ink could be made from fine silver powder or even mercury.

The production process for these inks was complex and costly, making them luxury items used only for important documents and valuable manuscripts. Moreover, these inks were not only expensive to produce but also to use. To bring out the precious metals, the writing materials had to be prepared with particular care.

Gold and silver inks also had the advantage of fading less over time compared to organic inks, which increased the durability of the texts. These inks reflected light in a special way, giving the manuscripts a radiant appearance that further enhanced their cultural and aesthetic value.

The use of these inks in the *"Codex Argenteus"* is an outstanding example of luxurious and artistic book production in late antiquity and the early Middle Ages.

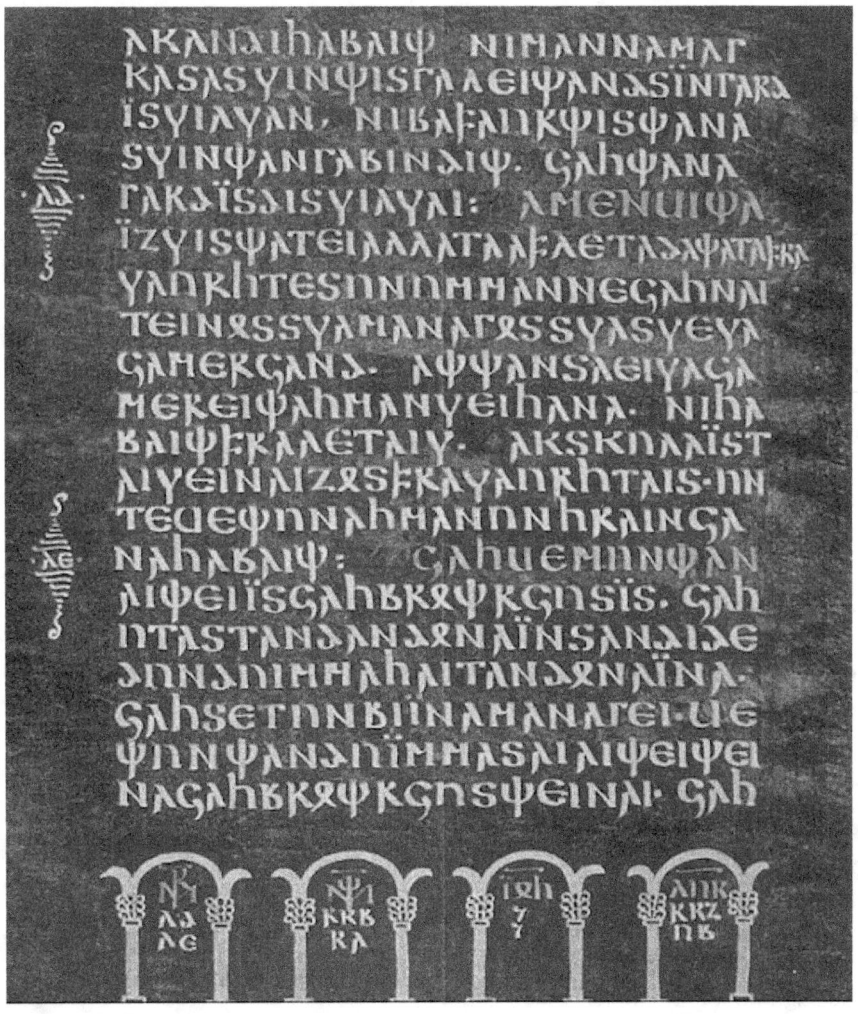

Fig. 150: Page from the Codex Argenteus written in silver and gold ink (Creative Commons)

Even more expensive today are certain branded inks for inkjet printers, whose prices per liter are sometimes exorbitant - though, to reassure you, they are nowhere near the cost of a liter of scorpion venom (~ €24 million). Behind these high prices (especially for colored printer inks) are not only manufacturing and development costs but also a clever business strategy: make the inkjet printer as cheap as possible to encourage purchase, and then profit later from the consumables (ink cartridges, specialty papers). However, this business strategy only works in part. There was a time when inkjet printers were only marginally more expensive than their set of ink cartridges, but the trade in refill kits and inexpensive inks has since flourished.

330. Inkjet printer printheads

However, if you want to obtain high-quality prints of photographs, it is better to use the printer manufacturer's original inks, as they are usually developed in conjunction with the print head. Such a print head, with nozzles measuring only a few micrometers (μm), from which tiny ink droplets are sprayed onto the paper under precise control, represents cutting-edge technology. There are several types of print heads, among which the piezo print head developed by Epson is particularly instructive in its operation. It utilizes a special solid-state physical effect discovered in 1880 by Jacques and Pierre Curie in tourmaline crystals, known as the "piezoelectric effect." This effect is a property of all ferroelectric crystals (which have a very high dielectric constant) and causes them to become electrically charged in the direction of one of their polar crystal axes when subjected to tensile or compressive stress. Certain ceramics, in particular, can be expanded and contracted by applying an alternating voltage at their resonance frequency using this effect. This principle is precisely what is exploited in a piezo print head (or more specifically, in a nozzle within such a print head).

A nozzle made of piezoelectric ceramic, when subjected to a voltage pulse, reduces its volume, causing an ink droplet to be expelled from the nozzle at high speed (typically 10 m/s). The subsequent voltage pulse increases the volume, allowing ink to flow into the nozzle from the tank through capillary action. This process creates a stream of ink droplets, achieving "firing frequencies" of up to 150,000 droplets per second per nozzle. By varying the pulse amplitude, the size of the ink droplets can also be controlled.

However, print heads can now be mass-produced at low cost using special etching processes similar to those used in electronic circuits. In this method, the ink is abruptly heated in a tiny chamber, causing it to expand and eject from the nozzle. As a result, print heads have become integral parts of the ink cartridges and are therefore disposable.

In contrast, piezoelectric print heads are reserved for high-end inkjet printers (which have now also replaced pen-based plotters in engineering and design offices), where precise alignment of the print head is crucial.

Fig. 151: Typical inkjet print head with the four ink tanks

331. Printing banknotes

By the way, did you know that modern inkjet printers can print banknotes quite passably (in terms of color quality, not security features) if they could? In the meantime, printers and photocopiers have integrated recognition software that either completely or at least partially prevents printing, so that the printer owner does not even have this obvious idea and makes himself liable to prosecution. Incidentally, this also applies to a certain number of image processing programs (e.g., Photoshop), which refuse to process euro or dollar bills, for example...

In this respect, the time-honored profession of counterfeiter still has a hard time despite modern technology - if he is merely a home worker and not the operator of a suitably well-equipped print shop. But even in this area of criminal gainful employment, there used to be - and still are today - truly gifted artists. One such artist, for example, was exposed due to carelessness (or stupidity, depending on how you want to look at it) in Cologne in 2006, as a number of newspapers reported at the time.

It started with an excavator driver at a garbage tip. As he was digging away, just like on any other working day, his excavator shovel suddenly unearthed a few sacks from the garbage (there were seven in the end), from which a large number of shredded dollar banknotes trickled out! Needless to say, the criminal investigation department immediately took an interest in these sacks and the circumstances of their discovery. In particular, their specialists for counterfeit money, sub-division "Dollar Blossoms", were immediately impressed by the high graphic quality of the finds (although they were obviously to be discarded as misprints). One of these specialists even expressed the opinion to a journalist that the counterfeits could only be distinguished from the originals by the printing paper used. And it was true. The forger was a master of his trade, respected as a lithographer and graphic artist ("Andy Warhol from Cologne") and even made it to an entry in Wikipedia with these skills. One day, he obtained a $100 banknote from the local savings bank, which he scanned (back then this was still no problem) and altered the serial number using a graphics program. After many attempts, he used the resulting image file to create high-quality printing plates for his private offset printing press. But by then the investigating authorities already had their eye on him, because he had disposed of documents with his name and address in the aforementioned bin bags along with the test and misprints. This meant he could be observed immediately and a trap could be set for him in peace and quiet by elegantly enticing a prospective buyer for the dollar flowers. The rest is quickly told. He was arrested at the very moment he wanted to sell fake dollars for real euros, his counterfeiting workshop was searched and broken up, and after the trial he was sentenced to several years in prison - because when it comes to money, the authorities know no mercy. And that has always been the case. Counterfeiting has existed ever since money has existed. Until early modern times, counterfeiters who were caught were almost always threatened with the death penalty.

With the advent of banknotes, counterfeiting almost developed into a shadow economy. It is estimated that around 1861, around half of the paper money in circulation in the United States was counterfeit. After the end of the Civil War, the state therefore went to great lengths to ensure that it did not lose control of the money. A special secret service tracked down the counterfeiting workshops and brought the counterfeiters to justice, which was not exactly squeamish in this respect. Towards the end of the 19th century, this "fiscal" problem was largely solved.

332. Real blossoms

The case of the Portuguese Artur Virgílio Alves dos Reis (1898–1955) demonstrates that flooding a country with counterfeit money can certainly lead to serious economic problems. His ingenious idea was to have counterfeit money printed by a legal money printing company using the same legal printing plates that were used to print the official "real" banknotes. The problem that had to be solved can be summarized in a nutshell as follows: He had to convince the banknote printing house ("Waterlow and Sons" in London), under great secrecy, with a plausible story and a large number of forged but notarized contracts, to print the desired number of banknotes (in this case, escudo bills with the likeness of Vasco da Gama) for his fictitious client - using the same printing plates with which the printing house had previously carried out an order from the Bank of Portugal. The whole thing was so cleverly arranged that the operation was completely secret and went off without a hitch. The only drawback (and what ultimately proved to be his undoing) was that the banknotes had the same serial numbers as the legal batch already in circulation in Portugal.

Fig. 152: There were two copies of this banknote with the same serial number, which ultimately led to problems and questions...

After the delivery of the 200,000 500-escudo banknotes, dos Reis managed to put a significant number of them into circulation in Portugal, which of course was noticed at some point. The number of counterfeits almost reached the number of legal banknotes with the same nominal value. However, random checks showed that all the bills were obviously genuine. The miraculous increase in money simply could not be explained - and the decision was made to hastily withdraw all 500-escudo bills from circulation. However, this led to great resentment among the population, which is why this action was quickly called off again. In the meantime, however, counterfeit money expert Luís Alberto Campos e Sá discovered that some of the serial numbers were obviously duplicated. And so orders were issued to the banks to sort all the corresponding banknotes in their possession by serial number. In this way, the extent of the crime slowly became apparent - and this quickly led to the trail of Artur Virgílio Alves dos Reis, who was arrested before he could flee abroad. He did not suffer the fate of medieval counterfeiters - who were often thrown into boiling oil - but he did receive 20 years in prison (15 of which he actually served). After his release from prison, he turned down offers from several large banks and settled down as a coffee planter in Angola, becoming a devout Protestant. Artur Virgílio Alves dos Reis died there in 1955, completely impoverished. In 1925, his "banknote bubble" brought Portugal as a state into great distress. This event was certainly also largely responsible for the fact that in 1932 the military dictatorship (since 1926) developed into a fascist state under António de Oliveira Salazar (1889–1970), which was only swept away by the "Carnation Revolution" in 1974.

Speaking of "carnations"...

333. Tulip mania

Another banking crisis had to do not with "carnations" but with "tulips", which went down in banking history as the bursting of the first really big financial and speculative bubble. It did not affect Portugal either, but the Netherlands, in 1637/1638.

Fig. 153: Illustration from a Flemish tulip book

And that's how it came about: The Netherlands is still known today as the land of tulips. If you travel through the flat country in spring (especially the area around Keukenhof), you will notice the colorful tulip fields everywhere. They bear witness to a special fondness of the Dutch for this ornamental flower, which originally came from Turkey (Ottoman Empire) and Iran (Persia) and which now - thanks to the diligence of Dutch gardeners - comes in an immense variety of colors and patterns.

After it arrived in Central Europe in the middle of the 16th century, where it adorned the first palace gardens (e.g., in Vienna) and its bulbs could be easily distributed or sold to interested parties, it quickly became a sought-after commodity. Many gardeners, especially in the Netherlands, began crossing different tulip varieties with each other to breed new, never-before-seen color and pattern variations. At some point, flower lovers began to spend more and more money on tulip bulbs. A craze developed in which wealthy citizens planted their own tulip gardens, formed associations and exchange groups, and a lively trade in tulip bulbs developed, which fetched higher prices the more sophisticated the shape of the flowers, their color, and their color patterns were. In some circles, tulip bulbs first became collector's items, then increasingly a new type of investment that could be easily increased through cultivation. At the beginning of the 17th century, the tulip trade became a rapidly growing economic factor in the Netherlands. The first color auction catalogs were printed (the so-called "tulip books"), some criminals financed themselves by stealing tulip bulbs (which in turn boosted the security trade), and - from around 1634 - speculators finally discovered the tulip bulb as an ideal object of speculation. Suddenly, financial jugglers began to buy up tulip bulbs - not to decorate their own gardens with colorful tulips in spring, but to sell them at auction for a profit. As early as 1623, individual bulbs of particularly rare tulips fetched up to 1,000 guilders. This corresponded to around six times the annual earnings of a simple craftsman at the time.

As prices rose, the tulip developed from a collector's item to a status symbol, which was a wonderful way to show off in certain circles (just like a Porsche today). A bulb of the same type of tulip, which cost 1,000 guilders in 1623 ("Semper Aurora"), already cost 5,000 guilders in 1629. And prices continued to rise. More and more wealthy people put their money into tulip bulbs, which they bought at the frequent auctions of the time. Soon there were tulip derivatives, share certificates on tulip bulbs were issued, and creative "securities" based on tulips were invented. If you had enough guilders (you could buy a whole house in Amsterdam for three tulip bulbs of a particularly rare color variety), you could also buy tradable subscription rights. You no longer enjoyed a blooming tulip but pressed it (without the bulb, of course) to show the buyer what he could expect from his purchase the following year. This in turn formed the basis for forward transactions, where it was possible to trade in bulbs that were still in the ground. Next to the variety labels, signs with the promissory notes of the future owners of the bulbs were stuck in the soil of the tulip beds. Of course, this couldn't go on for long. At some point, they were no longer able to find buyers who could raise enough money - the "tulip bubble" burst. And this is exactly what happened in 1637, when the auction proceeds fell further and further below the set starting price until finally, none of the tulip bulbs on offer fetched the amount for which they had once been purchased. Prices fell so rapidly that many of those affected lost their entire fortune from one day to the next - and if they had previously bought the tulip bulbs on credit, they were suddenly trapped in debt. This crash began to jeopardize the entire Dutch economy, prompting the government to decree that tulips were no longer to be regarded as objects of speculation, but only as normal goods that had to be paid for in cash.

The emergence of the price bubble associated with this "tulpomania" was probably facilitated by an inflationary monetary policy. This is because the money flows of gold and silver from their extensive colonial empire met in Holland to be converted here. This growing amount of money, which was particularly concentrated in the hands of rich merchants ("pepper sacks"), was simply looking for investment opportunities. And here it happened to be the tulip, which differed in shape, color, and pattern from all other flowers grown for ornamental purposes at the time. The inevitable crash certainly had a drastic impact on the national economy of the Netherlands at the time (there are reports of many bankruptcies in 1637/38). There was also an increase in suicides. Someone who had previously been "filthy rich" found himself in the gutter, as a well-known stock market speculator once put it. But the crash was probably still limited, as it was largely cushioned by the economic strength of the colonies. However, it can still serve as a textbook example of a stock market crash in which a speculative bubble bursts. You only need to replace "tulip" with "real estate" or "dotcom" and you have arrived in the present.

Holland is still the "tulip country" par excellence, although it is now under heavy pressure from foreign tulip producers, particularly from Africa. In 2002, for example, there was a sharp drop in prices, which caused a number of tulip producers in the Netherlands to go bankrupt. There is no doubt that Dutch growers have remained "the" tulip growers to this day. Every year, new varieties are added to the 1,200 or so that are regularly grown.

334. Tulip mosaic virus

Multicolored varieties whose petals are "flamed" in different colors (so-called "Rembrandt tulips") have always been particularly popular. When they first appeared in Holland, they made a significant contribution to the emergence of tulipomania. This is because they emerged "suddenly" and not deliberately through breeding. Tulip bulbs, which had initially only produced single-colored tulips, suddenly showed color changes up to the aforementioned "mottled" color patterns. What was not known at the time was that the tulips were suffering from a viral infection, similar to the way humans occasionally suffer from a flu infection (which is usually easy to see). This virus, which causes the "valuable" color and pattern change, is the tulip mosaic virus. It belongs to the large group of potyviruses - elongated protein envelopes that enclose RNA and are transmitted from plant to plant by aphids. Once they have penetrated a plant cell, they reprogram its protein synthesis apparatus, like all viruses, with the aim of producing as many copies of themselves as possible. In many plants, this results in mosaic-like damage patterns on their leaves, which is why they are also referred to as mosaic viruses (a frequently studied virus here is the tobacco mosaic virus, for example).

In tulips, on the other hand, the petals change color and even take on the flame-like hues that made tulip bulbs so valuable in their day. However, such a viral infection naturally also weakens the affected plant in the long term, so that some valuable varieties, such as the legendary "Semper Augustus," have since died out. At the time of tulipomania, up to 10,000 guilders - the equivalent of a small house - was paid for a "Semper Augustus" bulb. Today, you can buy "flamed" tulips for little money in DIY stores. However, these are genuine cultivars and no longer seriously diseased plants...

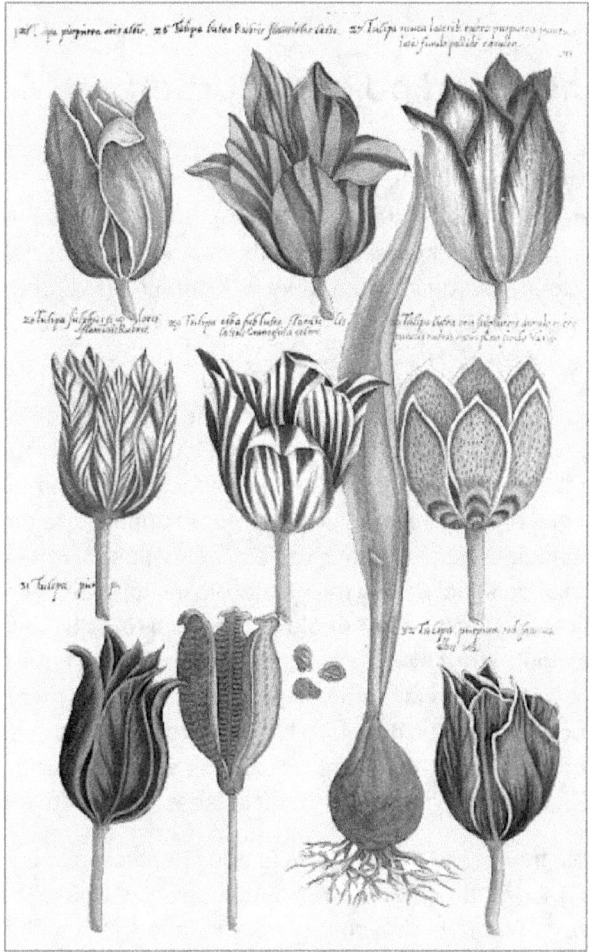

Fig. 154: "Rembrandt tulips" are characterized by unique colour patterns and nuances

335. Poisonous plants

However, it is not only plants that can become "seriously ill," but also humans and animals from certain plants, which are therefore simply called "poisonous plants." For example, it is not necessarily recommended to eat tulip bulbs or (even more dangerous) lily of the valley bulbs as a substitute for shallots. The lily of the valley, in particular, is considered especially poisonous due to its content of a large number of glycosides. It is not even advisable to drink the vase water. In any case, children should be discouraged from eating the beautiful red berries of the lily of the valley in summer.

Poisons have always played a prominent role in the history of humankind when it came to sending rivals, "enemies," disliked contemporaries, or even rulers of all kinds to the afterlife. The fear of poison attacks created the not entirely harmless profession of "taster," and people whose passion it was to kill others with poison were known as "poisoners."

336. Poisoners and poison murders

Some female personalities in particular had devoted themselves to this passion, such as the famous Marquise de Brinvilliers (1630-1676). But poisoners have always been in a bad position when the authorities get hold of them. And so, it is certainly not surprising that the aforementioned marquise did not die of natural causes. She was beheaded in front of a large crowd in Paris on July 17, 1676, after being subjected to a highly unpleasant water torture.

The "poison concoction" reached its peak during the Renaissance in Italy, in the city-states of Florence, Venice, and Genoa. At that time, it was possible to make an endless amount of money there with "poison-indicating" miracle cures: Amulets with precious stones, "snake fangs," and strong-smelling oils were supposed to protect against poisoning, as were candelabras made from eagles' feet, whose flames supposedly went out in the presence of poisons. During the Medici era in Florence, there are even said to have been state-licensed "poisoners" (if old writings are to be believed), from whom a Doge, for example, could virtually order poison murders via a catalog and price list. The prices were quite differentiated and almost affordable for a rich man. For example, the murder of a pope cost around 100 ducats, which in retrospect was a real bargain. Poison murders, on the other hand, which were associated with longer business trips, were more expensive. For example, you had to invest at least 500 ducats to poison a sultan.

If you look at the biographies of more or less famous Renaissance princes and ecclesiastical dignitaries living at the time, you will find quite a few whose lives were ended by poison. The House of Borgia is particularly notorious in this respect, and an opulent Franco-German television series has recently shed light on the subject. One of the most evil figures from this family was undoubtedly Rodrigo Borgia (1431-1503), who left the conclave in 1492 as Pope Alexander VI. He is said to have put aside large numbers of his real and suspected "enemies" with the infamous "Borgia poison." At the time, "poison" was not only considered an effective means of personnel policy in the Vatican. Today we know that "Cantarella" (the "Borgia poison") was a particularly effective arsenic-containing odorless and tasteless powder, the effects of which were ultimately experienced by Alexander VI himself and his son Cesare (1476-1507).

On August 5, 1503, they both sat together with Cardinal Adriano Castellesi da Corneto (1460-1521) in the cardinal's garden and enjoyed their dinner. Shortly afterwards, the cardinal became nauseous and developed a fever and chills. The next morning, Borgia's father and son showed the same symptoms. The news that the Pope had been poisoned quickly made the rounds in Rome. But contrary to expectations, both made a slight recovery in the following days, which did not last in the case of Pope Alexander VI, and he finally died on the night of August 17, 1503, to August 18, 1503. Contemporary sources report that the pope's corpse turned black, swelled up unnaturally, and emitted foul-smelling fluids. In short, the people concluded that the hated Borgia had been deservedly taken by the devil. However, the question of how this apparent arsenic poisoning occurred in the first place remains unanswered to this day. One entirely plausible theory assumes that the cardinal wanted to forestall his own assassination and accepted his own, but not fatal, poisoning just to send Alexander VI where he thought he belonged. After all, he was extremely rich, and the Borgias certainly had their eye on his fortune. However, it is also possible that the two Borgias wanted to poison the cardinal, but by mistake (or on purpose?) the poison was administered to them by one of the initiated servants. Be that as it may. Cesare and the cardinal survived the poison attack, but Rodrigo Borgia did not.

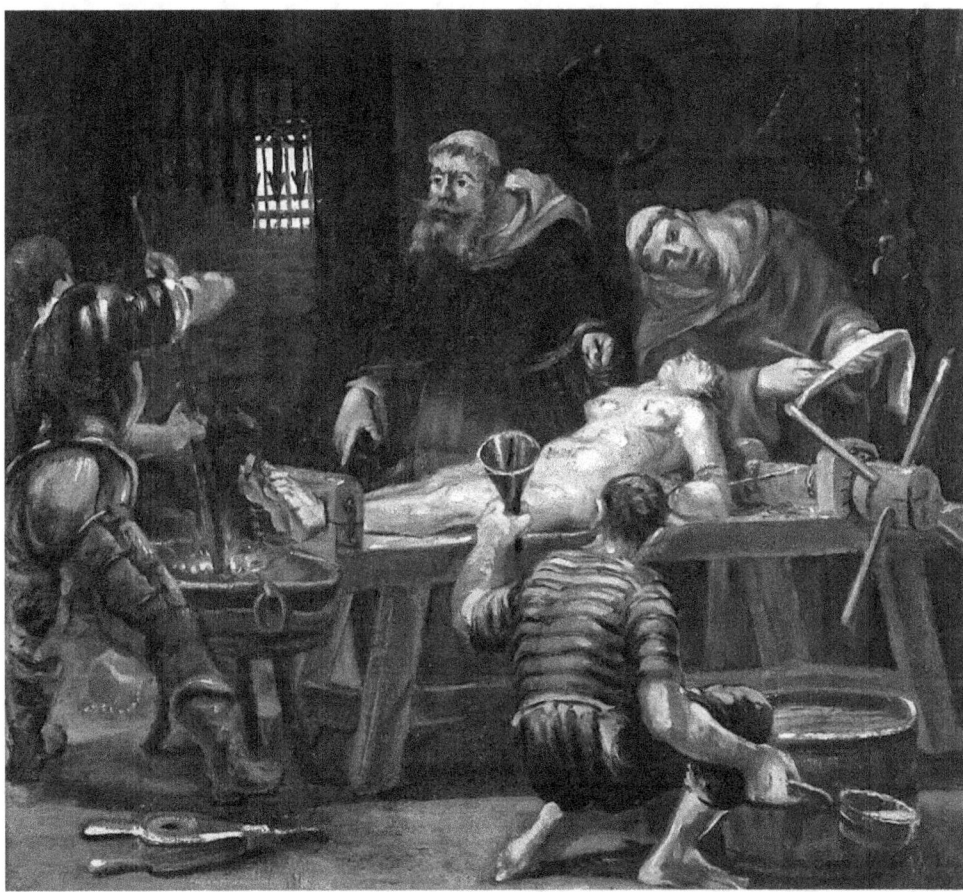

Ab. 155: The judicial authorities of the time were not exactly squeamish in their treatment of the poisoner Marquise de Brinvilliers (painting by Jean-Baptiste Cariven, Wikimedia)

Arsenic (an arsenic compound), popularly and affectionately known as "rat poison," was the poison of choice for many centuries if you wanted to eliminate someone inconspicuously. The appropriate dosage made it possible to achieve a more or less long "course of the disease," which significantly reduced the risk of detection for the perpetrator. There are several examples in history where caring wives gradually poisoned their weary husbands with small doses of arsenic while apparently and for all to see, sacrificially nursing them to the bitter end (the symptoms were very similar to those of cholera). The names of some of these ladies have survived to this day, of which the following are just a few examples for our own research: Florence Maybrick (1862-1941), Marie Lafarge (1816-1852), and Charlotte Ursinus (1760-1836).

337. The Marsh sample

Murder could rarely be proven against such poisoners until a certain James Marsh (1790-1846), a chemist by profession, appeared as an expert witness. He developed a method that could detect even the smallest amounts of arsenic clearly and reliably in court. As a result, arsenic became the poison of choice in the early 19th century (arsenic, in the form of its oxide, was available in powder form in virtually every household at the time, e.g., as a "fly stone" for systematic fly-killing or as a poison for mice and rats), and it quickly disappeared from the arsenal of poisoners, where it had previously held a particularly prominent place. This disappearance is connected to an 80-year-old domestic tyrant known as Georges Bodle, a farmer from Plumstaed in England, who died in agony in 1832 surrounded by his family, who were quite pleased about it. The neighbors and authorities quickly suspected that the death of the domineering elderly gentleman was the result of poisoning, which they blamed on one of the deceased's sons, the "younger" John Bodle. When it came to the inevitable jury trial, James Marsh, called as an expert witness, failed to convince the jury of the defendant's guilt with his evidence, which was evident to him, so the jury acquitted the defendant to his delight (a decade later, he confessed to the family's joint poisoning using the powder known as "poudre de succession").

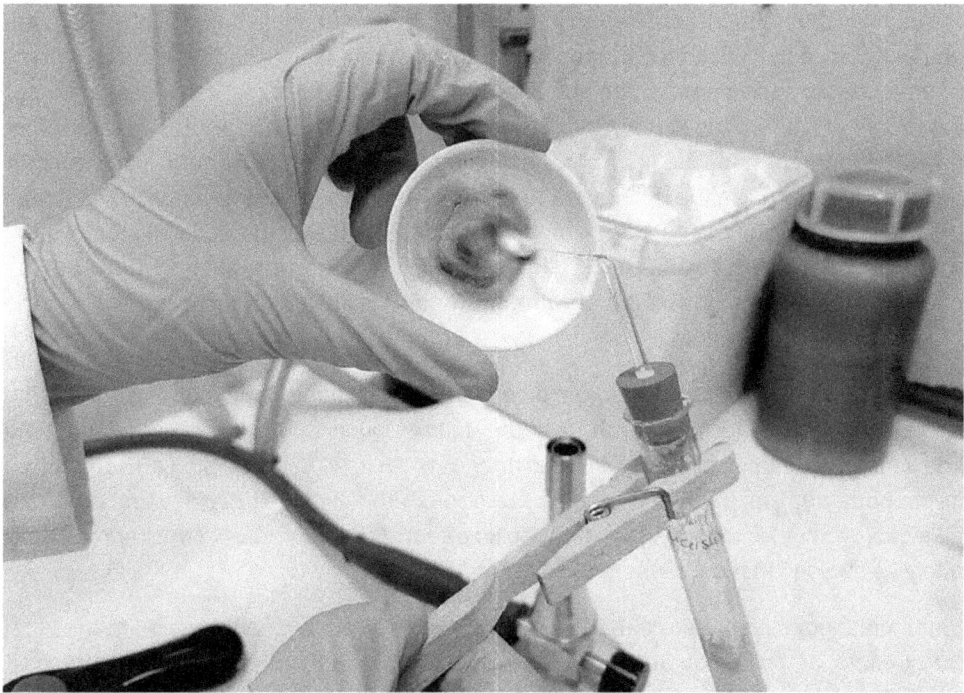

Fig. 156: Positive Marsh sample

This acquittal annoyed Marsh so much that he began to think about a new detection method based on the hydrogen arsenic discovered by Carl Wilhelm Scheele (1742-1786). Thus, he quickly invented the "Marsh test," which would subsequently cause all "poisoners" working with arsenic to fear the law (to put it in gender-neutral terms). The equipment he used was surprisingly simple. All he really needed was a glass flask with a glass tube extending from it and ending in a nozzle. In addition to the sample material (e.g., a piece of the victim's stomach or stomach contents), the glass flask contained some arsenic-free hydrochloric acid. If a little zinc is added, atomic hydrogen is produced, quasi in statu nascendi, which immediately combines with the arsenic to form arsine. The remaining hydrogen combines to form normal hydrogen molecules. This mixture then escapes from the nozzle and can be ignited. If you place a white porcelain bowl over the flame, it will either remain white (the victim has not been poisoned with arsenic) or it will turn black due to the deposited metallic arsenic (arsenic level, the accused has a problem). In this way, even the smallest amounts of this toxic substance can be detected.

338. Poisonous plants from the garden and from fields and meadows

But back to plant-based poisons. It's hard to believe, but fields and meadows, and even more so our ornamental gardens, are home to poisons that can make arsenic seem quite harmless (warning: polemic!). Here are a few examples. The striking dark berries of deadly nightshade (*Atropa belladonna*) should be avoided if at all possible. The Latin name of the genus, which tells botanists that it is a nightshade plant like the tomato or potato, contains the name of the Greek goddess of fate, Atropos, whose task is to cut the thread of life. You are at risk if you eat around 10 of these black berries as an adult or three to four as a child. They contain a highly effective mixture of various alkaloids that are absolutely lethal in the right quantities. These are nitrogenous plant bases, of which morphine is perhaps the best known. Atropine is also a highly effective alkaloid which, when taken, opens the pupils wide, which, as was once thought, enhanced the beauty of women ("belladonna" - beautiful woman). Today it is an ingredient in some eye drops but is also used in cardiology.

In May, laburnum (*Laburnum anagyroides*) blooms in parks and many gardens. The name is very fitting for this shrub from the legume plant family. However, few people know that it is deadly poisonous. Just 10 flowers or 15 to 20 seeds, which are later found in the pods hanging all over the bush, are enough to kill an adult. Here too, a certain alkaloid (cytisine) is responsible for the toxic effect. This ornamental shrub should definitely be avoided in places where children regularly play.[71]

Some poisonous plants are already recognizable as such by their name. I am thinking, for example, of the poison berry (*Nicandra physaloides*), the poison lettuce (*Lactuca virosa*), the pretty poison primrose (*Primula obconica*) and the poison buttercup (*Ranunculus bulbosus*). But nature is full of surprises, and the most dangerous plants are not always so easy to recognize. There are numerous inconspicuous plants that can be highly dangerous despite their harmless appearance. However, this knowledge is of little use if you don't know the plants behind these names.

---

[71] This also applies to the blue rain (Wisteria floribunda), although it is not quite as poison-ous as the laburnum

Fig. 157: Not everyone is aware that the pods of the laburnum contain deadly poisonous seeds, which is why this shrub must not be planted near children's playgrounds (Author)

Even the allotment gardener and the florist often have to deal with dangerous poisonous plants, although they are not always aware of this. Take, for example, the lily of the valley mentioned earlier (point 335). Drinking the vase water is enough to cause serious stomach and intestinal problems. Cyclamen, poinsettia (a spurge plant), lobelia, various arum plants, dieffenbachia (not recommended as a houseplant for families with children), and oleander are also poisonous. Oleander is particularly treacherous as it is often found in gardens and on terraces; even eating small amounts of its leaves or flowers can lead to serious poisoning.

We must also warn against some neophytes. For example, the giant hogweed (*Heracleum mantegazzianum*), which is sometimes referred to as Hercules' perennial or "Stalin's revenge." This umbellifer, which spreads rapidly along riverbanks, is particularly surprising due to its extraordinary size. A few splashes of its sap applied to the skin can cause terrible sunburn in the sunshine, which is why this weed should only be pulled out with protective goggles, a protective suit, or at night if possible (please take a shower afterwards). This phenomenon is caused by photosensitizing substances, which strip the skin's natural UV protection. In addition, the giant hogweed is also able to displace native plant species due to its enormous vigor and thus has a negative impact on biodiversity. However, its flowering also attracts many insects, which may please nature photographers (like the author)...

339. Ricin

Another poisonous plant that is found relatively frequently in allotment gardens due to its exotic appearance is the castor oil plant (Ricinus communis). Its beautifully marbled seeds, known as "castor beans," are used to produce an oil that is often used as lamp oil and as a laxative. More recently, it has also been used to produce various lubricants for a wide range of technical applications. In addition to the oil, the beans themselves also contain a glycoprotein (lectin), which is highly toxic and remains in the press residue. This is ricin. Just one milligram (!) of this potent protein is enough to kill a person. This is equivalent to eating about eight of the aforementioned castor beans, although the strength of the poisonous effect can be increased by chewing thoroughly.

The extreme toxicity of ricin had already prompted weapons developers to use this substance as a "warfare agent" during the First World War. Thankfully, these developments came to nothing for a variety of reasons. But then various secret services discovered this poison as an effective means of murder.

Fig. 158: An ornament to many an allotment or park, but extremely poisonous - the castor oil plant (Author)

340. The umbrella assassination

The so-called "umbrella assassination" of the Bulgarian writer and dissident Georgi Markov (1929-1978), orchestrated by the Bulgarian communist secret service, has attracted a great deal of attention in this regard. The name comes from the fact that a prepared umbrella was used as the murder weapon. Its tip was designed to act like a hypodermic needle, allowing a small metal ball to be injected into the victim - in the event of an "accidentally" staged collision in the crowd. This metal ball, with a diameter of approximately 1.5 mm, contained around 40 micrograms of pure ricin, which was released into the wound and killed the dissident within three days. Unfortunately (from the point of view of the Bulgarian secret service!), this projectile was discovered during the autopsy, and subsequent investigations were finally able to clarify the assassination and its background, although they were unable to apprehend the actual assassin (there is, however, a guess as to who this person might be). It is said that the KGB supplied the poison and the capsule.

Recently, one or more offshoots of the Islamic terrorist organization Al-Qaeda also seem to be interested in castor beans from the miracle tree. In 2011, the American secret service claimed to have discovered that some members of this terrorist group were increasingly buying castor beans. It is still not clear whether the bearded gentlemen wanted to press castor oil from them because they needed a cheap laxative, perhaps still marked by Ramadan (due to the binge after sunset), or whether the pressing residues were the actual object of desire. After all, it can be used very effectively to kill "infidels."

Incidentally, ricin is also a very effective insecticide. This is why you will hardly find any traces of caterpillars or other insect larvae feeding on the leaves of the miracle tree in the garden. The same phenomenon can also be observed in the natural habitat of this particular species of spurge, North Africa and the Middle East.

341. Jonah and the castor bean worm

This is contradicted by the fact that we can read the following about the "castor bean plant" in the Bible, in the Old Testament, more precisely in the Book of Jonah V 4:6-10:

"So the LORD God appointed a castor bean plant and it grew up over Jonah to be a shade over his head to deliver him from his discomfort. And Jonah was extremely happy about the plant. But God appointed a worm when dawn came the next day and it attacked the plant and it withered. When the sun came up God appointed a scorching east wind, and the sun beat down on Jonah's head so that he became faint and begged with all his soul to die, saying, "Death is better to me than life." Then God said to Jonah, "Do you have good reason to be angry about the plant?" And he said, "I have good reason to be angry, even to death." Then the LORD said, "You had compassion on the plant for which you did not work and which you did not cause to grow, which came up overnight and perished overnight."

According to the Bible, there seems to be a "worm" for which the castor oil plant is a tasty source of food despite the ricin contained in the leaves, and which also devours it so quickly that Jonah once had to do without it as a source of shade. As we now know, this "worm" is the caterpillar of an endemic moth that is related to our garden tiger moth (Arctia caja). Its scientific name is *Olepa schleini*, and it is only found in a narrowly defined area on the coast of Israel. But the really amazing thing is that it was only discovered as a "large butterfly" in 2005! There are approximately 2800 years between the events described in the Book of Jonah and the discovery of the bear moth.

Fig. 159: Caterpillars of Olepa schleini on a castor bean plant (Wikimedia)

The great theological insight behind this surprising discovery for all lepidopterists (butterfly experts) is: *"The Bible is right after all!"* - at least with regard to "worms" that gnaw on miracle trees. Europe and the Near East are an otherwise extremely well-researched area from a butterfly-historical point of view. Here (in contrast to the tropical regions of our world), there is only a chance of discovering a new species here and there from the species-rich group of small butterflies. The only problem is - how do you recognize a new species that has never been described before? It's not that easy. It's not enough to find some freshly caught and prepared animal in the identification book.

342. Taxonomy and taxonomists

Modern taxonomists (who are responsible for the description and naming of new species, among other things) must know the organisms they are working on very well, be able to draw on extensive reference collections, and also have unrestricted access to old and new literature in their respective fields. In short, the profession of taxonomist is based in an academic environment and is therefore concentrated in natural history museums and university collections. Unfortunately, biology has changed dramatically over the last half-century, and as a result, taxonomists have lost much of their public reputation to geneticists and molecular biologists. The problem is that for many animal and plant groups, there are virtually no specialists left today who can clearly identify them. This is particularly true of organisms whose diversity is hardly recognized anyway (such as the Aphidae, the "jumping plant lice" among insects). Butterflies and beetles are an exception here, as there are many amateur researchers who are hardly inferior in their expertise to academically trained taxonomists.

Collecting butterflies and beetles was once very popular. Particularly in the 19th and early 20th centuries, this type of person could often be found in the woods and fields ("with botanizing drum and butterfly net"). Today, their painstaking collections form the basis of our museum treasures.

343. Famous entomologists

There were quite a number of famous personalities who themselves amassed extensive collections of beetles and butterflies and did scientific work in these fields. Let's start with a gifted actor who was extremely well-known and famous during his lifetime: Ferdinand Ochsenheimer (1767-1822). His multi-volume work "The Butterflies of Europe", which he wrote in part together with his fellow actor Georg Friedrich Treitschke, can easily be viewed on the Internet in digitalized form and still bears witness today to the extraordinary diligence and knowledge of this lepidopterist of the first hour, who was a celebrated character actor in the Viennese court theater. Friedrich Schiller, who was able to experience him in 1801 as Talbot (the commander of the English) in his play "The Maid of Orleans", was certainly impressed by him.

The writer and philosopher Ernst Jünger (1895-1998) was a colepterologist (i.e. a beetle researcher). He is best known to literature enthusiasts for his war diaries from the First World War, which were published in 1920 under the title "In Stahlgewittern" and aroused great interest. Entomologists, on the other hand, particularly appreciate his work in the field of beetle science (colepterology). For example, a black beetle native to the Arabian region was named *Leptonychoides juengeri* in his honor. The best place to find out about his exciting life is the Ernst-Jünger-Haus in Wilflingen. His extensive collection of beetles and cicadas, which he amassed over the course of his long life, can also be found there.

Since 1986, the prestigious "Ernst Jünger Prize for Entomology" has been awarded in Baden-Württemberg every three years in memory of the famous but controversial writer, philosopher and amateur entomologist.

Fig. 160: Ferdinand Ochsenheimer was a celebrated theater actor by profession and one of the best-known lepidopterists of his time by avocation

Another well-known entomologist, at least among German butterfly collectors, was Arno Bergmann (1882-1960), whose collecting area included the Thuringian Forest and the Kyffhäuser. He was particularly interested in biocoenoses, the interplay between the plant world and the butterflies living there, i.e. what was later subsumed under the term ecology. He also left behind a major work, "Die Großschmetterlinge Mitteldeutschlands". However, he will live on in another term, the "Bergmann series". This refers to a fundamental series of spectral lines of alkali metals that lie in the infrared spectral range and are important for astrophysics, for example. Apart from his hobby as a butterfly collector, Arno Bergmann was a doctor of physics and later a secondary school teacher.

I can also think of the following entomologists, most of whom pursued their hobby professionally alongside their profession: Arnold Spuler (1869-1937), university professor (medicine) and politician of the "German National People's Party; Lionel Walter Rothschild, 2nd Baron Rothschild (1868-1937), banker from the Rothschild dynasty; the author of the Kinsey Report, Alfred Charles Kinsey (1894-1956), specialist in gall wasps."[72]

344. Lolita

And then, of course, there is the famous author of the extremely scandalous novel "Lolita" (1955): Vladimir Nabokov (1899-1977). The writer, who was born in St. Petersburg and later emigrated to the USA, is regarded by writers as one of the most influential storytellers of the 20th century and by entomologists as one of the best experts on American blue butterflies, a particularly species-rich family of butterflies. Some people still jokingly argue today about whether, in his case, it would not have been better if he had written a few fewer novels and published more about blue butterflies instead. Or whether he shouldn't have written a few more novels on the level of "Lolita" or "Ada" rather than spending a lot of time investigating bluebottles and their male sexual organs. But one thing can be said in any case: Nabokov only became wealthy through his writing, and only in his later years. And the fact that he was a butterfly specialist par excellence is hardly known to any readers of his novels or the audiences of his film adaptations.

Incidentally, the first film made by the famous director Stanley Kubrick (1928-1999) after he moved to England was "Lolita", the screenplay for which Vladimir Nabokov himself worked on. The task proved to be more difficult than expected, as the literary work was considered "pornography" at the time (a not insignificant factor in its popularity, by the way!) and they had to be damn careful that the film didn't end up on the Index and banned from being shown. But this task was accomplished with flying colors, as can be seen from the fact that the screenplay was nominated for an Oscar in 1963. As far as the film version is concerned (like the novel, it is about a parthenophilic love affair between an underage girl and a 40-year-old man), there was a certain curiosity at the premiere that is well worth mentioning. Alongside James Mason (1909-1984) as Humbert, the then 15-year-old Sue Lyon (1946-2019) played the character "Lolita". However, due to the very restrictive age rating of films, she was not allowed to attend the premiere in the USA - she was simply too young for such an "obscene" film, one of whose main roles she played...

---

[72] The author was acquainted with the following deceased entomologists of Upper Lusatia, to whom he owes much in terms of species knowledge and knowledge of the local nature: Hans Leutsch, Alfred Schmidt, Max Gube, Franz Rektor, Max Günther

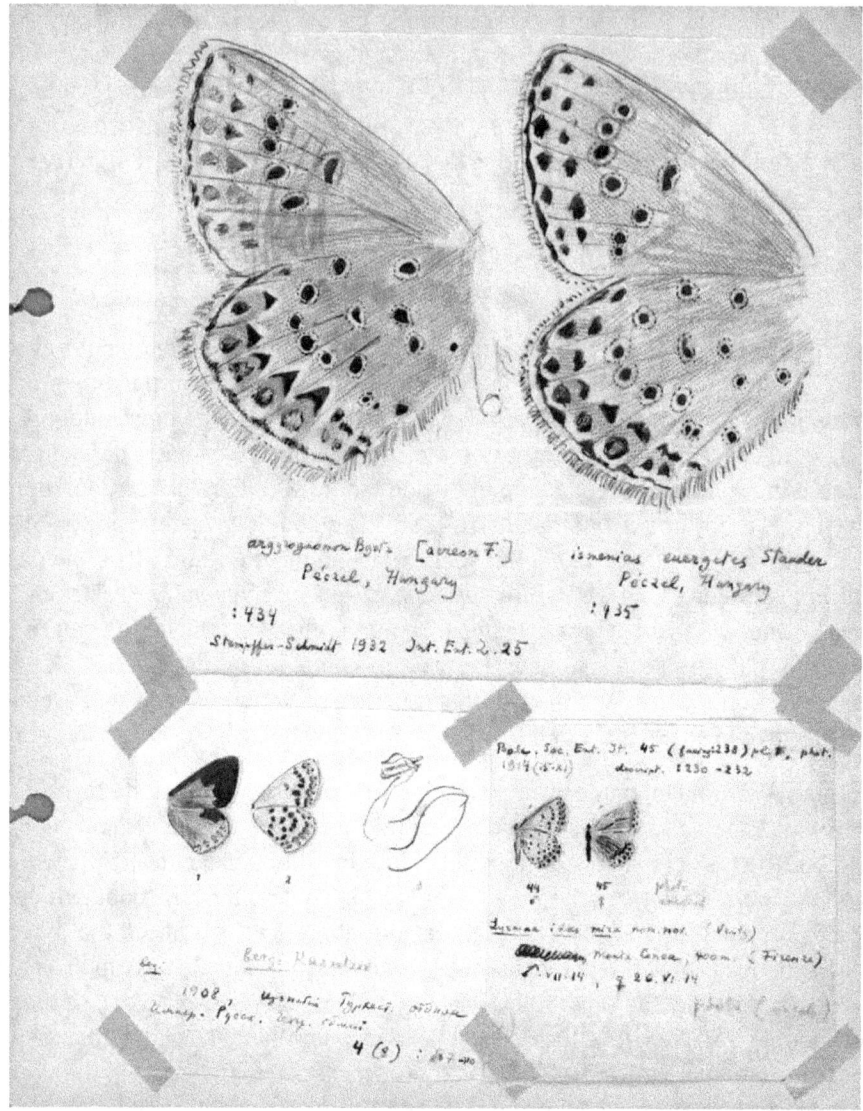

Fig. 161: A page from Vladimir Nabokov's sketchbook

345. Kubrick, Nixon and the man in the moon

The name of the brilliant director Stanley Kubrick (1928-1999) is also associated with a very special film, which was not made until a few years after his death and is an example of how conspiracy theories can be created with the means of an (apparent) documentary film, which are then believed by many people. I'm referring to the mockumentary *"Darkside of the Moon"* - better known in Germany under the title *"Kubrick, Nixon und der Mann im Mond"*.

Fig. 162: You can really see how much fun it was for the personalities involved in this mockumentary...

The term "mockumentary" refers to a film genre that depicts fictional events in the form of a documentary, often satirically or parodically mocking the claims of objectivity and cultural relevance of real documentaries. So what is this film about and why does it seem so credible? The content refers to the first landing of humans on the moon as part of the Apollo project (it is worth watching on YouTube) and is roughly in the tradition of the American science fiction film "Enterprise Capricorn" (1978), in which the public is fooled into believing a Mars landing. Allegedly, the first moon landing, in which Neil Armstrong (1930-2012) and Edwin Aldrin walked on the lunar surface on July 20, 1969, was faked by Kubrick in Hollywood style in absolute secrecy on behalf of the American government in order to be able to present viewers with a "studio version" à la "Enterprise Capricorn" instead of the original television footage of the moon in the event of failure. The apparent authenticity of this mockumentary is fed by a plausible-sounding story, interviews with well-known and highly placed personalities in NASA and the US administration (Edwin Aldrin, Henry Kissinger, Alexander Haig, Donald Rumsfeld, and Lawrence Eagleburger - to name just a few who should still be familiar, at least to the older generation) and a psychologically clever sequence of fiction and reality. If things hadn't been resolved at the end, this movie would have had the potential to serve as the basis for several conspiracy theories. However, the attentive viewer could have become suspicious during the so-called "documentary" if they had previously watched Stanley Kubrick's film *"Dr. Strangelove or: How I Learned to Love the Bomb"*. This is because the mockumentary mentions a certain "Dimitri Muffley", who is portrayed by Peter Sellers (1925-1980) in the aforementioned film - as the US president. Even the name "Jack Torrance" appears in the mockumentary - as a reminder, Jack Nicholson played him in "The Shining", a masterfully staged study by Stanley Kubrick about the interrelationship between reality and appearance, between reality and illusion.

346. Conspiracy theories

The more time that has passed since the moon landings in the 1970s, the more popular the conspiracy theory becomes - see "Kubrick, Nixon and the Man in the Moon" (point 345) - that the moon landings were all staged. This theory claims that NASA, with the support of the US government, faked the entire moon missions in a movie studio in order to win the race to the moon during the Cold War.

Supporters of this theory cite various arguments to support their claims. One frequently cited example is the analysis of photos and videos taken during the Apollo missions. Conspiracy theorists claim that shadows in these images are not parallel, which allegedly indicates artificial light sources. Similarly, the lack of stars in the sky in the photos is often cited as evidence that the images were taken in a studio with an insufficiently realistic backdrop.

Another point raised by conspiracy theorists concerns the flag of the United States "waving" on the surface of the moon. Skeptics believe that the movement of the flag as seen in the footage would be impossible, as there is no atmosphere on the moon that could generate wind.

Technological arguments also play a role. Some "theorists" claim that the technology of the 1960s was not advanced enough to carry out a manned mission to the moon. They argue that the risks and challenges, such as the Earth's radiation belt, were insurmountable and the astronauts would not have survived.

This conspiracy theory, which was virtually founded by the American author Bill Kaysing (1922-2005), was disseminated in the form of a television documentary that was first shown in the USA and then in a dubbed but uncommented form on German television (Spiegel TV, "Sind wir wirklich auf dem Mond gelandet?" - YouTube). These "documentaries" had something viral about them, so that NASA felt compelled to publish "counter-evidence." And this in turn strengthened the suspicion among conspiracy theorists that the manned moon landings were not entirely kosher. Interestingly, no attention was paid to the most important and plausible argument, namely that in the early 1970s there was a "Cold War" between the "West" and the "Eastern Bloc," which of course had an impact on space travel.

Assuming that Armstrong and Aldrin had not landed on the moon with their "Eagle" lander and that the television recordings had been faked a la "Capricorn One", this would have been easily recognized by the "Soviets," who were also skilled in radio and television technology. And they certainly wouldn't have missed such an opportunity to get one over on the "class enemy." Today, these arguments have all become irrelevant. In the meantime, probes such as the "Lunar Reconnaissance Orbiter" have photographed the Earth's moon with extremely high resolution (down to less than 50 cm/pixel) to such an extent that the tracks of the astronauts and the equipment left behind can be clearly recognized in the area of the landing sites of the Apollo missions and compared with their on-site images.

347. Why the manned moon landings are not fakes

To make it short and concise, let's compare the claims and their refutation. And lo and behold, all the arguments of the conspiracy theorists can be refuted quite quickly with mostly simple and plausible considerations:

Non-parallel shadows on the moon photos:

Claim: The shadows on the moon photos are not parallel, which supposedly points to artificial light sources.

Rebuttal: There is uneven terrain on the moon, which distorts the shadows. In addition, the sun as a point light source creates shadows that run differently depending on the terrain. This can also be observed on Earth.

Absence of stars in the photos:

Claim: The photos show no stars in the sky, supposedly indicating a studio shot.

Rebuttal: The moon's surface is very bright compared to the dark sky. In order to expose the surface correctly, the camera settings (short exposure times and small apertures) had to be adjusted, making the stars in the background invisible. This is a well-known photographic phenomenon.

Movement of the US flag:

Claim: The flag moves even though there is no wind on the moon.

Rebuttal: The flag was designed to fly as if in the wind. It was attached to a T-shaped pole to keep it upright. The movement of the flag results from the inertia created when the astronauts rammed the flag into the ground and rotated the pole.

Technological impossibilities:

Claim: The technology of the 1960s was not advanced enough to carry out a manned moon landing.

Rebuttal: NASA invested immense resources in the development of Apollo technology. This included advanced rocket propulsion technology, spacesuits and life support systems that were put through extensive testing. The success of Apollo 11 was confirmed by subsequent missions that brought back more scientific data and rock samples.

Van Allen Radiation Belt:

Claim: The astronauts could not have survived Earth's radiation belts.

Rebuttal: The astronauts crossed the Van Allen belts relatively quickly, minimizing their exposure to radiation. The spacecraft were also designed to provide sufficient protection against the radiation. The actual radiation dose received by the astronauts was far below dangerous levels.

Fig. 163: Landing site of Apollo 17, photographed by the Lunar Reconnaissance Orbiter

Independent observations:

Claim: The missions were not independently confirmed.

Rebuttal: Amateur astronomers and radio telescopes around the world tracked the missions and confirmed the signals from the Apollo spacecraft. Even the Soviet Union, then the USA's biggest rival in the space race, would have had the opportunity to uncover a fake, but did not do so because the data matched.

Moon rocks and scientific data:

Claim: The moon rocks brought back could be fake.

Rebuttal: The moon rocks brought back by the Apollo missions have been examined by scientists worldwide and confirmed as authentic. The composition and age of the rocks are consistent with what one would expect from lunar material and cannot be explained by terrestrial processes.

In this context, it is quite remarkable that the astronauts on the Apollo missions had a considerable amount of luck on their flights to the moon and back. Although space travel at that time was technologically impressive, it was also associated with great risks. One of the biggest, often overlooked risks was the danger posed by the sun.

During their stay in space, the astronauts were practically unprotected against cosmic radiation and solar activity in particular. If there had been increased solar activity during one of the Apollo missions, particularly in the form of a coronal mass ejection (CME), the situation would have become critical. Such events eject large quantities of high-energy particles into space, which could cause serious damage if they hit the space capsule directly and expose the astronauts to a lethal dose of radiation.

The space capsules used for the Apollo missions were not designed to protect against such extreme radiation bursts. A strong coronal mass ejection could have resulted in the crew being fatally irradiated. The thin walls of the Apollo capsule offered hardly any protection against the high-energy particles. This would not only have acutely endangered the astronauts' health but could also have affected the space capsule's electronics to such an extent that a safe return to Earth would have been impossible.

Fortunately for the Apollo astronauts, solar activity remained relatively calm during their missions. Looking back, this shows how much risk the early space missions took and how thin the line was between success and disaster.

In the meantime, this conspiracy theory has quieted down, although it still has a following who are not even convinced by the photos that have since been taken of the landing sites by probes of the "Lunar Reconnaissance Orbiter" type (because they are surely faked...).

348. A small panopticon of conspiracy theories

The conspiracy theory of the faked moon landings is probably one of the most beautiful in the gallery of conspiracy theories that enjoy general popularity. But their number is virtually unlimited and today, in the age of social networks, there is an audience for every one of them. We will therefore leave it at this point with a small but instructive selection and leave it to the reader to appreciate them. Perhaps you are also a fan of one or the other?

Here is a list, in descending order of popularity:

1. anti-vaccination movement
2. 9/11 as an inside job
3. moon landing was staged
4. new world order (NWO)
5. aliens in Area 51
6. chemtrails
7. illuminati control the world
8. QAnon (secret battle between Donald Trump and the "deep state")
9. flat earth theory
10. reptiloid aliens in positions of power

It is important to note that the popularity of conspiracy theories is difficult to measure and can vary greatly by region, time and social group. Furthermore, the popularity of a conspiracy theory cannot be equated with its "credibility" or "accuracy". Despite their popularity, almost all conspiracy theories have been scientifically refuted or have no basis in fact.

Nevertheless, the question remains as to whether there are also "conspiracy theories" that have subsequently turned out to be largely correct. And yes, they did (or do) exist...

349. "True conspiracy theories"

Here are a few examples of "theories" that started out as "conspiracy theories" and later turned out to be partially or completely true:

- **MK-Ultra:** the CIA's mind control program was long dismissed as a conspiracy theory until it was officially exposed in the 1970s.
- **Watergate affair:** The involvement of the Nixon administration in the break-in at the Democratic Party headquarters was initially considered implausible.
- **NSA mass surveillance:** Edward Snowden's revelations confirmed far-reaching surveillance activities.
- **Tuskegee Syphilis Study:** The unethical medical experiments on African-American men were conducted for decades before they were uncovered.
- **Operation Northwoods:** A never-implemented US military plan for false-flag operations against Cuba was later confirmed by declassified documents.

350. A pending case - the laboratory hypothesis on the origin of the coronavirus

In connection with the COVID-19 pandemic, there were indeed some claims that were initially considered conspiracy theories but were later at least partially confirmed. One notable example is the hypothesis about the origin of the SARS-CoV-2 virus that triggered the pandemic.

At the beginning of the pandemic, the idea that SARS-CoV-2 could have originated from a laboratory in Wuhan, China, was largely dismissed as a conspiracy theory. Many scientists and especially the media and politicians initially categorically rejected this theory. Over time, however, this hypothesis was taken more seriously as a possible explanation for the origin of the virus. Today, it is considered by many experts to be a plausible - albeit unproven - possibility alongside the natural origin of the virus.

Without going into too much detail, the timeline that led to this realization can be summarized as follows

- In December 2019, the first cases were detected in Wuhan, although it was initially suspected that transmission to humans took place at a wild animal market.
- From February to March 2020, the theory of a natural origin dominated, with bats seen as the likely source. The hypothesis of a laboratory origin emerged, but was largely dismissed as a conspiracy theory.
- From April to May 2020, the discussion about a possible laboratory origin intensified, although official bodies and many scientists continued to adhere to a natural origin.
- In the summer and fall of 2020, the debate in the scientific community intensified and calls for a more thorough investigation into the laboratory origin were raised.
- In 2021, the WHO conducted an investigation in Wuhan and classified the laboratory origin as "extremely unlikely", but this led to growing criticism of the investigation. At the same time, the US Secret Service resumed investigating the laboratory hypothesis.
- Between 2022 and 2023, both hypotheses, the natural origin and the laboratory origin, were considered plausible. The scientific community remained divided and agreed on the need for further investigation.

Currently (2024), the origin of the virus has still not been definitively clarified and the two main hypotheses continue to be debated.

If you are interested in the mechanisms underlying conspiracy theories, please refer again to the mockumentary "Dark Side of the Moon" (item 345). Although "Dark Side of the Moon" is intended as satire, it impressively shows how easily people can be influenced by well-presented but unfounded claims.

Particularly dangerous is the way in which seemingly credible sources are quoted and statements taken out of context are presented as proof of a greater, often secret truth.

This creates a deceptive basis of trust on which the entire theory is built. The film uses familiar conspiracy theory techniques, such as the manipulation of quotes and the misinterpretation of facts, to construct a convincing but ultimately false narrative. And this brings us to another phenomenon, propaganda.

351. Propaganda in the media and how to recognize it

The really "valuable" thing about "Dark Side of the Moon" is that it uncovers some difficult methods of disinformation that are used daily in various news programs, discussion panels, and reports, in which certain "truths" are consciously or unconsciously propagated. Particularly in times of crisis, the attentive observer can clearly see how, for example, in the mass media, the mandate to inform is increasingly replaced by propaganda controlled by various entities. The aim is always to establish a (certain) opinion among the majority of the population. This type of "program" can actually be inferred from its name because "propagare" means nothing more than "to spread" or "to disseminate." And it usually succeeds quite well, because hardly anyone masters or uses the simple antidotes to this poison, although they have already been revealed many times by a wide variety of authors, starting with Siddhartha Gautama (died around 544 BC): *"Believe nothing because a wise man has said it - believe nothing because everyone believes it - believe nothing because it is written - believe nothing because it is considered sacred - believe nothing because someone else believes it - believe only what you yourself have recognized as true."*

Believing "not" in something is the simpler exercise. Recognizing something as "true" yourself, on the other hand, is a much greater challenge, because it requires a certain amount of knowledge, common sense, and time to exhaust different sources of information (and of course access to these sources) and the ability to analyze. For example, to recognize the plausibility, or cui bono of a "media product". To do this, however, one must be aware of the most important methods of propaganda.

For example, it is noticeable in various television discussions that (a) there are often so-called claqueurs in the audience who clap particularly often at certain participants, encouraging the rest of the audience to do the same. This signals that these opinions are particularly important, correct, and authoritative; (b) the discussion is often completely unbalanced. For instance, during a television discussion about Richard Dawkins' well-known book "The God Delusion," which is critical of religion, the famous evolutionary biologist was pitted against two high-profile representatives of the official churches and a well-known Catholic layman. A similar method can be seen in discussions about the Ukraine conflict, where only one representative supporting the Russian position is invited, who also happens to speak poor German; (c) a representative of a "non-accepted" opinion either hardly gets a chance to speak, either because they are more polite than the eloquent "politicians," who usually emit sound bites, or when they do, they are cut off more or less elegantly by the moderator, especially when the conversation gets inter-esting.

Transforming a political talk show into a "tribunal" that serves only to discredit the in-vited person (e.g., with abbreviated and thus compromising "clips") has also been ob-served multiple times; (d) the same people are always invited, who often do not stand out due to their exclusive knowledge or insights but primarily use the talk show to raise their own profile. Of course, this can backfire when they meet intellectually superior or more knowledgeable guests. However, the selection process by program editors en-sures that this rarely happens. All these points, which can be observed regularly, ulti-mately lead to such talk shows no longer being watched, at least by the more educated citizens.

A similar pattern can now be observed in almost all opinion-forming press outlets - whether they are still published in printed form or mainly online. It begins with headlines that do not always do justice to the topic or even distort it. When reading the correspond-ing article, one increasingly notices a disconnect between the editorial text and the headline: the headline is a factual statement, while the editorial section is written en-tirely in the possible tense. The massive use of the subjunctive mood, in particular, should always raise suspicion. This suggests that the journalist is only reporting hearsay. However, there may also be an intention behind it - perhaps the journalist hopes that the reader will overlook the subjunctives and take the content at face value. If you discover such tactics, you should always (at least if you are interested in the topic) consult other sources (especially foreign ones). One goal should be to find the "original source" of the news.

Here, you may encounter another problem: modern journalism sometimes seems to be merely secondary-source journalism - as you can easily discover with Google. Many similar news items, for example, about the European or Greek financial crisis, or (even more instructive) the "refugee crisis" or the Ukraine conflict (to name just a few current topics), often stem from a specific report from one of many press agencies and are writ-ten less and less by journalists who work and research on the ground themselves.

Another issue that an attentive reader frequently encounters is the confusion between news and opinion (commentary). As informed citizens, we primarily want to be in-formed. We want to interpret events ourselves - sapere aude. If, on the other hand, you are interested in others' opinions, you can read their comments, which should be explic-itly marked as such. As informed citizens, we usually don't want to be presented with information that has already been filtered through an opinion mill. This approach feels condescending and therefore deeply propagandistic.

There are other typical characteristics that reveal a propagandistic agenda: these in-clude the omission of information, quoting out of context, negatively connoted person-alization ("Putin is to blame for everything!") - this is the "enemy image technique" - or the strategy of not allowing opponents to speak, or only doing so in a shortened or dis-torted manner. This involves ignoring the obvious, using euphemisms for one group and dysphemisms for another, and the overuse of the so-called "Nazi card" or its "neo-Nazi" variation as a last-resort argument when all other arguments are exhausted. In online media, comment sections have recently been blocked (with a few notable exceptions where editors have discovered that complete blocking is counterproductive overall) when propagandistic and controversial topics are discussed.

It is often more intellectually satisfying, and thanks to the possibility of hyperlinks to additional sources, often far more informative, to read the comments on a piece of media content than to consume the media content itself.

There is no doubt that distinguishing between news, opinion, and propaganda in a multimedia world is not easy. However, by adopting the few doctrines formulated by Buddha Gautama, you can learn to distinguish between reality and propaganda and maintain a critical mindset.

Images are a powerful tool of propaganda. They have the power to arouse emotions and shape opinions, often without people even realizing it. Visual representations can subtly but effectively change the perception of an event or a person. The selection, staging, and context of images play a key role in this. *"A picture is worth a thousand words,"* and propagandists use this to their advantage.

Photographs once were considered authentic. Today, this is no longer the case, as programs like Photoshop, artificial intelligence, and expert manipulators have mastered the art of altering images.

352. Image forgers and image retouching

However, there are now programs that can automatically detect all kinds of image manipulation, thus exposing image forgers. In this area, counterfeiters are also finding it increasingly difficult. Additionally, there is always the risk that image manipulation can backfire. For example, there are two official photos of the former Siemens CEO K.K. on the Internet. The otherwise identical photos show him in the 2004 version wearing an expensive Rolex watch. A year later, when Siemens began to downsize its fixed network division, "Communications" (1,350 jobs were to be cut), this picture of the Siemens boss no longer seemed appropriate, and the luxury watch was simply retouched out. When the deception was discovered, and Siemens' subsequent justifications seemed rather ridiculous, it turned out that the damage to the company's image was far greater than the manipulation was worth.

Even in the days of analog photography, altering photos to suit one's own purposes was a common tactic, especially for those in power. For example, two photos exist of Benito Mussolini (1883-1945) showing the Italian dictator on horseback - one with a servant holding the reins and one without the "reins holder." The latter, apparently retouched, became the "official" photo distributed.

The forgeries commissioned by Stalin are also legendary. The most famous shows him with Lenin and a number of other "revolutionaries"; the photo was taken on May 5, 1920, in front of the Bolshoi Theatre in Moscow, marking the departure of several tens of thousands of Red Army soldiers for the Polish-Russian war. Later, when Stalin's opponents,

Lev Borisovich Kamenev (1883, executed in 1936) and Leon Trotsky (1879, assassinated with an ice pick in exile in Mexico in 1940), fell out of favor, only the side of the photo without them was used for publication. When Stalin ordered the photo to be reworked into a grandiose painting in 1933, Kamenev and Trotsky were simply omitted, and the original disappeared into oblivion. Nowadays, the fact that it is no longer enough to simply manipulate photos may worsen the situation. Video clips also need to be cleansed of unwanted individuals (as in North Korea) or even altered to add compromising gestures, for example.

Fig. 164: When Lenin gave a speech in Moscow in May 1920, the two revolutionaries Leon Trotsky and Lev Kamenev stood on the steps of the podium (left). Later, however, Stalin made sure that the two disgraced revolutionaries "disappeared" from the photo.

353. The stinky finger or can you insult a speed trap?

But even that is no longer a major problem, as the "stinky finger" affair involving Greek Finance Minister Janis Varoufakis has shown (2015, for those who want to do some research). Incidentally, this gesture, which is considered obscene, was already in use in ancient Rome as *digitus impudicus*. At that time, however, it was not yet known that it could even be used to "insult" a speed trap. This was only established by German judges in 2000...[73]

---

[73] The reference number of the court ruling is: Bay ObLG AZ.: 5 St RR 30/00

354. Noah and the Flood

It is worth mentioning in this context that on the same day the aforementioned verdict was pronounced in Bavaria (February 23, 2000), Mozambique was hit by severe flooding caused by heavy and persistent rainfall. The floods resulted in approximately 700 deaths and set the already impoverished country back economically.

Incidentally, the so-called "Flood," which is described in detail in the Bible (Old Testament) and which has been made into a lavish Hollywood film ("Noah," starring Russell Crowe as Noah, 2014), is still regarded by some as the greatest flood in human history. This film adaptation is considered very realistic - at least as far as the biblical account is concerned. However, the fact that Noah was said to be 600 years old when he built the ark, while Russell Crowe, who portrayed him, was only 50, somewhat detracts from the authenticity it strives for. Few people are aware that there is an even older version of the Noah saga, in which Noah is not called "Noah" but "Utnapishtim," which is perhaps less suitable for a Hollywood movie.

355. Gilgamesh and the Atrahasis epic

It is part of the Epic of Gilgamesh, which in turn contains portions of an even older text, the Epic of Atrahasis. At over 3,800 years old, the latter is one of the oldest literary records of humankind. It was written down around the same time (or slightly earlier) as King Hammurabi (1792-1750 BC) had his legal codes chiseled into a diorite pillar in cuneiform script. In its entirety, the Atrahasis Epic (Atrahasis = "The Especially Wise One") is a Sumerian creation myth in which the story of the Flood is only one part.

The first part describes how the gods created humans, and the (incomplete) second part describes how the god Enlil wanted to get rid of them. To put it more simply, at some point, the god Enlil became annoyed by the noise they were making (he wanted to get a good night's sleep). So, he first sent a plague upon the people. But Atrahasis learned about it, consulted with his god Enki, and built a new place of worship, which eventually appeased the Sumerian gods. However, this peace did not last long - only about 1,200 years. Then Enlil first ordered the weather god Adad, and when he was unsuccessful, the fertility goddess Nisaba, to exterminate the people through a famine caused by drought. But Enki also revealed these plans to his priest Atrahasis, who in turn advised the people to only sacrifice to Adad and Nisaba, which promptly appeased them. When Enlil realized that these measures had also failed, he decided to drown the human race in a great flood. But again, Enki gave his high priest advice, which has been passed down on the corresponding cuneiform tablet:

Atra-Hasis opened his mouth and addressed his lord: "Tell me the meaning of the dream so that I may see where it leads!" - Enki opened his mouth and addressed his servant: "You say: What shall I seek? Pay attention to the message I want to tell you! - Wall, listen to me! Wall of reeds, pay attention to all my words! Destroy your house, build a boat! Give up your possessions! Save your life! - The boat you build shall be the same size (in all three directions). - Cover it like the Apsu, (the flood of water under the earth) so that the sun cannot shine into it. It should have a deck above and below. The rigging shall be strong, the pitch thick, and give the boat strength. I will let it come down on you here. - A multitude of birds, an abundance of fish. He opened the water clock and filled it. He announced to him the beginning of the flood for the seventh night"...

Fig. 165: Fragments of the 1st tablet of the Atraḫasis epic written in cuneiform script, which is kept in the British Museum. Missing passages were supplemented by, among other things, fragments of tablets from other versions of the same story, which were gradually found at often widely separated locations in Mesopotamia. (Wikimedia)

The third and last, but unfortunately only incompletely preserved tablet then tells of the actual Flood and the salvation of Atrahasis and his wife:

"The flood broke loose. - Its violence came upon the people like a battle. One could not see the other. No one was recognizable in the destruction. The flood roared like a bull. The wind howled like a neighing wild donkey. The darkness was black, no sun to be seen."

But then, from the gods' point of view, there was a major disadvantage to their actions: there were no more people to make sacrifices to them. And so, apart from Enlil, everyone was happy in the end that Atrahasis had managed to escape the flood with his boat.

According to the famous Sumerian King List (which also contains a reference to the Flood), Atrahasis appears under the name "Ziusudra" as the savior of mankind from the Flood. The same story then appears in a slightly modified form as a subplot in the Epic of Gilgamesh. There, on the eleventh tablet, you can read:

"Tube house, tube house! Wall, wall! - Reed house, listen, wall, understand! - Man of Shuruppak, son of Ubara-Tutu! - Tear down the house, build a ship, - Let wealth go, chase after life! - Give up possessions in exchange for life! - Lift all kinds of soulful seed into the ship! - The ship you shall build. Its dimensions shall be measured, - Let its breadth and length be equal; - You shall cover it like the apsû."

However, the recipient of this message here is not Atrahasis/Ziusudra, but Ut-napishtim. Surely not only a modern plagiarism detection program would find a certain similarity between the two texts.

And then there are some passages that have not survived in the Atrahasis epic, but which give a very vivid description of the preparations for the coming flood, the flood itself and finally the happy rescue of Utnapishtim and his traveling companions. A key line here is

"As the seventh day drew near, I let out a dove; The dove departed and came again: No resting place caught its eye, so it turned back. I let out a swallow; The swallow departed and came again: No resting place caught its eye, so it turned back. I let out a raven; the raven also went away; when he saw the water running away, he ate, scratched, lifted his tail and did not turn back."

When over 140 years ago, the cuneiform expert George Smith (1840-1876) leaned over the clay tablets found in the palace of Assurbanipal (king of the Assyrians between 669 and 627 BC) during excavations in Nineveh in the British Museum and read about the Flood, it must have gone right through the biblical scholars of antiquity. For in the first book of Moses, chapter 8, it states:

"Afterward he sent forth a dove from him, that he might know whether the waters had fallen on the earth. But when the dove did not find where its foot could rest, it came back to him in the box, for the water was still on the face of the whole earth. So, he put out his hand and took it into the box. Then he waited another seven days and again let a dove fly out of the box. It came to him at dusk, and behold, it had broken off an olive leaf and was

carrying it in its mouth. Then Noah realized that the waters had fallen from the earth. But he waited another seven days and sent out a dove, which did not come to him again."

It was the first time that an even older version of a Bible story could be identified - and that was a real sensation in the mid-19th century. What interests us today is the question of whether these Flood stories can be traced back to a geologically and historically verifiable flood event preserved in the collective memory of mankind, and, if so, where and when it took place. To cut a long story short, scientists have not yet progressed beyond the level of hypotheses. Based on the surviving texts, the Flood may have occurred in the Mesopotamian region around 2500 BC - as a significant and impressive, but likely localized, event. However, no geological evidence of extreme flooding in the Middle East that correlates precisely with this date has been found. Of course, this does not mean that the Flood is purely fictional. It is likely the result of real human experience, which was later exaggerated in tradition and transformed into a global catastrophe.

The explanatory models for the biblical Flood now range from volcanic eruptions, meteorite impacts, and tsunamis to climate change, without consensus on any one event. According to the sources, however, only weather-related "floods" seem plausible. This is because the Flood is described in the Bible as a global, worldwide inundation that rose for 150 days after 40 days of rain, and then subsided for another 150 days. The entire land is said to have been submerged for ten months, with the floodwaters covering even the highest mountain peaks. Only the crew of the ark was saved, eventually coming to rest on Mount Ararat in what is now Turkey.

356. The Ararat anomaly

In 1949, US Air Force reconnaissance planes were able to photograph a structure on the flank of the former volcano, which would later become known as the "Ararat Anomaly" (1995), when some people thought they recognized the outline of Noah's Ark. There is almost absolute certainty that this is a geological formation, but of course this has no major influence on the corresponding "conspiracy theories".

357. Dam burst on the Bosporus

Recently, the Flood has also been associated with an event that is said to have occurred around 5500 BC - the "dam break" on the Bosporus, in which vast quantities of water poured into the then only partially filled basin of the Black Sea within a very short period. With the end of the Würm (or Vistula) Ice Age, the melting of the inland ice led to a noticeable rise in sea levels, which naturally also affected the Mediterranean. At some point, the isthmus separating the Sea of Marmara from the Black Sea was breached and torn away, causing enormous volumes of water to flood into the Black Sea basin, which was over 100 meters lower at the time. By then, the peripheral areas of the Black Sea were already relatively densely populated, as evidenced by archaeological artifacts. People therefore had to leave quickly to escape the rapidly rising level of the Black Sea.

Some authors suggest that it was precisely this existential experience, passed down through generations, that led to the development of the Flood saga. Incidentally, during the so-called Cenozoic ice ages, there were numerous sudden flood events worldwide, but certainly, only a few people were affected (as there were not very many people at the time).

358. Great Spokane Flood

Some of the largest floods took place in North America in the area of the Columbia River Plateau (in the states of Washington and Oregon) between 13,000 and 15,000 years ago, creating the so-called *"channeled scablands."* The large number of valleys that have cut deep into the hard basalt of the Columbia Plateau differ from other valleys cleared by water or ice in their rectangular cross-sections, which led to a heated debate among geologists in the 1920s regarding their formation. In addition, ripple marks, which are unique in their size and shape, have been preserved in the area of sand deposits. The now dry soils of the channels have depressions in places, which can most easily be interpreted as scours. In addition, there are entire cascades of former cataracts, streamlined "islands," and large blocks of rock that have been transported over long distances and then left behind.

The geologist J. Harlen Bretz (1882-1981) was the first to make sense of these terrain features, sparking a scientific debate. Today, after more than 90 years of field research in this region, his ideas are generally accepted, and the formation of the scablands can be explained in simplified form as follows: At the end of the last ice age, a retreating glacier dammed the Clark Fork River, creating a huge glacial lake ("Lake Missoula") with a total area of about 7,700 km^2, a depth of about 600 m, and a water content roughly equivalent to that of today's Lake Michigan. Its shoreline can still be seen today, almost in fossilized form, on the slopes of the Sentinel Mountains near Missoula (Montana).

When the ice dam finally broke, around 15,000 years ago, the water masses of the glacial lake poured catastrophically into the Columbia River Plateau southwest of Spokane within a few days to a maximum of two weeks, reaching discharge rates of several km^3 per second and discharge speeds of up to 100 km/h. Such a flood was repeated about 40 times over a period of two millennia, with corresponding intervals during which Lake Missoula filled up again. This series of flood events is summarized under the name *"Great Spokane Flood."* The first and largest flood wave almost completely removed the loess cover of the Miocene Columbia River basalts before it began to work its way deep into the hard bedrock. The original loess cover remains completely intact only on some of the streamlined "islands," which were obviously not flooded. The valley itself was formed mainly by the enormous erosive force of retreating cataracts, as can still be observed today (albeit much more slowly) at Niagara Falls. In total, over 200 km^3 of rock were removed and transported in the scablands. As already mentioned, such floods were repeated several times, as the ice dam was constantly rebuilt by glacial advances. Today, you can still see the huge scree slopes and gravel banks at the lower reaches of the streams, which bear witness to the power of the flowing water.

Fig. 166: Landscape created by a catastrophic flood event in the area of the Quincy Basin (source http://hugefloods.com)

Such catastrophic emptying of glacial lakes, albeit on a much smaller scale, also occurred in other places during the ice ages. One example is the area around the Porcupine
River in northern Alaska. However, landscapes have also been discovered in the area of
the Aral Sea and the Caspian Sea that indicate extensive flooding events at the end of
the last ice age.

359. Jökulhlaups

The fact that glacial lakes can clear out entire valleys within a very short period when
they suddenly release their masses of water downhill can be studied very well in the so-
called Jökulhlaups ("glacier run"). This Icelandic term refers to the sudden emptying of
lakes that have formed under the ice sheet or glacier due to volcanic warming, for example. In geology, however, this term is also used generally for all glacier-related flood
events.

A particularly dramatic example is the large glacier run of 1996 in the Skeiðarárjökull
area in southern Iceland. The starting point was the Vatnajökull volcano, which is almost
completely hidden under an ice cap. When this volcano erupted once again in 1996, the
resulting heat caused the surrounding glaciers to melt rapidly, leading to the glacial
lakes filling up quickly and eventually breaking the ice barriers. Within a very short period, these lakes emptied, allowing huge masses of water to flow into the valley with incredible force.

The destructive power of this flood was immense. On the Skeiðarársandur sands to the south, scientists and local residents observed how scabland-like structures formed - an impressive reminder of past ice ages in which similar processes shaped entire land-scapes (point 358). The masses of water swept away blocks of ice up to ten meters in size, uprooted trees, and destroyed a massive steel bridge that was actually designed to withstand such forces of nature.

Here, two of the most powerful forces of nature, fire and ice, interacted fatefully, creating floods of devastating proportions that permanently changed life in this region and have also found their way into Nordic doomsday myths.

360. Doomsday scenarios

In this context, I am reminded of the well-known poem by Robert Frost (1874-1963), which is often quoted when it comes to doomsday scenarios. It is called "Fire and Ice" and was published in Harper's Magazine in 1920. Its message and magic unfold best in the English-language original:

> *Some say the world will end in fire,*
> *Some say in ice.*
> *From what I've tasted of desire*
> *I hold with those who favor fire.*
> *But if it had to perish twice,*
> *I think I know enough of hate*
> *To say that for destruction ice*
> *Is also great*
> *And would suffice.*

No interpretation is attempted here, only the question of the "end of the world". Some religious communities believe that the end is "near" - and I would even like to agree with them in a certain sense, because we are living in the last phase of the earth as a living planet. In order to understand this - and which also puts humanity's role in Earth's history into perspective - we need to take a closer look at the most important events in the history of our universe and Earth's history to date and those that are yet to come. And I mean "look at" in the sense of "visualize in terms of time". After all, our thinking apparatus can do little with periods of millions and billions of years, even if the corresponding annual figures flow smoothly over our lips.

361. The cosmic year

A year, on the other hand, with its twelve months, lies exactly within our imaginary horizon, so it makes sense to break down the history of our universe, which began 13.79 billion years ago with the Big Bang, to the length of a year. On this new time scale, January 1, 0:00, is the moment of the Big Bang, and December 31, midnight, is the "now."

According to our current understanding, space, time, and matter came into being at the moment of the Big Bang. This history can be divided chronologically into several phases: the Planck era, the GUT era, the quark and hadron era, the era of primordial element synthesis (creation of hydrogen and helium), and the radiation era. In our model, the radiation era ends approximately 11.4 seconds after the Big Bang - this is when the universe first becomes transparent (corresponding to around 300,000 years after the initial moment). Today, the photons decoupled in the process form the homogeneous and isotropic radiation field of the cosmic background radiation (point 156).

It took over three months for our Milky Way to finally form, with March 15 as the approximate date. Finally, on August 31, our Sun and, with it, the Earth formed from a collapsing interstellar cloud of gas and dust. The oldest rocks on Earth, such as those found in the Precambrian shields (e.g., the Acasta gneisses), formed around September 16. A few days later, on September 21, the first primitive life forms appeared on Earth in the form of simple prokaryotes. Biomineralizations known as stromatolites bear witness to them. A further month and a half on our timescale passed before the first complex cells with nuclei and other cell organelles emerged, with the cutoff date being November 9.

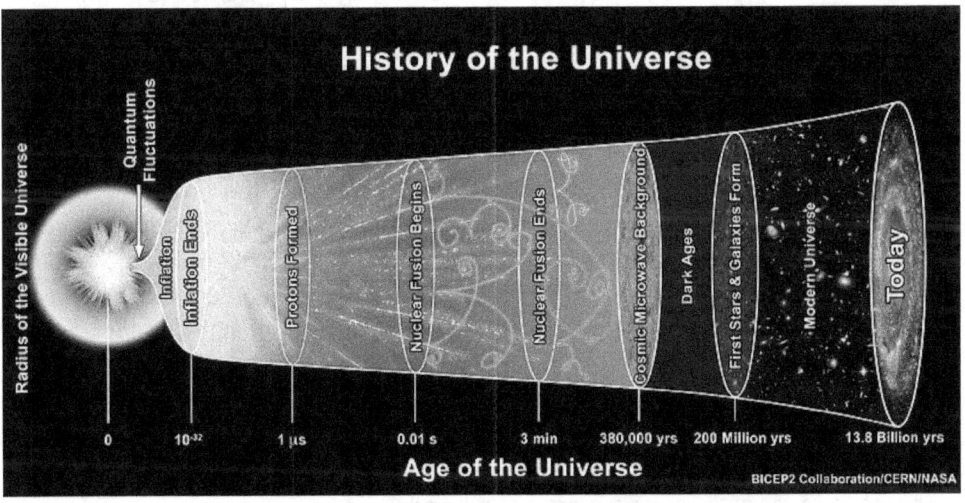

Fig. 167: Timeline of the universe - representation of the development of the universe over approx. 13.8 billion years (Creative Commons)

The first stable multicellular organisms began to colonize the Earth's oceans on December 5. On December 14, the event known as the "Cambrian Explosion" finally occurred, as virtually all of today's animal phyla and a number of other phyla that had already died out by the end of the day were created within a single day. If Pikaia, the ancestor of all vertebrates, had died out by chance, there would be no herring, sparrow, or humans to-day...

Ten days before Christmas, the great radiation of animal life began. It was still taking place in the oceans. It was not until December 20 that the first animals and plants slowly began to conquer the shores of the landmasses. December 22 marked the arrival of amphibians, and on December 23, the era of the reptiles began, which came to an abrupt end on December 30 at 6:24 a.m. The era of reptiles would likely have continued until New Year's Eve if a large meteorite had not hit the Gulf of Mexico (Chicxulub impact, point 238), resulting in the extinction of virtually all animals up to the size of a cat. This event marked the beginning of the great hour of the mammals that had evolved by December 26. The surviving small mammal species now found a planet virtually cleared of their predators.

After a brief respite, the mammalian era began, and in the early morning of December 31, New Year's Eve, at 6:00 a.m. and 5 minutes to be exact, the first "ape" appeared on the planet. Another eight hours passed before the first hominid, a group to which we belong, appeared. It was only at 10:24 p.m. that a particularly intelligent species among them discovered that stones can be used as tools, and another hour later (11:44 p.m.) that fire is very suitable for warming and preparing delicious food.

At around 11:55 p.m., it became quite cold in the northern hemisphere as the ice ages began. When it got warmer again, the first spark of culture began to emerge in humans - this being the name of the hominid who had mastered fire - by painting colorful pictures in caves, as seen in France, or carving figures from the ivory of previously hunted mammoths. It took only 2 minutes and 47 seconds for writing to be invented (December 31, 23:59:47). In the following seconds, great empires arose and fell until finally, two seconds before midnight, Christopher Columbus set off on his voyage to a new world and a new era - the modern age.

The Reformation, the era of witch hunts, Galileo Galilei's discovery of the craters on the moon, the moon landings, and the invention of the internet and the iPhone all occurred in the last second of our year. It must be noted that Homo sapiens was particularly active in the last tenth of a second of the year, having invented the steam engine, the automobile, harnessed electricity, developed computers and smartphones, and began the extensive exploitation of Earth's resources. In the "atomic age," humanity developed all the instruments necessary to potentially exterminate itself, waged countless wars, and triggered a mass extinction of animals and plants on an unprecedented scale, which is comparable to the catastrophic mass extinctions in Earth's history (point 232). Additionally, humanity was the first living being to leave the protective Earth and send research probes to the edge of the solar system and beyond...

362. The next cosmic year - what will it bring?

We can now seriously ask ourselves how many more seconds, minutes, hours, or days people on Earth will be granted as we look into the future. One second? - that's about 435 years - or perhaps ten seconds, which may seem like a very ambitious target given the current waste of resources. If we make it to one o'clock in the morning of the new year (which corresponds to approximately 1.5 million years and which the majority of nuclear waste disposal opponents take for granted), it would certainly be a great miracle. Much of humanity's continued existence will then depend on whether colonization of the solar system and, eventually, the Milky Way succeeds. Since we do not know what the future will bring (until then, several supervolcanoes will likely erupt on Earth), we can only speculate.

On the other hand, it is extremely unlikely that humanity will live to see January 2. It is also well known that the sun's luminosity is slowly increasing, so that by January 14, not only humans but probably all other higher organisms will have disappeared from the Earth's surface. By January 24, the last plant will have withered, and the last insect will have died, and by the beginning of March, there will be no more oceans, and the Earth will have completely turned into a desert planet. By mid-July, the sun will begin to expand into a red giant star, only to end up as an Earth-sized white dwarf just two weeks later, after a few intense "thermal flashes."[74] Of course, this now has all the time in the world to slowly cool down and fade away more and more. Whether the Earth will still exist as a planet is probable, but by no means certain.

The realization that follows from this "world history," which has now been made vividly tangible, is banal and may be unsatisfactory for some people: earthly life is only a fleeting, temporary marginal phenomenon in the infinite vastness of the cosmos. It begins with bacteria, ends with bacteria, and bacteria also dominate it in the few months in between (bacteria are undoubtedly the dominant terrestrial life form in terms of robustness, number of individuals, and biomass - approximately 1.5 to 3 kg of a human being consists of bacteria alone).

363. Man - a cosmic mayfly

In the cosmic calendar, humans can strictly speaking only claim the status of "mayflies", being born on New Year's Eve and dying out again in the course of the New Year. In this respect, those people who believe that "the end is near" are quite right. But even that is only relative, because in the cosmic calendar a human life - when it comes right down to it - is just ~two-tenths of a second long, and on this scale, it is still a long way to one o'clock in the morning of January 1st. Nonetheless, one thing should still be clear: the throne that humankind has erected for itself as the "image of God," the "navel of the

---

[74] For more details see my textbook "Physics of Stars" published by Springer (2024)

world," and the "conqueror of nature" is "cracking at the seams" (King Crimson). How-ever, the majority of humanity does not yet know this, does not want to admit it, or has simply not acknowledged it. Doomsday scenarios, on the other hand, have preoccupied people ever since they became aware of their finite existence on Earth.

364. Doomsday scenarios

Many religions not only contain a creation myth but also offer certain end-time scenar-ios, such as the "Last Judgement" or the "samvatta-kappa" in Buddhist doctrine, which pertains to the cyclical "kalpa." However, we are more interested in the "doomsday sce-narios" that can be substantiated by natural science, particularly those that could po-tentially lead to the extinction of humanity in its civilizational dimension. Several con-ceivable scenarios include: wars with nuclear and biological weapons; global epidem-ics (pandemics); social effects of overpopulation (such as ideologization and struggles for increasingly scarce resources); the collapse of biodiversity affecting food production (a side effect of resource plundering); globally impacting natural disasters such as the eruption of a supervolcano; uncontrollable climate change; an asteroid impact of a sig-nificant size; or the effects of a cosmic gamma-ray burst or a nearby supernova. Addi-tionally, there is the potential takeover of Earth by an artificial intelligence created by humanity itself (as Stephen Hawking [1942–2018] warned); and the destabilization of technical systems through cyber warfare, among other threats.

Most of the scenarios that threaten humanity are of our own making. In contrast, the probability of global threats triggered by natural disasters, which cannot be influenced by humans, is relatively low even over an average timeframe of a few hundred to a thou-sand years. The eruption of a supervolcano with global climatic effects remains the most likely natural disaster (consider the Yellowstone supervolcano, which is overdue according to volcanologists, and the Phlegraean Fields, which are increasingly active). "Climate change" is also often overestimated in this regard, as even in the most cata-strophic scenarios, there is still sufficient potential for adaptation to enable survival.

Since the development of atomic and hydrogen bombs, and their initially unlimited, later treaty-limited accumulation by the superpowers, along with the resulting, at least par-tially stabilizing strategy of "mutually assured destruction" (MAD), humanity has had the potential for self-extinction. This danger is less about the complete extinction of the spe-cies and more about the collapse of all civilizational and technological structures as a result of a nuclear exchange. Despite assurances from those responsible that arsenals are securely protected against unauthorized use, "securely protected" does not equate to 100% security.

Today, the risk is compounded by the integration of military nuclear capabilities into global information systems. This networking increases the likelihood that capable indi-viduals, such as cyber terrorists, could gain access to sensitive areas. There are numer-ous examples where hackers have breached supposedly "secure" information systems of banks, authorities, and industries to steal data ("phishing"), manipulate it, or even take control of technical systems. These incidents demonstrate the vulnerability of modern societies to such attacks.

Highly developed technologies open up new possibilities for terror and destruction with far-reaching and sometimes nearly uncontrollable consequences. Individuals with the skills and intentions to program computer viruses, splinter groups, fanatical terrorists, and even hostile state secret services could theoretically paralyze or significantly impair the vital infrastructure of entire cities and regions, such as the power grid or water supply, without being physically present, if they manage to overcome security systems.

However, the risk is not limited to military technology. As humanity has entered the scientific and technological age, new potential dangers have emerged that could threaten the existence of developed societies. Besides cyberwarfare, other technologies are coming into focus. Genetics offers the potential to develop deadly viruses through genetic engineering, while robotics and "artificial intelligence" could also be used destructively. Nanotechnologies, originally developed for medical or industrial purposes, could become tools of destruction in the wrong hands.

In short, modern science and technology, which have brought us immense progress and prosperity, also harbor unprecedented destructive potential if misused.

365. The "gray goo scenario"

In the context of nanotechnology, the so-called *"gray goo"* scenario should be mentioned. This scenario describes the theoretical possibility that self-replicating nanomachines could get out of control. These tiny machines, which in principle correspond to von Neumann probes[75] - but on a molecular level - could be able to replicate themselves from hydrocarbons, such as those found in crude oil. The basic idea would be to use such nanomachines to combat environmental disasters like oil spills. These machines could break down the spilled oil and use it as a building material to create copies of themselves. So much for the theory.

Currently, there are no scientific reasons that make such nanomachines impossible in principle. There is even a certain probability that such technologies will one day be feasible. However, the *"gray goo scenario"* becomes problematic when we consider possible errors in the programming of these machines. If a software error or design flaw enables the nanomachines to use not only hydrocarbons but also any other organic material for their reproduction, the consequences could be devastating. Since these machines multiply exponentially - similar to the famous chessboard problem where the number of grains on each square doubles - they could transform all organic life on Earth into a gray, amorphous mass within a short period of time. This "gray goo" would then be the end product of uncontrolled replication.

---

[75] A "Von Neumann probe" is a hypothetical concept for a space probe that can replicate itself by using raw materials from planetary or meteoritic material to create exact copies of itself

Although this scenario is primarily theoretical, it raises significant questions about the responsibility and ethics of developing new technologies. The best way to prevent such a scenario would be to avoid developing such nanomachines in the first place. However, it is questionable whether all scientists and developers would adhere to this precaution. Given the potential risks, it is crucial that safety protocols and ethical considerations are integrated into the research and development of such technologies to prevent unforeseen disasters.

The problem is that there will likely always be someone who views this as a challenge and pursues it, whether officially or covertly. This really does pose a danger, especially when considering "viruses" instead of nanomachines...

366. Designer viruses

And there are already examples of bioscientists who have succeeded in developing their own viruses using biotechnology methods. In 2002, for example, it became known that US scientists had produced artificial polio viruses - without using a host cell - using publicly available information and equipment purchased by mail order.[76] It took them two years of work, and in the end, their result was indistinguishable from a naturally occurring polio virus. This demonstration - which sparked significant controversy among their colleagues - opens the door to the possibility of creating other small RNA viruses in the laboratory that may have lethal properties for humans. The basic knowledge for this is provided by advanced university studies, the blueprints for suitable gene sequences are available online, and the ingredients and equipment can be sourced from mail-order companies.

To summarize, the American Academy of Sciences stated in the memorandum *"Making the Nation Safer: The Role of Science and Technology in Countering Terrorism"*:

"Even a few individuals with specialized knowledge and access to a laboratory could, without major cost or effort, produce a plethora of deadly biological weapons that could pose a serious threat to the U.S. population. They could also produce such biological agents using commercially available devices - which are also used to produce chemicals, drugs, food, or beer, and thus go unnoticed."

Such resulting risks urgently require new regulations and safety protocols at a global level. It is essential that governments, scientists, and international organizations work together to prevent the misuse of biotechnological innovations. This includes stricter controls on the dissemination of biotech knowledge and equipment, international agreements on the non-proliferation of bioweapons, and increased monitoring and cooperation in research safety.

At the same time, we must consider the ethical dimensions of these technologies. The ability to deliberately modify viruses or other biological organisms - such as the SARS-

---

[76] Cello, J., Paul, A. V., & Wimmer, E. (2002). Chemical synthesis of poliovirus cDNA: gen-eration of infectious virus in the absence of natural template. science, 297(5583), 1016-1018.

CoV-2 virus that triggered the COVID-19 pandemic - raises profound questions about human responsibility and the limits of scientific action. A responsible approach to bio-technological advances is therefore just as important as the technical mastery of these technologies. The scientific community and global society face the challenge of maintaining a balance between innovation and safety, ensuring that biotechnological advances serve the good of humanity and do not become a threat.

367. Brave new world

Scenarios that could lead to the virtual extinction of the human species can, of course, be spun further. The more complex a society is organized, the more dependent it is on technology, the more susceptible it is to attacks instigated by individuals or groups of individuals. And that should give us serious food for thought.

368. "Intelligence" as a limiting factor of technical civilizations

A "dangerous" question in this context is whether the specific human abilities summarized by the term "intelligence" might even, in a very general sense, be a limiting factor for technological civilizations, as has occasionally been more or less reasonably suspected.

Considering that human intelligence is the result of a long evolutionary history focused primarily on survival in a natural environment, it becomes clear that our cognitive abilities are not specifically geared toward the complex and often abstract demands of modern technological societies. Evolution has equipped us with an intelligence that allows us to create tools and adapt, but there are limits to this adaptability. Our cognitive structure is oriented toward solving short-term problems, recognizing immediate dangers, and managing social interactions rather than handling complex systems over the long term.

In a world increasingly dominated by technologies of our own making, our decision-making is often influenced by emotions, cognitive biases, and limited understanding. In a highly complex, globally connected world, this can lead to dangerously poor decisions. One example is how politicians address existential risks. Although humanity undoubtedly possesses the knowledge and technologies to address these risks, we often fail to take effective long-term action. This suggests that our intelligence may not be sufficient to cope with the enormous challenges of a technologically advanced civilization.

Another issue is the limitations of our moral and ethical development. While our technological capabilities have grown exponentially, our ethical framework and ability to cooperate globally have not developed proportionately. This imbalance results in the development of technologies whose long-term effects we may not fully control or understand. Artificial intelligence is a recent prominent example of this.

A look at the current global crises shows that humanity frequently struggles to deal responsibly with its own creations. From environmental degradation to the proliferation of weapons of mass destruction and the threat of biotechnological anomalies, these challenges highlight that human intelligence, as it exists today, may not be sufficient to ensure a sustainable future for civilization.

From this perspective, it will be intriguing to see whether humanity will survive the first second of the new cosmic year.

369. The end of the world is near - prophets and charlatans

According to prophets, religious scholars and a multitude of charlatans, the world has already ended umpteen times - even in our lifetimes. They all have the characteristic that they name a certain date, usually emphasized in some way, and when it is reached, nothing happens in the end. This was also the case on December 21, 2012, when the Mayan calendar (supposedly) ended. As nothing major happened on this day, the after-Doomsday parties that were organized in many cities were also well attended.

370. Nostradamus

Nostradamus, born as Michel de Nostredame in 1503 and living until 1566, is considered a special luminary in terms of prophecies, including those concerning the end of the world. His almanacs made him famous during his lifetime, as some of the prophecies they contained seemed to correlate with real events that had not yet occurred when he wrote the quatrains.

The unique aspect of these quatrains is their highly metaphorical formulation, written in a language (Old French) that can only be described as "bold" in terms of both spelling and grammar. Interestingly, they contain hardly any specific dates. This vagueness often leads Nostradamus experts to predict the exact date of future events using various methods, such as "counting letters," which can be quite curious. For example, there are claims that an earthquake with a magnitude of 9.8 is expected to strike California on May 20, 2015, or perhaps June 13 - the reader will know. But one question remains: How did Nostradamus know about the seismic Richter scale back then? Nostradamus experts have an answer: he was a prophet and could foresee events far into the future (up to the year 5352 AD). Additionally, Nostradamus is said to have predicted that in 2015, all language barriers would fall, enabling everyone to communicate with everyone - even animals. This would have delighted me and my cat Humpel, and the EU could have avoided the hassle of translating laws and decrees into countless languages. That would have been remarkable. Unfortunately, as we know, Nostradamus was wrong once again - or, more likely, his interpreters were.

Fig. 168: "And this mysterious book, written by Nostradamus' own hand, is it not enough of a guide for you?" (Faust I, Study)

Let's try our hand at interpreting a Nostradamus quatrain. Consider the translation of Quatrain - Century II, Verse 46 (if you want to look up the original online), a particularly clear verse:

"After a great catastrophe for mankind, an even greater one will come. It will rain blood, milk, famine, iron, and pestilence. There will be a fire in the sky, with a tail and sparks."

I believe the interpretation is quite clear: in 2012, 67 years after the catastrophe of the "Great War," an asteroid or comet would hit Earth, triggering the end of the world. Didn't think of it immediately? Take comfort. That's likely because you're not a true Nostradamus expert. Nevertheless, this example demonstrates that even true "experts" in Nostradamus exegesis can be wrong from time to time. In the case of Nostradamus, "now and then" is probably a serious understatement. Did you notice anything about the "end of the world" in 2012? I certainly did not.

The Nostradamus prophecies also provide strong evidence that it is much easier to fit events that have already occurred into one of Nostradamus' verses than to predict future events (which, unless they are banalities, are highly unlikely to ever occur). Nonetheless, many people believe in the nonsense that Nostradamus' followers claim to read from his writings today. It seems to be profitable for these "experts," as Amazon lists around 1,300 books on the keyword "Nostradamus" in German alone. Max Dessoir (1867-1947) was likely right when he wrote in his book Vom Jenseits der Seele (1920): *"The (real) miracle of Nostradamus is not his text, but the art of interpretation by his expositors."*

371. Exegesis and hermeneutics

There is nothing wrong with the method of exegesis or hermeneutics. Both methods are widely used and are indispensable for various subject areas, especially theology and jurisprudence. Exegesis is used to find out what a text actually "wants to say" in its historical and cultural context. It aims to decipher and interpret the original meaning of a text. In contrast, hermeneutics in the philosophical sense is concerned with the fundamental act of "understanding" a text. In methodological terms, hermeneutics involves explaining and interpreting texts in order to unlock their deeper meaning. Originally, both methods were developed to make the statements of sacred texts understandable, often in an allegorical sense. This means that they were intended to help decipher the hidden or metaphorical meanings contained in religious scriptures. Thus, both exegesis and hermeneutics have played a central role in providing deeper insights into significant texts and promoting their relevance and understanding.

372. The Revelation of John

Take the Bible, for example, and the last book of the New Testament, the "Revelation", better known as the *"Book of Revelation"*. If someone who is not particularly religious takes the trouble to read this text, the question inevitably arises: what is the author, who is apparently a Christian prophet, actually trying to tell us? The text is obviously in great need of explanation. Its mystical language and its context, which can only be fully understood within the historical framework of the time in which it was written, make an immediately obvious interpretation difficult.

This is where the work of exegetes, the interpreters of the Bible, begins. Depending on their approach, they interpret the "Revelation" differently. Some interpret it as a "critique of the present" in the context of the time of the Roman Emperor Domitian (51-96 AD), who was known for his brutal persecution of Christians. Others see the "Revelation" as a vision of the future, culminating in the end times, the "Last Judgment". In this view, the catastrophes "seen" by the prophet herald the final judgment and build up to this point, which is ultimately followed by the kingdom of God. This interpretation is particularly popular among Jehovah's Witnesses and is also often used artistically, for example in the US end-time film *"The Seventh Sign"* from 1988. However, the artistic value of this film is debatable.

Another interpretation of the "Book of Revelation" sees this work less as a prophetic prediction of future events and more as a drama of salvation. In this dramatic framework, the cosmic battle between the forces of absolute evil and the forces of good is depicted. This interpretation emphasizes that the book is a narrative representation of God's ultimate victory over evil, culminating in the establishment of His eternal kingdom. Here, the triumph of God, who is portrayed as a righteous judge and omnipotent ruler, is at the center of the narrative.

From a historical perspective, this interpretation can also be seen as a sharp social critique of the abuses and moral transgressions of the Roman Empire. John, the author of Revelation, addressed his message to the Christian communities in Asia Minor, which were still heavily influenced by Hellenistic culture at the time. These communities were not only exposed to cultural influence but also to political pressure and religious persecution from the Roman Empire.

John used symbolic and apocalyptic language to denounce the destructive influences of the Roman Empire and to warn the churches of its corrupting influence. He wanted to encourage Christians to remain steadfast in their faith and not give in to the pagan practices and moral corruption of Roman society. Revelation was intended to offer them comfort and hope in times of persecution by conveying the certainty that God would ultimately prevail and establish His righteous kingdom.

Through this interpretation, "Revelation" becomes an important document of resistance and encouragement for the early Christians. It shows how they were able to keep their faith in the midst of oppression and persecution and to hope for God's ultimate salvation.

A few phrases from the "Book of Revelation" have also made it into everyday language thanks to Luther's translation of the Bible. For example, the *"book with seven seals"* and *"the Alpha and the Omega"* from the quote: *"I am the Alpha and the Omega, the first and the last, the beginning and the end."* This "omega" has even found its way into scientific terminology - as the end point of history - namely as the "omega point".

373. The Omega Point Theory

Firstly, as the end point of all evolutionary developments per se (e.g., in the sense of the theologian and philosopher Pierre Teilhard de Chardin (1881-1955)) and secondly as the possible end point of a cosmological development of the entire universe.

The "Omega Point Theory" presents itself as a physical theory based on Einstein's General Theory of Relativity and predicts a so-called "Big Crunch" at the end of the development of our universe. Its occurrence depends crucially on the average density of the gravitationally effective energy (=mass) of the cosmos. If it exceeds a critical value, the cosmic expansion will eventually come to a standstill, and the cosmos will enter a collapse phase, which in turn will end in a "Big Crunch" after a finite period of time. This is precisely where the American physicist Frank J. Tipler locates the "Omega Point". Unfortunately, all observations show that there will (probably) never be a "Big Crunch" because the expansion behavior of our cosmos is aimed at an "eternal" expansion - the expansion rate does not decrease, nor is it constant, but is constantly increasing. Astronomers Saul Perlmutter, Brian P. Schmidt, and Adam Riess were awarded the Nobel Prize in Physics in 2011 for this discovery.

Fig. 169: Book cover of the German edition of Frank J. Tipler's book on the Omega Point Theory (the appendix is the rather interesting part)

So what does the "Omega Point Theory," which Frank J. Tipler developed in 1994 in his highly acclaimed semi-popular book "The Physics of Immortality: Modern Cosmology, God, and the Resurrection of the Dead," which quickly became a bestseller. In it, Tipler constructs a scenario with which he ultimately predicts the "resurrection" of each of us

- as a "simulation" in a cosmic computer - with the prospect of "eternal life," as promised by the Christian religion. Many parts of the book, insofar as they are not interspersed with theological vocabulary and refer to purely scientific aspects such as Poincaré's theorem of recurrence or the "heat death" of the universe, are also informative for a physicist and encourage reflection on the subject matter (note: it is not light fare!). However, the overall structure that the author builds on this seems rather crude and cannot even be taken seriously from a scientific point of view. The initial intention seems rather to have been that a kind of "content of faith" was given (the "Kingdom of God" after the "Last Judgment" according to the Apocalypse), and the author seems to have seriously asked himself (although it is hard to believe with such a technical luminary as Frank J. Tipler) how nature (the cosmos) must be constituted and what man must do so that this "content of faith" can be experienced "at the end of time." In any case, the blending of scientific argumentation and eschatological guidelines, such as those found in monotheistic religions, does not make reading any easier.

Briefly summarized and without going into details, Tipler's "Omega Point Theory" states the following:

- **Cosmological end of the universe:** Tipler postulates that the universe will end in a so-called "Omega Point" in the distant future. This point represents a state of maximum complexity and density in which the universe would end in a final collapse, similar to a Big Crunch.
- **Infinite computing power:** In the course of this collapse, the available computing power in the universe increases infinitely. Tipler argues that intelligent life forms or their successors (e.g. artificial intelligences) could use this infinite computing power to store and process any conceivable information.
- **Resurrection and immortality:** Infinite computing power would theoretically make it possible to simulate and recreate all "consciousnesses" that have ever existed. This would mean a kind of digital resurrection and immortality for all people.
- **Uniting science and religion:** Tipler attempts to combine scientific knowledge and religious ideas. In his theory, the Omega Point becomes a kind of scientifically based version of God, who is omniscient and omnipotent, as he has infinite computing power and controls the fate of the universe.

Even if he justifies all his points in a highly scientific and mathematical way, studying his train of thought may not lead to something that is "comprehensible." I think the attempt to combine cosmological theories with metaphysical ideas was a commendable try, but remains nothing more.

Incidentally, he followed this up in 2008 with a book titled *"The Physics of Christianity: A Scientific Experiment."* In it, he wants to show that all the "miracles" of Christianity can be explained by natural science - from the "virgin birth" to Jesus' walk across the waters of the Sea of Galilee to the dematerialization of Jesus' body. However, the "solutions" he offers are so grotesque (for example, he explains Jesus' walk across the water with a neutrino beam those forms under Jesus' feet and is directed "downwards" - could it be any simpler? How about a flyboard?) that at some point you stop reading. Some Christians, especially from the creationist scene, greeted it euphorically, but it was described by his fellow experts (e.g., Lawrence Krauss, who otherwise also likes to "speculate" and overshoot the mark) as utter nonsense, which does not do justice to the author's high professional reputation (he is, after all, a professor of mathematical physics at Tulane University in New Orleans, who has made recognized contributions to cosmology, for example).

374. Christian fundamentalism

With this book, Tipler reveals himself to be a follower of Christian fundamentalism, which refers to the Bible as the literally inspired word of God. This fundamentalism, which is particularly widespread in the so-called "Bible Belt" of the USA and is even dominant in the form of evangelicals, attempts to displace the modern scientific world view. This is done by building a counter-position on the basis of Bible verses, which aims to convince devout and simple-minded individuals.

In his book, Tipler takes up the basic principles of this fundamentalism and integrates them into his scientific theories. He interprets biblical stories and prophecies as scientific facts and attempts to harmonize them with his "Omega Point Theory", among other things. In doing so, he assumes that the biblical writings contain not only spiritual but also scientific truths that can be found in modern physics. This approach represents a fusion of religious belief and scientific theory, which is met with skepticism and criticism by many scientists.

Through his theory, Tipler attempts to strengthen the credibility of biblical scriptures and provide a scientific basis for religious beliefs. This is, of course, in stark contrast to the widespread scholarly consensus, which views the Bible as a historically and culturally significant, but not scientifically sound, document. Tipler goes further, however, by claiming that the events and prophecies described in the Bible are actually rooted in the structure of the universe and its physical laws.

This type of argument is also frequently found in the "intelligent design" movement. Like Tipler, adherents of Intelligent Design argue that the universe and the life within it did not come about through random processes, but that there is an intelligent cause or divine plan behind it. Intelligent Design attempts to use scientific methods and findings to prove the existence of a Creator and thus also opposes evolutionary theories and the scientific mainstream.

375. Intelligent Design

The problem was that a legal dispute made it difficult for evangelicals to exert influence on the secularized education system in the USA, so they developed a counter-position under the guise of science, known as "intelligent design," in order to introduce their views into the state education system via a back door. The method is extremely ingenious. By demanding, for example, that both Darwin's theory of evolution and the "theory of intelligent design" - whereby it is left open who the "designer" actually is (after all, anyone can think of that for themselves) - are taught on an equal footing, the evangelicals are trying to get a foothold in schools and universities.

"Intelligent design" is first and foremost a teleological position that claims that certain things in nature (the theory of evolution is just one particularly popular field) could be better explained by postulating the controlling influence of an omnipotent being outside of nature. To this end, the followers of "intelligent design" look for phenomena or problems that are not yet understood or clarified by science and offer an apparent solution in the work of an "intelligent designer." And once science has found a plausible and stringent explanation, they quickly turn to a new, as-yet-unsolved problem. They use scientific methods, publish in the appropriate technical language, and usually only hint in the background that only a supernatural influence remains to solve the problem. The word "God" or other religious vocabulary is avoided as far as possible so as not to give the impression of a "stopgap God," which is only ever used when gaps in a scientific line of reasoning need to be closed.

There seem to be several reasons why the ideology of "intelligent design" has chosen the theory of evolution as its most important field of activity. In the extreme, quasi-literal interpretation of the Bible, where there is a creation in six days, a Noah with his ark, and a flood that afflicts the whole world, it is of course quite a challenge to reconcile these views with what science has discovered. If the story of creation as recorded in Genesis is taken word for word as an unquestionable truth, then Darwin's theory of evolution, with constant change over great periods of time, is a natural enemy of these views. In this respect, the "official church" (Catholic and Protestant, with the exclusion of the "Evangelicals") is more pragmatic and allows faith to be "faith" and science to be "science"; indeed, it now recognizes the theory of evolution without any problems, as it is the domain of the natural sciences and is not affected by beliefs that place it on a different level (John Paul II, 1996). Evangelical movements cannot and do not want to accept this because it would also contradict their pronounced sense of mission and their claim to be the sole representatives of "divine truth." By means of the "theory" of "intelligent design," they attempt to establish a quasi counter-position to the theory of evolution, which is intended to give the impression that it also reflects a "scientific" position that is at least equivalent to the canonical theory of evolution in terms of its value. And that is precisely what it is not, because it attempts to explain certain phenomena using something that stands outside the natural sciences and is therefore in principle not accessible to falsification according to Karl Popper. The aim seems to be to discredit the scientific theory of evolution within the building of biology as a whole by giving the impression that it is itself controversial within this framework (which of course it is not, because *"nothing in biology makes sense except in the light of evolution"* - Theodosius Dobzhansky (1900-1975)).

Incidentally, "modern" forms of creationism no longer deny what is obvious, namely that living beings change over time. Otherwise, any chicken or dog breeders' association would be able to immediately reduce their views to absurdity.

376. Basic type theory

They have come up with something else - also with regard to the limited cargo space of Noah's Ark - in which they now claim that "God" only created certain basic types, from which all of today's living creatures then developed "after the Flood" through microevolution (and then definitely according to Darwin's principles).

If this view were to be followed consistently, creationists (which some do) would have to give up the "world age" of just 6,000 years according to their calculations. The Anglican theologian James Ussher (1581-1656) was able to date the exact time of creation to October 23, 4004 BC in a remarkable piece of painstaking work based on the sequence of generations in the Bible and the preliminary work of John Lightfoot (1602-1675). But let's get back to some aspects of "intelligent design" by asking ourselves what actually gives an "intelligent designer" away.

377. Irreducible complexity

The keyword here is "irreducible complexity." Before discussing this concept in more detail, it is worth mentioning one of the most influential books of the early 19th century with regard to questions that would later be associated with the concept of evolution. It was written by the English theologian and philosopher William Paley (1743-1805), was published in 1802, and bears the title "Natural Theology." What has survived to this day from this extremely witty book is the parable of the pocket watch that a walker finds by the wayside on his stroll. As he examines its mechanism in amazement, sees how one cog meshes with the other and how everything fits together harmoniously to form a whole, he comes to the conclusion that this watch could only have been created by a gifted watchmaker. He then uses the same reasoning in a completely plausible and comprehensible way to attribute the existence of plants and animals (which are no less, but rather even more wonderful than pocket watches) to the ingenious work of a creator. At that time, long before Darwin's On the Origin of Species (1859) was published, this was the only explanation for the existence of animals, plants, and humans that anyone could understand.

If the pocket watch were to be seen as a natural object created by "Darwinian evolutionary processes," then it would have to have "evolved" in many small steps, in whatever way. However, since individual stages of development are "selected" according to their suitability as a clock, it would have to be certain from the outset that a timepiece would emerge at the end of the development. In any case, each intermediate stage can serve as anything else, but never as a timepiece. Only the final interaction of all the parts really makes the watch a watch. We are therefore certain from the outset that a pocket watch must be an object of a conscious artistic product of a suitably skilled and talented watchmaker, dedicated to its goal or purpose, and not the end point of a natural evolutionary process analogous to biological evolution.

An object that consists of several parts, which only in their interaction make the purpose of this object possible, is called "irreducibly complex." If you take just one cogwheel out of a pocket watch, it stops and can no longer tell the time. Michael J. Behe, who introduced this term into the scientific debate in 1996, defines it as follows:

"A non-reducibly complex system cannot be created in a direct way (i.e., by continuing to improve the same initial function that continues to work through the same mechanism) by slight successive changes to less complex precursor systems, because any precursor to a non-reducibly complex system that is missing a part is, by definition, non-functional."

This concept is so interesting for the ideologues of "intelligent design" precisely because there are many biological organs and processes that are made up of many parts, which only develop their functionality (similar to a deaf watch) in their entirety. In the past, the human eye was often used as an example that everyone could understand. The retina, vitreous body, lens, and pupil obviously have to work together very precisely before a virtual image of the world can be created in the brain with their help. However, this example is not suitable for an "intelligent" eye designer. The evolutionary line of development of a vertebrate eye and an octopus eye (both lens eyes) can be traced step by step (and all the necessary intermediate stages can still be found today in recent animal species). In addition, the vertebrate eye (and thus also the human eye) is a misconstruction and the squid eye is an ideal construction of a lens eye. This is easy to recognize if you take a closer look at the "wiring" of the light receptor cells. In a vertebrate eye, it reaches into the inside of the eye and the nerve cord has to be routed to the outside via the "blind spot." No engineer in the world would think of wiring the pixels of a CMOS sensor for a digital camera in this way, for example. However, if you look at the evolutionary development of the eye from a light-sensitive pigment cell via a cup eye and a number of other intermediate stages to the modern vertebrate eye, it becomes clear that evolution cannot change a path once it has taken it, but it can iron out the "initial error" by optimizing it at a high level, so to speak.

The analogously constructed eye of an octopus, which, because it is "correctly" wired, does not need a "blind spot," shows that there is another way. An interesting question for a supporter of "intelligent design" in this context would therefore be what might have motivated the "intelligent designer" to develop the human eye (or, more generally, a vertebrate eye) in the same way and not differently, i.e., "correctly" in the engineering sense.

In any case, the human eye is not suitable for making its existence plausible. However, there are certainly objects whose gradual development and functional optimization cannot be explained so easily on an ad hoc basis. Michael J. Behe, for example, discusses the bacterial flagellum, which functionally represents a nano-ring motor consisting of several functional units that only works if all these functional units are present and correctly assembled. If one assumes the known evolutionary mechanisms, then the drive of the bacterium must have developed in individual steps that were random and directionless. This "irreducible complexity" of the "electrorotation motor" of a bacterium is indeed difficult to understand when viewed in the light of a step-by-step development. The flagellum only works when all the parts are in place, and only when it func-tions can it serve as a selection feature, which in turn can be used as a starting point for subsequent evolutionary development processes. This obviously does not apply to the individual functional units as individual objects. In fact, from the point of view of proba-bilities, it is completely impossible for a corresponding number of mutations to have oc-curred simultaneously by chance, which ultimately led to a functional flagellum. What is not considered in this argument, however, is the possibility that a functional change may occur in a developmental process. This is in the sense that the individual parts ini-tially perform a certain function (in the case of the flagellum, for example, in the form of a molecular pump that is built into the cell membrane), but in the course of evolution this function changes to a different one (into a flagellar motor). In the meantime, the evolutionary steps necessary for this have been elucidated, so that the frequently cited "design argument" for the bacterial flagellar motor no longer applies. The problem with "irreducible complexity" obviously lies in the fact that it refers to the current state of a corresponding object and says nothing about what previous states the object has passed through.

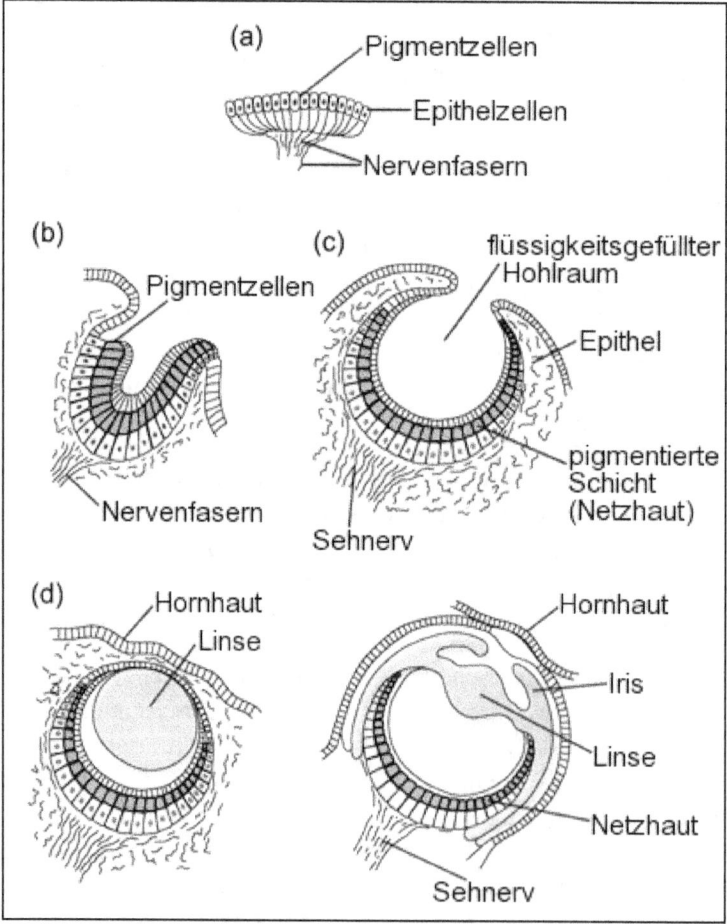

Fig. 170: Developmental stages that result in a highly complex cuttlefish eye through continuous evolutionary development

In summary, it must be stated that creationism, even if it comes in the form of pseudo-scientific "intelligent design," has nothing to do with science and therefore has no place in the state-organized education systems of secular states. Attempts to change this in individual states in the USA, for example, have always been fended off.

378. The Flying Spaghetti Monster (FSM) doctrine

Incidentally, as a result of this "defensive battle," a new and surprisingly open religion emerged in 2005 with a now astonishingly large number of members - and it happened like this: when it became increasingly apparent that the Kansas school board wanted to advocate the teaching of "intelligent design" alongside Darwin's theory of evolution, physicist Bobby Henderson felt compelled to send an ironic but entirely serious open letter to the said board. In it, he demands that - if "intelligent design" and therefore creationism - are to be taught in a public school, he also insists that "his" doctrine of the *"Flying Spaghetti Monster"* (FSM) should be taught on an equal footing. Should this not happen, he even announced that he would take legal action against the Kansas School Board.

Fig. 171: The parody of Michelangelo's "The Creation of Adam", touched by his noodly appendage, has become an iconic image of the "Flying Spaghetti Monster".

This fun religion, which has all the characteristics of a "real" monotheistic religion (with even some advantages!), spread quickly - even beyond the United States. Even in Germany, there is an offshoot of this religious community with a not inconsiderable number of followers. Its headquarters are in Templin in the Uckermark, where a "pasta mass" is celebrated regularly (I believe every Friday at 10 a.m.) with beer and pasta. Incidentally, after death, the faithful in "pasta heaven" have access to a beer volcano and a stripper factory, among other things. If that's not something else.

Incidentally, "creationism" is not just a variety of fundamentalist Christian sects.

379. Islamic fundamentalism

Darwinism has also become a red rag for Islam, which is why the Islamic publicist Adnan Oktar ("Harun Yahya," as he appears, is effectively his "artist's name") has published a particularly bright red book (at least in the English-language edition) weighing more than six kilograms in glossy paper entitled "Atlas of Creation," in which he proclaims the simple message in a way that is ridiculous for the expert and suggestive for the believer in Islam: Darwin is wrong! There is no need to go into the content any further, although his collection of pictures of fossils and plants and animals that exist today is quite respectable and impressive. After all, with this work he serves an area that has been little explored in Islam to this day because this doctrine of salvation has not dealt with real "science" for a long time. Strictly speaking, it has only been possible to speak of "Islamic creationism" since the 1970s, after Christian fundamentalists increasingly went public with this doctrine. The "philosophy" behind it can be summed up in the following sentence:

"All knowledge is already laid out in the Quran, and science must be measured against it. If the Quran is not compatible with science, science is automatically wrong."

This also shows a certain schizophrenia of teaching, e.g., in Iran. If the engineers there took this "doctrine" as seriously as their mullahs, nobody would have to fear that they could develop nuclear weapons sooner or later. It is just as schizophrenic when the terrorists of the "Islamic State" destroy cultural monuments that are thousands of years old because "there is nothing about it in the Koran" but arrange to do so using smartphones and the internet (and I haven't read anything about "smartphones" in the Reclam edition of the Koran either). And it seems downright grotesque to an enlightened Central European when a "sheikh" from Saudi Arabia with the most modern and expensive smartphone on his ear walks past you in the check-in hall of an airport, followed (at a suitable distance, of course) by his three (or more) women who are deeply veiled except for their eye slits. One is inclined to say that high-tech meets the spiritual Middle Ages...

380. Progressive role of Islam in history

And yet, there was a time when Islam played an extremely progressive role in the history of mankind. Fantastic buildings, such as mosques and palaces with their typical Islamic ornamentation, still bear witness to this today. The Moorish Alhambra on a hill near Granada in Spain is just one of many examples worth seeing.

The centuries under the dynasty of the Abbasid caliphs (between 750 and 1250 AD) are considered the "Golden Age," with the relatively peaceful century between 900 AD and 1000 AD on the Iberian Peninsula being the high point. The centers of Islamic knowledge and literature at this time were particularly in Persian Khorasan and Moorish-ruled Al-Andalus, with the Emirate of Córdoba (later Granada) as the intellectual center. Here, in a thoroughly cosmopolitan atmosphere, scholars from all scientific disciplines concentrated to recover lost ancient knowledge, translate it into Arabic (and thus preserve it), and generate new knowledge. Under the ruler (Caliph) al-Hakam II (915 AD - 976 AD), the great library of Córdoba was founded, in which up to 500,000 books and other written documents were stored and used.

Fig. 172: Abū Rayhān al-Bīrūnī, one of the most important scientists in the heyday of Islamic astronomy, geography and mathematics, on a Soviet stamp from 1973 worth six kopecks

Take, for example, the astronomer Abu l-Wafa (940-998), who was active in Persia at the time and who not only translated the main work of Claudius Ptolemy ("Almagest") into Arabic but also made a large number of mathematical discoveries (for example, in the field of spherical trigonometry). al-Biruni (973-1048) also worked in the province of Khorasan in ancient Persia, where he dealt with a variety of scientific topics, particularly astronomy and cartography. For example, he determined the diameter of the Earth using a measuring method he developed himself, whereby his value (12,679 km) differs only marginally from the modern value (12,756 km at the equator). He also penned a valuable work of history. The famous geographer and historian al-Bakri (1014-1094) worked in Córdoba. His main work, the "Book of Roads and Empires" (unfortunately, only fragments have survived), was for a long time a standard reference for merchants and adventurers who traveled the then-known world.

Alongside mathematics, astronomy, and geography, medicine was in full bloom during the "Golden Age" of Islam. The medical knowledge of antiquity was rediscovered and expanded upon, hospitals and sanatoriums were built, and the medical profession was held in high esteem. The most famous physician of that time was probably Ibn Sina (980-1037), who was better known in the Middle Ages and modern times under his Latinized name "Avicenna." His writings (in particular, the five-volume "Canon Medicinae") were used throughout the Middle Ages and were taught in the medical faculties of Christian universities. His knowledge was not limited to medicine. His great philosophical and scientific works have also come down to us, proving him to be a universal scholar of his time. Ibn Rushd (1126-1198), better known as Averroës, is another Arab polymath who had a significant influence on scholasticism, particularly through his commentaries on Aristotle. He was a jurist, physician, and philosopher and was simply called "the Commentator" in the Middle Ages.

Unfortunately, this "Golden Age" of Islam quickly came to an end when the Reconquista began in Andalusia and the Seljuks gained the upper hand in Persia. It is fair to say that this "Golden Age" has never been equaled in Islamic culture since. Nevertheless, relics from that time, when literature and science flourished, have survived to the present day, although we are hardly aware of this. For example, most of the proper names of stars in the sky come from the Arab world. And when we drink "alcohol," this word - like many words beginning with "Al" (such as algebra, alchemy) - also has an Arabic origin: "al-kuḥul" was the name given to the liquid made from wine by the alchemists, in which powdered cosmetics dissolved particularly well. Otherwise, it is well known that the consumption of alcohol is frowned upon (or rather "forbidden") in Islam, which has to do with certain verses in the Koran and their chronological order. The Koran is contradictory in this respect. Early verses (from the so-called Meccan period) still explicitly permitted the trade and consumption of intoxicating drinks. It was not until the late Medieval period that alcohol was considered "the devil's stuff," the consumption of which was forbidden under threat of corporal punishment. Now, in the canonical interpretation of contradictory verses, it is always the case that the most recent verse (viewed from the present) is the authoritative one (abrogation). And so today, a drunken stupor in an Islamic country that is inclined towards Sharia law can quickly lead to the punishment of flogging. In the Islamic calendar, we are only in the year 1436...

381. Iglauer Kompaktat

In the year with the same date, but this time in the Christian calendar, people in my hometown of Zittau (Sittaw), which was still Bohemian at the time, took note of the end of the Hussite Wars with great joy and relief, which had also brought much misfortune to the people of Upper Lusatia and neighboring Silesia. In the Iglauer Kompaktat, some of the demands of the Calixtines (the mainstream of the Hussite reform movement) were finally fulfilled (communion in both forms - host and wine) and Emperor Sigismund (1368-1437) was now also recognized by the Hussites as the Bohemian king and sovereign.

The compact was naturally a thorn in the side of the papal church and therefore only lasted a few years (until 1462). The religious community of the "Bohemian Brethren" eventually split off from the former Hussites and adopted a way of life based on early Christianity. During the time of the German Reformation, they turned more towards Lutheran doctrine, which led them to oppose the Catholic clergy even more.

382. Letter of Majesty from Emperor Rudolf II

Emperor Rudolf II's Letter of Majesty (1609) granted religious freedom to them and all other Protestant estates in Bohemia and Silesia, allowing them to build their own churches. However, this document (the only certified copy of which is in the old collection of the Christian Weise Library in Zittau) was not particularly valuable, as it did not ensure that the Catholics would tolerate Protestant churches. The destruction of the Church of the Bohemian Brethren in Klostergrab (1617) and the resulting Second Defenestration of Prague (1618) marked the beginning of the Thirty Years' War, the greatest catastrophe that ever befell the people of Central Europe. It also sealed the end of the Protestants in Bohemia, who were persecuted, prevented from practicing their religion, and forced into illegality. Their most famous bishop, the great pedagogue Johann Amos Comenius (1592-1670), finally had to leave his Moravian homeland. He died in exile in Amsterdam. It was not until the Peace of Westphalia in 1648 that Bohemian Protestants were granted equal status with Lutherans and Catholics under imperial law, which was reflected in a kind of private religious freedom.

Fig. 173: Copy of Emperor Rudolf II's letter of majesty from 1609

383. Bohemian brothers in Herrnhut

The Counter-Reformation continued in Bohemia, leading to the repeated expulsion of Protestant followers from Bohemia and Moravia until the mid-18th century. These expulsions were only formally ended with the Toleration Patent of 1781, issued by the Habsburg Emperor Joseph II. As part of these forced emigrations, some members of the Bohemian Brethren also ended up in Berthelsdorf in what was then Upper Lusatia. Here, they found refuge under the protection of Count Nikolaus Ludwig von Zinzendorf (1700-1760) and settled in the newly founded town of Herrnhut (a UNESCO World Heritage Site since 2024) from 1722.

It was in Herrnhut that the Herrnhuter Brüder-Unität (Unitas Fratrum), still known today and famous worldwide for its missionary work, was founded. This religious community now has over one million members who call themselves "brothers and sisters." Fortunately, however, not all of them live in Herrnhut...

Fig. 174: Herrnhut - View from the Hutberg, around 1731 (original in the Heimatmuseum Herrnhut, inv. no. 190)

As early as 1732, the members of the Moravian Brethren began to take their Christian faith out into the world. For example, they traveled to the Inuit in Greenland and to the indigenous people of Dutch Guiana (now Suriname) to preach the Gospel. Many of these missionaries combined their activities with extensive ethnological studies, the results of which are still of great value today.

Since 1878, souvenirs from these journeys have been collected in the "Ethnological Museum Herrnhut." The exhibits also include objects from James Cook's third voyage. The museum preserves and presents these valuable collections, which provide insight into the culture and life of the peoples encountered by the Herrnhut missionaries.

384. Wool and plucking science

What is probably less well known is that a man worked at the Herrnhuter Brüder-Unität who founded a completely new field of knowledge in ornithology, the "Gewöll- und Rupfungskunde". I am referring to Otto Uttendörfer (1870-1954). His passion from a young age was to research the feeding habits of our native birds of prey and owls.

Fig. 175 Otto Uttendörfer (1870-1954)

Birds of prey that hunt small birds (such as the sparrowhawk or the peregrine falcon) leave behind so-called "plucks" at their "plucking site," i.e., the feathers - which, being indigestible, they have to remove from their prey by the process of "plucking" before they can eat the bird. By collecting such plucks, assigning them to a specific species of bird of prey, and also determining the species affiliation of the prey bird based on its feathers, it is possible to make statements about its preferred prey. Birds, on the other hand, which swallow their prey - especially small mammals - whole, have to regurgitate the indigestible remains. These droppings are usually called "pellets" in birds of prey and owls. They almost always contain skeletal remains (e.g., skulls) of their prey, which can be used to identify the species of prey. Uttendörfer also collected and meticulously analyzed them. He also built a Germany-wide network of observers who provided him with plucks and pellets and their analyses. Their analysis resulted in Uttendörfer's main work, *"Die Ernährung der Deutschen Raubvögel und Eulen"* (1939), in which he systematically presents his findings on over 400 pages. After the war, his research earned him numerous academic honors.

Today, biologists like to use his method of analyzing birds of prey to study the small mammal population of an area in an economical way. Birds of prey and owls are better hunters than biologists are trappers. Fortunately, the populations of birds of prey and owls - with some exceptions - have been slowly increasing again in recent decades, as the corresponding monitoring programs prove. In particular, hunting pressure (e.g., on sparrowhawks and goshawks) has decreased significantly, and some species, such as the peregrine falcon, have been re-established in their native habitats through reintro-duction programs. I am thinking in particular of Saxon-Bohemian Switzerland and the Zittau Mountains, where it has been possible to observe them again for several years. Incidentally, they have now also discovered the street canyons of some large cities as breeding and hunting grounds (feral pigeons) - for example, in the center of Frankfurt am Main and even in Berlin, Alexanderplatz (not to forget Görlitz!). At the beginning of the 1960s, things did not look good at all for the largest and fastest native falcon in Germany. The dramatic thinning of the population was not so much due to illegal nestling removal or habitat destruction, but rather to an increasing lack of breeding success. The orni-thologists, who monitored the breeding sites on a voluntary basis to prevent disturbance and nest looting, noticed that the eggs often broke during incubation. Their shells had become thinner and more fragile.

385. Dichlorodiphenyltrichloroethane (DDT) and birdlife

The reason for this was an insecticide with the chemical active ingredient dichlorodi-phenyltrichloroethane, which has been increasingly used in agriculture and forestry since the 1940s and is better known by its trivial name, "DDT". At first glance, this chem-ical, known since 1874, had a seemingly ideal insecticidal effect. Even the smallest amounts killed insects, while it had little or no toxic effect on warm-blooded animals and plants, or only in higher doses. Moreover, dichlorodiphenyltrichloroethane is also relatively easy and extremely cheap to produce.

The Swiss chemist Paul Hermann Müller (1899-1965) was awarded the Nobel Prize for Medicine in 1948 for his discovery of the insecticidal effect of DDT (1939). At that time, DDT was the preferred insecticide, which was even used to powder the heads of children when head lice had nested in their mop of hair. However, the intensive use of DDT for pest control (especially in the USA) quickly showed side effects that had not been antic-ipated.

DDT is an extremely stable compound that hardly breaks down in nature and can there-fore accumulate in soil and water. Even more concerning was the fact that accumulation also occurs via the food chain. Insectivorous birds and subsequently predatory birds of prey were particularly affected. At a certain concentration, it acts as an enzyme that in-hibits calcium deposition in the eggshells of birds. The eggshells become more fragile, and breeding success declined. As a result, ornithologists in the USA noticed a dramatic decline in birds of prey such as the peregrine falcon and even the bald eagle, the national symbol of the USA, as early as the beginning of the 1950s. A correlation between the negative population trend and the steadily increasing use of pesticides containing DDT was quickly recognized. The same alarming symptoms were also evident in Europe. However, the negative effects were not limited to birds. Fish and mammals such as seals were also affected. DDT also gradually accumulated in humans and began to endanger their health. This upset environmentalists, and thus the first organized ecological movement emerged in the USA, which vehemently demanded a ban on the use of DDT and other chemical substances.

386. The silent spring

And it was precisely at this time that a book was published by a then still relatively unknown biologist named Rachel Carson (1907-1964), which quickly became extremely popular and therefore could not be ignored by politicians: *"Silent Spring."* It was published in 1962 and went on to become one of the most influential books of the 20th century. In it, she establishes a well-founded connection between the increasing use of DDT and the resulting consequences - such as a spring in which there is no more birdsong due to the lack of birds. It is clear that this book was not exactly greeted with enthusiasm by the chemical industry and its lobbyists, as well as a large number of farmers. Thus, a counter-campaign was quickly launched with the familiar methods that are still used with the same vehemence today: intimidation of the publisher through threats of legal consequences such as considerable claims for damages, denial of the author's professional competence, suspicion of a foreign-controlled conspiracy against the American food industry, suspicion of the author's "communist activities," claims that without DDT "bugs" and "caterpillars" would eat away the food of Americans, publication of crude counter-positions (such as "The Desolate Year," which depicts a world without pesticides in the darkest colors), media campaigns showing how important and how harmless insecticides are, etc. It was deliberately overlooked that Rachel Carson had not called for a ban on DDT, but merely urged a careful and well-considered use (especially with regard to the long-term consequences) of chemical substances in agriculture and forestry. She was well aware of the successes that had been achieved with the help of DDT, for example in containing malaria (through mosquito control) in Africa and Southeast Asia. Ultimately, however, the advocates for caution and environmental activists prevailed, as they had the better arguments - thanks in part to Rachel Carson. And what if the bald eagle, the heraldic animal of the USA, had become extinct? Thus, in 1972, an absolute ban on DDT came into force, which was subsequently adopted by many (now all) European countries.

These events surrounding DDT led to a political reassessment of the risks inherent in chemical products. In any case, the use of pesticides was a major experiment in which the negative consequences only became apparent in retrospect. And when DDT was detected in the bodies of penguins in Antarctica for the first time in 1965, it became clear that chemicals cannot be restricted to the area where they are used, but - if they are stable in the long term - ultimately spread globally.

Fig. 176: Rachel Carson (1907-1964)

This finding had far-reaching consequences for environmental awareness and the regulation of chemical substances. The realization that even the most remote areas of the world are not safe from the effects of human activity led to increased international cooperation and stricter laws to protect the environment. The example of DDT showed that nature is an interconnected system in which interventions in one place can have far-reaching and often unforeseen consequences elsewhere.

The global spread of DDT and other persistent organic pollutants (POPs) drew attention to the need for international agreements to control such chemicals. This eventually led to the 2001 Stockholm Convention, a milestone in international environmental policy that bans or severely restricts the production and use of the most dangerous POPs.

The story of DDT is an example of how scientific knowledge about the long-term consequences of environmental toxins can influence policy and lead to a rethink in the use of technologies. It reminds us that even when introducing new technologies, we must always consider the potential long-term effects on the environment and human health to avoid falling into similar traps again.

387. Ozone destruction by CFCs

Another instructive example of the unexpected consequences of human invention is the group of substances known as chlorofluorocarbons, better known by their abbreviation, CFCs. These compounds are purely artificial in nature and were specially developed as ideal coolants for refrigerators, as their boiling point can be precisely adjusted by their chemical composition. It was originally assumed that CFCs were largely chemically inert and would therefore not cause any environmental problems. Based on this assumption, they have been produced in huge quantities since the 1930s and inevitably released into the environment.

What had not been considered, however, was that the property of inertness (i.e., chemical stability and low reactivity with other substances in the environment) only applies to the troposphere. However, if CFC molecules reach the stratosphere, they are broken down by the sun's UV radiation. This produces reactive products, such as chlorine and fluorine atoms, which have a massive impact on the photochemistry of this layer of the atmosphere. This leads to the destruction of the fragile ozone layer, which protects us from the sun's harmful UV radiation.

The reaction chain for ozone destruction is mainly initiated by chlorine atoms. These split ozone molecules (O_3) into molecular oxygen (O_2) and mono-oxygen (O), whereby the chlorine atom acts as a catalyst and can destroy ozone molecules again in further reactions. A single chlorine atom can destroy several thousand ozone molecules before it is bound and loses its catalytic activity. This process leads to the thinning of the ozone layer, a phenomenon that has become known as the "ozone hole."

The fatal effect of CFCs on the environment was not recognized until the discovery of the ozone hole in the 1980s. The scientific findings ultimately led to international cooperation and the adoption of the Montreal Protocol in 1987, which regulated the gradual reduction and ultimately the complete elimination of CFCs. This agreement came into force in 1989 and obligated the participating countries to gradually reduce the production and use of CFCs.

In 2000, a near-global ban on CFCs came into force, which has contributed significantly to the recovery of the ozone layer. Nevertheless, the effects of these compounds will continue to be felt for decades to come due to their longevity in the atmosphere. The successful reduction of CFCs is an example of the importance of international cooperation and regulatory measures in environmental protection.

DDT and CFCs are all artificial substances that only become a danger in the long term and therefore initially attract little attention from the general public. The long-term negative effects of these substances on the environment were only recognized through extensive scientific research and the emergence of global environmental problems such as the hole in the ozone layer and the bioaccumulation of DDT in the food chain.

388. Catastrophic chemical accidents in Seveso and Bhopal

Catastrophic chemical accidents such as those in Seveso (1976) and Bhopal (1984) have had a profound impact on the risk assessment and regulation of chemical facilities and substances. The Bhopal accident resulted in over 3,800 deaths and an even greater number of injuries, highlighting the urgency of a new approach to managing chemical risks. These incidents forced industry and legislators worldwide to rethink their priorities and initiate comprehensive reforms focusing on both safety requirements and the environmental impact of chemicals.

Following these disasters, the focus in legislation has increasingly shifted to substance-based assessment. This means that the focus is not only on the expected benefit of a chemical product but also on its environmental compatibility, which must be proven. This reassessment has led to a significant tightening of safety requirements, which are monitored by strict control mechanisms, both in the production and use of chemical substances.

The rapid development of the chemical industry and the synthesis of new substances have made it necessary to investigate in detail how these substances behave in the environment once released and what effects they have on living organisms. This research includes investigating the toxicity, persistence, and bioaccumulation of chemicals in various ecosystems.

Fig. 177: Photo of the memorial plaque for the Bhopal disaster at Greyfriars Kirkyard in Edinburgh, Scotland. (Wikimedia)

The Seveso accident led to the introduction of the so-called Seveso Directives in the European Union, which aim to reduce the risks of major accidents involving hazardous substances. These directives require companies to take measures to prevent accidents and develop plans to deal with emergencies.

The Bhopal accident was one of the worst industrial disasters in history and sparked worldwide protests and demands for better safety standards. In India, this accident led to the introduction of stricter regulations for the handling and storage of hazardous chemicals and the creation of a National Environment Tribunal to resolve environmental disputes and enforce environmental laws.

The rapid development of the chemical industry and the equally rapid development and synthesis of new substances have made it necessary to investigate in detail how these substances, once released, behave in the environment and what effects they have on living organisms.

389. Ecotoxicology

The interdisciplinary research field of environmental chemistry and ecotoxicology, which was established towards the end of the 1960s, is concerned with the study of the effects of chemical compounds on the environment and living organisms. This specialized field arose out of the need to understand and manage the increasing pollution of the environment by synthetic chemicals. This discipline combines knowledge from chemistry, biology, ecology, and other sciences to investigate the complex interactions between chemical substances and environmental processes and to develop effective mitigation measures.

The number of known chemical compounds is estimated at around 6 to 7 million, and this number is constantly increasing due to targeted syntheses and discoveries, particularly in biochemistry. Only a small proportion of these compounds are used in the chemical industry. These chemical substances are found in a wide range of products, including pesticides, household chemicals, toiletries and cosmetics, food additives, pharmaceuticals, solvents, and disinfectants.

These chemicals are eventually released into the environment through improper disposal, leakage, or off-gassing. Some of these substances can be degraded in the environment, whether through natural processes such as biodegradation or artificial methods such as wastewater treatment. However, there are also substances that are highly persistent due to their structural properties and therefore accumulate in the environment over long periods of time.

A well-known representative of these long-term stable substances is DDT (dichlorodiphenyltrichloroethane), an insecticide that was widely used in the 1940s and 1950s (item 385). DDT is lipophilic, i.e., it dissolves in fats and oils and therefore accumulates in the food chain, a phenomenon known as bioaccumulation. This can lead to an increased concentration of DDT in organisms at higher levels of the food chain, including humans. Particularly problematic is the fact that DDT, once released into the environment, can persist for decades before it is fully degraded.

The accumulation of DDT and other *persistent organic pollutants* (POPs) in the human body can lead to adverse health effects. Some of these substances are known carcinogens, i.e., they can cause cancer. Others can disrupt the endocrine system, which can lead to developmental disorders, reproductive problems, and other health impairments. It is therefore of great importance to minimize the release of these substances into the environment and to understand their impact on human health and ecosystems.

In addition to the identification and monitoring of these pollutants, the development of safe alternatives and the promotion of environmentally friendly technologies are among the central tasks of environmental chemistry and ecotoxicology. Only a comprehensive and preventive approach can ensure the long-term protection of our environment and human health.

390. The Seveso poison dioxin

Of all environmental toxins, dioxin, or more precisely 2,3,7,8-tetrachlorodibenzodioxin (TCDD), has gained a certain degree of notoriety. Although it is of no economic use itself, it is formed from halogen compounds during certain combustion processes or is produced in large quantities as an intermediate product in the chemical industry. Even today, this substance would hardly be known among non-chemists if it did not have a catchy trivial name and if it had not been released in large quantities on July 10, 1976, in a chemical accident in the Seveso area in northern Italy. Since then, the dioxin TCDD has often been referred to as "Seveso poison." However, TCDD is also the active ingredient in the defoliant "Agent Orange," which the Americans sprayed in large quantities over the Vietnamese jungle during the Vietnam War (since 1965). The long-term effects of this act of chemical warfare (although in this case the term is not internationally recognized, as it was "only" a herbicide) are still present everywhere in Vietnam today: malformations in children, immune deficiencies, cancers...

But back to Seveso, the small Italian community in Lombardy, whose name nobody would know today if the aforementioned accident had not happened there. In the chemical plant there, where herbicides were manufactured, an unfortunate chain of circumstances led to an uncontrolled reaction during the production of 2,4,5-trichlorophenoxyacetic acid from 1,3,4,6-tetrachlorobenzene, which was accompanied by an increase in temperature and pressure in the production plant. This led to a partial dimerization of the trichlorophenol into the highly toxic TCDD, until finally, the reaction vessel exploded and the dioxin spread over a wide area in the form of a cloud. What followed was an act of concealment and initial appeasement of the residents of the surrounding area, who quickly felt the toxic effects of the dioxin in the form of more or less acute chloracne - and the company finally had to come clean. The farm animals that had eaten dioxin-contaminated grass also began to die off en masse. In addition, the leaves of trees and bushes at the site of the accident withered and dried up. As just one microgram of TCDD per kilogram of body weight leads to severe dioxin poisoning and this toxin only breaks down very slowly (when exposed to sunlight), the population around Seveso first had to be evacuated and then a large-scale decontamination was carried out. Among other things, the soil was removed and taken to special hazardous waste disposal sites.

Of course, there were also legal repercussions, and laws concerning dioxin were tightened worldwide to prevent such disasters in the future. Analytical methods have also been further improved in order to detect even the smallest quantities of this toxic substance in the environment. Today, a few nanograms per kilogram of body substance can be reliably measured (1 ppt (parts per trillion) = 1 ng/kg - "trillion" corresponds to our "trillion"). To visualize this tiny amount of substance, perhaps the following comparison can be made: If the earth's equator is given as 40,000 km, then one ppt corresponds to a length of 0.04 millimeters. If you carry out analysis with such precision, you will find that TCDD is present virtually everywhere: 3 ppt in soot from car exhausts, 40 ppt in road dust, 2 ppt in cigarette smoke, etc. Particularly high concentrations can be measured in the vicinity of waste incineration plants, which are now considered to be real dioxin emitters (up to 30 ppb - parts per billion, which is why they are subject to special supervision. The problem here again lies in the long-term stability of the substance and its accumulation in the environment and humans. While the clinical picture of acute TCDD poisoning is very well known, there is little reliable knowledge about the effects of even the smallest quantities. Consequently, everything should be done to avoid releasing dioxins into the environment. This is why waste incineration plants are required by law to operate at higher temperatures than the operators would like for economic reasons.

The use of "chemicals" in agriculture has caused yields to explode since the availability of artificial fertilizers and pesticides. This has secured the livelihoods of many millions of people. Insecticides help to contain insect-borne diseases such as the notorious malaria or sleeping sickness. In tropical and subtropical regions in particular, where such diseases are endemic, this has led to a significant improvement in public health and a reduction in mortality rates.

However, all this comes at the cost of increasing contamination of food with undesirable substances. Pesticide residues in food and contamination of soil and water sources are a serious problem that can have long-term effects on human health and the environment. Particularly in countries with less stringent environmental and food safety regulations, contamination can be significant, leading to increased exposure of the population to harmful chemicals.

The well-founded fear of this as well as an awakened "ecological awareness" among the affluent citizens of industrialized nations led to the prefixes "organic" and "eco" in front of food produced by "ecological forms of agriculture." The term "organic farming" refers to a form of agriculture that is particularly gentle on nature, characterized by little or no use of artificial fertilizers and pesticides and by the consideration of environmental and landscape protection concerns. This agricultural method promotes biodiversity, improves soil fertility, and helps to maintain the ecological balance. In addition, organic farming places great importance on species-appropriate animal husbandry and the avoidance of genetically modified organisms.

Of course, this is all very welcome. Nevertheless, it must be pointed out that less than 1% of the world's available agricultural land is farmed in this way - and a significant increase is not to be expected due to the steadily growing population, particularly in developing countries. The high productivity of conventional agriculture makes it possible to feed the growing world population, which is of crucial importance in regions where resources are already scarce. Switching to organic farming would lead to a decline in food production in many parts of the world, which could have disastrous consequences, particularly in poorer countries.

As sad as it is, "organic" and "eco" is just a market niche that can only establish itself in its pure form in affluent societies. This market niche particularly appeals to consumers who are prepared to pay higher prices for food produced under more sustainable conditions. Nevertheless, the influence of this niche on global agriculture remains limited. For most of its existence, mankind has not cared at all about "organic," as we would say today. In many regions of the world, human survival is still closely linked to maximizing agricultural yields, and environmental sustainability often plays a subordinate role. This presents humanity with the challenge of finding ways to ensure agricultural productivity while minimizing its environmental impact.

391. Environmental protection - an idea of the 20th century

The idea of environmental protection and nature conservation may have gained in importance in the 20th century, but history shows us that mankind has had a significant impact on the environment for thousands of years. Even in ancient times, human activities led to profound changes and sometimes irreversible damage to ecosystems.

Deforestation in the Roman Empire for timber and fuel led to massive forest loss and erosion. Similarly, the Sumerians turned fertile land into a salt desert by developing irrigation systems, which led to the long-term salinization of the soil. The collapse of the Maya empire is also partly attributed to environmental disasters such as droughts, which were exacerbated by extensive agricultural practices and deforestation. These examples illustrate that human intervention in nature can have profound and often negative consequences.

With the onset of the Industrial Revolution in the 19th century, the impact of humans on the environment increased even further. The increasing use of fossil fuels and industrialization led to accelerated deforestation and pollution. In Britain, forests were cleared on a large scale to meet the demand for fuel for steam engines and metal smelters. In Germany, the Harz Mountains and the Black Forest were severely depleted by intensive logging in Goethe's time. This deforestation contributed significantly to the alteration of the landscape and the deterioration of air quality.

Despite the lessons of history, mankind does not seem to learn from its mistakes. Today, it is mainly abject poverty and the greed for profit that are causing several dozen hectares of tropical rainforest to disappear every minute. The deforestation of these valuable ecosystems has serious consequences for biodiversity and the global climate. Those involved are either unaware that they are destroying one of the most important foundations of human life in the long term, or they are ignoring it. It is remarkable that humans can recognize and quantify the risk but simultaneously do little or nothing to reduce it. This shows that while human intelligence is capable of recognizing problems, human irrationality often prevents these problems from being solved.

Reason may prevail locally, but globally - driven by an ever-increasing number of people on this planet - destructive forces still prevail. It is nothing more than a myth that humans are "subjugating" nature. We are only walking a fine line with nature and cannot expect any preferential treatment from it.

392. Humans as the ultimate natural disaster

Its future history is just as uncertain as that of every other living creature on Earth. Or, as the Austrian sociobiologist Franz M. Wuketits once aptly put it:

"Humans are not only the cause of major natural disasters, but, strictly speaking, they themselves represent the greatest natural disaster currently afflicting the planet."

The extinction rates of plants and animals already exceed those of some of the major mass extinctions in Earth's history. Humans are a "large consumer species." They have spread globally, require more food, more land, and more energy than any other species and, as a result of their activities, are permanently reducing the diversity of life (to use a popular term in the ecology field) - which will one day be their undoing. And yet we are hardly aware of this indisputable fact. In my opinion, the ecologist R. Kinzelbach aptly characterized this in his book *"Ökologie - Naturschutz - Umweltschutz"* (Ecology - Nature Conservation - Environmental Protection), published back in 1989:

"In industrialized countries, most inhabitants do not necessarily have the feeling that they or their society are fundamentally doomed. It's only in the newspapers. If they didn't have their travelers, reporters, commentators, scientists, gurus, and alarmists - in short, that pluralistic spectrum of restless sensors - thanks to the Western ability to criticize and the highly specialized social structure, they might even think that the world is in order, buoyed up by the statements of their economic and state leaders."

If you follow current developments with a clear mind, you get an ambivalent feeling. Technological development is being driven forward in leaps and bounds. In the near future, cars will be driving completely autonomously on our roads. OLED displays are now so thin that they will soon be able to be adhered to the wall like wallpaper. More and more people are connected via social networks. Our research probes launched into space have explored the entire solar system and even passed by the dwarf planet Pluto, a kind of outpost of our planetary system. Nothing seems impossible anymore. What would have been a "miracle" a hundred years ago has now become an object of our desire to possess. Disasters and catastrophes only ever happen far away and do not (or only rarely) affect us. It's like the calm before the storm...

393. The thunder of guns can already be heard

In fact, the thunder of the guns can already be heard. The first dark clouds are gathering on the horizon, a harbinger of the storms to come. More and more people fleeing poverty, war, and religious persecution are crowding the borders of wealthy countries and seeking entry. They are looking for a better life, security, and stability, which are no longer available in their home countries.

The population explosion in emerging countries continues unabated, while the birth rate in developed societies is on the decline. This leads to an imbalance that increases social and economic tensions. In order to maintain or assert hegemonic claims, the major powers are opening up secondary theaters of war. However, this risky game is increasingly backfiring, as the example of the "Islamic State" shows. Here, modern technology and archaic ideology combine to form a dangerous mixture that does not bode well for the future. The Ukraine conflict is another example of such secondary theaters of war. The geopolitical tensions between Russia and the West (NATO's eastward expansion, the "color revolution") have plunged the country into a crisis that has far-reaching consequences for global security and stability.

The battle for influence, raw materials, and energy sources has long been in full swing. It is a battle that is being waged not only by states but also by multinational corporations, paramilitary groups, and even criminal networks. And we, the citizens, are right in the middle of it. We are witnessing these global changes and must rise to the challenge of actively shaping our future. Because ultimately, the fate of our world depends on the decisions we make today.

394. Pandemics

The world is now so interconnected that emerging infectious diseases with the potential of causing a pandemic similar to the Spanish flu, which raged at the end of the First World War (1918-1920) and caused more deaths worldwide than the actual fighting (estimated at between 25 and 50 million deaths), pose a real threat that is usually completely underestimated. In modern industrialized countries, we rely entirely on a functioning healthcare system and fail to recognize that it is not designed for a mass influx of sick people.

It is certainly true that pandemics have always accompanied mankind throughout its history. In the years 165 to 180 AD, for example, the Roman Empire was struck by an extremely deadly infectious disease, now thought to be a particularly virulent form of smallpox. The philosopher and Roman emperor Marcus Aurelius (121-180 AD) fell victim to it in Vindobona (a Roman military camp in the area of present-day Vienna) in 180 AD. This "Antonine Plague," which lasted for 25 years with brief interruptions and, according to conservative estimates, probably cost the lives of five million people, had a far-reaching political impact on the entire Roman Empire.

Fig. 178: Painting "The Plague in Ashdod" by Nicolas Poussin (1594-1665). The painting depicts a story from 1 Samuel in the Old Testament and primarily deals with the plague epidemic that ravaged Italy (particularly Rome) from 1629 to 1631.

Other pandemics that raged in waves, particularly in Europe, are summarized under the term "Black Death" (referred to as the "Great Pestilence" in the Middle Ages). This was probably mainly the plague caused by the bacterium Yersinia pestis. It wiped out at least a third of the entire European population in the middle of the 14th century alone. Infectious diseases (especially smallpox) were also extremely instrumental - to put it somewhat cynically - in the eradication of the indigenous peoples of North and South America during the conquest by European powers after 1492. Smallpox has since disappeared - thanks to the invention of vaccination by the British physician Edward Jenner (1749-1823). Since then, infectious diseases have largely (at least in perception) lost their terror. But this feeling is extremely deceptive in several respects. While bacterial infectious diseases can (still) be controlled relatively well with antibacterial drugs, this is more difficult or even impossible with viral diseases. Here, it is necessary to develop vaccines for preventive protection, which is generally very challenging. For a large number of viral diseases (including Ebola until recently), there is no vaccine.

Although most viral infections are relatively harmless, there are some viruses that can lead to pandemics. First and foremost is the highly variable influenza virus. An influenza outbreak detected too late with a flu virus strain that has a similar lethal effect to that of the Spanish flu could - as model calculations show - quickly escalate into a global catastrophe due to the high mobility of people. It is estimated that, without massive countermeasures, it would take just 60 days for all of the world's major cities to be affected.

Of all the predictable disasters that threaten a developed population largely concentrated in cities, a pandemic of an easily transmissible viral disease is the most likely. It is therefore essential to prepare logistically for such an event, as the primary aim is to prevent an outbreak. Once such a pandemic is underway, it is not only the health authorities in developing countries that are completely overwhelmed, but also the highly developed industrialized nations that quickly reach their limits.

The COVID-19 pandemic, which has kept the world on tenterhooks since 2020, has demonstrated impressively how vulnerable even the most prosperous countries are to the rapid spread of a new virus.

The Ebola outbreak (a hemorrhagic fever with a high fatality rate) in West Africa in 2014, which dominated the news channels for a while until interest in it waned, can also serve as a lesson in this context. The world would look very different today if the Ebola virus had had the same infectious potential as the Spanish flu influenza virus. So, we can count ourselves lucky that the epidemic was limited to a few countries in Africa and that the number of victims (even if that may sound cynical) was kept within limits at just over 11,000 (just to give you an idea of the scale, malaria alone claims well over a million lives worldwide every year).

395. Ebola

Ebola was first observed in 1976 in the Democratic Republic of Congo, where 318 people were infected in a town near the Ebola River, 280 of whom died. It is assumed that the "host animal" of this filovirus, notable for its characteristic filamentous shape, is fruit bats. These animals, which some Africans consume as "bush meat," could have transmitted the virus through direct contact with humans. This is how the virus entered the human body and was able to unleash its deadly effects - people bled to death from the inside out, as it were, which led to the term "hemorrhagic fever."

Since the first documented outbreak, numerous other Ebola epidemics have been observed, primarily concentrated in Central Africa. Although these outbreaks were all ultimately contained, each of them demonstrated the immense danger posed by this virus. The high lethality rate and the fact that there is no specific, widely available vaccine or antiviral therapy make Ebola a constant threat in affected regions.

The problem is that such viral infections often do not provide sufficient economic incentive for the pharmaceutical industry to develop specific virus inhibitors (antivirals) or vaccines due to their low case numbers. This development gap can lead to catastrophic consequences, as the 2014 Ebola outbreak in West Africa clearly demonstrated. In this case, however, several experimental vaccines were already at an advanced stage of development, including the active ingredient "rVSV-ZEBOV-GP" developed by Canadian scientists. In principle, this vaccine consists of vesicular stomatitis viruses[77] that are harmless to humans and have been genetically modified to produce a protein of the Ebola virus. The human body reacts to this protein by developing specific antibodies, which then provide protection against infection with the Ebola virus.

However, the development of such vaccines is anything but trivial. It requires extensive research and development work that can take years or even decades. The path from laboratory research to clinical trials to regulatory approval is long and complex. However, the Ebola outbreak of 2014 demonstrated that it is crucial to act quickly in high-risk cases and to make experimental therapies available as swiftly as possible. The subsequent rapid development and deployment of the rVSV-ZEBOV-GP vaccine in affected areas marked an important milestone in the fight against the Ebola virus and highlighted the importance of continuous research and international collaboration in combating dangerous infectious diseases.

396. COVID-19

COVID-19 was first observed at the end of 2019 in the city of Wuhan, in the Chinese province of Hubei. The disease is caused by the SARS-CoV-2 coronavirus, which belongs to the same family as the viruses that cause SARS and MERS. It is suspected that the virus jumped from an animal host, possibly bats or an intermediate host such as the pangolin, to humans. This zoonotic transmission is a crucial component in the development of new infectious diseases. However, there are also strong grounds for suspecting that the high virality of the SARS virus was artificially and purposefully (and largely secretly) created in a laboratory. Leading virologists now more or less support this theory.

The rapid and global spread of COVID-19 has led to a pandemic that has so far infected millions of people and caused significant health, social, and economic disruption worldwide. The virus is mainly transmitted from person to person by droplet infection but can also spread through contaminated surfaces. Symptoms range from mild respiratory symptoms to severe pneumonia and multi-organ failure, which is particularly dangerous for the elderly and people with pre-existing medical conditions.

Since the emergence of COVID-19, numerous measures have been taken to contain its spread, including lockdowns, mandatory masks, social distancing, and extensive testing and quarantine protocols. However, these measures did not always achieve the desired success and led to considerable social challenges and tensions. In Germany, for example, they were used by politicians to increasingly weaken constitutionally guaranteed basic rights and, in some cases, even to illegally suspend them.

A major turning point in the fight against the pandemic was the rapid development and approval of vaccines. Various research laboratories and pharmaceutical companies

---

[77] Vesicular stomatitis viruses (VSV) are a group of RNA viruses belonging to the Rhabdoviri-dae family. These viruses cause vesicular stomatitis in animals, especially cattle, pigs and horses, a disease characterized by the appearance of vesicles and ulcers in the mouth and muzzle area and on the claws

around the world began intensive work on COVID-19 vaccines at the start of 2020. The mRNA technology, which had previously hardly been used in vaccine development, proved to be particularly successful (although many experts no longer view it that way in retrospect). Vaccines such as those from BioNTech/Pfizer and Moderna are based on this technology, in which a small portion of viral RNA is injected into the human body to stimulate an immune response. However, whether the development of such RNA vaccines has opened a Pandora's box is currently the subject of public debate, in part because the various side effects of the corresponding vaccines have been investigated nationally and internationally.

Despite the rapid development and availability of vaccines, challenges remain. Virus variants, some of which are more resistant to existing vaccines, and vaccine hesitancy among the population continue to pose obstacles. In addition, economic and logistical difficulties, particularly in poorer countries, have slowed global vaccination progress.

The pandemic has also demonstrated the importance of a well-coordinated international response and strong healthcare systems. It has exposed weaknesses in the global health infrastructure and highlighted the need for investment in research and development, as well as in prevention measures and pandemic preparedness.

COVID-19 will go down in history as one of the most serious global health crises of the 21st century. The pandemic has not only challenged the medical and scientific community but has also had a profound impact on the daily lives of people worldwide. It is hoped that the lessons learned from this crisis will help us to be better prepared for future pandemics and to establish more sustainable and equitable healthcare systems worldwide.

397. The threat of new viruses and the limits of global pandemic preparedness

On the other hand, new, previously unknown viruses emerge every few years. Take, for example, various "swine and bird flu viruses," such as the current (2024) H5N1 virus, all of which represent a potential danger that should not be underestimated should they or their descendants develop into an epidemic or, even worse, a pandemic. Then, there may not be enough time to develop an antidote to prevent a catastrophe. Contingency plans based on logistical containment (such as quarantine measures) and regulating the availability of capacities in hospitals and any vaccines and medicines that can be used are therefore more important. However, it is worrying that although such emergency plans are mandatory in industrialized nations, populous emerging economies and developing countries cannot afford them or can only afford them to an insufficient extent.

As even the WHO has noted, humanity is poorly prepared for the emergence of new pandemics. This raises the question of whether a "killer germ" could emerge at some point in the future that has the potential to cause an uncontrollable global pandemic. Would such a "germ" be capable of wiping out the entire human race? The answer to this question is a clear "no." And for several reasons. Firstly, as a biological species, humans are globally distributed and as "omnivores," they can use a variety of foods. In addition, there is sufficient genetic diversity in the human population to ensure that there will always be individuals who are immune to a pathogen, whether a bacterium or a virus. In the event of an uncontrollable epidemic (such as the "great pestilence" in the Middle Ages), there may be a considerable loss of human life, with the collapse of social structures playing a significant role in addition to the actual infectious disease. However, this will not lead to the extinction of the Homo sapiens species. Even a "meteorite impact" below the "Chicxulub level" and a nuclear exchange between the major powers would not be able to achieve this - despite the radioactive contamination that would result in the latter case. As I said, this is only about humans as a biological species. However, living conditions can hardly be compared with those before such an "event," and it will take some time before "life" in our sense appears worth living again. We should therefore do everything we can to avoid both local and, in the worst case, global apocalypses.

398. Man-made climate change - the new warning sign

The warning sign that is currently being painted most clearly on the wall in this respect is "man-made" climate change. The adjective "man-made" is the real problem and the crux of the controversy, as "the climate" has always changed and will continue to do so in the future.

But let's start with the terminology, because "weather" and "climate" are often mistakenly lumped together, whether consciously or unconsciously, although there is a fundamental difference between "weather" and "climate": *Climate is what you expect; weather is what you get"* (Larry Gates). Weather forecasting deals with the "weather" that is highly likely to be expected in a region tomorrow and the day after tomorrow. Climate research is concerned, among other things, with the occurrence of cold and warm periods or, more precisely, with the mean values and deviations from the mean value of meteorological parameters for a certain region over such a long period of time that the statistical parameters mentioned make sense. In practice, a period of at least 30 years has been agreed upon for this averaging. For long-term studies, several of these 30-year averages are then averaged.

It becomes more difficult to derive statements about global trends (i.e., for the whole world) and not just regional trends (e.g., for Central Europe). Determining the temporal dimension of the climate, in particular, is an extremely difficult undertaking. Firstly, because "measurements" have only been available for a few hundred years, and secondly, because the derivation of climate parameters indirectly (tree rings, ice cores, geological markers such as sediments, isotope ratios) has many methodological weaknesses. Paleoclimatology is therefore only able to roughly reconstruct the Earth's climate development. The situation is better for the Quaternary (i.e., the last two and a half million years or so), and here in particular for the Holocene (the last 12,000 years or so) - at least as far as the Earth's northern hemisphere is concerned. Meaningful temperature curves for the mean annual temperature could be created for this period. They show the ups and downs of temperatures and thus the course of the so-called interglacials (warm periods) and glacials (cold periods), during which large parts of Eurasia and North America were covered by a mighty ice sheet. To be clear from the outset, we are (fortunately) living in an interglacial period today. The last great ice age, the Weichselian or Würm Ice Age, was not so long ago (it ended around 10,000 years ago).

Another critical aspect is the question of the role of humans in climate change. Many critics of the theory of anthropogenic climate change argue that natural factors such as solar activity, volcanic eruptions, cloud cover, and oceanic circulation patterns have a significant impact on climate and may be stronger than anthropogenic influences. There is evidence that climate models based primarily on anthropogenic emissions of greenhouse gases may not adequately account for these natural factors and therefore cannot accurately predict future climate development. This is particularly true with regard to the adequate inclusion of clouds in climate models, which has not yet been achieved satisfactorily. However, clouds are a decisive factor in the Earth's climate system (just think of their albedo).

Another point worth discussing is the question of the political and economic interests behind the climate debate. Critics complain that climate protection measures are often associated with considerable costs and economic disadvantages, especially for industrialized countries that have built their economies on fossil fuels. It is argued that some actors are using the climate debate to pursue their own political and economic goals, be it by promoting renewable energies or by introducing new taxes and regulations. According to critics, these measures could also lead to economic redistribution, from which certain sectors or regions benefit while others suffer.

Not to be forgotten is the criticism of the way in which climate research and climate policy are communicated. Some critics accuse climate research of exaggerated dramatization and scaremongering that is not always justified by the scientific data (e.g., the extremely unlikely extreme scenarios are always used in the press to argue the case, while the other, more likely scenarios are swept under the carpet). They call for a more sober and balanced discussion about the risks and uncertainties of climate change as well as the costs and benefits of climate protection measures.

Furthermore, the question is raised as to whether the current measures are actually having the desired effect or whether they are merely symbolic steps that do not address the actual causes and systemic problems. The "climate issue" has become deeply ideologically charged ("climate glue") and often lacks an objective debate on effective and pragmatic solutions.

399. Little ice age

As already mentioned, the last great ice age in Central Europe was not so long ago (it ended around 10,000 years ago). Our ancestors, who lived in the Northern Hemisphere between 1675 and 1715, also had to contend with particularly "icy" winters. Together with a similarly cold period between 1570 and 1630, this is even referred to as a "Little Ice Age". The Vikings under "Erik the Red" continued to raise livestock on the lush pastures of "Greenland" until around 1000 years ago, when a relatively sudden cooling forced them to abandon their settlements around 1350. The causes of natural climate change are manifold and make climate systems highly complex systems that defy simple description. I have already mentioned one very important aspect, the Milankovitch cycles, in another context (point 270).

Fig. 179: The painting "IJsvermaak" ("Ice Pleasure") by Hendrick Avercamp, which shows people on a frozen canal in the Netherlands in the harsh winter of 1608, contrasts with the situation today, in which the canals normally no longer freeze over in winter. Such artistic depictions of ice scenes are only known from the period between 1565 and 1640 (think of Pieter Brueghel the Younger).

In addition to the Milankovitch cycles, there are other important factors that contributed to the emergence of the "Little Ice Age." One hypothesis is that a series of volcanic eruptions played a role, including in particular the eruption of Huaynaputina in Peru on February 19, 1600, which cooled the climate worldwide. Volcanic eruptions are known to eject large quantities of ash and sulfur dioxide into the stratosphere, which reduce solar radiation and thus lead to a cooling of the Earth's surface.

Another factor was the low solar activity during the so-called Maunder Minimum (1645-1715), during which hardly any sunspots were observed. This period of low solar activity correlated with some of the coldest years of the Little Ice Age and is considered an important contributor to the cooling.

The effects of the Little Ice Age were far-reaching. In Europe, long, cold winters and short, cool summers led to crop failures and famine. Rivers such as the Thames in London and the canals in the Netherlands regularly froze over, leading to ice festivals and markets on the ice. The population suffered from the harsh conditions, and social and economic upheaval ensued. The effects were also felt outside Europe. In North America, settlers experienced particularly harsh winters, which affected agricultural production. In China, the cooling led to a shift in agricultural zones and political unrest, which ultimately contributed to the fall of the Ming dynasty.

The "Little Ice Age" clearly shows how sensitively human societies react to climate change. The complex interactions between the natural driving forces of climate and human reactions to them are therefore an important field of research in climatology. It also shows that climate change is not a modern phenomenon, but has always accompanied and challenged humankind.

400. Greenhouse effect

Furthermore, if you want to understand the change in the Earth's climate, you need to know that our "moderate" temperatures are the result of an effect called the "greenhouse effect". Most people assume that this effect is primarily caused by the trace gas carbon dioxide, but this is not entirely true. I will return to this later. Without this greenhouse effect, the Earth's equilibrium temperature would be almost exactly -18°C. This so-called equilibrium temperature value can, with reasonable accuracy, be interpreted as the Earth's mean annual temperature (although the concept of a planet's "mean annual temperature" makes little thermodynamic sense). However, since the actual mean temperature is +14°C, the difference of 32°C must be seen as the result of the aforementioned greenhouse effect. The physics behind this is more complicated than generally assumed and is not yet fully understood. Perhaps just this much: the Earth's atmosphere is almost transparent to solar radiation in the optical range; it only reaches the Earth's surface with minimal influence, where it is partly absorbed and partly reflected by the land and sea. This causes the surface to heat up and emit infrared radiation (thermal radiation, according to Planck's law) based on its temperature, for which the Earth's atmosphere is no longer transparent in certain wavelength ranges. The reason for this is certain gases whose molecules are not homo-nuclear like nitrogen or oxygen (the main components of the Earth's atmosphere). These primarily triatomic molecules are able to absorb the infrared radiation emitted by the Earth's surface (i.e., they become energetically excited - particularly in their rotational levels), which is why they are classified as "greenhouse gases". However, molecules excited in this way release the absorbed energy again after a short time - for example, through collisions with other molecules - or through the re-emission of the previously absorbed radiation. This time, however, the re-emission is no longer directed from the ground towards space, but isotropically in all directions, including back to the Earth's surface, which, according to common doctrine, receives slightly more radiant energy than directly from the sun. At this point, I have doubts about the common paradigm, as I cannot see how heat should flow from the colder atmosphere to the warmer Earth's surface. In the end, according to this theory, a new equilibrium temperature between solar radiation and terrestrial radiation is established (both contributions are always equal), which is greater than without the presence of greenhouse gases. That's the theory, which is far more complicated in detail.

So, what are these greenhouse gases? In order of their impact, they are water vapor, carbon dioxide, and methane, plus a dozen other gases of extremely low concentration. In this list, water vapor is the main cause of the natural greenhouse effect (its contribution to the 32°C is estimated at over 60%). This is followed by carbon dioxide, the concentration of which, again according to current doctrine, is considered a control variable of the greenhouse effect. It mainly enters the Earth's atmosphere (and the oceans) through volcanic eruptions and is primarily removed from the atmosphere through the weathering of rocks. Since the beginning of the industrial age, another source of carbon dioxide has emerged from human activity - in the form of carbon combustion for energy production.

401. Anthropogenic greenhouse effect

heir contribution to the overall greenhouse effect is usually referred to as the "anthropogenic greenhouse effect" - and this is the basis of the fear of a global climate catastrophe that needs to be prevented (around 6% of the total carbon dioxide in the Earth's atmosphere is of anthropogenic origin). The logic behind this can be explained as follows: Burning fossil fuels (coal, natural gas, crude oil) to generate energy produces carbon dioxide as a combustion product, the concentration of which in the atmosphere rises steadily (from 280 ppm in pre-industrial times to approximately 426 ppm today). This corresponds to 14 more CO_2 molecules per 100,000 air molecules. With this increase, the flanks of the already saturated carbon dioxide absorption bands increase, leading to more heat radiation being absorbed. This results in a slight increase in air temperature, which in turn leads to increased water vapor formation through evaporation. This intensifies the greenhouse effect (i.e., more water vapor in the atmosphere), leading to a further increase in the global average temperature. Observations as well as theory clearly indicate a correlation between the global mean temperature and the carbon dioxide concentration, with a lag of the CO_2 concentration relative to the temperature being postulated. By using climate simulation models to extrapolate into the future, taking this correlation into account, alarmist projections for the expected rise in temperature emerge. These projections are driving politicians, particularly in industrialized countries such as Germany. It is becoming increasingly apparent that nature does not necessarily adhere to the simulations and models of climate researchers.

402. Can climate simulations be trusted?

This is because a complex system, such as the Earth, can only ever be mathematically modeled approximately. It is not so much the algorithms based on the solution of the corresponding differential equations that are controversial, but rather the choice of initial and boundary conditions, as well as the choice of spatial resolution (referred to as the "grid parameters" - the Earth is surrounded by a three-dimensional grid in which the meteorologically relevant variables are calculated for each grid point at each time step).

The choice of initial and boundary conditions is crucial, as they define the starting points and limits within which the model should operate. Small uncertainties or inaccuracies in these parameters can accumulate over time and lead to significant deviations, affecting the accuracy of the forecast (an effect of "deterministic chaos"). Additionally, the spatial resolution of the grid influences the level of detail in the model: a finer resolution allows for more accurate local predictions but also significantly increases the computational effort.

One of the most important questions to consider when working with such "climate simulations" is the extent to which you can trust their predictive power over 10, 100, or even 1,000 years. Caution and skepticism are definitely warranted. Long-term predictions are subject to considerable uncertainty, as the climate system contains numerous feedbacks and non-linear processes that are difficult or impossible to predict. For example, small changes in ocean currents or land use can have a major impact on the global climate.

Another aspect is predicting future developments in human activities, such as greenhouse gas emissions, the use of fossil fuels, or land use changes. These factors are difficult to predict and can significantly influence the accuracy of climate models. The uncertainties in socio-economic scenarios also contribute to the need for caution in interpreting long-term climate model predictions.

In addition to uncertainties in the models themselves, there are also methodological challenges. Climate models must be regularly validated and calibrated by comparing their predictions with actual observations. These validation processes are complex and require careful analysis of discrepancies between the model and reality.

403. Climate simulations can neither be validated nor verified

But this could conceal a fundamental problem of climate simulations or numerical simulations of complex systems, which Naomi Oreskes, Kristin Shrader Frechette and Kenneth Belitz pointed out as early as 1994.[78] Their verification and validation is not possible - unless you wait until the prediction time to which the simulation refers has been reached. Then you can see whether the prediction has come true or not.

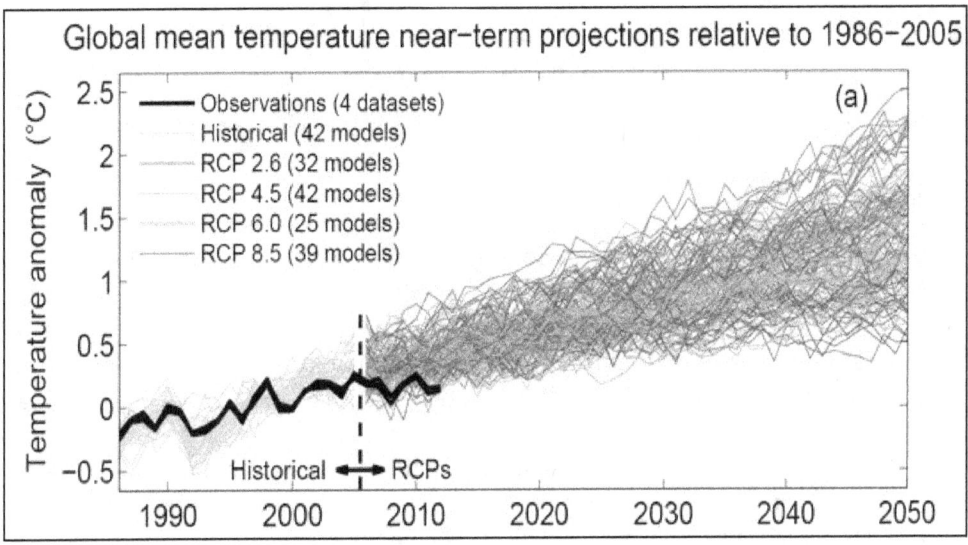

Fig. 180: Examples of simulations of various climate models with regard to the future development of the Earth's average global temperature. The "true" development is based more on the conservative models, while the media assessment is based more on the extreme models (source IPCC)

The problem with climate simulations, in particular, is that they have to average out an extremely complex, inherently chaotic system in which both the data basis is inherently incomplete and the knowledge about the system to be simulated is at best only approximately correct. The scientists who develop such climate models can therefore never be completely sure whether the model is overlooking any relevant factors. Of course, this does not mean that such numerical simulations of highly dynamic systems are useless per se. They are not, as they help to understand such systems and uncover their development trends. In other words, you can "play" with them to see what happens when this or that adjusting screw (e.g., the carbon dioxide concentration) is turned. However, we must not confuse the results of a simulation with reality, as is often the case with climate models. It must also be taken into account that calculations are always carried out with a fixed number of digits (accuracy) and the rounding errors that inevitably arise (and

---

[78] Cf. Oreskes, N., Shrader-Frechette, K., & Belitz, K. (1994). Verification, Validation, and Confirmation of. Science, 263, 4.

propagate) can certainly influence the system behavior as disturbances. The numerical stability of such simulations depends crucially on the numerical solution methods, which is why mathematicians have to invest a great deal of effort in their development. For all these reasons, great care must be taken when interpreting simulation results, especially when far-reaching political decisions depend on them.

From a risk management point of view, the results of climate simulations should be taken seriously, especially if different models lead to a similar trend (global warming). If it makes sense (which should be discussed), countermeasures should be considered. On the other hand, the "evidence" based on measurements, rather than simulations from various climate models, remains quite limited when viewed critically in the context of the Earth's natural temperature fluctuations.

404. Usurpation of climate change by politics

The usurpation of the actual (and ongoing) climate change by politicians has already led to some positive, but also several negative developments, the implementation of which would not have been possible without the specter of "global warming." If you take the models seriously, you have to decide whether you want to limit the anthropogenic share (in concrete terms, according to the current view, this means "turning the carbon dioxide screw") - which can only be a global task in which all carbon-burning nations must participate (as it stands, this is an illusion, at least regarding developing and emerging countries). In this respect and in the truest sense of the word, it is completely meaningless for Germany to act alone, as anyone who can look up the figures on the Internet and use a calculator can see for themselves. Either everyone pulls together or they don't. If all the major carbon dioxide producers (including China, India, the USA, and Russia) do not participate, the hoped-for effect will certainly fizzle out. And all those who have joined in will have wasted a large part of their efforts (which they have incurred with many disadvantages) (remember that the fear of the climate apocalypse has become a significant economic factor in industrialized countries - i.e., real money is at stake).

405. Positive effects of an increase in carbon dioxide concentration

But you can also turn the argument around and point out that a higher concentration of carbon dioxide in the atmosphere can certainly have positive effects. Not because of a possible increase in the greenhouse effect, but because of the improved conditions for plant or crop growth. Today, the carbon dioxide molecule is generally perceived as a "danger" to humanity. This perspective fails to recognize that all the "carbon" built into living organisms (including us) comes primarily from carbon dioxide in the air. If the carbon dioxide content of the air were to fall below approximately 150 ppm, it would have fatal consequences for plant growth - it would cease entirely. Richard Feynman (1918-

1988), the great physicist and Nobel Prize winner, once wrote a truism that still surprises many people today:

"Trees consist mainly of air. If you burn them, they become air again, and in the blazing embers, the blazing embers of the sun are released, which helped to transform the air into the tree. And in the ashes, we find the small remnant of the part that came not from the air but from the solid earth."

When the "energy maize fields" begin to sprout here in Germany in early summer, the carbon in their extremely fast-growing stalks comes from the carbon dioxide in the air, which they extract through the process of carbon fixation by photosynthesis with the help of sunlight. A higher concentration of carbon dioxide in the air would therefore tend to promote plant growth, as historical evidence from Earth's history suggests. There were periods when the carbon dioxide concentration far exceeded the 1,000 ppm mark, which was associated with extremely lush plant growth (for example, during the Meso-zoic era) (for comparison, today's value is approximately 420 ppm). It is therefore quite possible that the climate catastrophe will not materialize, and the Earth could become greener, deserts could shrink, and plants - as recent research has shown - might require less water for their growth. Yields could increase, and the Earth could support more people in a humane way.

What I am trying to say is that climate futurology, through its connections with politics and industry, has increasingly become an ideology in which proponents and opponents of the thesis of a developing (man-made) "climate catastrophe" are irreconcilably op-posed. It is time for pure science to reassert its role in this important area - through its tried-and-tested, ideology-free methods. Foresight can do no harm. However, the future should not be painted entirely black just because it benefits certain interest groups. The climate will continue to change, and humans will not have much substantial power to counter this. I believe there are currently other man-made problems that are more ur-gent to address.

But let's dwell for a moment on the benefits of increased carbon dioxide concentrations in the Earth's atmosphere in relation to plant life. Plants absorb carbon dioxide from the atmosphere (which, by the way, leads to a seasonal modulation of carbon dioxide con-centration) and, with the help of sunlight and water, convert it into glucose (grape sugar) in a highly sophisticated process, releasing oxygen molecules.

406. Photosynthesis

The importance of photosynthesis for life on Earth is fundamental. It can probably be regarded as one of the most important "inventions" of life ever. Without photosynthesis, life on Earth would most likely have remained small, primitive, and restricted to a few suitable habitats. The sun-driven production of energy-rich organic substances such as carbohydrates, lipids, and proteins made a new type of metabolism possible, which, in addition to the substances mentioned as food, could also use the oxygen produced during the light reaction as an oxidizing agent for an aerobic way of life. Heterotrophic (especially animal) life has always depended on plant life as the primary producer of biomass, regardless of its position in the food chain.

Fig. 181: A biochemical marvel - green leaves (copper beech). (Author)

In this context, it is interesting to analyze the energy and material turnover on Earth caused by photosynthesizing organisms (cyanobacteria, plants). The measure for this is the organic dry mass produced annually as part of phototrophic metabolism. It is approximately 115 gigatons (Gt) in terrestrial ecosystems and 55 Gt in marine ecosystems, net (i.e., minus material consumption through respiration). Limiting factors include the efficiency of photosynthesis (generally far below 5% net) and the carbon dioxide content of the Earth's atmosphere. The latter must be constantly regenerated (through volcanic degassing and respiration) so that the global carbon cycle does not lose its balance. It is estimated that the entire atmospheric carbon is turned over in the biosphere once in just 12.5 years. To break down the significance of these figures to a manageable level (photosynthesis drives virtually all biogeochemical cycles of terrestrial ecosystems), consider the following example: everyone knows the deciduous tree of our temperate latitudes - the European beech (Fagus sylvatica). A 100-year-old specimen has (at least in summer) around 200,000 leaves with a total surface area of approximately 1,200 m^2. They contain about 180 g of pure chlorophyll (the green "plant pigment" in which photosynthesis takes place), distributed across approximately 100 trillion chloroplasts in the leaf cells. On a sunny day, these chloroplasts are able to absorb around 9.4 m^3 of pure carbon dioxide from about 36,000 m^3 of air and subsequently produce approximately 12 kg of carbohydrates from it. The photolytic decomposition of the water required for this simultaneously produces around 9.4 m^3 of pure molecular oxygen - which corresponds to the daily requirement of two to three people. A portion of the 12 kg of carbohydrates is converted into lignin, which we commonly know today as "wood," via a complicated biochemical process. In short, beech wood not only contains aromatic substances important for "smoking" ham, sausages, and cheese but also - as Richard Feynman put it - solar energy. It can be partially "recovered" by burning the wood. In terms of organic waste and fast-growing "organic maize," this is the principle behind biogas plants, except that here the result of "fermentation," namely methane, is obtained and later burned, for example, in a gas engine connected to a generator to produce electrical energy.

407. Georg Imbert and the wood gasifier

However, a gas engine can also be used to power a vehicle, which brings us to an inventor that not even seventy years ago every reasonably educated German knew by name: Georg Christian Peter Imbert (1884-1950). He invented the "wood gasifier" around 1920 and a little later the "wood gasifier automobile." His "fuel" mostly consisted of wood shavings, which could be carried in a suitable container. And if the fuel threatened to run out, you just had to stop briefly to collect wood in the forest... But joking aside, the wood gasifier engine was a highly regarded source of propulsion at the time, especially during the war years, so it is well worth examining how it worked. After all, the Nazi regime hoped to use this "miracle weapon" to achieve the "final victory" despite the increasingly hopeless raw materials situation. The two basic technologies - the production of gas by carbonization of organic substances (which is known to produce carbon monoxide) and the gas engine - were of course well known in the 1920s. Even the first two- and four-stroke engines, developed by Nicolaus Otto and others around 1865, were "gas engines" that ran on illuminating gas. This "illuminating gas" was produced in large quantities during the manufacture of coke from hard coal, as coke was in turn needed in huge quantities for the production of iron and steel.

Georg Imbert's achievement was to improve and optimize the gas generator (i.e., the part in which "illuminating gas" is produced from wood chips, which in this case was referred to as "producer gas") so that it could be used to operate a light gas engine for a longer period. When stationary, it could power machines (such as threshing machines), or, if the system was integrated into a truck, it could drive around without gasoline...

When you see a wood gasifier for the first time, it looks a bit like a cannon stove. A fire glows at the bottom and wood chips are poured in at the top... But it's the exact inner workings that matter. Because in a "gasifier" not only a mixture of carbon monoxide (highly dangerous!), carbon dioxide, and hydrogen is produced, but also a variety of other substances, which in their entirety are simply referred to as "wood tar." This foul-smelling, viscous liquid must be permanently removed from the wood gasifier so that it does not get into the engine, which would otherwise quickly give up the ghost. The "gasifier" therefore also needs a gas cooler, a tar collection tray, and finally a post-cleaner, and everything should be designed as optimally as possible for a high gas yield. Ideally, the tar produced should be incinerated straight away to prevent large quantities of tar from being produced in the first place.

Fig. 182: Georg Christian Peter Imbert (1884-1950)

Georg Imbert had to make many attempts in this regard until he achieved "his" optimum design. Key elements here are "ring nozzle," charcoal bed, and air flow with reversal. Nevertheless, the energy content of the resulting gas remained quite low at ~1.2 kWh per cubic meter compared to pure methane (natural gas, approx. 10 times higher). The engine therefore had to achieve a high compression ratio (typically 1:9), which in turn made its design and construction more difficult.

Fig. 183: Imagine an S-Class car with a wood carburetor...

Even at the time of its invention, a wood carburetor engine was therefore hopelessly inferior to a gasoline engine in terms of performance. It was mainly installed in special trucks (of course, there were exceptions; there were also cars and even motorcycles with wood carburetors), which in turn were often purchased by trades that had a lot to do with wood themselves (e.g., furniture companies, joineries, etc.). However, this changed from the second half of the Second World War, when fuels became increasingly scarce. Here, the wood carburetor engine suddenly experienced a renaissance, which lasted a few years in Germany after the end of the war. After that, petrol and diesel were once again available in abundance, and the wood carburetor fared like the dinosaurs - it died out. And the name of its inventor was quickly forgotten.

So what does a trip in a wood carburetor car, such as an Opel Blitz truck, look like? Turning the ignition key and starting it was not an option. First, you had to heat up the gasifier. To do this, you put a layer of good charcoal about ten centimeters thick on the gasifier and a few shavings on top. A little methylated spirits helped to light the fire. This is a job that is easy for anyone who barbecues frequently. As soon as enough embers have formed, you can open the lid and fill the wood gasifier with wood chips or other chopped wood. Of course, you must not forget to put the lid back on at the end. As soon as the gasifier produces enough gas, it is time to start the engine. For this, there is the famous crank that you had to insert into the engine from the front - as you can sometimes still see in old movies. You had to turn the crankshaft by hand a few times so that the pistons could suck in the gas and the gasifier would get a draft - just like the chimney through the chimney. With a bit of luck, the engine will now start, and nothing stands in the way of a joyride.

However, you have to stop every 100 km or so to refuel with wood chips. Another 50 km further on, it's time to clean the gas purifier. And finally, after around two thousand kilometers, a thorough cleaning of the Imbert generator should be considered if you don't want the truck to stop in the botany at some point or the wood carburetor to catch fire and burn down.

What you might be able to expect from a paid truck driver is not at all acceptable for a private car (the appearance alone is horrifying). So it's no wonder that the wood carburetor car has hardly ever found any true friends. Incidentally, just as birds have evolved from the now-extinct dinosaurs, wood pellet heating has evolved from the wood gasifier. Somehow, Georg Imbert's invention still lives on. And if you take its technology as a basis (computer control, etc.), then perhaps even a wood gasifier car would have a chance again today, perhaps ecologically CO_2-neutral heated with organic corn pellets. Seen in this light, it is really surprising that this technology has not yet been adopted as part of the energy transition. The overall efficiency should be even better than if biodiesel were first produced from wood or other organic materials and then used as fuel.

408. The gyro car

While we're on the subject of ecologically sustainable drive systems, it's important to mention the "gyro car," a pioneering electric vehicle that had a brief but intriguing existence in the 1950s. This Swiss invention employed a large flywheel to store and utilize energy in a unique way.

The gyro car used a flywheel with a diameter of 1.7 meters and a weight of 1.5 tons, which was suspended between two sturdy frame supports. The flywheel was accelerated to approximately 3,000 revolutions per minute using an electric motor. This acceleration process took about 2 minutes from a complete standstill. The energy required to spin the flywheel was drawn from an external power source through current collectors on the roof, similar to the system used by trolleybuses.

Once the flywheel reached its operational speed, the bus could disconnect from its power source. It then relied on the energy stored in the flywheel to operate. This setup allowed the bus to travel approximately 6 kilometers at a top speed of 50 km/h before needing to return to the charging station for re-acceleration.

A total of 17 buses were produced with this drive system. Despite its innovative approach, the gyro car system did not see widespread adoption, but it remains a notable example of early attempts to harness mechanical energy storage for urban transit.

Fig. 184: A Gyrobus stop with Gyrobus in the 1950s (Creative Commons)

But obviously, this idea has not proved successful because none of these systems have been seen on roads anywhere in the world since 1960. However, there might be further developments in this area, similar to the evolution from wood gasifiers to wood pellet heating.

I prefer not to mention this idea to avoid stirring up any controversy: a flywheel next to every wind turbine as an energy storage solution. After all, it was able to store an incredible 5 kWh of energy in the Gyrobus, which speaks to a reliable energy storage system. However, this stored energy can be released unpredictably, especially if a flywheel breaks. In the past, particularly at the end of the 19th century, when flywheels were widely used as temporary energy storage devices or to stabilize machine speed, "flywheel explosions" occasionally occurred. Such events often led to serious accidents. Statistics available online show that between 1856 and 1883 alone, 16 people lost their lives in a total of 38 flywheel accidents in Germany. These figures illustrate the risks and potential dangers associated with the use of flywheels and show that, despite technical progress, safety aspects must always be considered critically.

409. The potter's wheel

The reasons for this were primarily material and bearing issues or excessive rotational speeds. Incidentally, the "flywheel" is a similarly ancient invention to the windmill (even older), with its origins in the potter's wheel. Its use can be traced back to 6000 BC in Mesopotamia. In Central Europe, the Neolithic period, the last post-glacial cultural stage of the Stone Age, is considered the era when ceramic products became increasingly significant. This shift occurred as people transitioned to a sedentary lifestyle and engaged in agriculture and animal husbandry. During this time, pottery was produced without a potter's wheel but was already fired in kilns to enhance its durability. The potter's wheel itself emerged only in the subsequent Bronze Age, when stoneware became increasingly rotationally symmetrical. More importantly, however, people remained extremely conservative for many centuries regarding the shapes and patterns used to decorate their pottery. An archaeologist who discovers pottery shards during excavations of ancient settlement sites can usually determine a rough date by visual inspection: if the shards belong to a "bell-shaped" cup, their maker was most likely a member of the "Bell Beaker Culture" (between 2600 and 2200 BC), which was widespread in Central Europe. Conversely, if the "beakers" resemble "funnels," it is highly probable they were made by a member of the "Funnel Beaker Culture" (between 4200 and 2800 BC). Both cultures overlap with the older "Bandkeramiker" and the younger "Schnurkeramiker," which largely coincide with the Bell Beaker Culture. The latter were known for pressing cords into the soft clay to create patterns and enhance the aesthetic value of their stoneware. Their craftsmanship can still be admired today in various museums with archaeological collections, such as the Bautzen City Museum in Upper Lusatia.

Fig. 185: "Zone Beaker" and 'Bell Beaker' (2500-2200 BC) from the Museum of Prehistory and Early History in Berlin (Wikimedia)

After the development of flint tools, the creation of utensils made from dried and, somewhat later, fired clay marked a particularly significant innovation in human development. In Mesopotamia, entire cities - cosmopolitan centers at their peak - were built from fired clay bricks. Even when these cities were buried under sand, large quantities of clay shards or tablets inscribed with mysterious characters revealed to early explorers, whom we might now call "amateur archaeologists," that a submerged city lay hidden beneath the rubble.

410. Ceramics

Objects made from inorganic, non-metallic materials such as loam or clay, which can be shaped and fired to produce durable vessels, figurines, tiles, or bricks, are now classified as "ceramics." This definition encompasses both traditional ceramics, which have been used for thousands of years, and modern high-tech ceramics, which have significantly impacted our lives.

For example, the hotplates on our electric stoves are made from special ceramics, as are the mirrors in our giant telescopes, which allow us to observe billions of light-years into space. The chisels on our machine tools are fitted with extremely hard ceramics that can cut through the toughest steels with ease. Even our broken teeth are now repaired with ceramic inlays and crowns rather than amalgam.

Ceramic plates protect spacecraft from burning up upon re-entry into the Earth's atmosphere. All high-temperature superconductors are, without exception, ceramics such as $YBa_2Cu_3O_{7x}$. Additionally, microelectronics would be inconceivable without specialized ceramics. These materials serve as insulators and substrates for electronic components, enabling the miniaturization and increased efficiency of modern electronic devices. Without these ceramic materials, many of today's technological advancements would be unimaginable.

411. Porcelain

The fact that "ceramics" encompasses more than just fired clay is illustrated by porcelain vases and figurines, which were often valued as highly as gold, especially during the Baroque era. In China, the art of firing porcelain reached its zenith in the 7th century. The slightly translucent, hard yet fragile porcelain was long considered a well-kept secret of Chinese ceramists. The Venetian trader Marco Polo (1254–1324) was the first to introduce porcelain vessels to Europe, where much like silk fabrics, they became highly sought after in Far Eastern trade. This demand created significant interest among the wealthy, leading to an increase in the export of these goods - primarily by ship - to Europe at the beginning of the 16th century. Notably, the Dutch "Vereenigde Oostindische Compagnie" (Dutch East India Company) played a key role, supplying European royal houses with porcelain from China and Japan, effectively establishing a monopoly. This resulted in extensive collections, including the renowned Ming dynasty vase collection in the Palace of Versailles (which houses the famous Hall of Mirrors). The desire to produce this type of "earthenware" locally grew, but the Chinese maintained their production secrets. Consequently, Europeans initially focused on creating increasingly refined imitations, distinguished particularly by unique glazes. The most famous of these is the so-called "Medici porcelain," produced in Florence from around 1575. Known also as "soft porcelain," it became a major export success for this northern Italian city-state.

Fig. 186: Chinese porcelain vessels from the Ming period (Wikimedia)

412. Ehrenfried Walther von Tschirnhaus and Johann Friedrich Böttger

The fact that Dresden (or Meissen) eventually became a center of European porcelain production is attributed to a sovereign with a keen interest in porcelain, Augustus the Strong (1670–1733), and a somewhat forgotten polymath, Ehrenfried Walther von Tschirnhaus (1651–1708), alongside Johann Friedrich Böttger (1682–1719). Tschirnhaus came from an old Upper Lusatian noble family with estates between Görlitz and Seidenberg (Tschernhausen, now Černohousy; Kieslingswalde, now Sławnikowice, the birthplace of E. W. von Tschirnhaus).

His primary interest since his grammar school days in Görlitz was mathematics. The "Tschirnhaus transformation," named after him, is still well known among algebraists today. After studying in the Netherlands and undertaking an educational journey across Europe, during which he visited notable scientists such as Isaac Newton in London, Gottfried Wilhelm Leibniz in Paris, Athanasius Kircher in Rome, and philosophers like Baruch Spinoza, he returned to Kieslingswalde in 1679.

Fig. 187: Ehrenfried Walther von Tschirnhaus (1651-1708)

In our context, his interest in glass production and its processing, which he learned about in Paris, is significant. There, large burning mirrors were used to melt alumina. To repeat these experiments, he began to produce burning mirrors from metal himself with the help of a mechanic. This was followed by further extended stays abroad, including the Netherlands and Paris. Back in Electoral Saxony, he continued his experiments with burning mirrors (one of them can still be admired today in the Mathematical-Physical Salon in the Zwinger in Dresden). He succeeded in liquefying asbestos, which was considered impossible at the time.

In 1697, he approached Elector Augustus the Strong to persuade him to increase the establishment of glassworks and stoneware factories. This was partly successful under the condition that he first had to organize a thorough search of Saxony for precious and semi-precious stones such as agate, amethyst, and others as possible raw materials. This marked the beginning of a phase of intensive searches for mineral raw materials in Saxony. In 1700, the Dresden glassworks known as the "Ostrahütte" was founded, which was managed by v. Tschirnhaus himself until his death.

At this time, a young alchemist named Johann Friedrich Böttger caught the attention of the rulers, first in Prussia and then in Saxony, who claimed to be able to produce gold coins from silver coins and demonstrated this convincingly to an audience (using sleight of hand, of course). As both royal houses were in need of money, Böttger became highly sought after, and he had to flee to Wittenberg (1701) due to a bounty from the Prussian king. There, he was kidnapped on the orders of Augustus the Strong and taken into "protective custody" with the unmistakable demand, "Make me gold!" Of course, as he did not have access to a heavy ion accelerator like the one at the GSI Helmholtz Center in Darmstadt today, this was an impossible task. Ehrenfried Walther von Tschirnhaus also suspected this, so after making the acquaintance of the talented Böttger in 1702 and becoming his boss by royal order in 1704, he set him to work on the development of porcelain. After many experiments, which initially involved increasing the temperature of the kilns used, he finally succeeded in producing "red stoneware" (now known as "Böttger stoneware") in 1706. This is an unglazed ceramic with a reddish color and high breaking strength that is low in water and completely sintered through due to the high firing temperature and is virtually pore-free. The most important "trick" in its production was "sintering," a process that v. Tschirnhaus studied for a long time. This involves heating a ceramic base material consisting of two components to such a high temperature that the main component remains solid, but the secondary component melts and fills all the pores, with the particles of the main component also sticking together like glass.

The first "real" hard porcelain was produced almost two years later, exactly on January 15, 1708, in the Dresden Jungfernbastei, as a surviving record proves. Tschirnhaus died at the end of that year, but the experiments to produce porcelain continued under the direction of Böttger until he was able to officially present the final "recipe" (a mixture of kaolin, feldspar, and quartz) and the associated firing technology in the spring of 1709.

Fig. 188: Decorative jug made of Böttger stoneware from 1710 (Wikimedia)

413. Meissen porcelain

This marked the actual birth of what was later known as "Meissen porcelain." The name comes from the fact that the first Saxon porcelain manufactory was established in the Albrechtsburg Castle in Meissen. This manufactory was founded in the summer of 1710 by electoral decree.

Albrechtsburg Castle in Meissen was the ideal location, as the castle with its spacious rooms offered sufficient space for the extensive production and was also well protected. The decision to establish the manufactory here was made by Saxon Elector and Polish King Augustus the Strong, who had a great interest in the production of porcelain, which was highly sought after in Europe at the time and was therefore often referred to as "white gold."

Since 1722, the two crossed swords have been the trademark of Meissen porcelain. This mark was introduced to clearly identify the products of the Meissen manufactory and protect them from counterfeiting. The crossed swords are still known worldwide today and stand for quality and craftsmanship.

Around the same time, other porcelain manufactories were established, not only in Germany but also in other European countries such as France (with the famous manufactory in Sèvres) and England (e.g., the Wedgwood manufactory). European porcelain production experienced a great upswing, spreading and developing rapidly.

Much work was put into the development of typical decors and glazes, which became the trademark not only of individual manufactories but of entire stylistic periods. The Meissen onion pattern, for example, was a very popular decoration that was developed in the 18th century and is still used today. The variety of designs reflects the creative wealth of the porcelain artists and the fashion trends of the time.

Today, porcelain production technology is largely perfected. Many porcelain products are now mass-produced on a quasi-industrial scale. Industrial production has considerably reduced manufacturing costs and made porcelain affordable for broad sections of the population. At the same time, research has shifted to the broad field of specialty ceramics. These modern ceramics often have little in common with what is commonly understood as "stoneware." They are used in high-tech applications such as medicine, electronics, and aerospace and are characterized by special properties such as high-temperature resistance, hardness, or chemical resistance. Innovations in this area underline the extent to which the porcelain manufactory's heritage has been continued and, at the same time, integrated into new, forward-looking technologies.

414. High-temperature superconductivity

The search for ceramic materials with the extraordinary property of so-called "high-temperature superconductivity" is of particular interest here. This is because superconductive materials open up the possibility of transporting electrical energy over long distances without any losses whatsoever - the dream of every energy engineer. The only problem is that, even today, "high temperature" in this context still means a temperature of less than -135°C (i.e., a temperature that is certainly not perceived as particularly "high" when exposed to it). This still severely restricts the possible applications of such electrical conductors, as at least liquid nitrogen (boiling temperature at -196°C) is required to cool them. A superconductor that would work at normal "room temperatures" would be ideal. But according to the current state of research, there is still a long way to go - if there are any materials with this property at all. So-called "metallic" hydrogen, of which the giant planet Jupiter consists of a high percentage, could be such a material. However, it cannot be produced in the necessary quantities under earthly conditions (pressure!).

415. What exactly is superconductivity?

Before I go into the significance of special ceramic materials in terms of superconductivity, I should briefly explain what "superconductivity" actually is. If we recall our school days or the time of our "apprenticeship" or "studies" (as far as they involved technology), then "Ohm's law" may still evoke a faint memory, at least as a concept. It states that at a constant temperature, the relationship between voltage and current for a given metallic conductor is always constant, with this constant referred to as electrical resistance (more precisely, "ohmic resistance"). It is a measure of how strongly the material inhibits the flow of current in a conductor. Perhaps the best way to visualize this is that the negatively charged free electrons, which are known to be present in every conductor, are accelerated by the applied electrical voltage but are constantly colliding with the atomic nuclei, which greatly impedes their movement. The conductor heats up because, during this process, the electrons transfer kinetic energy to their collisions partners, i.e., the atomic cores in the metal lattice. Just think of the filament of a light bulb, which starts to glow as soon as sufficient current flows. As a result of this interaction between electrons and the metal lattice, something like an average drift velocity is established, which determines how many electrons pass through the conductor's cross-section per unit of time, corresponding to the electrical current strength measured in amperes.

Experiments in the 19th century had already shown that electrical resistance is both material-dependent and temperature-dependent. When it became possible to liquefy gases such as oxygen, nitrogen, or even helium and then use them as coolants, people also began to take an interest in the behavior of electrical resistance at low temperatures. For example, physicist Gilles Holst (1886-1968) carried out experiments in the laboratory of the future Nobel Prize winner Heike Kamerlingh Onnes (1853-1926) to determine the electrical resistance of mercury, which he cooled with liquid helium (which his mentor had succeeded in producing shortly before). He noticed something astonishing: he observed the sudden and complete disappearance of the electrical resistance as soon as the temperature dropped below 4.2 Kelvin. His boss didn't want to believe it at first, but further experiments showed that this behavior was indeed real. Kamerlingh Onnes called this previously unknown behavior of mercury "superconductivity." Although this is not entirely accurate, Heike Kamerlingh Onnes is considered the discoverer of superconductivity. However, he did not receive his Nobel Prize (1914) specifically for this but more generally for his "investigations into the properties of matter at low temperatures" and the development of a method to liquefy the noble gas helium.

Most metallic superconductors have transition temperatures below 10 K. Only the intermetallic compound magnesium diboride becomes superconductive at a transition temperature of 39 K, making it the leader among metallic materials. In 1986, physicists Alexander Müller and Georg Bednorz discovered that lanthanum barium copper oxide loses its electrical resistance at a temperature of 35 K. A year later, a copper oxide compound (yttrium-barium-copper oxide) was discovered, which becomes superconducting at 92 K. This type of superconductor, which is not a metal but a ceramic material, was called a high-temperature superconductor. The two IBM researchers were awarded the Nobel Prize in Physics in 1987 for their work. This marked the beginning of an intensive search for further high-temperature superconductors with the aim of finding those with a transition temperature in the range in which nitrogen is liquid (between 63 K and 77 K). This is because liquid nitrogen is easy to produce and therefore the laboratory coolant of choice (as students, we used it to make delicious ice cream without any complications). As for ceramics, the current record for the transition temperature is held by thallium barium calcium copper oxide (TBCCO) at 138 K. It has existed since 1994 and has not yet been surpassed.

A major disadvantage of superconducting ceramics is that, like all ceramics, they are extremely brittle, and it is very difficult or impossible to produce flexible wires from them, e.g., for superconducting coils to generate extremely strong magnetic fields or for use as power cables. However, where they are unchallenged, such as in precision measurement technology, specifically "*Superconducting Quantum Interference Devices* = SQUID," they have now almost completely replaced conventional measurement methods (especially with regard to magnetic field measurements).

While there has been a generally accepted theory for "classical superconductivity" since 1957, this is only partially the case for high-temperature superconductivity, as it occurs in certain ceramic materials. Thus, there is still a lot of theoretical research to be done in this area.

416. Superfluidity

Superconductivity is linked to another amazing phenomenon: the superfluidity of liquid helium (He-4 below 2.17 K). Both phenomena cannot be explained by classical physics and offer fascinating insights into the world of quantum physics. Superfluidity was first discovered in 1937 by Pyotr Leonidovich Kapitsa (1894-1984, Nobel Prize 1978), John F. Allen (1908-2001), and Don Misner.

You are probably familiar with the light mill, an evacuated glass bulb containing an easily rotatable impeller with two or four blades, each of which is white on one side and black on the other. Now imagine that a nozzle is directed at these blades, through which air is blown. The wheel will then start to move and turn quickly, even in the dark. Now imagine that it is not air that is forced through the nozzle, but a liquid, a very special liquid, namely superfluid liquid helium. Then something really amazing happens. Everyone might expect the jet of liquid to drive the wheel even better than the air forced through the nozzle. However, nothing happens. Although the liquid helium splashes onto the blades, it cannot exert any pressure on them.

Depending on the experimental setup, superfluid helium has a vanishingly low or similar viscosity to "normal" liquid helium, which is paradoxical enough. This liquid can still move smoothly through capillaries. If, for example, a long test tube is placed in a beaker of superfluid helium, the liquid rises against gravity on the outside of the test tube. This effect, which creates a so-called "roll-in film," causes the test tube to fill from top to bottom until it is full or the beaker is empty. If this were to happen with seltzer, for example, we wouldn't believe our eyes.

In addition, there are a number of other effects that, in the case of superfluid helium, are diametrically opposed to our "classical world view." The most remarkable phenomenon is the absence of any internal friction, which means that a rotating container of superfluid helium gives the impression that the liquid inside it is standing still. This lack of friction means that the superfluid is not subject to the usual inertial forces and is therefore fundamentally different from normal fluids.

Another fascinating behavior is the ability of superfluid helium to flow through tiny pores and gaps that are impassable for normal liquids. This phenomenon is particularly evident in the so-called "helium fountain." Here, a part of a container containing superfluid helium is heated in such a way that a temperature difference occurs inside. Based on the second law of thermodynamics, one would expect the helium to flow from the warmer to the colder area. With superfluid helium, however, something surprising happens: it flows from the colder to the warmer region and rises like a fountain.

These astonishing properties of superfluid helium not only open up new perspectives for research but also call into question fundamental concepts of classical physics. Research into superfluidity thus offers a deep insight into the astonishing and often counterintuitive phenomena of quantum mechanics.

417. Wonderful world of quantum physics

This is where the wonderful world of quantum physics begins - in macroscopic terms: superconductivity and superfluidity can only be explained within the framework of quantum mechanics. You may learn something about this at school, especially if you take advanced courses. Otherwise, it is one of the supreme disciplines of any physics course, both in terms of the experimental side (I will only mention the double-slit experiment here) and - and to an even greater extent - in terms of the theory and the mathematics on which it is based (e.g., "Operators in Hilbert space"). The only thing that is "disturbing" for every physicist is that the theoretical description of quantum physical processes works excellently, but the "ideas" that one tries to form about these processes often seem to contradict common sense. Just think of the highly paradoxical behavior of superfluid helium (point 416). Thus, even today - despite the complete logical consistency of the mathematical apparatus required to describe quantum physical processes - the interpretation in terms of *"What is all this actually about?"* or *"What kind of reality is quantum theory based on?"* is still controversial. For example, superfluidity and superconductivity can certainly be explained using quantum theory, albeit not always in all details. In the case of superconductivity, John Bardeen (1908-1991), Leon Neil Cooper, and John Robert Schrieffer already succeeded in 1957, and their theory of superconductivity (known as the BCS theory) earned them the Nobel Prize in Physics in 1972. This theory deals with the quantum mechanical behavior of electrons, which, despite their identical charge (repulsion!), can unite to form pairs under certain conditions that are fulfilled in superconductors at low temperatures, fundamentally changing their behavior.

418. Fermions and bosons - the spin makes the difference...

Like most other elementary particles, electrons have a so-called intrinsic angular momentum, referred to as "spin" for short (imagine the electron rotating around its axis, which is nonsensical for a quasi-point-like particle, but this is just quantum mechanics). This "spin" is measured in units of a fundamental natural constant called "Planck's constant" (not "Planck's quantum of action") after its discoverer Max Planck (1858-1947). If the spin is a half-integer multiple of this constant, then we speak of fermions - electrons and also nucleons (i.e., the components of atomic nuclei, protons and neutrons) belong to this group. If an elementary particle or a system composed of fermions, such as the atomic nucleus of He-4, has an integer multiple of this (or zero), then it is referred to as a boson. "Fermions" are named after the famous Italian physicist Enrico Fermi (1901-1954), and "bosons" are named after the Indian physicist Satyendranath Bose (1894-1974). Both families of particles differ fundamentally in their statistical behavior, i.e., when they form a "gas" (like the free electrons in an electrical conductor) or a liquid (like the superfluid He-4).

419. Quantum mechanical state

How objects, i.e., electrons, atomic nuclei, or even the entire universe, are described in quantum mechanics can be challenging at first glance. The basic concept here is the "state" of such an object, which encompasses all "observable" properties of the object (or sometimes just a specific property such as spin, charge, location, momentum, or combinations thereof...). Due to its mathematical structure, it is also referred to as a wave function. What in classical mechanics describes the movement of an object (for example, a mass point), Newton's second law of motion ("force = mass times acceleration"), is in quantum mechanics replaced by the so-called Schrödinger equation, which is more complex to describe in words. It determines - analogous to Newton's second law for a mass point - the "motion" (better "evolution") of a quantum mechanical state (wave function) in space and time. The Schrödinger equation is a partial differential equation that describes the temporal development of the wave function of a system and thus forms the basis for predicting the dynamic properties of quantum systems.

You can at least admire it here, without me explaining it in more detail here:

$$ i\hbar \, \frac{\partial \psi}{\partial t} = \dot{H}\psi $$

While Newton's equation of motion is strictly deterministic (if the initial position and initial velocity (expressed by the particle's momentum) are given as initial values, as well as the acting forces, then the past and future positions of the mass point (or a system of mass points) can, in principle, be calculated for any point in time), the Schrödinger equation only gives probabilities of finding an object (if, for example, the position is used as a state) at a specific location. Only when a measurement process is triggered (in this case, a position measurement) can the exact location of the object be determined.

However, other properties - such as the velocity (expressed by the momentum) in this case - remain undetermined. This situation, which does not exist in classical mechanics and is one of the reasons for the many peculiarities of the microscopic world, is expressed by Heisenberg's uncertainty principle. This principle states that it is, in principle, impossible to simultaneously determine the position and momentum of a particle with arbitrary precision. The more precisely the position is measured, the less precise the measurement of the momentum becomes, and vice versa. This fundamental uncertainty is not a limitation of our measurement technology but a fundamental property of nature on a microscopic level and leads to a profound revision of our concepts of reality and causality as they exist in classical physics.

420. Pauli exclusion principle

The Pauli exclusion principle is a fundamental law of quantum mechanics that was formulated in 1925 by the Austrian physicist Wolfgang Pauli (1900–1958). It applies to all particles with half-integer spin (fermions, such as electrons, protons, and neutrons). Put simply, the Pauli exclusion principle states that two fermions can never occupy the same quantum state. A quantum state (see point 419) is described by a combination of properties such as energy, spin, and location. This means that in an atom, two electrons cannot exist simultaneously in the same "shell" with the same quantum mechanical properties.

Fig. 189: Wolfgang Pauli (1900-1958)

You can imagine the electrons in an atom as students in a classroom. Each student (electron) has its own place (quantum state). The Pauli principle ensures that no student can occupy the same place as another. If all the seats are taken, a new student has to move to a different class (higher energy state). This rule means that the electrons in an atom have to "get out of each other's way." Each electron has a specific place in the different energy levels that an atom has. This prevents all electrons from falling to the lowest energy level, which would make the atom unstable.

Another important effect of the Pauli principle is the so-called "degeneracy pressure." In a fermion gas, as occurs in white dwarfs or neutron stars, each particle has its fixed energetic position. As fermions cannot simply change their quantum state when all low-energy levels are occupied due to the Pauli principle, a pressure arises. This pressure, which does not depend on temperature but on fermion density, is called degeneracy pressure. White dwarfs are stars at the end of their life cycle that are stabilized by the degeneracy pressure of an electron gas. The electrons cannot be compressed any further as they already occupy all available quantum states. In neutron stars, extremely dense stars, it is the degeneracy pressure of the neutrons that prevents the star from collapsing. Here, too, the neutrons cannot all occupy the same state, which leads to enormous pressure.

421. Cooper pairs and superconductivity

Every "wire" (metal) contains a degenerate electron gas, which is set in motion when a voltage is applied and flows through the wire. In doing so, it collides with the metal atoms that form the metal lattice and induces vibrations that travel through the lattice like sound waves. These oscillations are known as "phonons." They can also be described quantum mechanically by a "state," so in certain contexts, it is possible to treat them as "particles" or "quasiparticles." When electrons collide with or interact with phonons, their kinetic energy is dissipated (causing the conductor to heat up), and the current flow is limited (resulting in ohmic resistance).

If you continue to cool a conductor, such as mercury, something remarkable happens. The thermal lattice vibrations (phonons) decrease more and more as the temperature falls. When the material-dependent transition temperature (4.153 K for mercury) is reached, an electron can excite a phonon by emitting energy, allowing a second electron to gain the same amount of energy by absorbing that phonon. This results in a polarization of the metal lattice, which overcompensates for the Coulomb repulsion between the two electrons of different spin orientation, leading them to experience a weak attraction. They then form a pair (more precisely a "Cooper pair," named after its discoverer Leon Neil Cooper) with a total spin of zero. This means that these "Cooper pairs" (which can be spatially separated in the lattice) are no longer fermions but "bosons." Bosons are not subject to the "Pauli exclusion principle," and this has dramatic consequences. They "condense" out of the electron gas, so to speak, meaning they ultimately all occupy the same quantum state (there is a certain analogy here to the formation of a Bose-Einstein condensate, though this cannot be explained in detail here). In this state, they are no longer able to interact with phonons (or with the ions that form the lattice). The Cooper pairs can now flow through the conductor without resistance because they all have the same momentum. This is referred to as a coherent state, which also applies to superfluid liquids (where, as mentioned earlier, internal friction disappears).

It is assumed that in superconducting ceramics, a similar but more complex mechanism of Cooper pair formation is at work, not based on electron-phonon interaction. The exact cause of pair formation in superconducting ceramics (more precisely ceramic oxides) is still unknown (or there are only assumptions, but no conclusive theory). It is therefore extremely difficult to determine the composition of new ceramics with this property in a targeted manner, i.e., based on knowledge about the mechanisms that lead to high-temperature superconductivity.

In 1987, the so-called "liquid nitrogen barrier" was broken concerning the transition temperature. This suddenly made technical applications of superconductivity possible that had previously only been dreamed of. Of course, it would be even better if an easily producible material could be found with a transition temperature in the "room temperature" range. Such a discovery would make highly effective superconducting drive systems, such as for maglev trains or more powerful computer systems, conceivable. In the latter case, clock frequencies of up to 100 GHz could be achieved using superconducting circuit logic (keyword "*Rapid Single Flux Quantum Logic*, RSFQ").

There is no question that superconductors could revolutionize the distribution of electrical energy via cable networks (keyword "ohmic loss"). Unfortunately, there has been no success in raising the transition temperature of ceramic high-temperature superconductors for around two decades. We can only hope that materials scientists will develop new compositions of superconducting ceramics and that physicists will support them with a plausible theory of high-temperature superconductivity in the foreseeable future. It is worth mentioning that organic superconductors have also been known since 1980 and that the discovery of fullerenes, a special carbon allotrope (allotropes are special structural forms of a particular element, such as graphite, diamond, or graphene in this case), has led to the discovery of new superconducting materials. Despite these advances, however, their common "disadvantage" for engineers remains their low transition temperature, which continues to severely limit their practical application.

422. Magnetic resonance tomography

Only very few people know that superconducting electromagnets have led to a revolution in medical technology - the "tube," as a magnetic resonance imaging (MRI) scanner is commonly known. This technology allows visualization and diagnosis of a patient's internal organs in the form of cross-sectional images without the need for invasive procedures (which are understandably perceived as unpleasant). MRI is an application of quantum physics and would not be possible without its principles. It relies on an effect known as "nuclear magnetic resonance," which is based on the fact that atomic nuclei can absorb and re-emit alternating electromagnetic fields. This procedure (known as MRI - magnetic resonance imaging for short) uses the combination of a strong magnetic field generated by superconducting coils with high-frequency pulses to excite the large number of hydrogen nuclei present in body tissue. By analyzing the relaxation radiation, proton densities are derived, which are then used to produce tissue sectional images via computer.

In contrast to X-ray examinations, MRI examinations are practically harmless for the patient, provided they do not have metal piercings (due to the magnetic field) or a pacemaker...

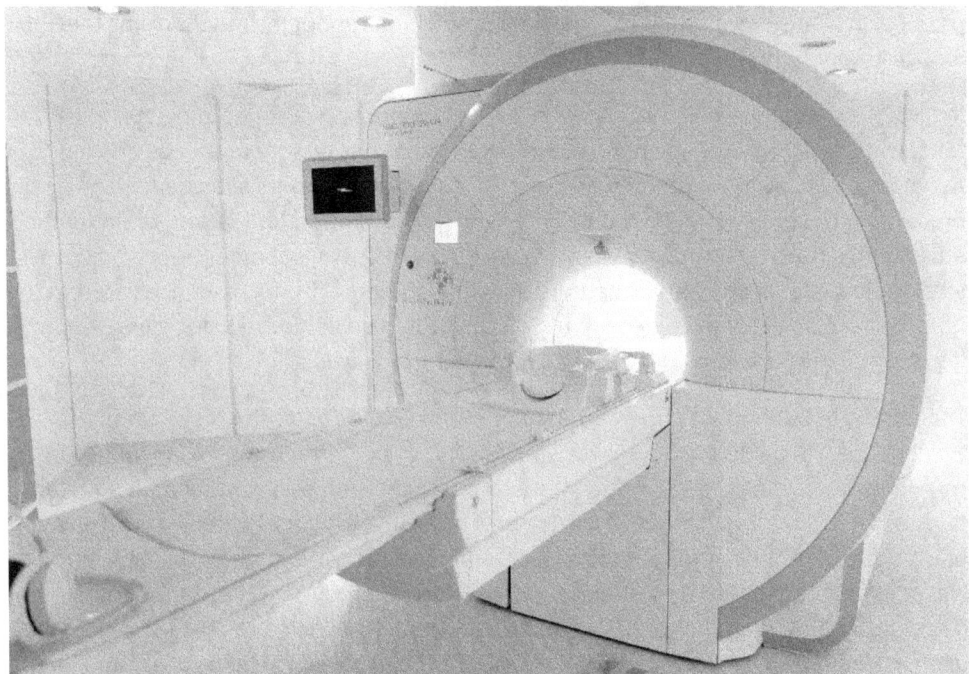

Fig. 190: Magnetic resonance spin tomograph

423. Computer tomography

Computer tomography (CT) is related to MRI but works with X-rays. This imaging procedure calculates body slices to create detailed cross-sectional images. A computer program reconstructs a three-dimensional image, called a computer tomogram, from numerous X-ray images taken from different angles. This method leverages the fact that X-
rays are absorbed to varying degrees by different types of tissue. By measuring the absorption not only in projection (as with a standard X-ray) but from multiple fluoroscopic
directions, it is possible to differentiate between spatially limited differences in density
(such as an organ) and greater layer thicknesses. Mathematically, this involves solving a
so-called "inverse problem," where the spatial distribution of X-ray intensities outside
the body is used to infer the internal structures where the radiation is absorbed. Key concepts for further study include "radon transform" and "Fourier analysis."

Despite their high purchase price, CT scanners and, to a slightly lesser extent, MRI scanners are now standard equipment in every university hospital and are used extensively.
However, when capacity issues arise, a bog body, a glacier mummy ("Ötzi"), or the mum-
mified remains of a once-powerful figure from ancient Egypt, known as the "Pharaoh,"
might be examined using the "tube" to investigate any ailments that may have afflicted
them.

424. The pharaonic murder - finally solved

Even ancient murder cases can be reexamined using computer tomography, as demonstrated by the gruesome murder of Ramses III. Ramses III, who lived from 1221 BC to 1156 BC, was killed in a brutal manner during a revolt instigated by his concubine Teje and his son Pentawer. Surviving court records, now housed in the Egyptian Museum in Turin, confirm this. The term "brutal" refers to the fact that he was slashed seven centimeters deep in the neck, a detail confirmed in 2012. Thanks to the preservation of ancient Egyptian rulers as mummies, this murder could be forensically investigated nearly 3,000 years later, as there is no statute of limitations on murder.

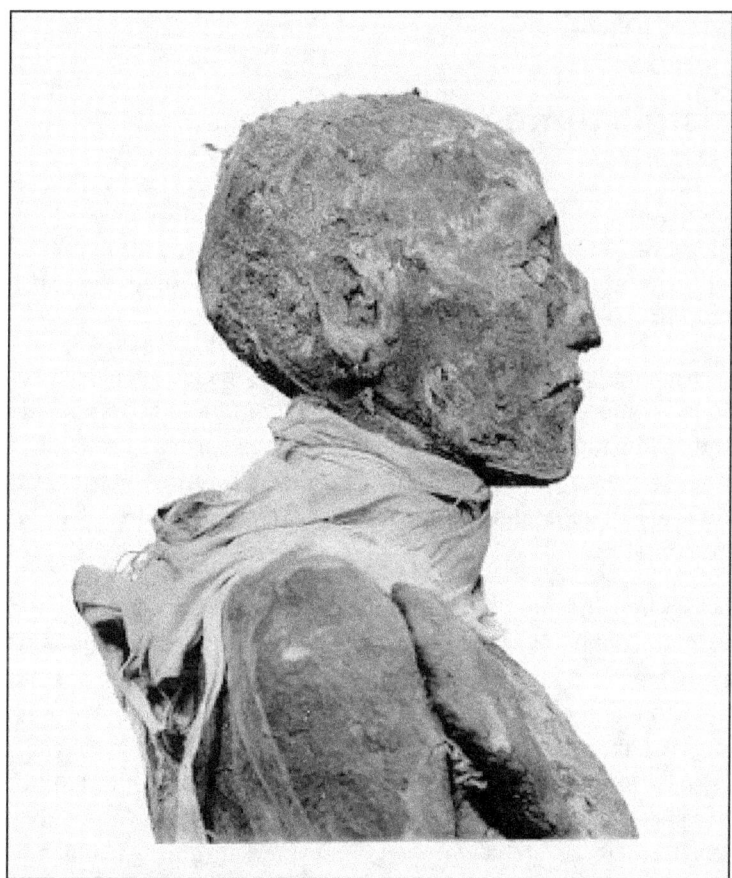

Fig. 191: Ramses III (1221 - 1156 v. Chr)

The mummy of Ramses III, recovered in 1881, is now in the Egyptian Museum in Cairo. It was subjected to a computed tomography (CT) examination, which revealed the seven-centimeter-deep cut in the pharaoh's neck. This cut, inflicted with a sharp knife, had been concealed by a scarf that was not removed for conservation reasons. This 2012 discovery provided definitive proof that Ramses III was indeed murdered in a palace revolt, or more specifically, a harem conspiracy.

However, the CT images also showed that Ramses III was seriously ill at the time of his death. He suffered from severe arteriosclerosis, which significantly impaired his health. It is possible that he would not have lived much longer without the murder, a factor the conspirators Teje and Pentawer might have overlooked. Their plot was uncovered shortly after the murder, and Pentawer was forced to commit suicide following a trial, a common practice at the time.

Interestingly, Pentawer's mummy was also identified, though, unlike his father's, it was poorly preserved. Modern genetics played a crucial role in this identification. Each mummy has a unique genetic fingerprint, allowing for the verification of family connections and solving historical mysteries even after thousands of years. Genetics thus helped identify the pharaoh's son as a conspirator and provided another piece of the puzzle in Egyptian history.

425. Seismic tomography

But back to computed tomography. Instead of electromagnetic waves (which include X-rays), seismic waves, or "earthquake waves," can also be used for computed tomography. Although "organs" cannot be detected, structures deep within the earth's crust and mantle can be imaged in three dimensions.

Seismic waves emanating from an earthquake source (epicenter) travel through geological formations at different speeds, depending on their material composition (density) and temperature. The time it takes for them to reach the earth's surface at a seismometer depends specifically on the path they take through the earth's interior, with the speed of propagation varying depending on the elastic parameters of the rocks, which in turn depend on temperature.

As you can easily imagine, the time from the epicenter to the seismometer depends on the direction in which the waves pass through spatially structured temperature inhomogeneities (e.g., a magma chamber). By determining the travel time for seismic waves that traverse the earth along different paths, the spatial structure of a material or thermal instability in the earth's interior can be calculated and visualized by solving an "inverse problem" - similar to methods used in computed tomography or MRI. In other words, seismic tomography aims to deduce the velocity distribution of the subsurface traversed from the travel times of the wave field measured at various stations to determine the geometric parameters of the corresponding fault structures.

Almost all parameters affecting the velocity of an earthquake wave are temperature-dependent. Therefore, the terms "cold" and "hot" can be used in this context as relative descriptors for "fast" and "slow" to describe the regions within the earth's interior through which these waves pass. The reference model is always an earth model that specifies the average seismic velocity as a function of depth. A deviation from this model, or an "anomaly," indicates that if the wave velocity is lower than specified in the model, the region must be warmer and the strength of the rocks lower than in the normal case. Conversely, if the wave velocities at the same depth are significantly higher than in the model, the opposite applies.

If you want to infer the cause (e.g., the shape of a corresponding geometric object) from observing "effects" (e.g., shadow images of an object from different projection directions), you must solve an "inverse problem" mathematically.

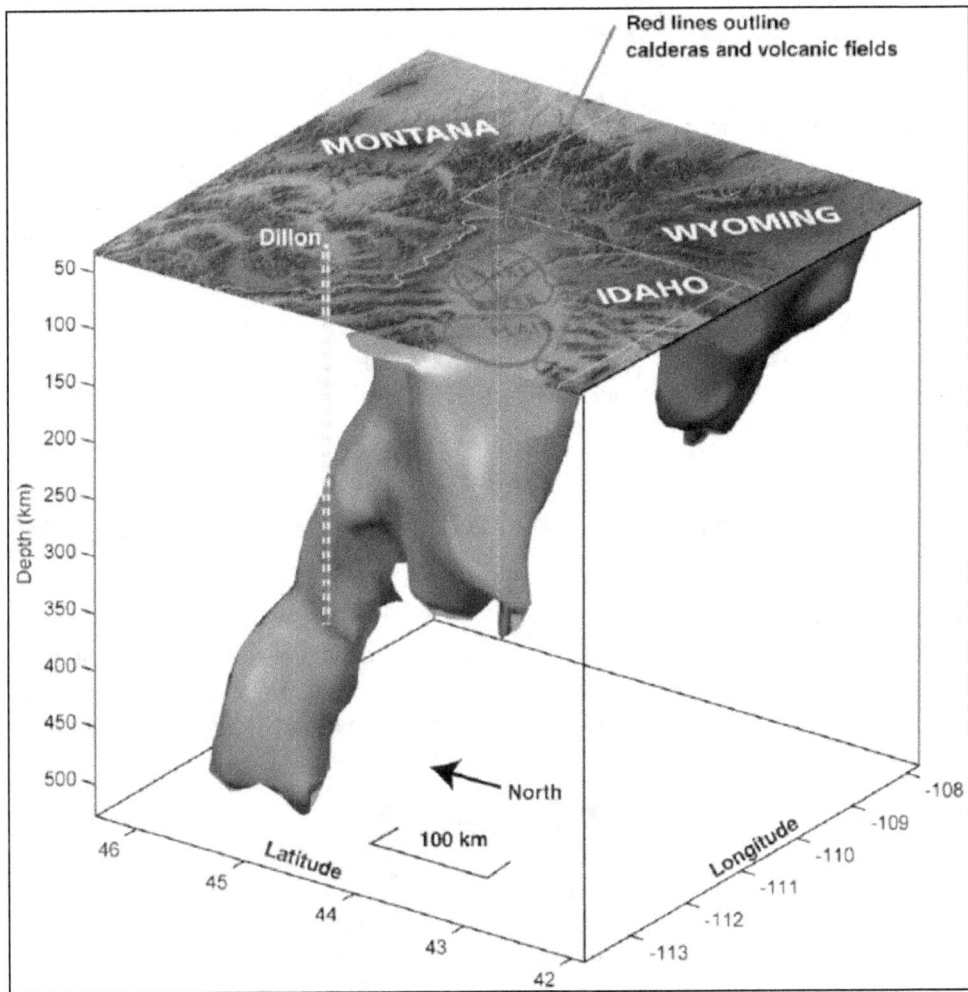

Fig. 192: Seismic tomography was used to visualize the magma chamber under the Yellowstone supervolcano (USGS, L.A. Morgan)

In seismic tomography, the causes are local temperature fluctuations inside the earth (which, as mentioned, are reflected in changes in the speed of seismic waves), and the effects are the arrival times of earthquake waves at various locations on the earth's surface, which deviate from a seismic model. An inverse problem consists of drawing conclusions from indirect observations (in this case, so-called "travel time residuals") about the physical or geometric properties that influence these observations (e.g., the size and shape of a magma chamber). The corresponding "direct problem," on the other hand, would involve mathematically tracing the paths of seismic waves from a given

417

model of a magma chamber to calculate the arrival times at specific points on the earth's surface - similar to the "ray tracing" used in computer graphics.

Mathematically, this is a very demanding task that requires the use of powerful computers due to the vast amount of measurement data that must be processed. The basis for this is the inversion of the so-called "Radon transform," which is very successful numerically using fast Fourier transforms. The detailed images of internal organs provided by modern computed tomography (CT) and MRI scanners serve as proof of this success.

Today, seismic tomography is one of the most advanced indirect geophysical imaging methods, offering a direct three-dimensional view of the Earth's interior. First used in the 1970s, it has provided crucial insights into the structure of the earth's body and the functioning of plate tectonics. It is now applied at various scales, including in the search for oil and gas deposits using artificial earthquakes. Geophysicists refer to this specialized geological exploration method as "local earthquake tomography."

It is fascinating to observe how the same mathematical process can be used to visualize tumors in the human body, forensic traces in mummies, and the size and shape of magma chambers in supervolcanoes. This illustrates the importance of mathematics as a fundamental science.

426. Johann Radon from Tetschen on the River Elbe

When Johann Radon (1887-1956, he came from Tetschen an der Elbe) published his groundbreaking article "Über die Bestimmung von Funktionen durch ihre Integralwerte längs gewisser Mannigfaltigkeiten" in the "Verhandlungen der Königlich-Sächsischen Gesellschaft der Wissenschaften zu Leipzig" in 1917, it was a fundamental mathematical article that had no direct application at the time.

Fig. 193: Johann Radon (1887-1956)

Radon, who worked as a mathematician in the fields of analysis and differential geometry, could hardly have imagined that his work would lay the foundation for one of the most important medical technologies of the 20th century: computed tomography. Without the Radon transform he developed, it would not be possible to reconstruct a three-dimensional image from the measured projections of an object - a process that has saved countless lives today.

This story should serve as an urgent reminder to those who underestimate the importance of basic research - especially politicians who believe that research without immediate practical benefits should be restricted. It is precisely basic research that provides the foundation for later innovations and applications. Without the seemingly "useless" knowledge gained through such research, many of today's indispensable technologies would not exist. Basic research must therefore be prioritized to ensure long-term progress in science and society.

427. Mathematics is not a natural science, but it is indispensable

Let's stay with mathematics. There is virtually no "culture" that has not had, or does not have, some form of mathematics (even if only for the sake of counting), at least to some extent. Cuneiform tablets from Mesopotamia already contain mathematical texts, such as methods for solving quadratic equations. Since then, mathematics has developed into a field of knowledge that is no longer comprehensible to a single person, and whose findings and methods have now found their way into all areas of scientific and engineering research, and even into the "social and economic sciences" such as sociology and economics, where it is impossible to imagine life without it.

Mathematics is characterized by its self-building nature, with the number of "theorems" and "proofs" constantly increasing. Algebra is based on arithmetic. Geometry builds on arithmetic and algebra. The most important and indispensable discipline for scientific and technical applications, analysis in the form of differential and integral calculus, relies on arithmetic, algebra, and geometry. And so it continues. Mathematics resembles a vast family tree, with branches that keep growing and strengthening. Everything that has been "proven" in mathematical research remains. This is the key difference from the natural sciences (though mathematics is often counted among them), where every insight must be measured against experience, empiricism, and experimentation.

Since the inception of mathematics, mathematicians have been drawn to formulating problems that represent significant challenges. Some are very simple and easy to state, such as Fermat's Last Theorem or Goldbach's Conjecture, which asserts that *every even number greater than 2 can be expressed as the sum of two prime numbers.* The former was proven by Andrew Wiles in 1995, while number theorists are still working on Goldbach's Conjecture today.

428. Hilbert's problems

The first significant collection of mathematical problems, whose solutions are considered groundbreaking for the development of mathematics, was presented by the famous Göttingen mathematician David Hilbert (1862–1943) at the Second International Congress of Mathematicians in Paris in 1900. This collection, known as the "23 Hilbert Problems," was not only a list of mathematical challenges but also a vision for the future of mathematics. In these problems, Hilbert identified the central challenges that, in his opinion, could lead to fundamental progress in mathematics. Hilbert's problems cover a wide range of topics, from number theory and geometry to mathematical physics and logic. They were formulated to have a significant influence on mathematical research and thinking in the 20th century and to shape the work of the mathematical community for decades.

Of the original 23 problems, 15 are considered solved today. This means that well-founded and generally accepted solutions have been found for these problems, often leading to new mathematical theories and methods. However, three of the problems remain unsolved and continue to pose a major challenge for mathematicians. These unsolved problems are still the subject of intensive research and discussion, as their solution promises fundamental insights into various areas of mathematics. In addition, five of the problems are considered unsolvable in principle. This assessment usually results from the fact that the questions in these cases are not formulated precisely enough to be answered unambiguously. The realization that some of these problems are unsolvable also has profound implications for understanding the limits of mathematical methods.

One of the best-known of the three unsolved problems is the Goldbach Conjecture. This conjecture, which states that any even number greater than two can be expressed as the sum of two prime numbers, is a prime example of a seemingly simple mathematical problem that has proven extraordinarily difficult to prove. Despite numerous efforts and significant progress, particularly through analytic number theory, no definitive proof of this conjecture has been found to date. The Goldbach Conjecture remains a fascinating and challenging problem that keeps mathematicians around the world engaged.

429. Millennium problems

There is also such a compilation of mathematical problems for the 21st century, known as the "Millennium Problems". You can even earn some extra money as a mathematician by solving them, as the Clay Mathematics Institute has offered prize money of one million dollars for each solution to these seven problems. We have already seen one of them - it involved finding exact solutions to the Navier-Stokes equations, which are fundamental to fluid mechanics (point 28). The others are

- the proof of the conjecture of Birch and Swinnerton-Dyer,
- the proof of Hodge's conjecture,
- the solution of the P-NP problem,
- the proof of the Poincaré conjecture (solved in 2002 by Grigori Jakowlewitsch Perelman),
- the proof of the Riemann conjecture,
- the investigation of Yang-Mills' equations.

They will not be explained in detail here. If you are interested in them and perhaps even think that you can contribute something to their solution, you can find out about them yourself on the Internet, where all the relevant documents can be viewed on the Clay Institute website.

Now a few brief remarks on the importance of a good knowledge of mathematics for those aspiring to become scientists or engineers. Many first-semester physics students are surprised to find that the first two years consist mainly of mathematics lectures, and that university mathematics is presented very differently from what they were accustomed to in school. In fact, in the first four semesters, it makes virtually no difference whether you study mathematics or physics. Pursuing a career in physics means first learning the "official language" of this highly demanding science, which is mathematics. However, this does not mean you need to be particularly adept at "mental arithmetic" (in the past, many sales clerks were often better at this than some seasoned professors of mathematics or physics), but rather that you must develop a deep mathematical understanding to use this "tool" effectively for solving problems. This can only be achieved through persistent practice. Therefore, as a student, you should not be surprised if you have to work through dozens of exercise sheets and extensive practical reports every week in addition to reviewing lectures to succeed reasonably well. Much more important than talent are diligence, a high tolerance for frustration, and the determination to work on a problem until you find "your" solution. And if it turns out to be incorrect, simply sit down again to understand the mistake. On the other hand, if you are simply seeking a fun student life, STEM subjects are definitely not suitable unless you are an "overachiever."

Fig. 194: Grigori Perelman explains part of his proof of Poincaré's conjecture on the blackboard to his eager audience

If you want to mitigate the shock that the transition from high school mathematics to university mathematics causes for many students, I recommend searching the Internet for videos of mathematics lectures (e.g., on the Tübingen multimedia server TIMMS) and watching them. If you can adapt to this, you should seriously consider studying a subject that is "math-heavy," because the intellectual rewards that await you at the end are substantial. Otherwise, you may struggle to get through the first few semesters.

Now a few words about the Internet.

430. Joseph Weizenbaum's "dunghill"

Joseph Weizenbaum (1923-2008) once described it as a "great heap of dung in which many pearls and diamonds are hidden." Anyone who learns to find these "pearls" and "diamonds" can use the Internet as a never-ending source of knowledge.

Let's just stick to the subject of "lectures." The term dates back to the time of scholasticism, when books were still very rare. In a lecture, a topic is presented orally on a blackboard with chalk or, more recently, via PowerPoint "multimedia," which has the advantage that viewers can follow the lecturer's explanations with two senses (in contrast to reading). The disadvantage is that lectures can generally only be attended by enrolled students, and you usually have to pay for (public) lectures - and they usually take place at locations that are currently inaccessible and usually at inconvenient times. In principle, the Internet now makes it possible for anyone interested to attend lectures and talks without having to leave their home. Lecture notes are now also available to everyone free of charge on the web and can replace expensive textbooks. In short, there is no longer any reason to remain ignorant.

I myself use a tablet PC to watch a lecture, a presentation, or a BBC documentary online in the comfort of my armchair (or bed, if you prefer). The important thing is to learn as early as possible to separate the 90% "crap" of the internet from the real pearls and diamonds, according to Weizenbaum, and to concentrate on the latter and exclude the others from consideration as far as possible.

Joseph Weizenbaum, probably the most prominent and profound critic of the modern information society, also wrote the following quote, which can also be seen as part of a modern school policy agenda:

"The highest priority of schools is to teach pupils their own language so that they can articulate themselves clearly and distinctly: in their silent thoughts as well as orally and in writing. If they can do this, then they can also think critically and interpret the signals with which their world floods them. If they can't do that, then they will be victims of the clichés and stereotypes that the mass media spouts for the rest of their lives."

The ability to express oneself orally and in writing, and to do so as flawlessly as possible, seems to be rapidly disappearing in the age of text messaging and email, and not just according to my own observations. Alexander von Humboldt (1769-1859) and many of his contemporaries were still able to write books by hand in such a way that they could be typeset 1:1.

431. AI - phrase threshers and bullshit generators

Today, it is not only here and there that we see articles and essays in which ignorance or lack of content is camouflaged, for example, by grandiose and meaningless foreign words. If you want to make excessive use of this method, you should refer to a product of "artificial intelligence" a la Weizenbaum, the "bullshit generator" or "phrase thresher." By entering the appropriate nouns, you can easily create grammatically correct but largely meaningless sentences that also sound extremely "learned." By using these tools intelligently, entire passages, for example for "party programs" or "speeches," can be generated with little effort - at least sometimes you get the impression that this is really the case in practice when you read or hear them...

Nowadays, "bullshit generators" can be easily "programmed" by AI. Politicians, in particular, can now use them without any problems to generate content-heavy and impressive speeches that will provoke a storm of applause. The following prompt is for you to try out yourself (suitable for ChatGPT, among others):

You are a bullshit generator and a great phrase-monger before the Lord. I will give you a topic below. Your task is to formulate a short political speech that sounds highly intellectual but is otherwise completely devoid of content. Do you understand?

Now give a topic and let yourself be surprised! Of course, you can improve the prompt further to create even more impressive speech templates.

Finally, here's what I think is a particularly nice example from so-called "scientific prose," which I found some time ago. It comes from linguistics, a science that is particularly committed to human expression.

Here is the quote:

"The 'kaleidoscopic polemic' (W. Lepenies) about the relationship between a structuralism oscillating between methodology and panstructuralist ideology and Marxism has dominated the theoretical scene in France in recent years and, in the works of Althusser's school, has led to the attempt to catalyze a 'scientified' version of Marxist theory by assimilating the results of linguistics, cybernetics and psychoanalysis, which in turn is interpreted from the perspective of linguistic aporias, to humanist interpretations of Marxism."

Everything clear? A dissertation in such a style cannot really be graded anything other than *"summa cum laude"*.

432. Bicycles and cycling

Freedom of movement is something completely different from freedom of purpose - in the physical sense. The lack of the latter in modern (urban) people has given a new lease of life to an invention that was long hated by motorists: the bicycle. The sales figures (around four million last year in Germany alone (2023)) make cars look old by comparison.

With the introduction of the mountain bike in the 1990s, its popularity has steadily increased, both as an everyday item (commuting to work, inner-city transportation) and as a piece of sports equipment and means of transportation for leisure activities. In its classic form, it is technically largely exhausted, "high-tech" in many details, and unrivaled from an energy point of view (range and speed with an average drive power of ~150 W). It should therefore not be uninteresting to tell you something about its history, which can probably be traced back to the year 1801. In this year, a Russian farmer from the Urals by the name of Yefim Artamanov is said to have demonstrated something like a "pedal scooter" at the Tsar's court, but this can no longer be clearly proven. The actual "invention" of a two-wheeled vehicle goes back to the Baden forestry official Karl Friedrich Christian Ludwig Freiherr Drais von Sauerbronn (1785-1851), who took his first ride on it in 1817. His wheel, which he named "Veloziped", was regarded by his contemporaries as more of a curiosity than a serious means of passenger transportation capable of development. One year after this memorable ride, a Parisian newspaper wrote about the locomotion machine, which was now called the "Draisine":

"This machine will not be of much use; for one can only make use of it in well-kept avenues or parks.... The vehicle is good for children to play with in the garden..."

In a certain sense, the author was even right here, because what we would call a sensible road today only existed very, very occasionally at that time. But von Drais had nevertheless struck a chord with the times with his invention, and so a large number of other "inventors" and engineers began to work on it, some of whom came up with some truly adventurous designs.

The fact that "bad roads" no longer bother us much when cycling today (at most they annoy us, unless you ride a racing bike) is due to the invention of the air-sprung tire (Pneu, patented in 1888). Before that, "bicycle wheels" looked like "carriage wheels," i.e., they had a wooden rim that was fitted with a metal tire to hold it together and protect it. The first important question that engineers had to deal with was a) the drive technology and b) the size of the wheels. Anyone who has ever broken a chain on a bike ride can understand that the drive principle used by Drais for his "Draisine" is not ideal. The first idea was therefore to equip the front wheel with a pedal drive. This required the largest possible front wheel and a much smaller rear support wheel. The size of the front wheel was limited by the position of the saddle, as optimum power transmission to the pedal crank had to be guaranteed. In this case, one pedal crank revolution was equal to one wheel revolution and the distance covered corresponded to one wheel circumference. This meant that higher speeds could be achieved compared to the velocipede, and it was also more robust against uneven roads. This bicycle, known as the "penny-farthing," appeared around 1845 and enjoyed a certain popularity until it was replaced by better designs. In terms of evolutionary biology, the front wheel became bigger and bigger, the rear wheel smaller and smaller, "mounting" became more and more difficult, and the consequences of accidents for the rider more and more threatening. Quite simply, the penny-farthing had reached an evolutionary dead end, similar to the American sabre-toothed tiger at the end of the last ice age. While the latter is known to be extinct, the penny-farthing is being produced again today in small numbers and is often used as an eye-catcher at show events.

The first bicycles, which looked something like today's bicycles, were built from around 1861 by the Frenchman Pierre Michaux (1813-1883). They also still had a front-wheel pedal drive. However, the biggest drawback was the difficulty of steering, as the pedal drive directly on the front wheel made the vehicle unstable. An American, a certain Hemmings, said to himself around 1869, "why have two wheels when you can have one big one," and invented the first monocycle. The rider sits inside a larger wheel, which consists of two rings connected by spokes. The inner ring serves as a kind of guide rail for the actual chassis with saddle and the drive mechanism, which consisted of a drive wheel with pedals (the patent literature knows of a multitude of drive variants, including a motor). There was no steering and no brakes, which required a certain amount of courage on the part of the driver. Nevertheless, news of the "Flying Yankee Velocipede" also reached Europe, where the Leipziger Illustrierte Zeitung (1882) reported on it as follows:

"Although the new vehicle does not appear to be suitable for practical use as a means of transportation due to the considerable diameter of the outer wheel, it nevertheless offers an interesting exercise for friends of sport and is also capable of industrial utilization as an attractive spectacle, especially for those who are not very familiar with the laws of mechanics."

Fig. 195: You don't necessarily need two wheels to have fun on a motorcycle... (Creative Commons)

Of course, such a bicycle did not catch on, but the principle has certainly survived, as evidenced by some exotic-looking motorcycles (so-called monowheels).

However, the bicycle only became really useful with the introduction of the bicycle chain and rear-wheel drive, which happened around 1885 (by John Kemp Starley (1854-1901)). In principle, this marked the invention of the "low wheel" (two wheels of equal size), which has stood the test of time to this day. First solid rubber tires, then pneumatic tires (1888, John Boyd Dunlop (1840-1921)) increased comfort, and gears (first hub gears, then derailleur gears from 1930) finally made the bicycle a widely used and serious means of mass transportation.

"Cycling races" have been held since 1869, cycling as an Olympic discipline since 1896, and the famous 'Tour de France' has been held - with interruptions due to the war - since 1903. The "hourly world record" with a racing bike is currently 56.792 km (2022, Filippo Ganna) and 144.17 km (2016, Todd Reichert) for fully clad recumbents. When riding in the slipstream, speeds of up to 296 km/h have already been achieved using muscle power alone (2018, Denise Mueller-Korenek). This alone underlines the fact that the bicycle is "the" invention in the vehicle sector that is most effective at converting drive energy into kinetic energy. As a reminder, the average drive power is approx. 150 W (=0.2 hp). Christopher Froome, the winner of the 2015 Tour de France, achieved a drive power of around 425 W "on the mountain," which is already within the limits of what is just about humanly possible.

The innovation of the new millennium in terms of bicycles is to artificially increase their drive power by using battery-powered electric motors. Such "pedelecs" (pedal electric cycles) or e-bikes are delighting bicycle dealers with steadily increasing sales figures, as a hub motor that can be switched on when required helps older people in particular to remain mobile when cycling (the old cyclist's saying comes to mind: *"You can best tell how you are going downhill when you are going uphill..."*)

433. Electromobility

Electromobility is the political order of the day here in Germany anyway: as is well known, one million electric cars were to populate German roads by 2020. In 2015, the number of battery-powered cars had already surpassed the 19,000 mark (excluding hybrids). By the beginning of 2023, there were already more than one million purely electric cars on German roads.

The acceptance of electric cars has definitely increased, even if they often travel shorter distances on a single charge than diesel or petrol vehicles and sometimes have problems in cold weather. However, charging times have improved rapidly, and the charging infrastructure is constantly being expanded. The higher price of electric cars is offset by government subsidies and lower operating costs.

Fig. 196: Advertising for an electric car: Perfection of Style and Service

Pedelec fans are already convinced, as more than two million electric bikes have now been sold in Germany. However, the success of e-bikes has not been transferred to the automotive sector to the same extent. It was a strategic mistake on the part of the government not to include e-bikes in the target of one million "electric cars" by 2020. If this had been done, politicians would have been able to report that their plans had already been exceeded.

The main problem affecting the acceptance of electric vehicles is still the efficiency and cost of the battery technologies used. However, there are promising advances such as the development of solid-state batteries and other innovative technologies that can improve range and charging times. Despite these advances, the question of whether battery production can be made sustainable and environmentally friendly remains an extremely critical issue. The production of batteries requires the extraction of rare raw materials, which is often associated with environmental and socio-political problems.

434. Battery technology: the key to electromobility

Battery technology plays a decisive role in the acceptance of electric cars, as it significantly influences the range per charging cycle and the duration of a full charge. Alongside price, these factors are the most important criteria for consumers considering an electric car. The service life of the battery, measured by the number of possible charging cycles, is also an important aspect.

Fig. 197: Lithium-ion batteries are the most popular battery systems for electromobility today (Creative Commons)

Batteries that are to serve as the primary energy source for a car obviously have to meet special requirements. The key metric here is energy density, i.e., how much energy can be stored per unit of volume or mass and recovered for use. Diesel fuel, for example, has an energy density of 43 MJ/kg. The best batteries today achieve an energy density of just 4.7 MJ/kg (aluminum-air batteries which cannot be used as rechargeable batteries for chemical reasons) in experimental setups (i.e., far from practical usability). Lithium-ion batteries, which have an energy density in the range of 0.5 MJ/kg, are state of the art for use in motor vehicles. This may explain why electric car batteries are so large, so heavy, and also so expensive.

To understand the challenges of battery technology, you need to know the electrochemical processes that take place in batteries. Batteries consist of several galvanic elements in which chemical energy is converted into electrical energy. They consist of two different metal electrodes (anode and cathode) and an electrolyte that forms a redox system. The voltage they deliver depends on the materials used, the electrolyte, and the temperature. The trick is to find the right components for a secondary cell and to create the optimum physicochemical conditions for its operation. This is anything but trivial, as two centuries of battery research have shown.

The principle of batteries has been known since the experiments of Luigi Galvani (1737-1798) and Alessandro Volta (1745-1827). Today, you can build a battery yourself with simple means: two galvanized nails, some sheet copper, wire, and two lemons from the supermarket are all you need. You can measure the voltage with a light-emitting diode and a voltmeter. The citric acid serves as the electrolyte, and the metal electrodes generate a potential difference that causes the diode to light up.

Lithium-ion batteries are the most advanced energy storage devices in electromobility. They are available in different variants, such as lithium metal oxide accumulators, lithium polymer accumulators with liquid organic electrolytes, lithium manganese accumulators with many charging cycles, and lithium-air accumulators, which have a high development potential. IBM hopes to achieve a theoretical energy density of around 40 MJ/kg with lithium-air batteries, which would be a real breakthrough. A typical lithium-ion battery today, with an energy content of 180 MJ (50 kWh), weighs around 400 to 500 kg and enables a range of around 150 km. Lithium-air batteries could significantly increase the range due to their higher specific energy density if they are successfully developed.

Despite many challenges, there is reason for optimism. Battery technology is constantly evolving, and there are promising approaches to solving existing problems. Electric cars already have a satisfied clientele as second vehicles for short distances. With further development work, they could soon be a real alternative to conventional mid-range cars. The future of electric mobility looks promising, and battery technology will play a central role in realizing this vision.

But back to the lithium-ion batteries that can be found everywhere today.

435. Lithium

They have brought honor to a chemical element that a few decades ago was almost exclusively of interest to nuclear weapons manufacturers (as lithium deuteride). It is known to be found in the third position (atomic number Z=3) in the periodic table, in the first main group, the alkali metals. It was formed mostly during the first minutes of the Big Bang as part of primordial nucleosynthesis (alongside deuterium, primordial helium, and a small amount of beryllium). In principle, it cannot be fused within stars, but it can be destroyed very effectively. It is therefore quite remarkable that it ranks 27th in the table of elemental abundances in the Earth's crust (similar to zinc or tungsten, estimated proportion ~0.006%) and is therefore present in greater absolute quantities in the Earth's crust than lead, for example. However, because lithium compounds are rarely found in abundance, the extraction of this very light alkali metal is difficult and cost-intensive, which naturally impacts battery prices.

In nature, lithium occurs in two stable isotopes. The nucleus of the more common (92.4%) lithium-7 consists of three protons and four neutrons, while lithium-6 has only three neutrons. The theory of element formation (nucleosynthesis) during the Big Bang makes fairly accurate predictions regarding the proportions of hydrogen, deuterium, helium, and lithium (all other elements are only produced in the course of stellar evolution), which agree quite well with the observed proportions, with the exception of lithium. Surprisingly, in light of primordial nucleosynthesis, the observed amount of lithium (lithium-7) in the atmospheres of old stars located in the halo of the Milky Way (globular clusters) appears to be too low, which constitutes the content of the so-called "cosmological lithium problem." However, recent measurements on halo stars seem to confirm that this "problem" is only a misconception. Rather, it seems to be the case that lithium, which is specifically heavier than hydrogen and helium, slowly sinks into the central regions of these stars over the many billions of years of their existence, where it is then destroyed during nuclear processes. The result is the observed depletion of this element in the stellar atmospheres, which is about 2/3 of the primordial value. However, the last word on this matter has not yet been spoken, as there are also several other serious explanatory models for this phenomenon. In any case, it is interesting to know that the lithium in our lithium-ion batteries is very old (exactly 13.787±0.020 billion years), because it was created in the first few minutes after the ominous time zero that we call the "Big Bang."

436. Big Bang Theory

The realization that "our" world emerged less than 14 billion years ago in a still inexplicable way from something known as the "Big Bang" is now explained in detail in the so-called standard model of cosmology. It is now so well empirically proven that alternative views such as the steady state theory once founded by Hermann Bondi (1919-2005), Thomas Gold (1920-2004), and Fred Hoyle (1915-2001, who coined the term "Big Bang") are no longer seriously pursued today.

The "Big Bang Theory" is, as the name suggests, a mathematical-physical construct (theory) that was created to explain and justify our observations, as far as they relate to the history of the development of the cosmos as a whole, with as little contradiction as possible. The theoretical background in terms of space, time, and gravity is provided by Einstein's general theory of relativity. What "happened" and "happens" in this model is determined by thermodynamics and quantum theory in the form of the standard model of elementary particle physics. Both theories, Einstein's theory of gravitation and quantum theory, restrict our picture "downwards" with their very own areas of validity because there is still no recognized quantum theory of gravitation, at most a few approaches worthy of discussion. This prevents us from glimpsing the true cause of what is generally perceived as the "explosion" of a singular state, the "Big Bang." But with our physical theories, we can at least approach the time "zero" with some certainty down to seconds (which is astonishing enough). It is the point in time from which "physics as we know it" begins to function. In truth, however, as the well-known cosmologist Hubert Reeves (1932-2023) once put it, the Big Bang is merely our horizon in time and space, which we regard as the zero point of our history out of convenience and for lack of anything better: "We are explorers before an ocean: we cannot see if there is anything beyond the horizon." Empirically, we have already approached this horizon to within a few minutes of the Big Bang - in the form of the results of the aforementioned "products" of primordial element synthesis (more precisely, the primordial hydrogen/helium ratio). A view of the tiny inhomogeneities, quantum fluctuations, in the plasma expanding with space during the Big Bang is provided by the cosmic background radiation (also known as 3 degrees Kelvin radiation), which is being measured ever more precisely by satellites (e.g., "Planck") and which gives us insights into both the exact spatial structure, the age of the universe, and the early formation of structure in the cosmos. In terms of time, it takes us back to ~380,000 years after the "Big Bang," when the cosmos became quasi-transparent (this epoch is called the "age of hydrogen recombination") because the electromagnetic radiation field was now able to decouple from matter. This radiation field, which today corresponds closely to that of an ideal black body (Planck radiator) with a temperature of 2.725 K, fills cosmic space homogeneously and isotropically and is best observed in the microwave range. Alongside the observation of the expansion of space (visible in the distance-dependent escape velocity of galaxies, Hubble's law), it is one of the most important empirical pillars of the Big Bang theory. Both observations, the cosmic expansion and the isotropic background radiation, are actually "the" proof that our cosmos must have had a dense and hot beginning.

When it was established in 1929 on the basis of observations made first by Vesto Slipher (1875-1969) and then systematically by Edwin Hubble (1889-1953) that our universe was expanding, it was immediately clear that it must have had a smaller expansion "earlier." The Russian mathematician and meteorologist Alexander Friedmann (1888-1925) had already deduced precisely this kind of behavior from Einstein's gravitational field equations. This discovery led to a paradigm shift in our understanding of the universe: a static cosmos was replaced by the image of a dynamic, constantly changing universe that originated in a dense and hot state, the so-called Big Bang. Or to put it another way: theory and empiricism clearly point to the existence of the Big Bang, in which "our" cosmos was created.

People usually think of the Big Bang as an explosion "in space," but this is wrong if you look at it in the light of the general theory of relativity. The Big Bang not only encompassed the material content of the cosmos accessible to us (represented today by the Hubble sphere) but also the entire physical space itself, which also expanded. This space, which has been continuously expanding ever since, is probably infinite in size, and all the matter and energy of the universe are distributed within it.

Unfortunately, the geometric concepts on which these findings are based are beyond imagination, which is why the metaphor of an "explosion" is often used to illustrate the event. This metaphor, although physically imprecise, provides a simple way of conveying the complex mathematical and physical relationships without having to resort to the precise but inaccessible formulaic language of mathematics. After all, it was this "explosive" event that gave rise to everything we know - including ourselves.

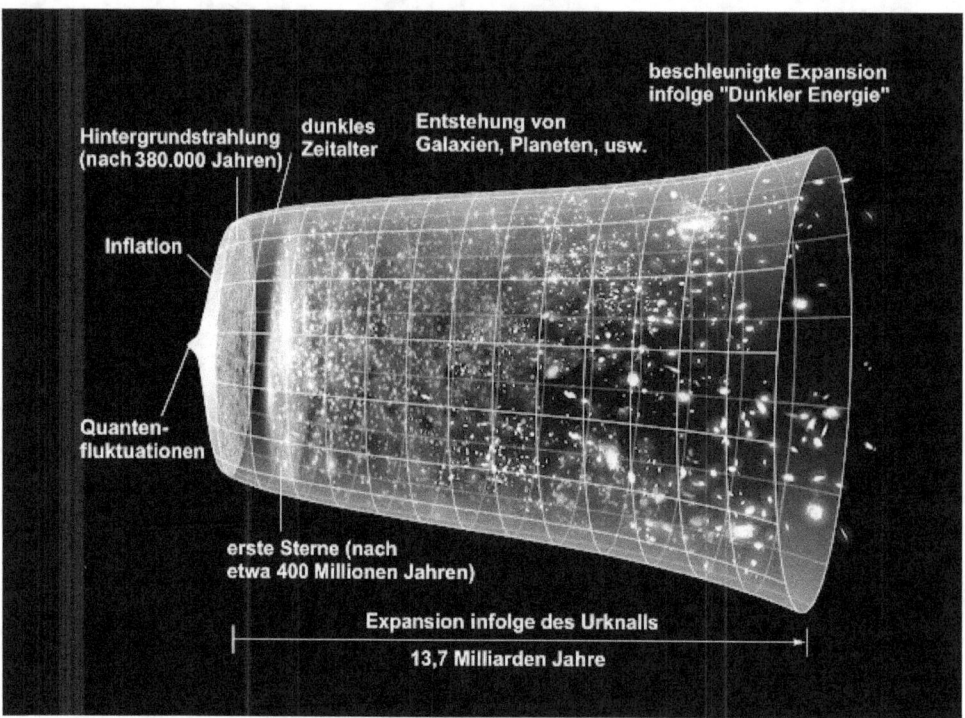

Fig. 198: Space-time diagram of the development of the manageable cosmos according to the standard model of cosmology (Creative Commons)

437. Why is there anything at all and not rather nothing?

However, as far as we approach the answer to the question of what exactly happened during the Big Bang, the old question of metaphysics, *"Why is there anything at all and not rather nothing?"* (in the formulation of Gottfried Wilhelm Leibniz (1646-1716)), remains unanswered. A question of this kind is inaccessible to empirical science, as it is located on a higher, more general level. It is, in short, philosophical in nature. Martin Heidegger (1889-1976) therefore even described the question *"Why is there being at all and not rather nothing?"* as the central question of metaphysics, which cannot be answered with logical arguments alone.

A key term here is the concept of "nothing". It has many aspects of meaning, which in the case of the above statement can be reduced to the non-existence of being.

Fig. 199: Sokrates (469 v. Chr. – 399 v. Chr.)

The ambivalence between what is and what is not was already a central theme in the philosophical thinking of the pre-Socratics (the name given to the natural philosophers who worked before the Athenian philosopher Socrates (469 BC - 399 BC)). These early thinkers attempted to understand the basic structures of reality and to clarify the relationship between what exists and what does not exist. With the realization that the non-existent is neither recognizable nor explorable, as taught by Parmenides of Elea (between 520 BC and 460 BC), philosophical considerations shifted to the closer definition of being and its most essential property, existence.

Parmenides thus set a decisive emphasis that deeply influenced later philosophy, especially in the doctrine of "ontology," i.e., the science of being, which deals with the question of what it means to be. Other key concepts that are important in this context include "reality" (Wirklichkeit), the "problem of identity," and the "problem of universals," which goes back to Plato. This latter problem concerns general concepts such as number, man, God, color, etc., and the question of whether these really exist or are merely "human constructions" - a topic that was controversially discussed in the so-called "universals controversy" and still leads to different views in various "ontologies" today.

However, this should not be the central topic here. At this point, we want to return to the "Big Bang," an event that, in a certain sense, seems to bring existence out of nothing and thus sheds a whole new light on the metaphysical discussions of the pre-Socratics.

438. The ominous "nothing"

When dealing with cosmology, it is increasingly common to read that the world was created from "nothing," without this ominous "nothing" being specified in more detail, either philosophically or physically. In modern physics, the term "quantum mechanical vacuum" has become established for this "nothing," which, as we know today, is fundamentally different from the "mathematically conceivable vacuum," i.e., the absolute void. Even in antiquity, "nothing" was identified with another concept, namely "emptiness" (which, of course, is a philosophical limitation as soon as "space" is also granted an independent existence). For the atomists, this was the space between the atoms. For their (the atomists') doctrine, this space, which could perhaps be described as a "microvacuum," was necessary in order to be able to postulate the movement of indivisible atoms at all.

In modern physics, however, the quantum mechanical vacuum is by no means an absolute void, but rather a place of constant activity and fluctuation. This vacuum is filled with virtual particles that appear for a short time and then disappear again due to Heisenberg's uncertainty principle. These fluctuations are not just theoretical constructs, but have measurable effects, such as the Casimir effect,[79] in which two metal plates lying close to each other experience a measurable force of attraction caused by quantum fluctuations in the vacuum.

---

[79] The Casimir effect describes the attraction between two close, unequal plates in a vacu-um caused by quantum fluctuations of the electromagnetic field, resulting in a measurab-le force that arises as a consequence of the confined modes of the virtual photons within the gap between the plates.

From a philosophical point of view, "nothing" has always been a difficult concept. Aristotle argued that "nothing" could not exist, as existence always implies a form of being. The scholastics of the Middle Ages took up this idea and discussed it against the background of theological concepts such as creation out of nothing (creatio ex nihilo), a central concept of Christian theology, which emphasizes the omnipotence of God and the radical dependence of creation on its Creator.

For ancient atomists such as Leucippus (5th century BC) and Democritus (ca. 460-459 BC), however, emptiness was not an absolute absence of being but a necessary condition for the movement and interaction of atoms. Without this emptiness, they argued, atoms would be static and immobile, and there would be no change, no movement, and no diversity in the world. This idea of emptiness as a necessary space for movement differs fundamentally from the modern physical concept of the vacuum, but it also shows how long humanity has struggled with the idea of "nothingness."

In today's cosmology, the idea that the universe could have originated from nothing is seriously considered. One hypothesis states that the universe originated from a quantum fluctuation of the vacuum. This idea is deeply rooted in quantum mechanics, which posits that on the smallest scales, defined by Planck time and Planck length, space and time themselves are subject to quantum mechanical uncertainty relations. This uncertainty could have enabled the spontaneous creation and expansion of a small, dense, and hot universe.

This view naturally has far-reaching philosophical and physical implications, as it calls into question our understanding of existence and raises fundamental questions about the nature of space, time, and matter. The term "nothing" thus becomes a complex and multi-layered concept that goes far beyond the mere absence of matter. It becomes fertile ground for interdisciplinary discussions between philosophers and physicists, who together try to uncover the deepest secrets of the universe.

The ominous "nothing" therefore remains a fascinating mystery. It challenges us to question our notions of reality and existence and, at the same time, opens up new ways of understanding the fundamental structure of the universe. Thus, "nothing" remains an essential challenge and source of inspiration for human thought, which is always striving to fathom the deepest secrets of existence.

Fig. 200: Nothing

439. The "nothing" as "emptiness"

As it was impossible for most ancient philosophers to conceive of absolute emptiness, it was either denied (*horror vacui*) or filled with a "fifth substance" by Aristotle. This fifth substance (alongside fire, earth, water, and air), the "*quinta essentia,*" was supposed to be immaterial, structureless, and free of the opposites contained in the aforementioned elements (e.g., warm and cold, solid and liquid). It was believed to evenly fill the entire universe (ether) and also be the actual cause of the movement of the celestial bodies. For this reason, there is no place for the vacuum in Aristotle's philosophy, whereas for the atomists, the existence of the void is a necessity to postulate the movement of atoms as an inherent property of matter.

Aristotle's denial of the vacuum can be seen as a direct consequence of his conception of motion. His concept of dynamics is based on the key concepts of force, velocity, and resistance. Force is the actual cause of a movement process. However, a body only moves as long as the cause, i.e., the effect of the force, continues. From the moment the force ceases to act, the movement should also cease. Galileo Galilei (1564-1642) was only able to demonstrate about 2000 years later with his discovery of the law of inertia that this assertion did not correspond to reality. According to Aristotle, the air that opens in front of the flying arrow and closes behind it is the actual cause of its movement. Once this statement is recognized as correct, it follows that there can be no movement without air. The air or the "finer" ether thus becomes media without which movement is virtually impossible. A completely empty space is therefore unthinkable in the context of peripatetic physics, even purely kinematically.

440. The vacuum

In this way, the ether became a fixed concept in the natural sciences for the next two millennia. The question of the existence of a "vacuum" only became topical again at the beginning of the modern era, not so much for ideological and philosophical reasons, but more for purely practical considerations. Even before Galileo's time, miners knew that normal suction pumps could only lift water to a height of around nine meters. For this reason, cascades of pumps had to be built to drain mines. The analysis of these frequently used devices brought the vacuum, as a synonym for "airless space," back into the focus of research.

Evangelista Torricelli (1608-1647), probably Galileo's most famous student, succeeded in creating a vacuum for the first time in 1644. He filled a 90-centimeter-long glass tube with mercury, sealed it with a finger, and then inverted it into a container also filled with mercury. After removing his finger, he observed that a free space around 14 centimeters long formed above the mercury column. Torricelli hypothesized that this space was completely empty and therefore represented a vacuum. Further experiments confirmed his hypothesis. For example, it was demonstrated that the empty space above the mercury column could be entirely filled with water, proving it to be truly empty. Despite these clear experimental results, opinions diverged at this point. René Descartes (1596-1650), for example, was a staunch opponent of the view that the world is divided into a "corpuscular part" and a vacuum. Through astute reasoning, he argued that space and substance form a unity. For him, there was nothing in the world but matter in motion. This perspective is also expressed in the following quote:

"In this, however, many seem to me to be mistaken, that although they assume a fluid in heaven, they imagine it as an empty space that offers no resistance to the movements of other bodies, but also has no power to take them with it. For there can be no such emptiness in nature, and it is common to all liquids that they do not resist the movements of other bodies only because they themselves have a movement in them, and because these movements easily take place in all directions with such force that in a certain direction they necessarily take all the bodies contained in them with them, as far as no other cause holds them back, and they are solid, resting and hard, as can be seen from the previous."

Based on the thesis of a fluid-like ether filling the entire cosmic space, Descartes developed a vortex theory in his work "Principia philosophiae" (1644), according to which the movement of the celestial bodies is caused by the movement of the ether. The fascinating aspect of his theory was that it plausibly explained the movement of the planets without invoking remote forces.

The great significance of Cartesian ideas in the history of science lies less in their factual content. More importantly, Descartes emphasized the fundamental equality of earthly and cosmic laws and highlighted the material unity of the entire world. In this context, one can certainly see analogies between his substance-filled cosmos and the quantum mechanical vacuum of contemporary physics.

In the decades following Descartes, experiments with the vacuum became increasingly popular. Vacuums created by pumping out hollow bodies are usually referred to as "technical vacuums." They are of great practical importance. The famous Magdeburg mayor and councilor Otto von Guericke (1602-1686), for example, perfected the air pump and used it to create vessels empty of air. His most famous experiment is likely the one with the Magdeburg hemispheres. First conducted in 1654 based on his suggestion, it clearly demonstrated the enormous power of air pressure. Von Guericke had the air pumped out of the interior of two adjoining copper hemispheres, and it was found that even two teams of eight horses were unable to separate the hemispheres. This impressive experiment is still often demonstrated today in a slightly modified form (without horses) in physics lectures. Further experiments of the time also showed that light can propagate in a vacuum, but sound cannot. Life is also not possible in a vacuum. The observation that both light and magnetism propagate in a vacuum (Robert Boyle, 1627-1691) and the long-range theory of gravitation postulated by Isaac Newton (1643-1727) indicated that the vacuum is not completely devoid of properties after all. It was necessary to explain how certain effects could propagate through empty space.

441. The "world ether"

In this way, various concepts of a space-filling universal ether emerged, which would statically fill Newton's absolute space and simultaneously serve as a medium for the transmission of long-range effects. With the help of the ether, the intention was ostensibly to reduce the long-range forces occurring in physics to the more obvious short-range forces. The fact that gravity, for example, propagates through empty space without the mediation of a medium seemed completely absurd even to Newton. Nevertheless, the ether hypothesis remained purely formal in his mechanics. The problem of interaction could not yet be solved by the physics of the 17th and 18th centuries. Dubious substances such as the heat substance (phlogiston) or the light substance were proposed as alternatives to Newton's theory of action at a distance. They were supposed to be imponderable and components of the universal ether.

In the 18th century, interest in the ether diminished. Many scientists of the time had come to terms with the remote effect of gravity. Newtonian mechanics, with its axiomatic structure and astonishing predictive power, became the epitome of a scientific theory.

In addition to mechanics, optics was particularly advanced during this time. Targeted experiments and theoretical investigations led to the rediscovery of the idea laid down by Christiaan Huygens (1629-1695) in his "Traité de la lumière" in 1678, according to which light is a wave process. He systematically explained the rectilinear propagation, refraction, and reflection of light. In 1690, Huygens also developed an astute, but unfortunately only phenomenological, theory of birefringence without recognizing the fundamental property of light behind this phenomenon. Since Newton favored the corpuscular concept in his "Opticks" (1704), the wave theory struggled to gain acceptance due to his authority. However, this changed in 1801 when Thomas Young (1773-1829) successfully explained the interference phenomena observed at the double slit using the wave theory. Experience showed that light can propagate through transparent solids, liquids, and gases as well as through a vacuum. At that time, however, scientists were only familiar with elastic waves. Thus, the question arose of why light waves can propagate in bodies of high elasticity as well as in bodies of low or vanishing elasticity. To answer this question, it was necessary to introduce a hypothetical light ether that would fill both the microvacuum and the macrovacuum (i.e., the empty space between celestial bodies). On one hand, this light ether was imagined to be so extremely thin that it could hardly influence the movement of the planets around the sun. On the other hand, it was also supposed to be elastic enough to allow light to propagate through it in the form of longitudinal ether waves, similar to sound waves in the air.

The first significant setback to the idea of the elastic ether came from Augustin-Jean Fresnel (1788-1827), who revisited the question of the polarization of light in his work. He experimentally established that no interference occurs when light rays polarized perpendicular to each other interact. This observation could only be explained if light has a transverse character. In other words, light is not a longitudinal wave, like a sound wave, but rather a transverse wave, similar to those in solids. The proof of the transverse nature of light waves implied that the light ether must behave like a solid. However, this clearly contradicted the observed movement of the planets around the sun.

The discovery that light exhibits transverse wave characteristics posed a dilemma for the ether concept. Solid-state properties had to be attributed to the ether for it to serve as a medium for light. To preserve the ether hypothesis, it was necessary to endow the light ether with new, sometimes contradictory properties. Ultimately, only a few mechanical properties could be attributed to it without theoretical contradiction. It was conceived as a quasi-rigid substrate at rest, filling the entire cosmic space, enabling the propagation of light, and representing an absolute reference system. Despite the growing internal contradictions, scientists continued to explore the ether theory. Scholars such as Augustin-Louis Cauchy (1789-1857), Franz Ernst Neumann (1798-1895), and George Gabriel Stokes (1819-1903) worked on this theory, albeit unsuccessfully.

The situation became even more complex with the new findings in the fields of electricity and magnetism. It had long been known that electricity and magnetism operate through a vacuum. To describe these effects adequately, the English physicist Michael Faraday (1791-1867) introduced the concept of electric and magnetic fields. He laid the experimental foundation for the synthesis of these forces (Faraday's law of induction), which James Clerk Maxwell (1831-1879) then developed into a comprehensive theory of electromagnetism. Maxwell's equations describe the structure of the electromagnetic field, which exists in reality and represents a new type of matter (in the physical sense) alongside substance.

Maxwell's theory was the first real field theory in the history of physics. However, interpreting it within the framework of mechanical ideas proved challenging, if not impossible. The persistence of the ether concept led to the creation of seemingly uncanny models involving ether vortices and associated ether pressures and tensions. These models eventually led to the realization that electromagnetic phenomena could not be ultimately explained by mechanics.

One unexpected consequence of Maxwell's equations was the prediction of electromagnetic waves. The theory predicted that an accelerated electric charge emits electromagnetic radiation, which propagates at the speed of light and transports energy. Electromagnetic waves in the form of radio waves were experimentally demonstrated later by Heinrich Hertz (1857-1894). It also became clear that light is an electromagnetic wave phenomenon, which provided a new foundation for optics. The direct proof of the light ether was still pending. Considerations by Maxwell and Hermann Helmholtz (1821-1894) suggested that it should be possible to experimentally determine Earth's movement relative to the assumed stationary ether. The corresponding experiment was carried out in 1881 by Albert Abraham Michelson (1852-1931) and, in a significantly improved form, in 1887 with Edward Williams Morley (1838-1923). Using a precisely constructed interferometer (the Michelson interferometer), they attempted to measure the difference in speed between light beams aligned with and perpendicular to Earth's motion. If light were an ether wave, the speed of light should add to the speed of the Earth in the direction of its motion. However, the experiment, and all subsequent experiments with increasing accuracy, consistently yielded negative results.

442. Overcoming the "world ether"

There is simply no "ether wind", and therefore no "light ether". Instead, the result was a realization that was difficult to accept: the speed of light in a vacuum is completely independent of the observer's state of motion. While the non-existence of the ether with its exotic properties was quite acceptable, the resulting constancy of the vacuum speed of light led to new conflicts. It turns out that the so-called Galilean transformations - one of the cornerstones of classical mechanics - lose their validity. Speeds that are close to the speed of light, approximately 300,000 km/s, can no longer simply be added together. "Time" itself changes and depends on the state of motion of the observer (reference frame). We owe this truly revolutionary insight to Albert Einstein (1879-1955).

The key experiment on the way to this realization was the famous Michelson-Morley experiment in 1887. The two physicists, Albert A. Michelson (1852-1931) and Edward W. Morley (1838-1923), wanted to prove the existence of the "ether", which was imagined as a carrier medium for light waves. It was assumed that the Earth moves through this ether, which should lead to a measurable "ether wind".

443. The Michelson-Morley experiment

Michelson and Morley developed an extremely precise interferometer to measure the speed of light in different directions. They expected the light to move faster in the direction of the Earth's motion through the ether and slower in the opposite direction. This would have shown up in the form of shifts in the interference pattern.

Fig. 201: Experimental setup of the Michelson-Morley interferometer at the Western Reserve University in Cleveland (July 1887). Top left Albert A. Michelson, below Edward W. Morley (Creative Commons)

To their surprise, however, Michelson and Morley found that there was no difference in the speed of light, regardless of the direction of the measurement. This meant that the expected ether wind did not exist.

In 1905, Albert Einstein published the "Special Theory of Relativity" (without knowing the results of the Michelson-Morley experiment), in which he fundamentally revised the ideas of classical physics about space and time. This also finally removed the foundations of the ether hypothesis. The vacuum proved to be truly empty in the classical sense. What remained were particles and fields. The further history of the vacuum was determined by quantum physics, which emerged around the same time.

444. The quantum vacuum

In particular, the development of relativistic quantum field theories led to a new and completely unexpected concept of the vacuum. A new "ether" was discovered, so to speak, which no longer has anything to do with its classical model, but nevertheless exists and is measurably noticeable in experiments. It is created by removing all real elementary particles - i.e., protons, neutrons, electrons, photons, and neutrinos (to name but a few) - from an area of space. The vacuum created in this way is known as a physical vacuum or "quantum vacuum" for short. It represents the maximum experimentally achievable "void," so to speak. It differs from the mathematical vacuum in that it is still permeated with fields that are in their energetic ground state. The mathematical vacuum, on the other hand, is merely an abstraction. In principle, it cannot be created. The "quantum vacuum" is the maximum achievable "nothingness" of nature.

Classically, it can be assigned properties such as a finite electric and magnetic field constants, the square root of whose product is the reciprocal of the vacuum speed of light. Quantum mechanically, it represents the lowest stable state that an area of space can assume when it is penetrated by physical fields. It is important to know that in quantum field theories, each elementary particle is described by a field. Such a quantum field in a vacuum state means that no real elementary particles are present, i.e., the so-called zero-point energy of the field (which is always greater than zero) is not sufficient to create "real" particles. However, it is sufficient for so-called "virtual" particles and antiparticles to be constantly created, but these disappear again within a time limit, which is given by the Heisenberg energy uncertainty relation. The "zero-point fluctuations" predicted by quantum theory show that the quantum vacuum only appears "empty" on average. To use an analogy, it behaves like the surface of the ocean as seen from a space station. The closer you get to its surface as you descend, the stronger its waves appear.

If you could resolve the physical vacuum down to a length scale of the order of the Planck length ($\sim 10^{-35}$ m), you would see something like a "quantum foam" of the highest dynamics - a kind of bubbling lake of particles and antiparticles of all kinds. The quantum vacuum therefore virtually contains the entire particle spectrum of all known and yet-to-be-discovered elementary particles. For this reason alone, it is expected that the physical vacuum has a direct influence on observable parameters, such as the electric charge of an elementary particle. And this is precisely the case, as a large number of experiments have shown.

Under certain conditions, the virtual particles can "materialize," i.e., become real particles. This happens, for example, when elementary particles collide in particle accelerators or in the immediate vicinity of the event horizon of a black hole. In the latter case, this is known as "Hawking radiation" (named after its discoverer Stephen Hawking (1942–2018)), which ultimately means that even black holes do not live forever.

An important property of the quantum vacuum is that it lacks spatio-temporal references, or, to put it another way, an observer must never be able to determine their own state of motion relative to the "zero-point radiation" of the quantum vacuum.

445. The Unruh effect

This requirement leads to interesting consequences, as Canadian physicist William G. Unruh discovered back in 1976. He devised a "thought experiment" which I would like to briefly present here without using too many technical terms, as it shows how real thermal radiation can be generated from the quantum vacuum.

Imagine an elevator car made of an ideally electrically conductive material (this restriction is necessary so that no electromagnetic radiation can penetrate the box from "outside" - keyword "Faraday cage"). This cabin should be located somewhere in the cosmos, far away from interfering gravitational fields. There is an absolute vacuum both inside and outside the box. The question William Unruh asked himself was: "What happens when this elevator box is accelerated uniformly?"

As can already be shown in the context of classical electrodynamics, the ideally conducting car floor emits an electromagnetic wave at the exact moment when the acceleration begins, which propagates toward the ceiling at the speed of light. There it is reflected, transferring a small part of its energy to the ceiling atoms. In the process, photons are emitted whose spectrum is purely thermal and obeys Planck's law of radiation. The result of the acceleration process is a highly diluted photon gas made up of real photons inside the elevator car.

It is not at all difficult to extract these photons from inside the elevator car and thus remove them. All you have to do is cool the cabin down accordingly. In this way, a vacuum is created inside the cabin, but it is clearly different from the vacuum outside the box. The clearest way to see this is to install a photon detector inside and outside the elevator and compare their readings. While the inner detector is at rest relative to the inside of the box, the detector mounted on the outside moves at an accelerated rate relative to the surrounding vacuum. It must therefore react to the zero-point fluctuations of the quantum vacuum and thus indicate the presence of real photons, while the detector inside the elevator box shows no deflection. This means that the "inner" vacuum must also differ energetically from the "outer" vacuum. The outer vacuum is defined precisely by the fact that its energy disappears on average. If one accepts this definition, then it is reasonable to assign a negative energy to the vacuum inside the elevator box. An approximation of both vacuum energies is only possible if the "extracted" photons are reintroduced into the box. Then the "outer" vacuum would no longer differ from the "inner" vacuum, and both would have the same energy.

It follows from these considerations that, for an accelerated observer, a blackbody spectrum is superimposed on the spectrum of the "zero-point radiation," which leads to the zero-point fluctuations becoming stronger. He therefore gets the impression that he is moving in a radiation field with a temperature that depends on the magnitude of the acceleration. An interesting exercise for a budding theoretical physicist is therefore to calculate how strongly the elevator must be accelerated in completely empty space (without extracting the photons) so that its interior turns into a comfortable oven... (solution in m/s^2).

Incidentally, this scenario corresponds quite exactly - because of the equivalence between gravity and acceleration - to the conditions in the region of the event horizon of a sufficiently small black hole. If one uses the value for gravitational acceleration there, then one also obtains a corresponding Unruh temperature and an associated thermal radiation field, the aforementioned Hawking radiation.

Such and similar considerations make it clear that the cosmic void is by no means a simple "void" in the conventional sense. Rather, in the light of modern quantum field theories, this void proves to be a highly dynamic substratum of the world. This "quantum vacuum" can be imagined as a kind of new "ether," which, however, is in no way related to the classical concept of the ether. In order to understand the "beginning of the world," it is therefore essential to include quantum calculus in a suitable way in the description of this "beginning" (which for us means a beginning in time).

This challenge is taken on by so-called "quantum cosmology." It has set itself the goal of solving the singularity problem of classical cosmology in particular, which is characterized by an infinitely small and dense state of matter at time zero. Although quantum cosmology is still highly hypothetical, there are justified hopes that it will one day be able to explain the "Big Bang" physically and comprehensively. It is based on so-called "quantum gravity," a theory that integrates gravity in terms of quantum physics. Numerous theoretical approaches to quantum gravity currently exist, but a fully developed theory is still pending. This should be kept in mind when reading about concepts such as brane worlds, loop quantum gravity, ekpyrotic universes, or similar theories. These concepts are part of the dynamic and ever-evolving field of modern theoretical physics that strives for a deeper understanding of the cornerstones of our universe.

446. The emergence of our universe from the quantum vacuum

The idea that our universe could have emerged from a quasi-timeless quantum universe (perhaps together with many others) through a vacuum fluctuation cannot be entirely dismissed. As early as 1973, physicist Edward Tryon (1940–2019) showed that such a scenario is plausible. He noticed something that Pascual Jordan (1902–1980) had already noticed two decades earlier: a cosmos with a flat (i.e., Euclidean) geometry has a total energy of zero. In this case, the sum of all the individual energies of the particles that make up the cosmos is exactly equal to the amount of their mutual gravitational energies (which are known to be negative).

However, if this is the case, then the Heisenberg energy uncertainty principle can be applied. This states that the product of energy and time must always be greater than or equal to a special natural constant, Planck's quantum of action. From this, it can be argued as follows: *The more the total energy of the universe approaches zero, the longer the uncertainty relation allows this fluctuation to exist in time. If the energy value is exactly zero, then the associated quantum fluctuation ("universe") will exist forever.*

Considerations of this kind feed the assumption that there could be a universe beyond the Big Bang, which could be described as a quantum universe. This quantum universe could be characterized by the absence of temporality, possess additional spatial dimensions, and have the potential to generate entire "classical" universes from within itself, which in turn could form a kind of multiverse.

Physics and mathematics have no fundamental objections to such speculations. However, the crucial question remains as to how these ideas can be verified. This poses considerable difficulties, as we are causally trapped in "our" universe. Perhaps the only thing that remains is the stringency of a future theory that postulates a clearly defined, but in principle untestable initial state. If such a theory plausibly and on the basis of the laws of nature logically and comprehensibly explains what must have happened 13.78 billion years ago for a universe full of galaxies, stars, and life to develop to the complexity of consciousness and intelligence, then one could trust the premises of such a theory, even if direct verification is not possible.

However, none of the theories of quantum gravity proposed so far have achieved this status. Although these theories offer interesting perspectives and raise many profound questions, it remains a challenge to definitively confirm or disprove their validity. In the meantime, we can only hope that future advances in theoretical physics and mathematics may one day enable us to answer these intriguing questions.

447. Cargo Cult Sciences

The famous physicist and Nobel Prize winner Richard Feynman (1918–1988) critically questioned the problem that arises here in his book *"Surely You're Joking, Mr. Feynman!"*, which is well worth reading. He argued that a science that operates outside the realm of meaningful experimental or observational verification is not actually a "real" science but rather a form of "cargo cult science." This is certainly true for some aspects of cosmology (for example, the "ekpyrotic universe" according to Steinhardt and Turok, which is based on the yet-unverified string theory in the form of its generalization to "branes"). Therefore, such theories should not necessarily be taken too seriously but rather regarded as hypotheses in the truest sense of the word. Popular scientific presentations of the subject, in particular, often convey a certainty that does not actually exist. Cosmology is a complex field in which even experts can struggle to distinguish hard facts from pure speculation. As Lev Davidovich Landau (1908–1968) said about cosmologists in 1962: "Cosmologists are often mistaken, but never in doubt," which probably hits the nail on the head.

But back to the term "cargo cult science," which also applies to some mainstream research fields - though only to some. Real hardcore cargo cult science can hardly be distinguished from nonsense, if at all, and is therefore not pursued by serious scientists. As the attentive reader will have noticed, I have already mentioned at least three examples: hell topography (point 179), astrology (point 58), and homeopathy (point 249). All of these "specialist fields" claim to operate scientifically.

448. Cargo cult

And yet, on closer inspection, what they produce is just nonsense. So, what is the "cargo cult" that Richard Feynman explicitly referred to in his graduation speech at Caltech (California Institute of Technology) in 1974?

Anyone who has seen the impressive film "Memories of the Future," based on Erich von Däniken's book of the same name (this is cargo cult science made into a movie), may recall the opening scene. South Sea islanders are shown sitting next to dummy planes made of wood and straw, eagerly awaiting the military pilots who brought them one or two achievements of Western civilization during World War II. Back then, airplanes regularly landed and took off from makeshift airfields, and they saw ships dock and disembark, bringing items like chocolate, useful tools, radios, and other goods they had never seen before. But then the war ended, and the military left. The planes stopped landing, and no more ships docked. The supplies ceased, and the islanders missed all these amazing and useful things. So they built themselves headphones and microphones out of wood, sat in bamboo huts, and pretended to make contact with the planes by radio. They stood on deserted airstrips and gesticulated with wooden paddles to instruct the planes that weren't coming. They did everything exactly as they had seen the soldiers do - with anticipation, but without success. Because the planes didn't land and the ships didn't appear either. This type of "cult" has since been called the "cargo cult" by ethnologists.

Erich von Däniken wanted to show how primitive peoples react to the arrival of "extraterrestrials", which he claimed were responsible for many "god cults" around the globe. Richard Feynman used this term to describe pseudo-scientific concepts that try to appear serious and scientific to the outside world, but on closer inspection turn out to be a sham. Their proponents may wear white lab coats, but what they do is the same as the South Sea islanders: they wait for airplanes that never arrive...

Fig. 202: This is no longer possible today - South Sea islanders are waiting for the achievements of Western civilization as an example of the cargo cult (Creative Commons)

Not only in the natural sciences is it important to avoid deceiving yourself and others by fudging your results or adapting them to the zeitgeist; by forgetting to point out things that speak against the published research results, or by stylizing hypotheses into theories that they are not. You always have to be prepared to discard your research results if they ultimately turn out to be wrong, no matter how good they may seem.

Nevertheless, the current scientific community is not immune to cargo cult practices, as several recent scientific scandals have made clear. Whether some topics in gender studies should also be considered as such is an open question. If Richard Feynman were still alive, he would likely have exposed some of the views in current "climate science" as "cargo cult science," because it is precisely in this area that the close intertwining of science and politics is evident, mediated by state research funding. This can be seen, for example, in the fact that studies that do not align with expectations of "man-made climate change" are quickly dismissed or ignored by other scientists (or politicians) instead of being openly falsified using scientific methods or, if this is not possible, being recognized as potentially correct.

449. The Piltdown Man

One of the most famous scientific forgeries in the history of science was the "Piltdown Man." The skull fragments, found in 1912, were long regarded as the "missing link" between apes and humans until they were exposed as a relatively well-made fake in 1953. Prior to this, in 1856, a primitive human skull was discovered in the Neander Valley near Düsseldorf, and it was assumed that the cradle of mankind was in Europe. This theory, which fit the spirit of the times, became obsolete when an obviously much older prehistoric human skeleton was found in Southeast Asia.

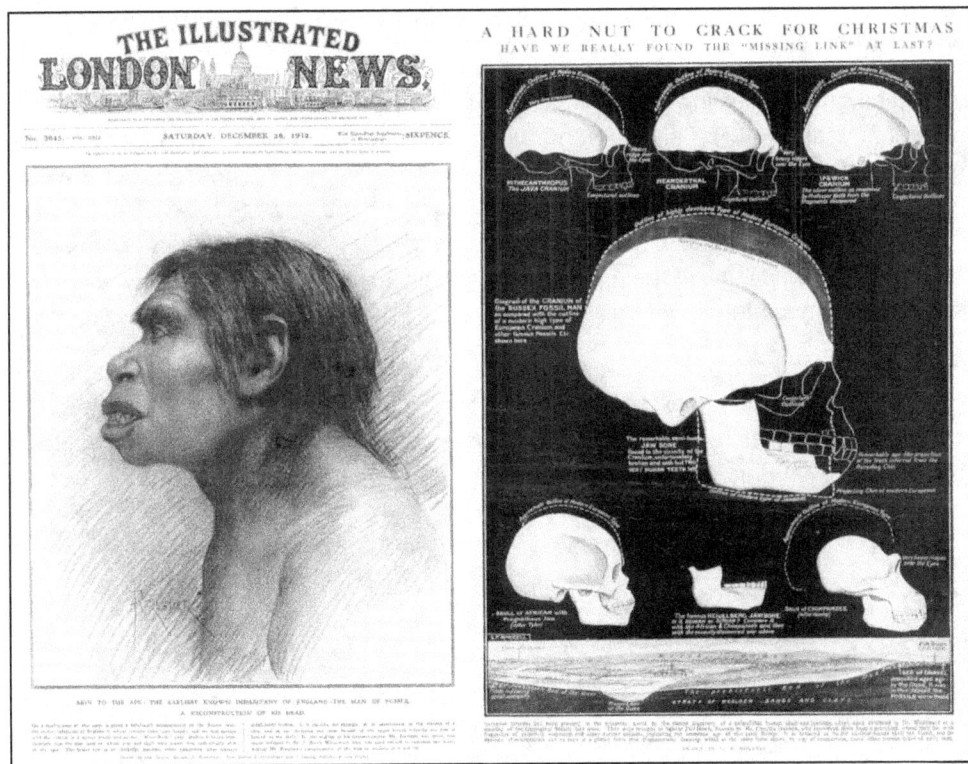

Fig. 203: Dawson and Woodward's findings are published in the Illustrated London News on December 28, 1912

At the end of the 19th century, educated circles began to accept the theory put forward by Charles Darwin (1809-1882) that humans and apes had a common ancestor (often misinterpreted as "humans descended from apes"). For this reason, every discovery of prehistoric human bones was seen as a kind of sensation, and the finder could be sure of a high scientific and social reputation. One such find was made in 1912 in a gravel pit near the village of Piltdown in the county of Sussex, England, by the amateur archaeologist Charles Dawson (1864-1916). This find consisted of a fragment of skull, a lower jawbone, and a number of teeth. A reconstruction revealed an ancient-looking creature that resembled an ape and was given the scientific name Eoanthropus dawsoni. This fossil was examined in detail and deemed genuine, and the English were proud that the "Man of the Dawn" was obviously British. In the anthropological literature of the first half of the 20th century, this "species" played an important role in the reconstruction of the then still quite clear tribal development of man. Although some scientists expressed doubts from the outset, particularly in comparison with Neanderthal skulls, these were largely ignored by English researchers.

In 1953, during a re-examination of the bones, archaeologist Kenneth Oakley and his colleagues realized that the finds came from different locations and had been adapted to the environment of the gravel pit at Piltdown. The skull was a human skull from the Middle Ages, the lower jaw was a 500-year-old lower jawbone of an orangutan, and the teeth once belonged to a chimpanzee. The bones and teeth had been artificially prepared to look like ancient bones. In addition, changes were made so that telltale features were not even visible.

In general, the unmasking of the Piltdown Man was greeted with relief by the research community, as it put an end to some unpleasant controversies. The only remaining question was who had committed the forgery. There are two theories: Either it was a farce that got out of hand and the author no longer had the courage to resolve it, or perhaps it was the discoverer, Charles Dawson, himself, who had already attracted attention for other frauds. As all the protagonists are now dead, the question of who was responsible will probably never be answered.

However, science can also degenerate in a terrible way when it panders to certain ideologies. A particularly bad example of this is the "racial science" of the "Third Reich," a pseudo-science with blood on its hands. Classifying it as "cargo cult science" would be more than trivializing. It shows how "science" can be misused politically to provide a "justification" for an inhumane ideology.

450. Lysenkoism

In the Soviet Union during the Stalin era, pseudo-sciences were also cultivated as long as they served the ruling ideology - in this case, Stalinist communism. An outstanding representative of this was the agronomist and biologist Trofim Denisovich Lysenko (1898-1976), whose teachings are known today as "Lysenkoism." He was undoubtedly one of the most highly decorated scientists in the Soviet Union: a three-time Stalin Prize winner, decorated seven times with the Order of Lenin, hero of a) the "Soviet Union" and b) "Socialist Labor," multiple deputy of the Supreme Soviet, and - one could almost say "naturally" - a member of the Presidium of the USSR Academy of Sciences. And yet, a significant part of his work and findings were "cargo cult" in the best sense of the word.

His area of expertise was agriculture, and his theoretical foundation was a modified form of Lamarckism - a theory of evolution that had virtually no followers outside the Soviet Union at the time - and his reputation essentially grew out of his relationship with Josef Stalin, whose repressive regime he exploited to silence not only competitors in the scientific community.

Fig. 204 Trofim Denissowitsch Lyssenko (1898-1976)

Lysenko believed (similar to Lamarck) that acquired characteristics are hereditary. This aligned very well with the heresy of "scientific communism," which assumed, among other things, that a new breed of human being, the "socialist personality" - or, more precisely, the "Soviet man" - could be created through education and indoctrination, with characteristics propagated at the time passed on from generation to generation. In this respect, "classical genetics" with genes was, of course, profoundly "unsocialist" and "bourgeois," which quickly made its adherents in the Soviet Union of the Stalin era suspicious in communist circles. And once caught in the clutches of Lavrenti Beria (1899-1953) and his henchmen, this quickly meant the loss of career opportunities, including exile to Siberia or even death. Lysenko was one of those who used this incomparable apparatus of repression to oust opponents who outmatched him in terms of character or intellect, and to push himself into their positions. It is therefore easy to understand why, after Stalin's death in 1953, Lysenko's star began to fade as investigations into his crimes commenced. Under Nikita Khrushchev (1894-1971), he was eventually dismissed as president of Lenin's Academy of Agriculture, and his teachings, now recognized as bizarre by both the state and party leadership, were gradually banned from

449

school and university curricula. Lysenkoism, which had once plagued Soviet agriculture like mildew, had had its day.

Lysenko was a very hard-working scientist with innovative ideas who approached the breeding of new plants pragmatically, but ultimately based his work on incorrect assumptions. He was particularly interested in developing new grain varieties. He carried out experiments to produce "summer wheat" from "winter wheat" (which, as we know, must overwinter once), whose life cycle would be limited to a single summer and could be expected to produce higher yields - particularly in climatic zones where its cultivation is typically risky. Believing that acquired characteristics were inherited, he moistened winter barley grains, allowed them to germinate, and then stored them in cold storage. He later sowed them as spring barley and claimed that he had bred "spring barley" from "winter barley," whose yield exceeded that of normal spring barley by more than 30%. He believed that in this way, he had transformed "winter barley" into "spring barley." He called this breeding method "Jarowization." However, this phenomenon is actually explained by a process known to botanists as "vernalization" and has nothing to do with the inheritance of acquired characteristics.

According to Soviet figures, the area planted with Jarovized grain in 1938 was around 18 million hectares, and the reported yield increases were quite remarkable. These initial successes established his reputation, and a personality cult began to develop around him. He was hailed as a "genius" in the press, which in turn inspired him to develop a universally valid theory to explain this new breeding method. At this point, he departed from the path of serious science. It proved to be a great advantage for him that his simplistic ideas appealed to Stalin. His successes were highly exaggerated, while his numerous failures were covered up, and his critics were silenced. In his theoretical writings, he dismissed chance in evolutionary processes, justifying his stance with quotes from Lenin and Stalin.

According to Lysenko, hereditary factors represented something like a "concentrate of environmental influences" to which an organism had been exposed over generations. Based on this entirely false premise, he eventually developed a kind of "new biology," which was rooted in "dialectical materialism" but had little to do with biology itself. His ideas even found their way into Soviet school textbooks. These books claimed that Lysenko and his colleagues had succeeded in, for example, turning pine trees into spruce and wheat into rye, and that apple trees would soon flourish in the cold steppes of Siberia, yielding bountiful harvests. Lysenkoism is perhaps the prime example of how, under the influence of political ideology, science can be virtually replaced by dogma. You might even notice parallels to modern times.

451. Mitschurin hat festgestellt...

However, Lysenko's teachings were not entirely of his own making. Many of his ideas can be traced back to the great fruit grower Ivan Vladimirovich Menshurin (1855-1935), whose practical successes (he developed fruit tree grafting to perfection, among other things) are still undisputed today. However, his theoretical ideas, which were strongly based on Jean-Baptiste de Lamarck (1744-1829) and adopted by Lysenko, were simply wrong. Unfortunately, only a few mocking verses about him have survived, which are still quoted today from time to time: *„Mitschurin hat festgestellt, dass Marmelade Fett enthält, drum essen wir auf jeder Reise Marmelade eimerweise..."* (Mitschurin has discovered that jam contains fat, so we eat buckets of jam on every journey...). Or, when there was little or no butter to buy in the early years of the GDR, the rhyme circulated among the population: *„Mitschurin hat festgestellt, dass die Butter Gift enthält. Um das Volk halt zu gesunden, ist die Butter nun verschwunden..."* (Mitschurin has discovered that butter contains poison. To keep the people healthy, the butter has now disappeared...)

452. Epigenetic processes

Interestingly, around 50 years after Lysenko's death, there are indications that certain acquired characteristics (e.g., some disease patterns) are passed on to offspring under specific circumstances. The functioning of these so-called "epigenetic processes" is definitely linked to genes and their inheritance mechanisms, but not to the actual genetic information stored in the DNA molecule. The decisive factor here is special epigenetic factors, which are able to regulate the activity of genes by including or excluding them, or groups of them, in relation to gene expression in the cell. These factors are not located in the DNA itself but in certain proteins in the chromosomes, more precisely in the histones. These histones represent the "coils" around which a DNA molecule winds in its condensed state in eukaryotes. If these "coils" are chemically altered, this has a corresponding effect on gene expression. For example, if a living organism experiences stress, such as from changing environmental conditions, lack of food, or exposure to toxins, certain histones can be permanently chemically marked, influencing the behavior of body cells by permanently switching certain genes on or off. It appears - and experiments have now shown - that such epigenetic markers can be passed on to offspring. There are even well-founded hypotheses that certain diseases, such as diabetes or obesity, are, to some extent, epigenetically determined. Research into the relevant mechanisms is therefore of great importance in terms of public health policy. For hereditary and evolutionary biology, epigenetics provides a new framework for explaining hereditary biological abnormalities.

Let's take the honeybee as an example from the animal world. It is well known that this insect looks the same in the early larval stage. Those larvae that are fed a mixture of honey and pollen by nurse bees develop into sterile worker bees. The larvae, on the other hand, that are fed with royal jelly undergo a transformation and develop into egg-laying queen bees. The genes of the worker bees and those of the queen bee are completely identical. It appears that the diet - in this case, the honey-pollen mixture - leads to the explicit deactivation of certain developmental genes.

The chemical mechanism that acts specifically on the histones is called methylation. This refers to the "attachment" or "removal" of methyl groups from certain histone proteins, with the result that gene expression can be controlled. Remarkably, special pharmaceuticals can be used to specifically influence this process, offering new therapeutic approaches for certain diseases. Initial successes have been achieved, for instance, in detecting cancer cells where the genes that normally prevent pathological cell growth are switched off. Furthermore, as has only been known for a few years, epigenetic dysregulation plays a crucial role in the development of certain immunological and neurological diseases as well as various growth disorders. These include rare conditions such as Silver-Russell syndrome (a specific form of dwarfism), which is categorized under the term *"genomic imprinting."*

453. Great little queen - Matilda of Flanders

It is interesting to note that there are virtually no short-statured people among the crowned heads of world history. The reason for this may lie in their rarity, as height was not a decisive parameter in dynasties that were preserved through succession. Thus, it may come as a surprise that the wife of King William I (around 1027-1087) and later Queen of England, Matilda of Flanders (around 1030-1083), was only 1.27 meters tall, while her husband is said to have had an exceptionally handsome figure.

Her exact height is known because her tomb in the abbey of Sainte Trinité in Caen was opened in 1961, and her bones were measured. However, during her lifetime, her short stature was never mentioned in written records. There is simply no reason why a "great queen" could not be short...

Matilda of Flanders was a remarkable woman. She was born around 1030, likely in what was then Flanders. Her marriage to William the Conqueror, then Duke of Normandy, took place around 1050. This union was not only a dynastic alliance but also a political one that helped consolidate William's rule. Matilda was a loyal and energetic partner to him.

On May 11, 1068, Matilda was crowned Queen of England, presumably in Winchester, England's first capital. Her coronation took place two years after the famous Battle of Hastings, in which William defeated the English troops and thus seized the throne.

Matilda was not only queen but also the mother of ten children, including two future kings, William II and Henry I. Despite her many duties as a mother and ruler, she was known for her charity and support of the church. She promoted the building of churches and monasteries and was a major patron of Sainte Trinité Abbey in Caen, where she was later buried.

Historians know surprisingly little about her personal life and daily activities. The surviving sources tend to focus on her role and importance as a queen and mother. There is evidence that she was educated and intelligent, qualities that assisted her in the politics and administration of Normandy and England.

Matilda's life and work show that "height," in the literal sense, says nothing about a person's importance and influence. Despite her small stature, she was a great queen whose influence and legacy continue to this day. Her short stature, which only became known centuries after her death, does not detract from her impressive legacy and historical significance.

454. The Bayeux Tapestry

This does not apply to her husband, who considers the famous "Bayeux Tapestry" a monument that has survived to this day. It depicts scenes of the conquest of Great Britain by the Norman Duke and later King of England William I, known as "the Conqueror," on a strip of fabric approximately half a meter wide and around 68 meters long, featuring elaborate, colorful embroidery.

The iconic pictorial history begins with the meeting between Harald Godwinson, Earl of Wessex, and the English King Edward and ends with the famous Battle of Hastings on October 14, 1066. This battle resulted in the victory of the Normans over the Anglo-Saxons and the death of their king Harald II (1022-1066). It then took just five years for the Norman duke, who was crowned King of England in the same year, to bring the whole country under his rule.

Fig. 205: Excerpt from the "Bayeux Tapestry" (Creative Commons)

The "Bayeux Tapestry," which is sometimes also called the "Tapestry of Queen Mathilda" (although it was probably commissioned by Edith of Wessex (1029-1075)), is of historical and cultural significance in many respects. This textile is even worth mentioning for its contributions to astronomical history...

455. Halley's comet

Approximately in the middle of the tapestry, at scene 33, you can see a "comet with a tail," which some figures are pointing at with their fingers. Today, we know that this is the oldest depiction of Halley's Comet, which made a perihelion passage in 1066 and came as close as 15 million kilometers to Earth (which, by the way, is exceptionally "close" for Halley's Comet). In any case, it must have been an extremely conspicuous celestial phenomenon at the time, much like how comet Hale-Bopp was for our generation in the spring of 1997. For the English, who confronted the Normans at Hastings, it was undoubtedly seen as an unlucky omen. William I, on the other hand, probably viewed it as a lucky sign or a "heavenly message of victory."

Comets have always been among the most spectacular celestial phenomena and have always been observed with great interest. In the Middle Ages and early modern times, however, "guest stars," as they were also known, were mainly regarded as harbingers of bad luck and were often blamed for many misfortunes in the world.

Fig. 206: Depiction of Halley's Comet on the Bayeux Tapestry. On the right, King Harald II (Creative Commons)

456. Comet pamphlets

The so-called comet pamphlets from that time, which still lie dormant in some archives and libraries, are well-known. There are reports of *"terrifying wrath rods of God"* and *"hair stars bringing pestilence"*. It seems that in those days, all the evils of the world that could befall a person were thought to be caused by "hair stars", which translates as "comet". In the town archives of the sixth town of Löbau in Upper Lusatia, for example, there is an entry for the year 1472:

"On the day of Agnetis a terrible comet appeared, which was visible in the sky for 6 weeks, whereupon war, famine and pestilence ensued in many places."

As this comet (its official name is C/1472 Y1) was particularly bright and therefore easy to see with the naked eye, there are many entries about it in old annals. It was observed very closely by the astronomer Johannes Müller, known as Regiomontanus (1436-1476). He recorded the results of his observations in a paper that was only published posthumously in 1532 by a certain Johannes Schöner (1477-1547) in Nuremberg. Interested readers can find it today in digitized form on the Internet under the title *"Problemata XVI de cometae (1472) magnitudine longitudineque ac de loco ejus vero"*. It is considered to be the first scientific description of a comet.

Fig. 207: Typical comet pamphlet from the 17th century (Creative Commons)

But as I said, comets were still considered to bring bad luck well into modern times, as the following impressive words on a comet leaflet from the time of the Thirty Years' War show:

"All comets bear witness to us - A great deal of misfortune, gloom and danger, - And never has a comet's appearance - Be without evil meaning. - Achterley misfortune in general arises, - When a comet burns in the air:

1) *Much fever, sickness, pestilence and death,*

2) *Hard times, scarcity and great famine,*

3) *Great heat, drought and barrenness,*

4) *War, robbery, fire, murder, riots, envy, hatred and strife,*

5) *frost, cold, storm wind, bad weather, water shortage,*

6) *The ruin and death of many high people,*

7) *Fire and earthquakes at many an end,*

8) *Great change of regiment,*

But if we repent from the heart, - God will also turn away all misfortune and pain."

Here we can see that at that time the old Aristotelian interpretation of comets as a phenomenon of meteorology and not of astronomy was still used to explain them: Comets are vapors from the earth that have ignited in the uppermost layers of air.

In 1577, the famous astronomer Tycho Brahe (1546-1601) was able to track a very bright comet for more than two months at his observatory in Uranienburg using special sounding instruments (the telescope had not yet been invented at the time). The aforementioned Löbau town chronicle also reports on this "Great Comet of 1577":

„Den 12. Novembr. biß aufn 6. Jan. 1578 hat man abermahl einen großen erschröcklichen Cometstern am Himmel gesehen, im 16 grad des Steinbocks, nicht weit von Saturno, welcher nach Untergange der Sonnen erschienen, und seinen Schwantz, so sich in die länge auf 39 gradus soll erstrecket haben, nicht schnurschlecht von der Sonnen, sondern seitwarts derselben gekehret, daß sich ihrer viel darüber verwundernde entsetzet, maßen dergleichen und von solcher Länge wenig Cometen gesehen worden, ...".

(Den 12. Novembr. bis aufn 6. Jan. 1578, a large, faint comet star was seen in the sky, at 16 degrees of Capricorn, not far from Saturn, which appeared after the setting of the sun, and its tail, which is said to have extended to 39 degrees in length, did not turn away from the sun, but to the side of it, so that many people were astonished at it, as few comets of this kind and of such length have been seen...).

Of course, Tycho Brahe was not frightened by such a comet, but soberly noted that the horizontal parallax of this celestial body was less than 15 minutes of arc. From this he concluded that comets orbit outside the Earth - and apparently even far behind the Moon. However, he was still unable to find a qualitative explanation for the phenomenon. It was not until more than 100 years later - in the meantime Johannes Kepler (1571-1630) had discovered his planetary laws and Isaac Newton (1643-1727) had used his law of gravitation to explain the movement of celestial bodies - that Edmund Halley

(1656-1742) was able to take comet research a big step forward again by discovering that comets are periodic, i.e. recurring phenomena. His research object, the "Great Comet of 1682", is now called "Halley's Comet" in his honor. However, he was unaware that there was even an illustration of it in the form of an embroidery on an old English tapestry. This comet was of course also followed with anxious hearts in Löbau in Upper Lusatia, and this time the annalist reported on it in particular detail:

„D. 23 Jul (1683. d. A.): Zu Abend um halbweg 10 Uhr, sahe man zum erstenmahl einen neuen Cometen, er war aber dermahln sehr kleine, mit einem kurzen Schwanze, durch den Tubum kondte er fein deutlich undt helle betrachtet werden, mit bloßen gesichte mochte er etwan einem Sterne Vierter größe gleichen. In folgenden Tagen spürte man, daß er rückgängig war undt zwar sehr schwaches Lauffes, sintemahl er nicht viel über einen halben grad desTages verrichtete. Er blieb lange Zeit klein und es war nicht zuerkennen, ob er ab- oder zunahm. Endlich sah man doch, daß er mercklich zunahm, sonderlich war er am 28. Aug: als er Beim SiebenGestirn stund, als der halbe Mond groß iedoch blaß undt ohne eines frischen Kern. So lange, als wir ihn sahen, nahm sein Lauff stets zu, biß er zulezt auff 5 grad kam, Erstlich war er (...) zwischen dem Kopffe des großen Behren undt dem Fuhrmanne, hernach ging er durch den Fuhrmanne undt kam nahe unter der Ziege hin, ferner zum Sieben gestirn undt ward im Kopffe des Wallfisches zulezt gesehen. - Nachdenklich ists, daß eben dieser Comet damahls erschien, als die Türcken Wien belagert hatten, Man sollte wohl meinen, er hatte den Türcken Krieg bedeutet undt sonderlich der Stadt Wien solch unglück; Aber es kann nicht seyn, Das unglück war schon da, durffte nicht erst bedeutet werden. Vielmehr wollte ich sagen: Dieser Comet verkündiget das Türken groß unglück. Denn ob es wohl erstlich, da er noch klein war, gegen Norden stund, so hatten wir ihn doch hernach, da er am grösten erschien, undt am stärcksten lieff, also auch vermuthlich der Erde am nechsten war, gegen den Türckenkrieg zu zusehns, als eben diese Gott undt Christus verhaßete Menschen, oder vielmehr Teuffel ihr 30 tägige Fasten Ramadan hielten. Gott trieb sie bald darauf durch die Waffen der Christl. Potentaten, am 20 Tage ihres Fasten von Wien, die sie so lange geängstiget hatten undt verliehe den Christen eine überaus herrlichen Sieg ...".

(D. 23 Jul (1683. d. A.): In the evening at half past 10 o'clock, a new comet was seen for the first time, but it was very small at that time, with a short tail, through the tube it could be seen clearly and brightly, with the naked eye it might have resembled a fourth magnitude star. In the following days it was felt that it was receding, although it was very faint, as it did not move much more than half a degree a day. He remained small for a long time and it was impossible to tell whether he was losing or gaining weight. At last it was seen to increase considerably, especially on August 28, when it stood by the seven stars, half the size of the moon, but pale and without a fresh core. As long as we saw him, his course increased until he finally came to 5 degrees, first he was (...) between the head of the great bear and the carter, then he passed through the carter and came close under the goat, further to the seven-pointed star and was seen at the head of the whale. - It is thought-provoking that this very comet appeared at the time when the Turks were besieging Vienna. One might well think that it had meant war for the Turks and especially such misfortune for the city of Vienna; but it cannot be so, the misfortune was already there and could not be meant first. Rather, I wanted to say: This comet announces the

Turks' great misfortune. For although at first, when it was still small, it was towards the north, we had to watch it afterwards, when it appeared the largest and ran the strongest, thus presumably also the closest to the earth, against the Turkish war, when these very people who hated God and Christ, or rather devils, kept their 30-day fast of Ramadan. God soon drove them from Vienna, which had frightened them for so long, through the weapons of the Christian potentates on the 20th day of their fast and gave the Christians an exceedingly glorious victory ...).

It is interesting to note, among other things, that a telescope ("Tubum") was obviously already being used for the observations described. A few years earlier, Georg Samuel Dörffel (1643-1727) from Plauen had been able to prove from his observations of the "Great Comet of 1680" that this comet moved on a parabolic orbit with the "sun" as its focal point. He was thus able to show that the two comets of 1680 and 1681 were one and the same, once before and once after the perihelion passage. His work "Dissertatio de Cometa" made this otherwise forgotten astronomy enthusiast (he was a clergyman by profession, most recently superintendent in Weida, Thuringia) world-famous among comet researchers.

With the telescope as an aid, the search for new comets subsequently developed into a favorite pastime of many astronomers and astronomy enthusiasts, which, if successful, led to high esteem and recognition not only in specialist circles. The Frenchman Charles Messier (1730-1817), for example, discovered 19 comets in the course of his life,[80] but it was not these comet discoveries that made his name famous. It was actually a by-product of his observational activities that made him really famous, his widely known catalog of nebulae (1784), which is still familiar to every astronomer and amateur astronomer today.

Another example from the area around the town of Löbau in Upper Lusatia should also be mentioned: Ernst Wilhelm Leberecht Tempel (1821-1889) from Niedercunnersdorf, where his birthplace can still be visited today. He is considered the discoverer of 21 comets. These include such prominent ones as 55/Temple-Tuttle, the original comet of the Leonid meteor stream, and 9P/Tempel-1, which was visited by NASA's "Deep Impact" probe in 2005 and even bombarded with a large copper projectile.

Astronomers with more theoretical knowledge developed new and fast calculation methods (at the time, the logarithm table and the head was the little man's computer), especially for parabolic orbits. The Bremen physician and amateur astronomer Wilhelm Olbers (1758-1840, he discovered six comets himself) should be mentioned here in particular, whose work "Abhandlung über die leichteste und bequemste Methode, die Bahn eines Kometen aus einigen Beobachtungen zu berechnen" (Treatise on the easiest and most convenient method of calculating the orbit of a comet from a few observations) considerably simplified the determination of comet orbits.

---

[80] As far as "amateur astronomers" are concerned, the Australian William Ashley Bradfield (1927-2014) has almost caught up with him, as he discovered a total of 18 new comets

457. Comet research

While the orbital properties of comets were increasingly understood, significant progress on their physical nature was made only in the 20th century. All previous attempts at interpretation were rather vague. For example, the discovery that many short-period comets have their aphelion near the orbit of Jupiter led to the hypothesis that comets are celestial bodies ejected from Jupiter from time to time, though this hypothesis did not explain exactly how this ejection process works.

Fig. 208: Periodic comet 13P Olbers. The background stars appear as lines because the comet's nucleus was tracked to eliminate its motion (author)

The highlights of comet research in the 19th century include Friedrich Wilhelm Bessel's (1784-1846) explanation of comet tails as a result of dust escaping under the influence of a repulsive force emanating from the Sun and Giovanni Schiaparelli's (1835-1910) proof that comets are connected to meteor streams. Bessel's dust theory lasted for a very long time, as it came quite close to the truth in some essential aspects. In 1900, Svante Arrhenius (1859-1927) demonstrated that Bessel's repulsive force could be explained by the radiation pressure of sunlight. This idea was later developed further by Karl Schwarzschild (1873-1916) and Arthur S. Eddington (1882-1944), who in 1910 (incidentally, the year Halley's Comet was once again visible) proposed a theory explaining how dust particles can be expelled from a comet when it is close to the Sun. In 1868, Giovanni Donati (1826-1873) and William Huggins (1824-1910) observed a comet through a spectroscope for the first time and found that its light was significantly different from ordinary reflected sunlight. The molecular bands predominating in comet spectra could only be explained more than half a century later with the help of quantum theory. The famous "dirty snowball" model of a comet nucleus (Fred Whipple, 1906-

2004) and the theory of the "comet cloud," which surrounds the solar system as a spherical shell from the outside (Jan Oort, 1900-1992), date back to 1950. Another major discovery that should not go unmentioned is Ludwig Biermann's (1907-1986) explanation of comets' plasma tails as the result of the interaction of comet gases with a particle radiation permanently emanating from the Sun, now referred to as the "solar wind."

In 1986, an event occurred that truly inspired comet research - the return of Halley's Comet, once observed by Matilda of Flanders, King Harald II of England, William the Conqueror, and Edith of Wessex, to name just a few. Even though its visibility was limited this time, it was still a media event and the target of a whole armada of interplanetary space probes, among which the ESA comet probe "Giotto" is particularly noteworthy. Almost 30 years later, the comet lander "Philae" is standing on the surface of comet 67P/Churyumov-Gerasimenko, albeit somewhat tilted, waiting for sunlight so that it can once again "phone home" - though it has yet to succeed.

It is remarkable how much mankind has advanced, at least technically, in the almost 1000 years since the Battle of Hastings.

458. Problem Progress

The only thing that hasn't changed is the noise of battle. Even "progress" is not an absolute quantity and, strictly speaking, depends on the context in which the term is used. It is most often used in the sense of "technical progress," where it is most appropriately applied. Here, it describes the transformation of an economy based on the products it provides and their continuous qualitative improvement, driven by targeted research and development. It usually implies social advancements "for the better," although this must also be viewed in the context of general living conditions and the social environment (what use is technological progress to a person who cannot participate in it?). On the other hand, it is not self-evident that a society will develop into a "technological society" on its own. In fact, this is rather unlikely because, as many examples in human history show, traditions and their maintenance play a major role in people's coexistence and thus hinder development in the direction of "progress."

Fig. 208: Reductive evolution

"Progress" in the technical sense is initiated during revolutionary phases, which are often followed by long periods of stagnation (e.g., the transition from the Stone Age to the Bronze Age or from the Bronze Age to the Iron Age) or by developments that parallel social progress, such as the Industrial Revolution. The Industrial Revolution, which began in England in the second half of the 18th century, is still ongoing, as evidenced by advancements like LED-based energy-saving lamps.

An interesting question in this context is whether increasing hostility towards technology and science, combined with anti-progress ideologies, can halt this development.

459. Belief in progress

The term "belief in progress" is closely linked to the idea of inevitably steady technological and economic advancement. Historically, this belief has often led to significant achievements and improvements in various areas of life. However, with the development of the atomic bomb, the release of persistent toxic substances in chemical accidents, and the large-scale environmental destruction caused by technology - one of the worst inventions in this respect being undoubtedly the hand chainsaw - the image of progress has begun to darken. These developments have made many people realize that progress is not inherently positive.

This critical view of progress is often suppressed because technological progress is closely tied to another central concept: economic growth. Politicians and decision-makers often view growth as a prerequisite for technological and social advancement. However, it is now widely recognized that infinite growth in a world with limited resources inevitably leads to a dead end. Yet, this insight is only slowly permeating political consciousness, as "no growth" is often perceived as a threat. In reality, it is the "dictates of growth" that force individuals and companies to consume and produce more and more.

Modern economic systems and laws literally compel this constant growth. However, upon closer inspection, it becomes clear that the only way to address the problems caused by "growth" - such as resource scarcity, debt crises, trade wars, and overpopulation - is to decouple progress from growth.

This decoupling requires a shift in many areas of society. First, technological progress must be directed more towards sustainability and resource conservation. Innovations should be evaluated not only for their economic benefits but also for their ecological and social impacts. Second, the concept of prosperity should be redefined. Prosperity should be understood not only in material terms but also in terms of quality of life, health, and environmental awareness.

Recently, there have been increasing efforts in this direction. Concepts such as the circular economy, the sharing economy, and sustainable business models are gaining importance. These approaches aim to use resources more efficiently and minimize the ecological footprint. The idea of "degrowth" - the deliberate renunciation of growth in favor of sustainability and quality of life - is also gaining popularity.

Blind faith in progress has led to considerable problems in the past, necessitating a fundamental rethink in politics, business, and society. This is the only way to address the challenges of the future and preserve a world worth living in for future generations.

460. Sustainability

The magic word that has haunted political and economic literature for several decades is *"sustainability."* It originates from a principle developed early on in forestry by Hans Carl von Carlowitz (1645-1714), which states that only as much wood should be taken from a forest as will regrow in the same amount of time, and applies this principle to general economic life. The variety of meanings that this concept has acquired culminates in the demand that economic activity should always be geared towards preserving resources for future generations. The capitalist mode of production, however, clearly contradicts this principle, which is why it can only be enforced politically through appropriate regulatory conditions.

Colloquially, this term has unfortunately become a euphemism and a battleground for ideologues of all stripes, used in various contexts to support different agendas. The natural vagueness inherent in the term has undoubtedly contributed to this conceptual confusion. For example, some view the production of biofuels from maize or palm oil as a sustainable method to meet "sustainable" and "climate-neutral" fuel needs, while others see it - rightly so - as an unprecedented environmental destruction through agricultural monocultures that undermine biodiversity, landscapes, and food security. Similarly, while some see the proliferation of wind turbines in Germany as a step toward a "sustainable," "decarbonized" (another new buzzword!) and "climate-friendly" energy economy, others - again, rightly so - view it as a sustainable destruction of cultural landscapes through visual intrusion and the killing of thousands of birds, which some green ideologues accept as collateral damage, while simultaneously risking a secure energy supply due to the volatile nature of wind power. Ironically, the same ideologues who advocate wind farms with fervor may oppose traditional songbird hunting in Mediterranean countries - justifiably so - while some view nuclear energy as an existential threat to humanity (rather than the nuclear arsenals of major powers), while others - certainly justifiably - consider it one of the few options for alleviating global energy poverty without consuming vast amounts of land (consider the "energy density of primary energy sources"). There are even those so ideologically committed that they oppose research into inherently safe nuclear reactors, while others see nuclear power as a bridge to a low-energy economy until the possible advent of energy generation through nuclear fusion.

Thus, when people talk about "sustainability," one should listen carefully and try to discern what is meant beyond the term itself. Currently, this term obscures rather than clarifies the complexity inherent in nature and society. It often masks a multitude of interpretations that muddle rather than illuminate the actual complexities.

Indeed, the concept of "sustainability" appears ill-suited to advancing human societies, especially when demographic developments - particularly rapid population growth - are ignored. The uncritical acceptance and popularity of the term may also reflect a deeper phenomenon: widespread mistrust in particularly affluent and saturated societies. This mistrust is directed both at humanity's past achievements and at the possibilities for future developments. It seems as if a supposed escape from the uncertainties of the modern world is being sought in "sustainability" - a return to supposedly simple and natural solutions that frequently fail to address complex realities.

Simultaneously, "sustainability" is increasingly being stylized as a moral imperative that is rarely questioned. This often leads to the exclusion or suspicion of alternative approaches and technologies that might be more effective or efficient. The debate thus becomes ideological, focusing more on enforcing certain worldviews than on actually solving problems.

Given these developments, the term "sustainability" should not be adopted uncritically but should be continually scrutinized and contextualized. This is the only way to prevent it from becoming an empty phrase that creates confusion rather than clarity.

461. Fears for the future

And not without reason. Financial crises are shaking the global economy, wars are flaring up everywhere, and they are leading to an unmanageable refugee problem. The issue of poverty remains unresolved, though efforts to address it are generally on the right track. People are alarmed by reports of new diseases and epidemics, and fears of apocalyptic scenarios such as the "climate catastrophe" are spreading everywhere. Almost all central debates of the present are characterized by anxieties about the future. To alleviate these fears, instruments are being taken out of the technological "moth box" and stylized as "future technologies." At the same time, real, groundbreaking technologies are demonized by certain circles and ultimately banned, even though they could potentially alleviate the misery.

The priority should be to advance science and technology and use them in ways that enable more people to lead happy and dignified lives. Perhaps it is time to think not only about "growth," but also about "standstill" - not in the sense of freezing existing conditions, but rather as a cautious adjustment to the unsatisfactory living conditions that still afflict hundreds of millions of people worldwide. The goal should be to reach a level where no one has to go hungry and where the wealthy are willing to make concessions to their standard of living to allow the less fortunate to lead a decent life.

If humanity wants to survive in the medium term, it must confront this task and not risk its material and intellectual resources for questionable "progress" or armed conflicts in which only a small percentage of people are involved - as is unfortunately the case today. Instead, we should focus on building a just and sustainable society that can meet future challenges and offer everyone a perspective.

Beyond ideologies and religions, the real meaning of our existence is the pursuit of happiness, as the Dalai Lama (Tenzin Gyatso) once said. But it is an ancient realization that true individual happiness cannot thrive or grow on the unhappiness of others.

462. Global conflicts and the need for a paradigm shift

Recently, two conflicts in particular have heightened global fears about the future:

The Ukraine conflict has reignited fears of a third world war. The direct confrontation between Russia and the West over the Ukraine theater of war (with NATO, which does not see itself as a party to the conflict but does little to bring it to a diplomatic end) has brought the world to the brink of a nuclear abyss, although few people are aware of this. This conflict has not only exacerbated geopolitical tensions but has also strained the global economy through energy crises and disruptions to supply chains. The fear of further escalation and the possible use of nuclear weapons has increased global insecurity.

At the same time, the ongoing conflict between Palestine and Israel, particularly following the events of October 2023, has highlighted the complexity and apparent insolubility of long-standing geopolitical tensions. This conflict exemplifies how historical trauma, religious differences, and territorial claims can lead to a seemingly endless cycle of violence. The global impact of this regionally limited conflict - ranging from diplomatic tensions to fears of terrorism - underscores the interconnectedness of our world and the need for global solutions.

These conflicts illustrate that fears about the future are not just abstract concepts but have a very real and immediate impact on the lives of millions of people. They also highlight the urgent need for a rethink in global politics - shifting from confrontation to dialogue and understanding.

Against this backdrop, as already mentioned, the term "standstill" is gaining importance. This is not about resignation but about a conscious realignment of priorities. Instead of an uncontrolled race for resources and power, we need a global consensus on peace, justice, and sustainability.

The solution does not lie in isolationism or nationalism but in increased international cooperation. We must recognize that global challenges - whether climate change, pandemics, or armed conflicts - can only be overcome together.

Education plays a key role in this. By promoting critical thinking, intercultural understanding, and ethical awareness, we can raise a generation capable of addressing the complex problems of our time.

Ultimately, it is about finding a balance between technological progress and human well-being, between economic growth and environmental sustainability, and between individual freedom and collective responsibility. Only in this way can we shape a future where fears are replaced by hope and where the pursuit of happiness does not come at the expense of others but contributes to improving the lives of all.

463. Bloch's concrete utopias

It is therefore time once again to examine "concrete utopias", to paraphrase Ernst Bloch (1885-1977). The challenge lies in the fact that the future is inherently uncertain, and predictions about specific developments can only be made - with limited certainty - over a manageable time horizon. Futurology, the study of potential future developments, frequently falls short when it extends beyond this manageable time frame.

Fig. 209: Ernst Bloch (1885-1977)

At the beginning of the 20th century, for example, a "horse dung problem" was predicted for the future. Around 1900, there were around 100,000 horses living in New York alone, which were indispensable as draft animals. They produced around 1,400 tons of horse manure every day, which had to be continuously removed from the streets. As plausible projections by urban planners showed, this quantity would continue to increase in the course of steady economic growth, so that eventually complete and timely removal and transportation would no longer be possible. A rough estimate showed the pessimistic result that a layer of horse manure about three meters high could be expected on all New

York streets by 1950. Thank goodness this horror scenario was averted by the invention of the automobile (the success of which was not predicted by futurologists at the time).

Even today, it is difficult to attest to the "scientific nature" of futurology when it attempts to make statements about future developments that go beyond a time horizon of perhaps ten years (keyword "climatology"). Just think of the triumph of the computer, which was not foreseeable at the time when Steve Jobs (1955-2011) presented the first "Apple". As we all know, unexpected political and technological developments can quickly derail a forecast for the future. Nevertheless, future models are not pointless per se. In fact, they are extremely important in relation to certain issues, especially if they are based on basic parameters such as population growth and key economic indicators. The future development of the earth's climate is also undoubtedly one of them, regardless of whether it contains a significant anthropogenic component (and is therefore at least theoretically controllable within certain limits) or not.

However, as soon as we begin to develop a concrete idea of a desirable future, we move into the realm of utopia. "Utopia" originally refers to a place that can be translated as 'nowhere'. This term is always used when one wants to create a positive contrast to the existing social conditions - an ideal world that is supposed to overcome the shortcomings and injustices of the present. The early utopians already attempted to do this.

The Marxists took up this idea and developed what they considered to be a "scientific" doctrine by extrapolating the historical development of humanity, especially in the West, towards a classless society. In this vision, the ultimate goal is a kind of "communist Elysium", a state of complete social harmony and equality. This gave the Marxist utopia an eschatological character that is otherwise only known from monotheistic religions. In this interpretation, utopia is not only seen as a theoretical possibility but is even stylized as the inevitable goal of socio-economic developments. This idea of an inevitable, positive future has shaped the thinking of entire generations and remains a powerful element of political and social movements to this day. However, it is often overlooked that such utopias, as inspiring as they may be, can in practice often lead to ideological dogmas that do not fully grasp the complex and often contradictory reality of human societies.

If we think of a utopia as something desirable, something worth striving for, something that can be achieved beyond mere castles in the air, then we arrive at the concept of a "concrete utopia" coined by Ernst Bloch. It is closely linked to the "principle of hope", which is also the title of Ernst Bloch's main philosophical work, in which he points out possible paths of social development in relation to the real, which have nothing to do with building "cloud cuckoo land", but rather show instructions for action in order to make the desirable a reality. For Ernst Bloch, the path is the essential thing and its end point remains nebulous and secondary but nevertheless visible to the extent that it remains a desirable goal.

464. Romantic utopianism

With the "romantic utopians", on the other hand, the desirable "final goal" is given as a "utopia" and the path to it remains nebulous.

Thomas More (1478-1535) was an English liberal politician. In his work "*Utopia*", written in 1516, he describes what he saw as a liberal ideal society, which has a thoroughly modern feel to it. In his "*Solar State*" (1602), Tommaso Campanella (1568-1639) designed a utopian state that corresponded to his vision of a papal universal monarchy with the abolition of private property (which he identified as the cause of all social ills). Henri de Saint-Simon (1760-1825) in turn, two hundred years later, coined the magic word "*l'Industrie*" - only the members of the community who produce useful goods and provide useful services (as opposed to the "parasitic elements", by which he primarily meant the nobility) are valuable to society.

Today, we need to think about how humanity can prevent self-inflicted local and global catastrophes by envisioning a society that exists in balance with nature, does not overburden the planet, does not exterminate itself, and yet remains committed to scientific and technological progress. We must also identify ways to achieve this balance, including: limiting population growth, implementing a resource-saving circular economy, halting the destruction of the planet's biodiversity, prioritizing politics in conflict situations, reducing the potential for accidental self-destruction, and overcoming poverty and ignorance through models of equal participation in minimum social standards and global thinking, among other strategies. This is primarily a matter for politicians. However, it remains to be seen whether we can trust politicians who are involved in inscrutable networks of interests and who are generally expected to think more short-term, from election period to election period. The diverse conflicts of interest between nations also make it more difficult to tackle global problems.

In any case, the 21st century will play a key role in the history of humankind in terms of the number and scale of problems to be solved. On the other hand, the prospects for addressing these problems may be better than generally assumed. Since humanity is the originator of these problems (and has recognized this self-critically), it is also within our power to solve them. The opportunity lies in our creative potential, combined with the new technical possibilities offered by modern means of telecommunication, for example. These technologies enable the instantaneous exchange of information, the acquisition of knowledge (if one chooses to pursue it), and the exertion of influence on political processes beyond national borders through the global networking of millions of people via social networks. Internet platforms such as Wikileaks have shown that it is no longer so easy for governments and secret services to operate in secret, and that they are therefore subject to increased public scrutiny.

We don't know what the world will look like in perhaps 50 years' time. But we can certainly help today to ensure that the world changes in a more positive direction in the future (unfortunately, as of 2024, it does not currently seem to be heading in that direction).

But back to "Utopia". This "nowhere" gave rise to the genre of "utopian literature", from which the genre of "science fiction" later developed.

465. Brave new world and "Big Brother is watching you"

Three outstanding works of fictional social criticism were particularly feared in totalitarian states (dictatorships), which is why they were usually not accessible there. The first is *"Brave New World"* by Aldous Huxley (1894-1963) from 1932, in which he develops a gloomy fiction of a totalitarian state that is largely non-violent ("world government"), but instead consolidates its power base through "human breeding" and ideological indoctrination. On the list of the 100 most important books of the 20th century published by the Parisian daily newspaper "Le Monde", this novel is ranked 21st, immediately followed by "1984" by George Orwell (1903-1950). It was published in the summer of 1949 by the London publishing house "Secker & Warburg" and was to be Orwell's last work, as he died of tuberculosis just one year later. Unfortunately, he did not live to see the great success he enjoyed with this gloomy depiction of a totalitarian society (*"Big Brother is watching you"*).

The gloomy vision of "Oceania", where "Big Brother" rules, a "Newspeak" (*"War is peace; Freedom is slavery; Ignorance is strength."*) clouds people's minds and manipulates their innermost thoughts and feelings, was particularly feared in the communist regimes, as there were obviously easily recognizable parallels here (for those who have not yet noticed, the "Newspeak" is celebrating its primeval state in Germany today). Who, if not Josef Stalin himself, should be hiding behind the "thick black moustache" that "Big Brother" wears? And the little man with the tangled white hair and goatee who calls himself Emmanuel Goldstein in the novel - weren't these the features of Leon Trotsky, Stalin's later archenemy?

The first German translation of 1984, which had previously attracted significant attention in the United States and the United Kingdom - including politically motivated attention - was published in 1950. It is therefore not surprising that the novel was banned in the GDR and could not even be found in the "poison cupboards" of university libraries. By 1978, there were even the first convictions related to this book: anyone caught reading it or lending it to others was guaranteed several years in prison. This circumstance "elevated" the novel to the extent that it became virtually compulsory reading for dissidents in the GDR. I myself read it as a student in the mid-1980s in the form of a poor-quality photocopy of a West German paperback edition, but without really being able to comprehend the "dangerousness" of the ideas it presented, which were supposed to justify several years in prison.

Fig. 210: Man hat manchmal wieder den Eindruck, dass man aufpassen muss...

After the fall of the Berlin Wall, the rapid development of modern information technologies and their applications led to a shift in the interpretation of Orwell's work as a "negative utopia." Today, "Orwell" or 1984 is often immediately associated with the concept of a total surveillance state, which has now become technically feasible. The NSA scandal and data retention (allegedly for noble purposes, of course!) are just two key examples in this context. However, we should not forget that Orwell's novel is not a prophecy but a warning.

The third book I would like to recall here is also by George Orwell: "*Animal Farm*" (1945). It was particularly hated by those in power in the Eastern Bloc countries because the power structures of the "dictatorship of the working class" were particularly clear to see in the parable told. The commandment "*All animals are equal - butsome are more equal than others*" has since become a catchphrase to express in a cynical way that the revolutionary principle of "equality" (Égalité) cannot be realized in any hierarchically organized form of human coexistence.

466. Thomas Hobbes' theory of the state

The principle of "equality" is a fundamental concept in state theory and was used by Thomas Hobbes (1588-1678) as the starting point for his considerations on the social contract. By postulating a natural state of man in which every individual is considered equal and free, but under which anarchy and lawlessness prevail and in which every person is a wolf to every other person, Hobbes derives man's natural right to enforce his egocentric nature by force against others, even in the face of resistance, which he views as an expression of individual freedom. According to Hobbes, to escape this state of nature, individuals must ultimately enter into a contract of submission to a sovereign whose absolute power enables them to live together peacefully for the sake of self-preservation, thereby relinquishing their right to self-determination. In his state-theoretical work Leviathan (1651), Hobbes describes in detail how a relationship of dependency arises in the form of a social contract, which provides the possibility of creating balance through codified rights and laws, ultimately evolving into the institution of the state.

Fig. 211: Cover of the first edition of Thomas Hobbes' "Leviathan" (Creative Commons)

The state's task is to counteract anarchic passions by restricting individuals' natural freedoms in an agreed-upon manner and by punishing violations of law and order: "I surrender my right to rule myself to this person or society on the condition that you also surrender your right over yourself to him or her." In this framework, the sovereign is the guarantor of rights and the monopolist of force. As long as the sovereign acts rationally and reasonably, he guarantees his subjects' rights to life and liberty (limited by natural law) and is able to "compel all citizens to peace and mutual aid against foreign enemies."

Since Hobbes, equality has been primarily, but not exclusively, understood as "equality before the law" and has thus - since 1775 (with the Constitution of the United States of America) - become a component of many constitutions.

467. Liberty, equality, fraternity and the planet Neptune

The word "equality" then reappeared in the famous slogan of the French Revolution of 1789: "Liberty, Equality, Fraternity," although it was not declared the official slogan of the Second Empire until 1871 by French Emperor Napoleon III (1808-1873). This slogan was originally coined by Maximilien de Robespierre (1758-1794) in a speech to the National Guard and was later adopted by the people. Incidentally, Robespierre was the political mastermind of the French Revolution who gave opponents of the revolution the choice of either changing their convictions or facing death (keyword: guillotine). He thus became one of the founders of the "Reign of Terror" (la Terreur, June 1793 to July 1794), to which he himself ultimately fell victim.

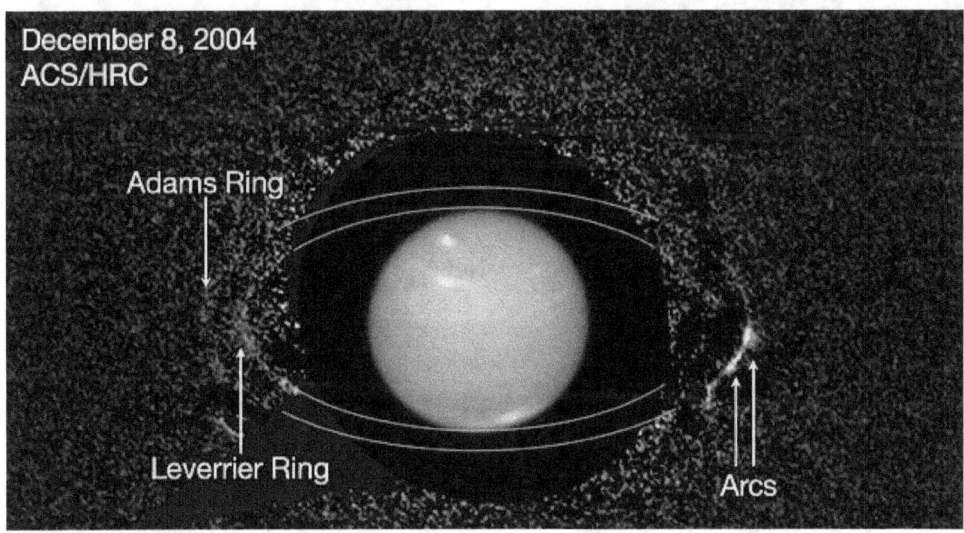

Abb: 212: Recording of the "arcs" in the Neptune ring by Voyager 2 (SETI Institute)

Modern "gender research" has also studied this slogan intensively, and it has been suggested that it would be more accurate to say "liberty, equality, solidarity" instead of "liberty, equality, fraternity." You can find out why for yourself. What is probably even less well known is that the terms "Liberté, Égalité, Fraternité" are also astronomically significant. When the Voyager 2 space probe flew past the planet Neptune in 1989, it captured a number of images of the ring system that had recently been discovered with terrestrial telescopes. A total of five regular rings could be detected, although they were not as impressive as those of the planet Saturn.

The outer ring, named "Adams," is located at a distance of approximately 62,900 km from the center of the planet. It was completely closed at the time of the flyby but shows three conspicuous and two less conspicuous thickenings, with a length ranging from 1° to 10° within a segment of about 40°. The three brightest of these arcs were named "Liberté," "Égalité," and "Fraternité," which is unusual for astronomical objects, following the slogan of the French Revolution. For a while, they could even be observed from Earth using giant telescopes. The origin of these "arcs," on the other hand, is difficult to understand. They obviously have something to do with Neptune's moon Galatea and other, as yet unknown, moons in the Neptune system.

Fig. 213: It is extremely difficult to make Neptune's rings visible with terrestrial telescopes - but this is no problem for the James Web Space Telescope (it works in infrared). The "arcs" once discovered by Voyager 2 are no longer clearly visible. (NASA, JWST)

Galatea has a diameter of 158 km and orbits the inner edge of the Adams ring around Neptune once every 0.429 days. According to some scientists, the gravitational effects and the motion of these moons lead to special resonance zones in the ring, where the lighter ring particles, which probably consist of ice, tend to accumulate. Ice has a higher albedo than rock, which makes these regions appear brighter than the rest of the ring. Observations with the 10-meter Keck telescopes in Hawaii in 2002 and 2003 revealed that the "arcs" do not appear to be stable objects. All the arcs except "Liberté" have more or less disintegrated since the visit of Voyager 2, and "Liberté" is also in the process of disintegrating. Its position relative to the "Fraternité" arc has also changed significantly. These observations are currently difficult to explain with existing theories. The moon Galatea likely plays an important role in this dissolution process. In any case, these ring structures are changing faster than previously expected.

In addition to the unusual outer ring, Neptune also has four other rings, named "Galle," "LeVerrier," "Lassell," and "Arago" from the inside out. These are the names of astronomers who were involved in some way in the discovery of this large planet at the outer edge of our planetary system. As for the rings, "Galle" and "Lassell" are relatively wide, while "LeVerrier" and "Arago" are quite narrow. They consist mainly of micrometer-sized dust particles and, at least in parts of the Adams ring, ice particles.

468. Galileo Galilei was the first person to see the planet Neptune

Incidentally, the first person to observe the planet Neptune was Galileo Galilei (1564-1642), the famous Italian naturalist who is the subject of Bertolt Brecht's work "The Life of Galileo." Galileo also discovered the craters of the Moon, the phases of Venus, sunspots, and the four "Galilean moons" of Jupiter using his self-made telescope. Of course, he did not know at the time that the small star he noted in his notebook next to Jupiter and its moons at the beginning of January 1613 was actually the planet Neptune, which would only be officially discovered over 200 years later.

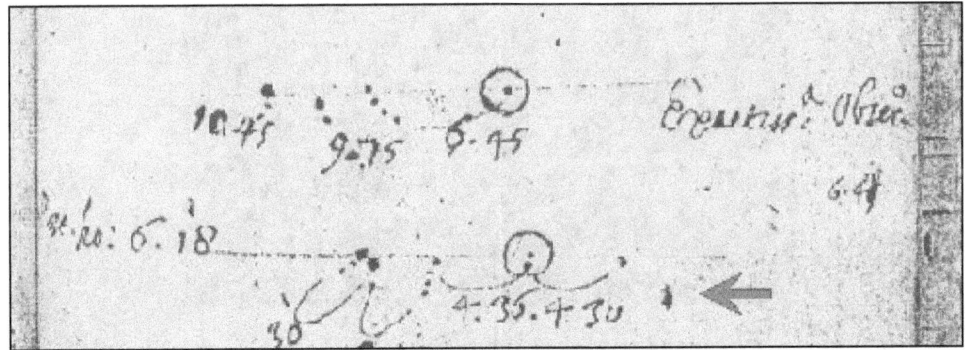

Fig. 214: In the early morning of January 6, 1613, Galileo Galilei sketched the position of the moons of Jupiter that he had discovered. What he didn't know was that one of them was not a moon of Jupiter at all, but the planet Neptune... (Creative Commons)

This early observation of Neptune is remarkable for several reasons. On the night of January 3 to January 4, 1613, one of the rare occultations of Neptune by Jupiter occurred. Both planets were in opposition to Earth, and their opposition paths intersected in such a way that Jupiter was able to pass in front of Neptune. This event lasted about 9 hours and 45 minutes. Galileo observed through his telescope during this period but did not notice this extraordinary conjunction. However, some of his sketches of Jupiter and its moons show a small star, which can now be identified as Neptune.

469. The real discovery of Neptune

The actual discovery of Neptune is still regarded today as a special milestone in the history of astronomy and a triumph of celestial mechanics. It is the planet that was predicted on paper, purely by calculation, and then promptly found - just two full moon diameters away from the calculated location. This fact alone is reason enough to describe the story of its discovery in more detail here.

Fig. 215: Urbain Jean Joseph Leverrier (1811-1877)

As previously mentioned in the chapter on Uranus, after its discovery by Wilhelm Herschel (1738-1822), several observations (in the form of position determinations) were found that predated the actual time of discovery. After the discovery, the movement of this planet was meticulously tracked with the utmost precision, so by the early 19th century, there were enough observations to determine its orbital elements accurately and to calculate ephemerides from them. However, as early as 1811, some astronomers noticed that the calculated positions deviated from the observed ones by up to 20 seconds of arc.

In 1821, Alexis Bouvard (1767-1843) undertook a further analysis of the observation data but could not resolve the existing discrepancies. Friedrich Wilhelm Bessel (1784-1846), the famous astronomer from Königsberg, initiated the discussion about whether these discrepancies were caused by another, unknown planet beyond the orbit of Uranus. Although he worked on the problem, he was unable to solve it. Later, around 1845, two then young and relatively unknown mathematicians, Urbain Jean Joseph Leverrier (1811-1877) in France and John Couch Adams (1819-1892) in England, began to tackle the "Uranus problem," both assuming from the outset that the deviations in Uranus' orbit were caused by an unknown planet. Their goal was to determine the position of this unknown planet as precisely as possible so that it could be located with a telescope.

At that time, there were no calculators or computers to solve such problems as we have today. Instead, astronomers relied on the now forgotten logarithmic tables, which the "celestial mechanics" of the time memorized and used to carry out complex calculations based on Newton's theory.

Adams' results were available in the fall of 1845, and Leverrier's in early 1846. Adams' predictions led to the observation of the new planet the following year, specifically on August 4 and 12, 1846. However, the observations were not analyzed due to the lack of valid star charts for the region in question, resulting in the missed opportunity. On September 23, 1846, Johann Gottfried Galle (1812-1910) in Berlin received a letter from Leverrier asking him to search for the planet Leverrier had predicted. That very evening, Galle and his assistant Heinrich Louis d'Arrest (1822-1875) began their observations and, after less than an hour, promptly identified a star not marked on the new Berlin Academic Star Charts. After a day of further verification, they confirmed the proper motion of this "star" with their nine-inch refractor, leading to the discovery of Neptune. Leverrier gained fame, while Adams was somewhat overshadowed. Today, we know that the Frenchman deserved greater recognition, as recent research in the archives shows.

Just one month later, in October 1846, the British scientist William Lassell (1799-1880) succeeded in discovering Neptune's first moon, which was named Triton. With data from Voyager 2, we now know what Triton's surface looks like. Today, a total of 14 moons of Neptune are known. Interestingly, when the records of the French astronomer Joseph Jérôme Le Français de Lalande (1732-1807) were later examined, it was found that he had noted the position of Neptune in 1795 - mistaking it for a star - twice in his observation diary. While Galileo had mistaken it for a moon of Jupiter, de Lalande had confused it with an ordinary star.

When physicists study celestial mechanics today, they are awestruck by the men and women who performed the most extensive numerical calculations in the 19th century using only their heads, paper, and pencils. Their primary tools were extensive logarithm tables, which alleviated some of the difficult multiplications and divisions. Calculations that modern computers can perform in a fraction of a second with a corresponding program sometimes took months to complete, and it was crucial to avoid miscalculations and carefully manage error propagation.

470. Talented ladies as "human calculators"

For the often monotonous calculations required for perturbation calculations - such as including the influences of large planets on the orbit of a planetoid or comet in orbit determination and ephemeris calculation - astronomers were happy to employ computationally skilled women, who effectively took on roles similar to today's computers. Some of these women independently produced significant research results and were recognized accordingly in the scientific community. For example, Luise de Pierry (1746 to after 1807) was the first woman to teach astronomy at the Sorbonne in Paris starting in 1789. She was particularly interested in calculating solar and lunar eclipses. The results she obtained were largely incorporated into the work of Joseph Jérôme Le Français de Lalande, who also acknowledged her contributions. His illegitimate daughter, Marie-Jeanne de Lalande (1769-1832), also worked as an astronomer. Her most notable work was a star catalog of approximately 10,000 stars, which was published in 1799.

471. A clever woman corrects Newton

Particular mention should be made of Marquise Émilie du Châtelet (1706-1749), who, despite her short life (she probably died of childbed fever after giving birth), is considered one of the most important female scientists of the Enlightenment period. She was not only a close friend of Voltaire (1694-1778) but also his partner, and she even corresponded regularly with King Frederick II of Prussia. At the suggestion of Voltaire, who himself had written a popular book on Newtonian mechanics around 1737, she undertook a French translation and intellectual analysis of Newton's main work, the *"Principia Mathematica."* In this process, she began to adapt Newton's difficult-to-understand mathematical calculus into Leibnizian notation, which had become increasingly popular, significantly improving the comprehensibility of Newton's work. Another lesser-known fact is that the concept of "living force," which we now know as "kinetic energy," is essentially attributed to du Châtelet, who corrected a major error made by Newton. Newton had erroneously assumed that kinetic energy was proportional to speed, whereas it is actually proportional to the square of the speed.

A large part of her philosophical reflections has also become a significant part of Enlightenment history. She fervently supported views that largely aligned with those of Voltaire. She criticized the theologies of revelation in general and even mocked the creation story of the Old Testament - a bold stance at the time - writing: "How amusing that the first three days [of the creation story] were limited by evening and morning before the sun was created on the 4th day...".

Marquise Émilie du Châtelet was undoubtedly one of the most educated and well-connected women of her time. Her work was recognized with respect by the male-dominated scientific community of the era. It is unfortunate that she remains overshadowed by her companion Voltaire in terms of the reception of her works.

Fig. 216: Marquise Émilie du Châtelet (1706-1749)

472. Diderot and his encyclopedia

Their lifetimes also encompass that of another great Frenchman whose legacy endures on Wikipedia, Denis Diderot (1713-1784). He is rightly considered one of the foremost encyclopedists and a key pioneer of the French Enlightenment of the 18th century. As a great organizer and editor, he succeeded in inspiring almost all the intellectual giants of his time to contribute to a project - a comprehensive collection of knowledge, a "circle of knowledge" (*enkyklios paideia*) - where terms and phrases were arranged, explained, described, and contextualized. Little is known about the early part of his life, up until around 1745. In that year, he published his *"Philosophical Thoughts,"* still anonymously, which marked him as a man who would decisively shape the century and establish it as the *"siècle des lumières,"* the "age of light."

The idea of an all-encompassing encyclopedia did not emerge out of nowhere but was inspired by a previous model. In 1728, a two-volume work entitled *"Cyclopaedia, or Universal Dictionary of Arts and Sciences"* was published in London by Ephraim Chambers

(1680-1740). The Paris publishing house Le Breton recruited the young Denis Diderot as a translator into French, a role he accepted due to his need for a steady income and his relatively modest demands for payment. Eventually, the idea emerged to not just translate but to create a significantly more extensive work. In 1747, d'Alembert (1717-1783), then an important mathematician and philosopher and a member of the French Academy of Sciences, joined the project. With his help, other notable contributors were enlisted, including Charles-Louis de Montesquieu (1689-1755), Bernard le Bovier de Fontenelle (1657-1757), and Voltaire (François-Marie Arouet, 1694-1778).

In 1749, the project faced a crisis when Diderot was arrested and imprisoned in the fortress of Vincennes - partly due to his *"Philosophical Thoughts,"* which irritated the authorities. He was released after one hundred and two days in prison following a confession and the intervention of his publishers, who had already lost the money they had advanced.

In 1750, Diderot published the *"Prospectus of the Encyclopaedia,"* outlining the meaning, purpose, and structure of the planned work. The main goal of this prospectus was to attract subscribers who would financially support the ambitious project. The first volume of the encyclopedia was finally published on July 1, 1751, provoking strong reactions, particularly from the clergy and especially the Jesuits, who saw the work as a threat to religious and moral order. Paradoxically, these attacks increased the encyclopedia's visibility and attracted more supporters.

The year 1752 became known as the year of the censors. The censors disapproved of the Enlightenment articles in the Encyclopaedia and eventually had them banned by royal decree. The enforcement of this ban fell to the royal director of publishing, Chrétien-Guillaume de Lamoignon de Malesherbes (1721-1794). Instead of enforcing the ban rigorously, de Malesherbes secretly warned Diderot of the impending confiscation. Diderot quickly responded by secretly handing over the manuscript to de Malesherbes and retreating into hiding to avoid persecution.

Meanwhile, de Malesherbes officially instructed the police to confiscate the manuscript from the publisher Le Breton. As expected, this attempt failed miserably. Through a combination of courage, cunning, and loyalty, Diderot and his supporters managed to preserve the work and secure its continued publication, which allowed the Encyclopaedia to gain even greater distribution and influence in the subsequent years.

The Encyclopaedia also gained many supporters within government circles. In particular, Madame de Pompadour (1721-1764), the educated mistress of Louis XV, used her influence to have the ban lifted just three months after it was imposed, despite some resentment at court. The French government even officially approached d'Alembert and instructed him and Diderot to continue the project. The third volume was published in November 1753. By the seventh volume, which appeared in November 1757, the letter "G" had been reached. This was the same year that Robert-François Damiens (1715-1757), an "unsuccessful assassin" by profession, attempted to assassinate Louis XV, drawing the attention of the famous executioner Nicolas-Charles-Gabriel Sanson (1721-1795) and his nephew Charles Henry Sanson (1739-1806).

The execution of Damiens was particularly brutal (quadripartite) due to the attempted regicide. More details can be found in the *"Diaries of the Executioners of Paris,"* Volume 1 (e.g., in Google Books), which dedicates an entire chapter to the assassination of Louis XV. Following this attempt, laws regulating the writing and printing of books and journals in France were tightened. Clerical circles used these laws against the "encyclopaedists," attempting to blame them indirectly for inciting Damiens' crime. Initially successful, this effort led to d'Alembert's resignation as co-editor and the withdrawal of the royal license for the "Encyclopaedia" in March 1759. The project seemed to be at an end, but the authors continued to work illegally under de Malesherbes' benevolent tolerance, ultimately delivering the complete, seventeen-volume work to their subscribers in 1765. In 1772, the eleven illustrated volumes were also completed.

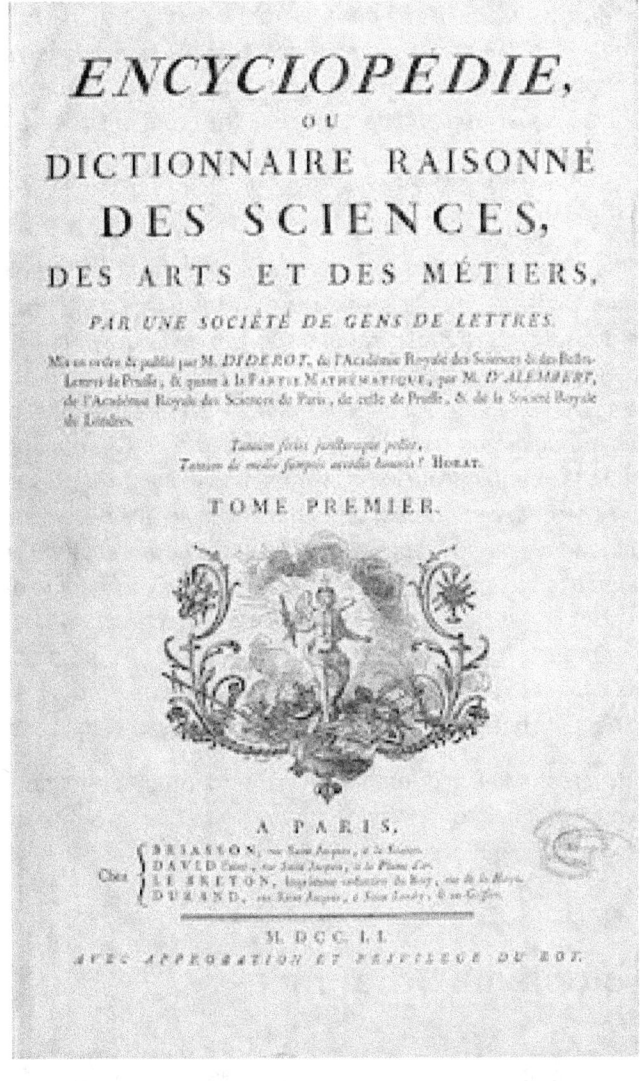

Fig. 217: Title of the "Encyclopedie ou Dictionnaire Raisonne des Sciences des Arts et des Metiers", Volume 1 - a series of books that advanced the Enlightenment proclaimed by Emmanuel Kant...

At the height of its distribution, the publishing house employed over 200 people, including paper manufacturers, printers, bookbinders, and engravers. A complete edition of the work cost around 900 livres in 1772, which was roughly 1.5 times the annual earnings of a well-paid silk worker in Lyon. Commercially, the encyclopaedia, with contributions from about 160 individuals, was a resounding success for the publisher. For generations, this work and its numerous successors served as the primary source of knowledge, consulted and referenced widely. Even today, it is still enlightening to read a few articles to understand what people knew and thought in the 18th century. For example, consider the term "madness" as discussed in the seventh volume:

"If one departs from reason unwittingly, because one is devoid of ideas, one is stupid; if one departs from reason knowingly, though with regret, because one is the slave of a violent passion, one is weak; but if one departs from it confidently, namely in the conviction that one is following it, one is, it seems to me, mad. These, at least, are the unfortunates whom one imprisons and who perhaps differ from other people only in that their madnesses are of the rarest kind and do not fit into the order of society... (d'Aumont)".

Or here is another term from the category "cuisine", as dealt with in the fourteenth volume - *sauerkraut*:

"The French mutilate this word to "choucroute". It is a dish that is popular everywhere in Germany; it is based on sour cabbage, hence its German name. Sauer means acid, kraut means cabbage. If you want to make sauerkraut, you first cut white cabbage into very thin slices; the Germans have a board for this purpose that resembles a slicer and is equipped with a sharp knife. If you grate the cabbage on this type of slicer, it is cut into thin slices, which are collected under the slicer in a trough. Once a sufficient quantity has been collected, the cabbage cut in this way is placed in barrels, layer by layer, sprinkled with salt and a few juniper berries; once the barrel is full, it is covered with a board and a weight is placed on top so that the cut cabbage is compressed. The whole thing is placed in a cellar and left to ferment for a few weeks. When you want to eat the cabbage, you wash it and cook it with salted meat, sausages, partridge and - depending on your preference - other meats. This ragout is much appreciated by the Germans; it is served at the table of the richest as well as the poorest. Foreigners have little taste for it, but this ragout seems to be quite useful for sailors on long journeys ...".

Until recently, encyclopaedias in printed form were important collections of knowledge and reference works which, in one form or another, whether as a massive "Brockhaus" or a small pocket encyclopaedia, could not be missing in any household of the educated middle classes.

473. Encyclopædia Britannica

One of the oldest encyclopaedias still in existence today is the 32-volume Encyclopædia Britannica, which contains around 44 million words - 300 times more than the small book you are reading now. Created in Scotland around 1768, it is still maintained to this day. However, in keeping with contemporary trends and technological advancements, it has been available only in digital form since 2012. With the advent of online encyclopaedic resources, the printed, expensive, and space-consuming versions have virtually lost their raison d'être. The innovative Wikipedia project (which, despite its many strengths, also has weaknesses, particularly in more politically sensitive areas) has transformed the encyclopaedia from an elitist publication into a widely accessible source of knowledge (provided one has internet access).

The Encyclopædia Britannica was initially published in Edinburgh by the society *"A Society of Gentlemen in Scotland"* and has undergone numerous editions since then. Throughout its history, it has been revised and updated several times to reflect the latest scientific and cultural developments. The eleventh edition, published in 1910-1911, was particularly significant for its comprehensive and scholarly articles.

Fig. 218: Nowadays, it is rarely found as an ornament on your own bookshelf, but almost only in larger libraries - the Encyclopædia Britannica

Over time, the Encyclopædia Britannica became a symbol of academic and intellectual authority, attracting contributions from many renowned scientists and experts, including Albert Einstein and Marie Curie. The digital transformation in 2012 allowed Britannica to keep pace with the rapid advancement of knowledge and offer regularly updated content. This shift has also contributed to the democratization of knowledge by providing easy access to high-quality information to a global audience.

In addition, the digital version offers interactive features that were not possible in the print edition, such as multimedia content and direct links to additional sources. Despite the competition from freely accessible online encyclopaedias, the Encyclopædia Britannica remains a respected reference work and an important part of the modern knowledge landscape. Its adaptability to new technologies and media ensures its continued relevance in the digital age.

Beyond the major encyclopaedias, countless "small" and highly specialized, sometimes even unusual, encyclopaedias have been created over time.

474. Wrynecks and weathercocks

In the years following the French Revolution, an anonymous observer noted that many former royalists had suddenly become ardent republicans and were now advancing their careers under the new regime. In early 1990s Germany, the term "Wendehals" (from the bird "Wryneck") was coined to describe such individuals. The anonymous author of the *Dictionnaire des girouettes* used the term "weathercock" to illustrate people who, like a weather vane, always turned with the prevailing wind. Published in Paris in 1815, the year of Napoleon's defeat at Waterloo, this dictionary detailed the careers of individuals who adeptly adapted to political upheavals to survive unscathed. The degree of opportunism was indicated by stylized weathercocks or flags next to each name. The more flags, the more frequently the individual had changed their political stance.

Among the numerous notable names listed, Charles-Maurice de Talleyrand-Périgord (1754-1838), Napoleon Bonaparte's renowned foreign minister, earned sixteen flags. Napoleon himself, while in exile on Saint Helena, remarked: *"What convinces me that there is neither a punishing nor a rewarding God is the fact that decent men are always unhappy and wretches are always happy. You will see a Talleyrand die in his bed."* And indeed, Talleyrand did.

Joseph Fouché (1759-1820), Napoleon's infamous Minister of Police, received twelve flags. Fouché's career exemplified the art of opportunism; he changed his allegiances as circumstances dictated, making him the archetype of the turncoat. Stefan Zweig (1881-1942) chronicled Fouché's life in his 1929 biography, available on Project Gutenberg, with a certain admiration for Fouché's ability to shift his convictions to suit the moment. Initially allied with the moderate Girondists, Fouché later joined the radical Mountain Party and advocated for the king's execution. Anticipating the fall of Robespierre, he joined forces with Jean-Lambert Tallien (1767-1820), only to be arrested and imprisoned. Fouché soon ingratiated himself with the Directory, the final revolutionary government, particularly with Paul Barras (1755-1829). Despite their reservations about him, he secured the position of Minister of Police, amassing considerable wealth. During the imperial era, he was honored extensively. However, when Napoleon's second marriage caused a scandal, Fouché fell out of favor. Predicting the emperor's downfall, he aligned himself with the Bourbons. When news broke that Napoleon had escaped from Elba and was marching towards Paris *("The bad wolf docked in the Bay of Juan"* was the headline in the BILD newspaper of the time), Fouché fled Paris for safety. Upon Napoleon's return and his reentry into Paris *("His Imperial Majesty entered the Tuileries, surrounded by his loyal subjects"* read the contemporary BILD headline), Fouché returned as well, flattering the emperor and was quickly reinstated as Minister of Police during the "Hundred Days."

Fig. 219: Joseph Fouchè (1759-1820) 🐦🐦🐦🐦🐦🐦🐦🐦🐦🐦🐦

Sensing that Napoleon's reign would not last long, he betrayed the planned Belgian campaign in a conspiratorial manner to the British Field Marshal Arthur Wellesley (1769-1852), better known as the "Wellington" (he was the 1st Duke of Wellington). When Napoleon was finally defeated at Waterloo, Fouché became President of the Provisional Government and thus finally broke with the Bonapartists. The situation continued. When Louis XVIII, a Bourbon, finally returned to the French throne, he once again became Minister of Police - a former Jacobin who had once voted for the death of the Bourbon King

Louis XVI... One can imagine that this *"lexicon of weathercocks"* caused quite a stir in France in those days. A certain Beuchot felt compelled to restore the nation's honor. His elaboration on this topic resulted in a book that can be found in the digital library Gallica under the title *"Dictionnaire des immobiles"* - i.e. a "lexicon of character feasts". After all, it contains 38 pages of text... So the "character festivals" at the time of the "First Republic" couldn't have been very far off - probably also thanks to the invention of Mr. Joseph-Ignace Guillotin (a doctor), to whom so many "character festivals" fell victim in the turmoil of the French Revolution...

The term "revolution" is always used when there is a fundamental and lasting change in a system within a comparatively short period of time. This "system" does not necessarily have to be a social system such as a monarchy. Science and technology also experience revolutions. In science, they are usually simply called "paradigm shifts".

475. Paradigm shift in science

This term comes from the scientific theorist Thomas S. Kuhn (1922-1996), according to whom a "paradigm" should essentially describe the framework conditions in which a scientific theory is embedded as an explanatory model of reality. If it is necessary to change this framework, this is referred to as a "paradigm shift." A typical example of this is the transition from the geocentric world view to the heliocentric world view, which was carried out by Nicolaus Copernicus (1473-1543) from Ermland. Under both paradigms, it was possible to calculate the positions of the planets in the sky for future points in time with an accuracy that was satisfactory for the practice of the time. However, Copernicus was able to show that under his premise the explanatory model could be logically simplified to a certain degree without the results becoming worse.

Paradigm shifts often occur in science. However, here it was linked to a change in worldview that called religious doctrines into question. The explosive nature of this only became apparent after Copernicus' death, when his ideas slowly began to spread in the West. This was due to the fact that his main work, published in 1543, received little attention from his contemporaries due to its mathematical nature, although the Catholic Church was also very interested in it - in a positive sense (Pope Paul III (1468-1549) had it reported to him in the Vatican Gardens). One reason for this was the "calendar misery" (i.e., the divergence between church holidays and the calendar due to an incorrect year length), the solution to which became increasingly acute and finally led to the "Gregorian calendar reform" in 1582, incorporating Copernican ideas. It was not until reformatory church teachers such as Philipp Melanchthon (1497-1560) and Martin Luther (1483-1546) recognized the ideological explosiveness of heliocentrism and attempted to stop its spread by preventing disputations at universities or ridiculing the Copernican doctrine *("Es ward gedacht eines neuen Astrologi, der wollte beweisen, daß die Erde bewegt würde und umginge, nicht der Himmel oder das Firmament...,* Luther's Table Talk") (A new astrologer was thought of who wanted to prove that the earth was moved and revolved, not the heavens or the firmament...). The Catholic Church, on the other hand, only became aware of the work in connection with Giordano Bruno (1548-1600) and Galileo Galilei (1564-1642) and, after thorough examination and in compliance with church law, placed it on the Index of Prohibited Books in 1616 (until 1835). But - from the Church's point of view - it had done no good. Once the Copernican ideas were recognized as true, the emancipation of science from the church began - with the visible success today that a space probe flew past Pluto some time ago (15 July 2015), photographing its surface and the surface of its moon Charon, and anyone with Internet access could take part in it - if they were interested...

When it comes to Nicolaus Copernicus, people always think of his astronomical contributions to the establishment of the heliocentric world system. But it is always forgotten that he also made significant contributions in the field of economics and political economy.

476. Nicolaus Copernicus as a great economist

At the beginning of the 16th century, when Europe was facing numerous economic challenges, Nicolaus Copernicus's fiscal ideas came to the fore. One of his most significant achievements in this field was the formulation of an early version of "Gresham's Law." This economic principle states that *"bad money drives good money out of circulation."* Copernicus observed that inferior coins, which were made of less valuable metals or had a lower precious metal content, led people to hoard more valuable coins. He set out these observations in his work *"Monetae cudendae ratio"* (1526), which dealt with the problem of coin debasement and its effects on the economy.

He recognized that the progressive debasement of coins led to inflation and economic instability. Through his astute analyses and the formulation of clear principles, he made a significant contribution to the understanding of monetary theory. His findings helped to explain the behavior of currencies in circulation and formed the basis for later economic theories and economic reforms.

In addition to theoretical analysis, Nicolaus Copernicus was also practically active in coinage reform. He was involved in the coinage policy of Prussia and Poland and actively participated in the development of reform proposals to stabilize and improve the monetary system. The aim of these reforms was to improve economic conditions by creating a stable and uniform currency. Copernicus's work in this area demonstrates a deep understanding of economic dynamics and the need for sound monetary policy.

Copernicus's economic contributions are particularly remarkable when one considers that he lived at a time when economics was not yet an established scientific discipline. His work was written in a context in which economic issues were often closely linked to political and social challenges. Nevertheless, he succeeded in formulating principles that have universal validity and were later developed further by important economists such as Adam Smith (1723-1790) and David Ricardo (1772-1823).

The integration of his economic insights with his extensive education and diverse activities makes Copernicus an outstanding example of interdisciplinary thinking in the Renaissance period. His ability to make connections between different fields of knowledge and to find practical solutions to complex problems distinguishes him as one of the great thinkers of his time.

Nicolaus Copernicus therefore not only made great contributions in the field of astronomy, but he is also regarded as a pioneer of economics and political economy. His work shows that his intellectual curiosity and analytical mind went far beyond astronomy and significantly influenced the development of economic theories. Copernicus's economic insights remain a testament to his perceptive mind and his contribution to the development of modern science.

477. What did Copernicus look like in reality?

Of course, we know what Nicolaus Copernicus looked like. This is because a number of portraits of him have been preserved. His physiognomy is now so well known that most educated people have no difficulty recognizing him in pictures.

It is generally known that Nicolaus Copernicus was buried in Frombork Cathedral after his death on May 24, 1543. However, the exact location of his tomb was forgotten over the centuries. Since 1580, when an epitaph was erected in his honor and later destroyed, archaeologists have searched in vain for his remains. Various attempts in 1802, 1909, and 1939 were unsuccessful.

It was not until 2004 that local historian Jerzy Sikorski revived an earlier hypothesis by former regional historian of Warmia, Hans Schmauch (1887-1966), that the grave must be located near the Holy Cross altar, the fourth altar column on the right. This prompted the bishop in charge to commission a team led by Polish archaeologist Jerzy Gassowski to search for it. In the summer of 2005, his team discovered the remains of thirteen graves near the altar, some of which were badly damaged. Among them were the mortal remains of a man aged 60 to 70 years, which could be attributed to Copernicus.

A facial reconstruction carried out in November 2005 using the skull showed a striking match with contemporary portraits of the astronomer, including features such as a broken nose and a scar above the left eyebrow. A DNA analysis was originally planned for final identification. However, the search for living relatives of Copernicus in the maternal line was unsuccessful, as only one of his sisters had descendants, but her maternal line extended only back to the 18th century.

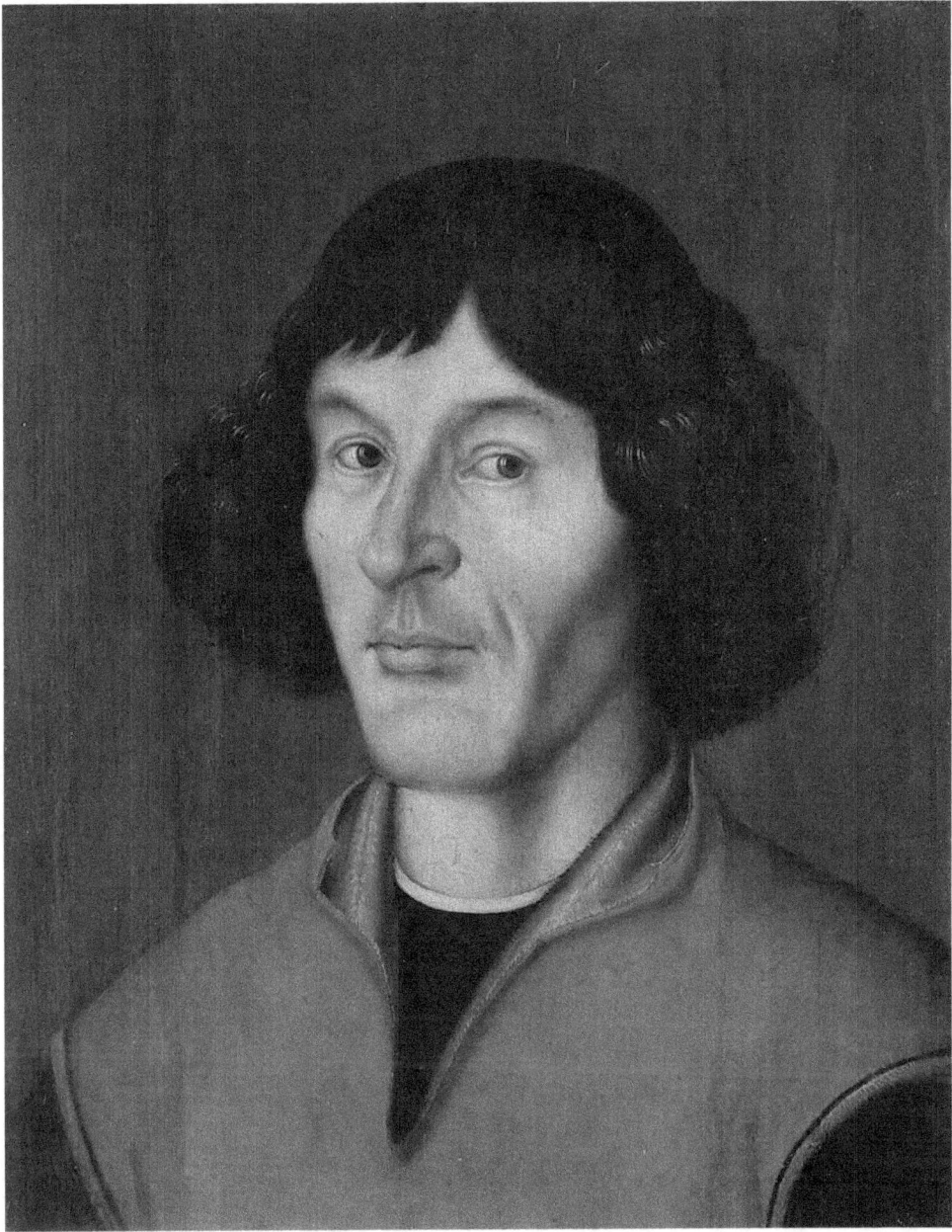

Fig. 220: Nicolaus Copernicus (portrait from the town hall in Thorn, around 1580)

Fortunately, nine hairs were found in a book that once belonged to Nicolaus Copernicus and which ended up in the library of Uppsala University after the Polish-Swedish wars in the 17th century. Four of these hairs could be used for genetic analysis. On November 20, 2008, Polish archaeologist Jerzy Gassowski and Swedish DNA expert Marie Allen announced that the DNA analysis of two hairs from the book and one tooth from the skull showed with high probability that they could be attributed to the famous astronomer.

However, this result was met with skepticism, as the DNA analysis also revealed that the skull belonged to a person with light (blue or gray) eyes. This contradicted most historical color portraits of Copernicus, which consistently depict him with dark brown eyes and dark hair. The discovery was only officially announced on November 3, 2008, after further investigations. Forensic expert Dariusz Zajdel from the Polish Central Forensic Laboratory used the skull to reconstruct a face that matches the known self-portraits of Copernicus astonishingly closely. It was also established that the skull belonged to a man who had died at around 70 years of age - the same age as Copernicus at the time of his death.

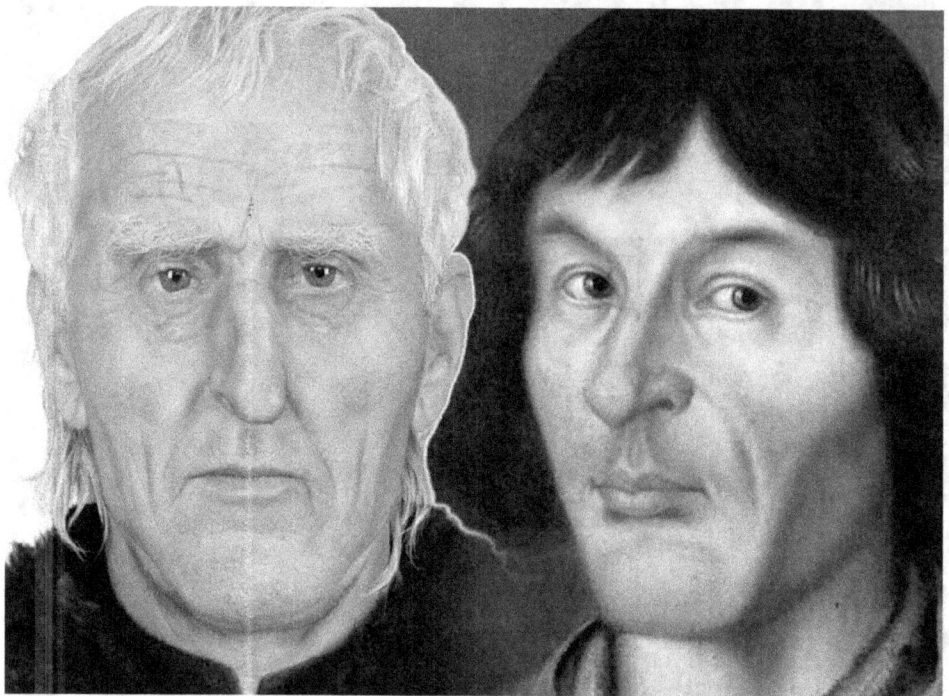

Fig. 221: Facial reconstruction of the 70-year-old Nicolaus Copernicus based on the skull found in Frauenburg Cathedral compared to the portrait shown in Fig. 220 (Source: Central Forensic Laboratory of the Police Headquarters)

Although the grave was in poor condition and not all the remains of the skeleton were found (including the missing lower jaw), the DNA analysis finally enabled an almost certain identification. On May 22, 2010, the previously excavated remains were solemnly reinterred in Frauenburg Cathedral. An inscription on the fourth pillar on the right-hand side of the cathedral now commemorates the great astronomer with the words:

"† Nicolaus Copernicus - natus 19.02.1473 Thoruniae - defunctus 21.05.1543 Frauenburgi - astronomus - heliocentrismi artifex - canonicus warmiensis."

The ceremonial reburial of Copernicus in Frauenburg Cathedral in May 2010 was a significant historical event. His new black granite tomb, with the depiction of his heliocentric model of the solar system, is a reminder of his incomparable contribution to science and his great significance in the history of mankind.

478. Facial reconstructions of historical personalities

In recent years, artificial intelligence (AI) technology has made remarkable progress, particularly in the field of generative image models. One of the most fascinating applications of this technology is the ability to create quasi-photorealistic images of historical figures such as Julius Caesar or Marie Antoinette. These images are often so detailed and realistic that they are almost indistinguishable from real portrait photographs.

Fig. 222: Gaius Julius Caesar and Marcus Aurelius, facial reconstruction by Allessandro Tomasi using AI on the basis of ancient portrait busts (© Allessandro Tomasi)

The key to this impressive ability lies in the so-called "generative" AI models. These models are actually artificial neural networks that have been trained to generate new and plausible image content. Prominent examples of such models are "Generative Adversarial Networks" (GANs) and "Variational Autoencoders" (VAEs). These systems learn from extensive data sets of portrait photos how characteristic features of a person, such as facial features, hairstyle, or clothing, are related to each other.

In order to create a realistic image of a historical personality, these models often use text descriptions, sketches, busts, and paintings as a basis. The models can then use these templates to synthesize new, photorealistic images. In doing so, they draw on the previously learned patterns and combine them to create a coherent overall representation. Additionally, some systems offer interactive editing tools with which the user can adjust the appearance of individual parts of the face. Advanced processes such as "latent space interpolation" also make it possible to generate new intermediate versions from several initial portraits. This can be used, for example, to visualize changes in a person's age. The AI uses the display characteristics of the reference images to generate new, plausible variants.

The quality of AI-generated images has increased enormously in recent years and has now reached a photorealistic level. However, these processes are not without their challenges. Inaccuracies in the training data or technical limitations of the models can lead to artifacts or distortions in the results. Despite these challenges, facial reconstruction of historical figures using generative AI models offers a fascinating opportunity to bring history to life visually. Incidentally, there are entire compilations of such images on YouTube, including one of the famous Lisa del Giocondo (1479-1542) by Nathan Shipley (Fig. 223).

479. A painting in the Seine...

Many people will be familiar with her portrait - it was briefly seen floating in the Seine during the opening ceremony of the 2024 Olympic Games, for example, after it was removed (or stolen) from the Louvre (by the "Minions," I suppose symbolically). What exactly this was supposed to signify was probably not clear to most viewers. But at least it was funny to watch, unlike some of the other scenes from the opening ceremony...

Who was Lisa del Giocondo, born Lisa Gherardini on June 15, 1479, in Via Maggio in Florence? She is probably better known by the name "Mona Lisa," under which she was painted by the famous Renaissance artist Leonardo da Vinci (1452-1519). For this reason, many art historians have studied her, which might not otherwise have been the case. Their research revealed the following biography:

Lisa del Giocondo was born as the daughter of the landowner Antonio di Noldo Gherardini and his third wife, Lucrezia del Caccia. Her father came from an old, wealthy noble family and owned several estates in Chianti. Lisa had three sisters and three brothers.

Fig. 223: AI-generated facial reconstruction of the Mona Lisa by Nathan Shipley (© Nathan Shipley)

At the age of 15, Lisa married the 30-year-old cloth and silk merchant Francesco di Bartolomeo di Zanobi del Giocondo on March 5, 1495. Lisa was Francesco's third wife; her dowry amounted to 170 guilders and a farm. The couple had five children and another daughter who died shortly after birth. Lisa also raised Bartholomeo, Francesco's son from his first marriage.

Francesco died of the plague in 1538 at the age of 73. After his death, Lisa moved to the Monastero di Sant'Orsola, where her daughter Marietta lived as a nun. Lisa died there on July 15, 1542, at the age of 63.

Leonardo da Vinci probably began work on the "Mona Lisa" in 1503 and worked on the painting for several years. It is assumed that the portrait of Lisa was intended as a sign of the Giocondo family's esteem and social advancement. However, there is no concrete evidence beyond the purely professional relationship between the artist and his model.

Although the identification of the "Mona Lisa" as Lisa del Giocondo is not entirely uncontroversial, she remains a fascinating figure of the Renaissance and a symbol of the timeless beauty and mystery surrounding Leonardo da Vinci's masterpiece.

480. Make lenses to see the moon big

Leonardo da Vinci was undoubtedly the most versatile artist, scientist, and engineer of the Renaissance and left a lasting impression through both his artistic masterpieces and his scientific research. His contributions in the fields of optics and astronomy are particularly noteworthy. Leonardo's diary entries in his famous notebooks provide a deep insight into his studies and discoveries in these fields.

For example, the *Codex Atlanticus* contains the sentence: *"Make magnifying glasses (occhiali) to see the moon large."* Some therefore believe that he came very close to inventing the telescope.

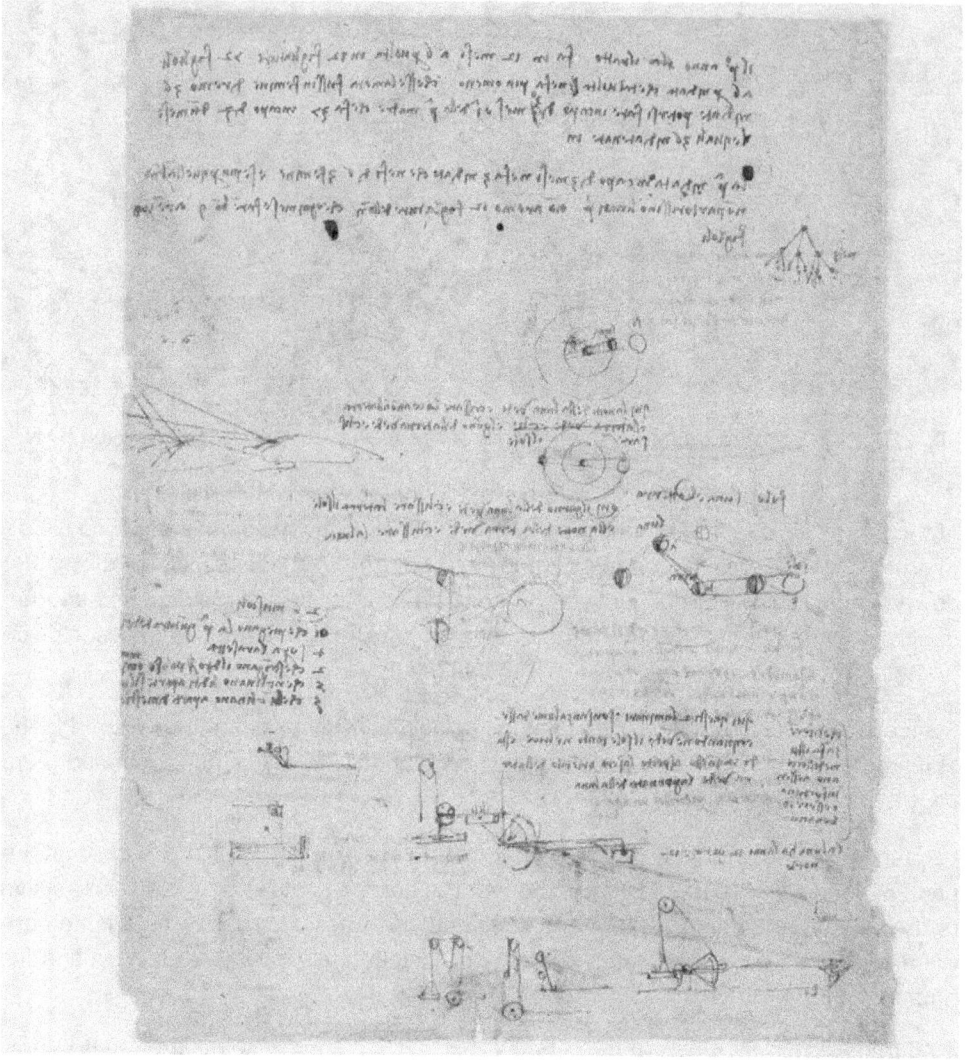

Fig. 224: A page from the Codex Atlanticus on astronomical topics (Creative commons)

Leonardo was deeply fascinated by the nature of light and its interaction with matter. In his notebooks, he describes in detail the properties of light and its behavior in various media. In contrast to the popular idea at the time that the eye emits rays of light to see objects, Leonardo argued that light emanates from objects and falls into the eye. This view, which corresponds to the modern concept of the propagation of light, was revolutionary for his time. He studied the reflection and refraction of light and observed exactly how rays of light are reflected from different surfaces. His studies on reflection led him to the realization that light is reflected at an angle that corresponds to the angle of incidence - a principle known today as the law of reflection.

Another central aspect of Leonardo's optical research was the study of the anatomy of the eye. He carried out extensive anatomical studies to better understand the human eye and drew detailed sketches of its structure. His notes show that he understood the function of the lens and the retina and recognized the eye as a receiver of light. Leonardo also recognized the importance of the pupil and described how it adapts to different light conditions. These detailed sketches and notes testify to a deep understanding of human visual perception that was far ahead of his time.

His studies also contributed significantly to his understanding of spatial representation. In his notebooks, for example, he describes the principles of linear perspective and how objects appear smaller with distance. Among other things, he investigated atmospheric perspective, which is the change in color and clarity of distant objects due to air turbidity and light scattering. These findings helped him to create more realistic images with greater depth. An excellent example of this is his famous "Mona Lisa," in which the background is depicted in soft, blurred tones to convey the impression of distance.

Another important topic in Leonardo's optical studies was the camera obscura, a forerunner of the modern photographic camera. He described how the camera obscura worked and how it projected an inverted image of the surroundings onto a surface. Leonardo recognized the importance of this phenomenon for the understanding of vision and used it to refine his theories on optics. His notes show that he regarded the camera obscura as a model for the human eye and applied its principles to the functioning of vision.

481. Camera obscura and Renaissance painting

The *camera obscura* (Latin for "dark chamber") is an optical device that played an important role in the development of painting, particularly during the Renaissance. The principle was already known in ancient Greece and China, but the Renaissance revolutionized its use and understanding.

The camera obscura consists of a dark chamber or box with a small hole or lens through which light enters. The light projects an inverted image of the outside world onto the white wall opposite.

The Renaissance was a time of revival of scientific and artistic interests. A decisive moment for painting was the discovery of linear perspective, to which artists such as Filippo Brunelleschi (1377–1446) and Leon Battista Alberti (1404–1472) made a significant contribution. They developed techniques with which three-dimensional spaces could be realistically depicted on two-dimensional surfaces.

Renaissance artists used the camera obscura primarily to improve the accuracy and detail of their works. One of the first and best-known users was Leonardo da Vinci (item 480). In his notebooks, he described in detail how the camera obscura worked and how it could be used to study light and shadow. Although there is no concrete evidence that

Leonardo used it directly for his paintings, his notes show that he understood the principle very well.

Another example is the Italian painter and mathematician Giovanni Battista della Porta (1535–1615), who described the use of the camera obscura in painting in his work "Magia Naturalis" (1558). He showed how this device could be used to create complex architectural perspectives and draw precise proportions.

Michelangelo Merisi da Caravaggio (1571–1610), a Baroque master known for his dramatic use of light and shadow, may also have used the camera obscura. The precise lighting and photorealistic detail of his works suggest that he may have used optical aids. There is speculation that he used the camera obscura to plan his compositions and determine the position of light sources to achieve dramatic effects.

In the 17th century, the Dutch masters also used the camera obscura, notably Jan Vermeer (1632–1675). Vermeer's paintings are characterized by exceptional clarity and detail, which has led many art historians to believe that he may have used the camera obscura in the composition of his scenes. In particular, David Hockney and other researchers argue that Vermeer and his contemporaries may have used optical aids such as the camera obscura to achieve the complexity and precision of their works. Vermeer's "Maid with a Milk Jug" and "View of Delft" are examples where the exact rendering of light and shadow could indicate such a technique.

482. The secret of Jan Vermeer's "Letter Reader at the Open Window"

Jan Vermeer van Delft, the mysterious master of the "Dutch Golden Age," has fascinated art lovers and art historians for centuries. His works, of which only around 35 are known, are characterized by extraordinary lighting, intimacy, and enigmatic everyday scenes. Vermeer, who was born in Delft in 1632 and died there in 1675, left posterity an oeuvre of unparalleled quality and depth, and every museum that is allowed to display a work by him is proud to do so.

One of his most famous paintings, "The Letter Reader at the Open Window," was painted around 1657–1659 and is one of the key works of his early creative phase. It can be seen today in the Old Masters Gallery in Dresden's Zwinger Palace. The painting shows a young woman in a light-flooded interior, intently reading a letter. The scene radiates the contemplative calm so typical of Vermeer. But the work holds a surprise that would only come to light centuries later.

The story of "The Letter Reader at the Open Window" took an unexpected turn when researchers discovered during X-ray examinations in the 1970s that an overpainting was concealed beneath the visible surface of the painting. On the originally blank wall in the background, Vermeer had painted a large image of Cupid, which was later painted over. The discovery was a sensation for the art world, but it was not until 2017 that the extensive restoration work began to uncover the original version. As the undertaking was considered delicate from the outset, the pros and cons were discussed with leading international art historians and restorers before the decision was made to proceed.

The uncovering of the hidden Cupid painting, which was completed in 2021, fundamentally changed the interpretation of the work. The naked Cupid on the wall, a well-known symbol of love, lends the scene a new dimension. It suggests that the letter the young woman is reading must be of a romantic nature. This revelation fits seamlessly into Vermeer's well-known preference for "pictures within a picture," which often carry symbolic meanings and comment on or complement the main content.

Fig. 225: Jan Vermeer's masterpiece "Brieflezend meisje bij het venster" (ca. 1657-59) before and after uncovering the overpainting (Staatliche Kunstsammlungen Dresden)

The overpainting of the Cupid painting was obviously not carried out by Vermeer himself but only some time after his death. This raises interesting questions about the reception and handling of works of art in earlier centuries. It is possible that a later owner or dealer found the painting disturbing or, more likely, wanted to adapt it to contemporary tastes.

In any case, the rediscovery and uncovering of the original state of "The Letter Reader at the Open Window" has significantly expanded our understanding of Vermeer's work. It shows how the artist created complex visual and symbolic layers in his paintings to tell stories and convey emotions. The presence of Cupid enhances the intimacy of the scene and invites the viewer to speculate on the thoughts and feelings of the reading woman.

This discovery joins a series of other discoveries that have refined our image of Jan Vermeer in recent decades. It underscores the need to examine even seemingly well-researched works of art repeatedly using new methods. At the same time, it sheds light on the complex fates that paintings can experience over the centuries.

For the understanding of Vermeer's oeuvre as a whole, this revelation means a reassessment of his artistic intentions and working methods. It confirms his mastery of composition and his ability to create deeper layers of meaning through subtle details. "The Letter Reader at the Open Window" now presents itself as an even richer and more multi-layered work that sheds new light on Vermeer's genius. For this reason alone (among many hundreds of others), it is worth traveling to the former electoral residence on the Elbe, to Dresden...

483. Canaletto view from Dresden

If you then head to the right bank of the Elbe below the Augustus Bridge, you can enjoy the famous Canaletto view. Afterwards, it is a good idea to admire the 1.33 x 2.37 meter painting of the same name by Bernardo Belotto (1722-1780) in the "Old Masters" gallery, which was painted in 1748. This was just five years after the completion of the Frauenkirche, which has since become Dresden's landmark alongside the Zwinger and the Hofkirche.

Fig. 226: "Canaletto view" of Dresden's old town from 1748 (Dresden State Art Collections)

Prior to this, Bernardo Bellotto (1747) had been invited by Elector Frederick Augustus II (1696–1763, who as "Augustus III" was also King of Poland) to his residential city of Dresden, where he commissioned him to paint some so-called "vedute" (city views).

Bellotto, who was born in Venice and honed his craft under the guidance of his famous uncle Giovanni Antonio Canal (1697–1768), better known as "Canaletto," brought his masterful technique and unique style to Dresden (Bellotto later always signed his paintings "Canaletto" with the consent of his teacher). The creation of the city view later known as the "Canaletto View" is a prime example of Bellotto's ability to combine architectural precision with atmospheric light and detailed city views. As we know, he used a camera obscura (point 481) to help him accurately capture the perspective and proportions of the cityscape. This technique enabled him to combine topographical accuracy with painterly detail. Today, photographs taken at the painter's location can be superimposed almost congruently over the painting.

The "Canaletto View of Dresden" shows a wide perspective across the Elbe with an unobstructed view of the magnificent buildings of the old town. The outstanding buildings include the Frauenkirche, the Hofkirche, and the Residenzschloss. Bellotto also captured the hustle and bustle on the Elbe and along the riverside streets, which clearly shows the dynamism and liveliness of the city (and which it has not lost to this day).

The careful composition and the play of light and shadow lend the painting depth and realism, making it an important historical document and artistic masterpiece. It offers not only a visual treat but also a valuable insight into the urban life and architecture of the time.

The painting was created during a time of great cultural prosperity in Dresden, which was partly due to the influence of the Saxon electors and the Polish kings as patrons of art and architecture. Bellotto's works from this period also contributed to the detailed reconstruction of Dresden's historic cityscape, after Dresden was almost completely destroyed by Allied bombers under the command of Arthur Travers Harris (1892–1984) shortly before the end of the Second World War, when the city was full of refugees from Silesia...

484. Bomber Harris

Arthur Travers Harris, also known as "Bomber Harris," was commander-in-chief of the British Bomber Command from 1942 to 1945. Harris was an advocate of the strategy of area bombing, which aimed to destroy the enemy's civilian infrastructure and thus break the morale of the population. This strategy, set out in the Area Bombing Directive in 1942, led to numerous area bombings of German cities, including Hamburg, Cologne, and eventually Dresden. Harris argued that such attacks were necessary to weaken the German war effort and shorten the war. Critics, on the other hand, accused him of causing unnecessary suffering and being disproportionate. In this respect, the bombing of Dresden is considered the most debated attack.

On the night of February 13–14, 1945, British and American bombers carried out a massive attack on Dresden. Within less than 24 hours, around 800 British and American bombers dropped approximately 3,900 tons of high-explosive bombs and incendiary bombs on the city. The raids caused a firestorm that destroyed large parts of the historic baroque city and claimed the lives of an estimated 25,000 people (this figure is highly controversial and probably a lower bound).

Fig. 227: Dresden 1945 (Peter Richard, Deutsche Fotothek)

Dresden was known for its cultural treasures, including the Semper Opera House, the Zwinger Palace, and the Frauenkirche. The city also had little military significance, which led to further controversy over the justification for the attack. However, Harris and his supporters emphasized that Dresden was an important transport hub and site of war industry.

In Germany, the bombing of Dresden was long regarded as a symbol of the horrors of air warfare and the destruction of German cities. The GDR government used the event for propaganda purposes to denounce the cruelty of "Anglo-American imperialism." After the reunification of Germany, a more differentiated discussion of the bombing began. Historians emphasized the need to view the events in the context of National Socialist crimes and the war as a whole. The annual commemorative events in Dresden, especially on February 13, are an expression of collective remembrance and an opportunity to reflect on the horrors of war and the importance of peace.

In England, Arthur Harris was celebrated as a hero during the war, and his strategy was considered decisive for the victory over Nazi Germany. After the war, however, a critical examination of the moral implications of strategic bombing began. Harris himself remained a staunch advocate of his tactics until his death in 1984 but never received the same posthumous recognition as other British war heroes.

In the US, the bombing of Dresden was less controversial, as American involvement was seen as part of a wider strategy to support the Allied war effort. Nevertheless, there were also critical voices here that questioned the humanitarian costs of such attacks. In recent decades, a differentiated historiography has developed in both countries that critically scrutinizes the strategic decisions and their consequences. Historians such as Frederick Taylor and Antony Beevor have helped to paint a balanced picture that does justice to the complexity of the decisions made at the time and the tragedy of the victims.

The destruction of Dresden left a city in ruins, but the reconstruction is an impressive testament to human will and strength. In the years following the war, Dresden was slowly rebuilt, and many of the magnificent baroque buildings that had made the city famous were restored. Particularly noteworthy is the reconstruction of the Frauenkirche, which had stood for decades as a ruin and memorial against the war. After the reunification of Germany, it was faithfully reconstructed with support from all over the world and reopened in 2005. Now the Canaletto view of Dresden is complete again. The city has finally recovered from the horrors of war and has developed into a modern city that has retained its historical and cultural identity. The reconstruction is not only a symbol of the city's rebirth but also of hope and peace in Europe.

485. "World peace" - the utopia of modernity

Unfortunately, it is true that armed conflicts, as not only European history shows, must largely be regarded as "normal" - as a continuation of politics by other means. This insight goes back to the great military scientist Carl von Clausewitz (1780-1831), who literally wrote in his great but unfortunately unfinished work "*On War*" (1832): "*War is a mere continuation of politics by other means*". He put it somewhat more precisely in his "Sketches", in which he noted:

"*War is nothing but a continuation of political intercourse with the interference of other means, in order to assert at the same time that this political intercourse does not cease through war itself, is not transformed into something else, but that it continues in its essence, however the means may be shaped which it makes use of.*"

This view has profound implications for our understanding of war and peace. It suggests that wars often seem inevitable because they are seen as instruments of politics to achieve goals that do not seem feasible by other means. One need only think of the reaction of the Russian Federation - "Putin" for short - to the refusal of the NATO states to stop their eastward expansion and agree on a mode of joint and undivided security between NATO and Russia. According to the Western narrative, the resulting "Ukraine war" from 2022 onwards is a war of aggression by Russia and, according to the Russian narrative, a defensive war of the breakaway Ukrainian provinces against the Kiev regime - "Selenskyi" for short. In reality, however, it is clearly a proxy war between the declining hegemon USA and the BRICS states demanding a multipolar world order, with Russia as an important player and its supporters.

However, this does not mean that peace is impossible, but rather that peace must be actively shaped by resolving political conflicts by non-military means. There is a qualitative difference compared to all other wars in world history, and this lies in the potential ability of nuclear powers to destroy a large part of humanity and the natural environment using nuclear weapons. The associated deterrent ("MAD" = Mutually Assured Destruction) is certainly effective but is increasingly being questioned by some political circles these days. This should be worrying - but it obviously isn't. Most people in highly developed industrialized countries (thank God!) know wars almost exclusively as a multimedia spectacle that takes place far, far away. The mentality behind this, which Johann Wolfgang von Goethe (1749-1832) describes so beautifully in his "Faust I", is obviously timeless (point 62).

Whether the ideal state of lasting "world peace" will ever seem achievable is questionable, given the more than 250 armed conflicts in the world since the end of the Second World War, in which the USA has been involved 44 times and Russia (including the Soviet Union) 10 times (as of 2024). And yet there have certainly been periods in world history in which (at least regionally) peace has lasted longer.

486. The Pax Romana

The "Roman period of peace," the "Pax Romana," is an extremely remarkable and largely peaceful epoch in European history. It lasted for more than two centuries, beginning with the reign of Emperor Augustus in 27 BC and ending around 180 AD. This exceptionally long period of relative peace and stability transformed the Roman Empire into a stronghold of economic prosperity and cultural flourishing. It followed extremely turbulent times in the late Roman Republic when, after years of bloody civil wars, Emperor Augustus (63 BC - 14 AD) succeeded in consolidating his power and implementing far-reaching reforms. He created a centralized administration, reformed the tax system, and established a professional army. These measures laid the foundations for a long period of peace, the beginning of which has gone down in history as the "Augustan Age."

During the Pax Romana, as already mentioned, the Roman Empire experienced a period of unprecedented stability and prosperity. The political order, which was shaped by a series of mostly capable emperors, allowed trade and the economy to flourish. An extensive network of trade routes connected the most remote corners of the empire and promoted the exchange of goods, ideas, and cultures.

The Romans used this time of peace to build up an impressive infrastructure. Roads ran through the empire like veins, connecting cities and facilitating trade and troop movements. Aqueducts supplied the growing cities with water and magnificent public buildings bore witness to Rome's power and wealth.

However, the Pax Romana was not only a time of material prosperity. It also brought with it a period of cultural prosperity. Artists, writers, and thinkers found ideal conditions for their work in this stable environment. Poets who are still known and revered today, such as Virgil (70 BC - 19 BC), Horace (65 BC - 8 BC), and Ovid (43 BC - 18 AD), historians such as Livy (59 BC - 17 AD) and Tacitus (58 -120 AD), as well as philosophers such as Seneca (4 BC - 65 AD), enriched the intellectual life of the empire with their works and left behind a cultural legacy that continues to have an impact today. In Alexandria, Claudius Ptolemy developed a geocentric mathematical theory of the movement of celestial bodies that was to endure for well over 1,000 years. With his work *"Geographike Hyphegesis,"* he was also the greatest geographer working in the West at the time.

Despite its name, the Pax Romana was of course not free of conflict. On the borders of the empire, especially in Germania and against the Parthians, there were repeated clashes. There were also challenges internally, as the Jewish uprisings in Judea show. Never-theless, the Roman institutions managed to overcome these crises and keep the empire stable overall.

The gradual end of the Pax Romana began with the reign of Emperor Commodus (161 - 192 AD) towards the end of the 2nd century. His incompetent government ushered in a phase of instability that ultimately led to the crisis of the 3rd century. It roughly covers the period between 235 and 284 A.D. It was a time of intense political, military, and economic turmoil in the Roman Empire, which brought the Roman Empire close to the brink of collapse. This era was characterized by unprecedented political instability, frequent changes of emperor, and short reigns, which led to constant uncertainty. More than twenty emperors ruled during this period, many of whom had come to power through military usurpation and were often overthrown or murdered just as quickly. These constant power struggles weakened the central power and undermined the authority of the empire.

At the same time, the Roman Empire was confronted with considerable external threats. Hostile peoples such as the Germanic tribes, including the Franks and Alamanni, invaded the western provinces, while in the east the Arab-born Sassanids posed a serious threat. The empire's military resources were overstretched and the constant fighting led to fur-ther destabilization. Economic problems further exacerbated the crisis. Constant wars and civil unrest led to a decline in agricultural production and trade activities. Inflation and coin debasement caused widespread economic insecurity. The tax burden on the population became overwhelming, and many people became impoverished or fled to the countryside to escape taxation.

It is easy to imagine that the social and cultural impact of this period of crisis was con-sid-erable. The constant insecurity and threat led to a decline in urban life. The cities, once centers of trade and culture, fell into decline and the population retreated to more rural, easier to defend areas. The social structure changed, and many people sought protection and stability with powerful local landowners, laying the foundations for the feudal system that was to follow.

The crisis finally came to an end with the reforms of Emperor Diocletian (284-305). Dio-cletian famously introduced the tetrarchy, a system of quadripartite rule in which the empire was divided into four administrative areas, each under an Augustus and a Cae-sar. These reforms strengthened military and administrative efficiency and brought back a certain stability. Diocletian's economic reforms, including the introduction of a new tax system and the stabilization of the currency, helped to restore economic order. He also introduced social and legal reforms that strengthened the empire in the long term. However, the former prosperity of the Pax Romana could no longer be achieved.

487. The lingering lessons of the "Pax Romana"

The time of the Pax Romana offers valuable lessons for today's world, which is once again being shaken by crises. It demonstrates how important stable and intelligent lead-ership is. Emperor Augustus and his successors ensured a long period of stability with strategic measures. Strong institutions ensured order and continuity, which modern governments should consider essential for order and prosperity.

Economic integration and infrastructure, such as trade routes and public buildings, fos-tered prosperity. Investment in infrastructure and cooperation is therefore crucial to economic success. The cultural heyday also underlines the importance of education, culture, and science for an innovative society.

The decline of the Pax Romana is a reminder to be vigilant against political instability. Frequent changes of power and internal tensions led to uncertainty. Mechanisms for the peaceful transfer of power and conflict resolution are therefore necessary. Military threats highlight the importance of a strong defense policy. Economic difficulties and social change underline the need for economic resilience and social cohesion. Social inequality must be reduced and social security systems strengthened.

The reforms of Emperor Diocletian illustrate how important adaptability and willingness to reform are in times of crisis. Governments and organizations should be prepared to make the necessary changes to ensure stability and success. Overall, the Pax Romana teaches that peace and prosperity can be achieved through wise leadership, strong in-stitutions, economic integration, cultural promotion, and adaptability.

Although the Pax Romana came to an end, its influence remained. The political and legal structures created during this period had a lasting impact on the further development of Europe and were rediscovered during the Renaissance. The achievements in art, litera-ture, and science set standards that are rightly admired to this day.

488. Roman decadence

However, the Pax Romana also laid the foundations for a social phenomenon known as "Roman decadence," which was particularly evident in the late phase of the Imperium Romanum (and signs of which can already be seen locally in Europe). It refers to a departure from traditional Roman virtues such as discipline, a sense of duty, and simplicity in favor of a lifestyle characterized by luxury, corruption, and moral neglect (only as far as the wealthy ruling class was concerned, of course).

This decline was closely linked to the conditions of that period of peace. The long period of peace and the unprecedented economic prosperity of this era led to an accumulation of wealth in the hands of the upper classes. The wealthy Romans flaunted their wealth with lavish parties, opulent villas, and expensive clothing in order to enhance their social status. These excesses thus became a status symbol and a sign of power, which led to a gradual loss of the traditional Roman values mentioned above.

The political stability of the Pax Romana also led to a concentration of power and wealth in the hands of relatively few, which in turn encouraged intrigue and power struggles within the elite. Corruption and bribery were the order of the day, as many officials and politicians pursued their own interests rather than the common good. This internal disintegration ultimately undermined the efficiency and authority of the government, with corresponding effects.

The loss of military spirit was also a decisive factor in the resulting decline. Security within the empire led to complacency and neglect of military discipline. Dependence on mercenary troops increased, which in turn had a negative effect on the loyalty and effectiveness of the army. The dwindling military strength made the empire more vulnerable to external threats and internal uprisings.

Economic problems exacerbated the situation. The luxury and extravagance of the elites stood in stark contrast to the growing poverty of the lower classes. The tax burden became overwhelming and drove many people into poverty or to flee to the countryside to avoid paying taxes. The resulting social tensions, in particular, contributed to the destabilization of the empire.

The cultural decline was also particularly noticeable. The once flourishing cities, centers of trade and culture, increasingly fell into decline as the population migrated to more rural and easier-to-defend areas. The social structure changed, and many sought protection and stability with powerful local landowners, laying the foundations for the feudal system that was to follow.

All these developments took place over several centuries, beginning around the late 2nd century AD and culminating in the final fall of the Western Roman Empire in 476 AD, the year in which the last Western Roman Emperor, Romulus Augustulus (c. 460 - c. 507), was deposed by the Germanic military leader Odoacer (c. 433 - 493).

489. The Decline of the West

Empires undergo a development that leads from rise through a more or less long period of prosperity to decline. The first person to recognize this and provide examples was the philosopher of history Oswald Spengler (1880-1936). In his major work *The Decline of the West*, which was published between 1918 and 1922, Spengler developed a comprehensive theory on the cyclical progression of cultures and civilizations. He viewed every culture as an organic being that goes through fixed phases of birth, growth, maturity, and decline.

Unlike many other historians of his time, Spengler saw history not just as a linear sequence of events but as a succession of cultures, each going through its own life cycle. Within this framework, he compared the Roman Empire with other great civilizations and described the transition from cultural prosperity to decadence and finally to decline.

For Spengler, this decline symbolized the inevitable conclusion of the cyclical course of every culture. The Roman Empire, which had once flourished through its cultural and military strength, had now become a rigid civilization that was no longer able to cope with internal and external challenges. For Spengler, the final fall of the Western Roman Empire was therefore the logical end of such a cultural cycle.

Fig. 228: First edition of the "Decline of the West" by Oswald Spengler

In *The Decline of the West*, Spengler also drew parallels between the decline of the Roman Empire and modern Western civilization. He saw similar patterns of decadence, materialism, and cultural exhaustion in his own time (German Empire). He therefore warned that Western civilization was also in a phase of transition and could experience a similar decline to the Roman Empire if it did not renew its creative and spiritual energies.

Although Spengler's work still enjoys a certain popularity among interested circles, it is also subject to critical scrutiny. Some historians and sociologists, for example, argue that his theory is too deterministic and underestimates the ability of societies to renew and change. Others emphasize that history is not necessarily cyclical and that every historical context is unique. In any case, it is worth reading this book again - especially if you have some classical education - and considering its premises.

490. The Clash of Civilizations

One person who did this a long time ago was Samuel P. Huntington (1927-2008). In his book *The Clash of Civilizations and the Remaking of World Order*, Huntington, an eminent American political scientist, took up the idea that cultures and civilizations play a central role in shaping and changing the world order. In his work, published in 1996, Huntington put forward the provocative thesis that after the end of the Cold War, world politics would no longer be determined by ideological or economic conflicts, but by cultural and religious differences.

The core thesis of Huntington's book is that the main lines of conflict in the future will run along cultural and civilizational lines. He identified several major civilizations - including Western, Islamic, Confucian, and Hindu - and argued that interactions between these civilizations would be increasingly characterized by mistrust, competition, and conflict. Huntington saw Western civilization in a similar stage of cultural exhaustion and decadence as Spengler had described for the Roman Empire and warned that Western values and institutions would be challenged by the rise of other civilizations.

Huntington's book was widely discussed among experts and met with both approval and fierce criticism. Supporters, for example, praised his clear and bold analysis of post-Cold War geopolitical realities and his emphasis on the role of culture and religion in international politics. Critics, on the other hand, accused Huntington of overemphasizing cultural differences and underestimating the potential for dialogue and cooperation between civilizations. Some warned that his theses could lead to a self-fulfilling prophecy by influencing the perception and behavior of actors on the international stage.

The relevance of Huntington's core statements for today is still controversial. In a globalized world where cultural identity and religious affiliation are constantly causing tensions, his predictions seem to be borne out in many ways. The increasing tensions between the Western world and Islamic countries, the growing importance of China and India on the world stage, and the internal cultural conflicts in multicultural societies reflect the patterns described by Huntington. And the current dangerous flare-up of tensions between Western-oriented Israel and the Islamic states (Iran) and terrorist organizations (Hamas, Hezbollah) that support the Palestinians seem to prove him right here too.

Nevertheless, the question remains open as to whether these conflicts are actually inevitable or whether they can be overcome through better intercultural understanding and cooperation. Like Spengler, Huntington presents a challenge that encourages the reader to reflect on the deeper cultural and civilizational structures that shape our world. Both works offer valuable insights into the dynamics of cultural rise and decline and invite readers to apply the lessons of history to the present in order to shape a better future.

In today's political science, both Spengler and Huntington play only a minor role, as globalization and increasing economic interdependence have relativized the importance of cultural and civilizational differences. Instead of deterministic models that predict the downfall of civilizations, the focus today is on complex, interdisciplinary approaches that take economic, political, and social factors equally into account. Nevertheless, the work of Spengler and Huntington offers valuable perspectives for understanding historical patterns and cultural dynamics. Their theories are a reminder that cultural identity and social values continue to play an important role in world politics, even if they are no longer the only driving forces.

491. How long will humanity continue to exist?

On a meta-level, the problems discussed here raise the very interesting question of the extent to which societies of rational and technologically developed beings are capable of surviving in the cosmos on their own once they have reached the stage where the possibility of self-destruction exists.

This question ultimately leads us to profound reflections on the nature of intelligent civilizations and their long-term survivability. Our biological and psychological traits are known to have evolved over millions of years primarily to survive in a prehistoric environment hostile to humans. These traits may not be optimally adapted to the challenges of a high-tech civilization. Instincts and behaviors that were once essential for survival can therefore become dangerous in a high-tech society.

This is because people tend to prioritize short-term benefits over long-term consequences. This tendency, known as 'temporal discounting,' could prove highly problematic in the existential risk world we live in. The challenge in this case is to develop mechanisms that encourage us to think and act for the long term.

The ability to work together effectively as a global community and make decisions is becoming increasingly important. Our political systems and institutions are often not optimized for this global level. Effective global cooperation and governance could therefore be crucial to managing existential risks. As the aim here is to balance conflicting interests, this undertaking is very risky and crisis-prone (keyword "armed conflicts").

Furthermore, the rapid pace of technological progress could overtax our ability to adapt and develop ethical frameworks. There is always the danger that we develop new technologies faster than we can understand and regulate their possible consequences. One example of this is the roll-out of artificial intelligence, which we are currently witnessing. Technology ethics play an important role here to ensure that innovations do not become uncontrollable.

A key factor for the long-term survival of a society could also be the ability to survive and recover from crises. This may require the development of backup systems on a planetary or even interplanetary scale and could include alternative habitats, resources, and technologies available in the event of a disaster.

These considerations suggest that intelligent civilizations may need to fundamentally change in order to survive in the long term. This may require an expansion of cognitive capabilities, a redesign of social structures, or even a transcendence of biological nature. Many philosophers of science now believe that biotechnology, nanotechnology, artificial intelligence, and other disruptive technologies could make such a transformation possible.

In theory, humanity could exist for millions or even billions of years if we overcome global challenges and evolve technologically (which, by the way, is highly unlikely for many reasons). However, according to the famous physicist Enrico Fermi (1901-1954), the fact that, despite ever-improving astronomical observations, not a single sign of extraterrestrial civilizations has yet been found points to a fundamental problem. This problem, known as the "Fermi Paradox," poses the question of why, if the universe is so old and vast, we have been unable to discover any traces of other intelligent life forms...

492. The Fermi paradox

The Fermi Paradox is based on a simple but profound question posed by Enrico Fermi in the 1950s: *"Where are they all?"* This question refers to the apparent contradiction between the (then-assumed) high probability of the existence of extraterrestrial civilizations in the universe and the lack of evidence or contact with such civilizations.

Fermi posed this question during a conversation with colleagues (including Edward Teller (1908-2003), Herbert York (1921-2009), and Emil Konopinski (1911-1990)) over lunch at Los Alamos National Laboratory in 1950. The discussion centered on UFO sightings and the possibility of extraterrestrial life. Fermi noted that, given the enormous number of stars in the Milky Way, many of which could have planets on which life could evolve, and the assumption that some of these life forms would be intelligent enough to undertake interstellar travel, our galaxy should have been colonized by such civilizations long ago. But despite this probability, there is no convincing evidence for the existence of extraterrestrial civilizations. This discrepancy between high probability and lack of evidence is what constitutes the Fermi Paradox.

In detail, Fermi's argument (which can be reproduced mathematically on the famous "beer mat") was as follows:

- There are an estimated 100-400 billion stars in our galaxy, many of which are sunlike stars.
- Many of these stars probably have planets, some of which are suitable for the formation of life.
- On these planets, intelligent life capable of interstellar space travel could evolve.
- Even with a very conservative estimate of the rate of spread of intelligent civilizations, the entire Milky Way system would have to be colonized within millions of years.
- Yet we see no evidence of such colonization or contact with us.

One possible explanation for the Fermi Paradox is that technologically advanced civilizations ultimately destroy themselves. This could happen due to a variety of factors, including wars, environmental disasters, uncontrolled technologies, or other existential risks. Another explanation could be that such civilizations deliberately hide or do not want to make their existence known in order to protect themselves from potential threats.

But the solution could also be even more banal, namely that the probability of "life" emerging on some extrasolar planet and developing into an "intelligent" life form capable of space travel is much lower than astronomers and astrobiologists, for example, usually assume (or rather "believe"). And we now want to investigate this thesis a little more closely.

493. Yes, where are they, the aliens...

The empirical basis for answering the question of whether there is life in general and extraterrestrial intelligence in particular anywhere in the universe other than on Earth, particularly in our Milky Way, is, strictly speaking, more than meager. On the other hand, we are gaining more and more clarity about how life works in detail, how it could have originated on Earth (here, too, a definitive answer is still pending), and according to which principles it has developed over millions of years. Biologists, particularly evolutionary biologists, have provided important insights into this. One of their most important and, for many people, rather disturbing findings is that biological evolution has no goal (even if it might appear so in hindsight). This is because the game of life is being played out anew every moment, and each species must constantly prove itself in order to survive - or become extinct. According to current knowledge, humans, as the "crown of creation," are nothing more than a chance product of an evolution that only appears logical and consistent in retrospect. Humans are only one representation of many millions of evolutionary lines that have survived to this day, such as the peacock butterfly, the house sparrow, the poppy, the slipper animal, the sequoia, or the oryx antelope. They have all successfully survived the struggle for existence; otherwise, they would not exist. But whether they will survive in the future is just as uncertain for them as it is for humans as biological objects. Through their special abilities, they have even created the potential to catapult themselves out of the history of life. A "dangerous" question in this context is whether what we commonly refer to as "intelligence" is perhaps even a limiting factor for technological civilizations, as has occasionally been suspected here and there, more or less justifiably.

Fig. 229: Typical appearance of an "alien" (generated by AI)

In any case, the fact is that to this day (2024), there is still not a single serious proof (indication) of the existence of extraterrestrial or non-terrestrial life. The question of whether the phenomenon of "life" in the cosmos is rather common or rather rare remains unanswered, and everything you can read about it is ultimately speculation. The "research" of Erich von Däniken, which he published a long time ago, does not help either. It is extremely exciting and suggestive but unfortunately only pseudoscience.

Even modern astrobiology, a scientific discipline that is still searching for the object of its research, has not yet been able to provide any concrete evidence of extraterrestrial life. However, it has gained many useful insights into how life could emerge and thrive under extreme conditions and what chemical and physical conditions are necessary for this to happen. These findings open up new perspectives on the possible forms and locations of life in the universe. Depending on how one interprets this information, one can come to the conclusion that intelligent life forms are either common or extremely rare in the cosmos. It could even be that we as humans are "Alone in Space" - the only intelligent species in an immense universe. But the truth is that we don't know.

494. The emergence of intelligent life [81]

The emergence of life is, of course, initially a basic prerequisite for the development of "intelligent" life, whereby we understand "intelligent" in this context to mean at least as intelligent as we humans. However, it is important to emphasize that intelligence exists in nature in many different forms. Dolphins, ravens, and bonobos also show impressive cognitive abilities in their own way - and of course, my tomcat "Humpel," whose behavior often indicates remarkable intelligence.

Despite these examples, the evolution of "intelligent" life, as we define it in relation to humans, is by no means inevitable according to everything science has gathered so far. On Earth, it has taken about 3.6 billion years for a species capable of cultural and technological advancement to evolve. During this long period, no other species - except for hominids - has even come close to a similar development.

It is remarkable that even humanity, which is the dominant intelligent species on Earth today, almost disappeared from the scene 70,000 to 80,000 years ago. This is evidenced by a so-called *"genetic bottleneck,"* which indicates a dramatic reduction in the human population. The cause was probably the eruption of the Toba supervolcano on Sumatra, whose catastrophic climatic effects caused the population of Homo sapiens to shrink to a few thousand individuals (point 320). This event shows how fragile the development of intelligent life can be and how much it depends on external circumstances. Despite these setbacks, humans have managed to recover and build the civilization we know today.

On the other hand, of course, no one knows exactly at what level of organization and under what selection conditions consciousness (as a prerequisite for something like culture) can develop at all, since in this respect we naturally only have the singular case of earthly life as an example. In the context of the synthetic theory of evolution, no "ultimate goal" (for example, in the form of self-conscious, thinking beings of any kind) can be derived. And there is something else to think about: the time factor.

---

[81] The argumentation is based on the statements that the author developed in his two specialist books "Planetologie extrasolarer Planeten" and "Astrobiologie" (Springer Spektrum 2014 and 2015)

Here, too, the example of the development of life on Earth suggests that the emergence of multicellular life itself is perhaps not so self-evident. It took almost 2 billion years from the beginning of chemical evolution (i.e., 4.1 - 3.8 billion years ago) to the appearance of the first metabionts (*gabonionts*) in the middle Pre-Proterozoic (i.e., 2.1 billion years ago). After a brief diversification phase of a few million years, they appear to have quickly died out again.

The first "permanent" multicellular organisms only date back to the Ediacaran, and from then on the development to humans only took about 600 million years... But even this development is, in retrospect, anything but predetermined. Or as Stephen Jay Gould (1941-2002) once put it: If *Pikaea*, the oldest known ancestral form of the cephalochordates (*Cephalochordata*), from which all animals with internal skeletons (*vertebrates*) evolved, had died out in the Cambrian, there would be neither herring nor humans today. After this point in time (542-488 million years ago), hardly any new animal phyla emerged (the taxonomic unit "phylum" represents the basic anatomical structure of living organisms). The development of life would have been completely different, and today animals with an exoskeleton might be the most complex living creatures on Earth. They would certainly have been able to form "states" (like ants and termites, for example), but whether they would have undergone technological development with the handicap of a rigid exoskeleton is left to the imagination of science fiction writers. And one more question is allowed in retrospect. What if the dinosaurs had not been wiped out by a meteorite impact 65 million years ago? Would mammals then have undergone a massive radiation a few million years later, from which humans eventually emerged? Would "intelligent" dinosaurs (or their descendants) perhaps be living on Earth today instead of us? Or, as Stephen Jay Gould (1941-2002) put it in one of his many clever essays:

"Had not the celestial body destroyed their (the dinosaurs', ed.) flourishing diversity, they might still be alive today (Why not? They had been doing well for 100 million years, and only 65 million years have passed since then in the history of the earth). If dinosaurs were still around, mammals would almost certainly be small and insignificant (as they were during the hundred-million-year reign of the dinosaurs). And if mammals are small, limited in their capabilities and not endowed with consciousness, they certainly don't give rise to humans who can express their indifference. Or who name their sons Peter. Or who marvel at heaven and earth. Or who ponder the nature of science and the proper relationship between fact and theory. They would be too stupid to try; too busy getting their next meal and hiding from the evil velociraptor."

Such "What if...? - questions" - for example, *"What if no celestial body had hit the earth 65 million years ago?"* - help to develop a feeling for probabilities, even if no reliable statistical data is available. In this case, it is about the probability that intelligent beings will emerge at some point in an evolutionary tree, which will undergo a cultural-technological development and thus be able to communicate interstellar.

In order to be able to seriously estimate the frequency of "intelligent life" in the cosmos, which has undergone a cultural-technological development (i.e., forms an "extraterrestrial civilization"), some specific questions must first be answered. However, without further "examples" of such civilizations, this is hardly possible or only possible with considerable uncertainty. And these questions relate less to astronomy than to biology. With an estimated potentially habitable planets in our Milky Way alone, there should be enough places where the miracle of life has taken root.

The first important question in this context is under what conditions and in what time frame multicellular life develops from unicellular life. The second question concerns the emergence of "intelligence," a special cognitive ability that leads to self-reflection and consciousness above a certain level of complexity and, under certain circumstances, enables the emergence of civilizations. However, intelligence alone is not enough. Assuming a dolphin had the same (or even higher) cognitive abilities as modern humans, the path to a technological civilization would be blocked due to its physical limitations, which restrict its ability to actively change its environment.

Intelligent behavior has developed many times in the animal kingdom - not only in warm-blooded animals such as birds or mammals. The example of the dolphin shows that further abilities are necessary for a living being to be able to shape its environment in a purposeful and conscious manner. This includes the development of a syntactic-grammatical language (with corresponding vocal abilities) and its writing in order to effectively preserve information independently of the individual - a fundamental prerequisite for culture. Equally important is the ability to change the environment with one's "hands" according to one's own ideas, which creates the potential for setting in motion a technological development based on mathematics and science. The probability of this happening can hardly be seriously estimated. For even the emergence of a civilization is by no means a guarantee that it will take the path to science and technology (in the modern sense) in the course of its existence, as many examples on Earth show...

Another question that plays an important role in this context (especially with regard to a possible solution to the Fermi paradox) concerns the average lifespan of a technological civilization. For example, the question arises as to whether such civilizations can perish after a relatively short time due to internal conflicts or scarcity of resources.

Some general considerations can be made about the emergence of the characteristic "intelligence" in an evolutionary tree, which must naturally be based on the example of the Earth. It is remarkable, for example, that this specific characteristic in its human form only appeared relatively late in the possible existence of the terrestrial biosphere. In around 500 million years, conditions on Earth will become increasingly worse until life on land and eventually in the oceans becomes impossible. The reason for this is the gradual increase in the sun's luminosity on its way to becoming a red giant star. If we assume that there are about 4.2 billion years (100%) between the emergence of life and the extinction of complex life, then the characteristic "human intelligence" only emerged when life on Earth had already completed about 85% of its possible period of existence.

Obviously, the time span from the emergence of life to the development of an intelligent species capable of cultural and technological development has little to do with the lifetime of a star in the main sequence phase. As Stephen Jay Gould has impressively shown, what we regard as "progress" in biological evolution is rather an extension of an asymmetric distribution of complexity, represented by the number of species (ordinate) over this property at a given time (e.g., in the present or in the Triassic). Since the development of life is not teleologically predetermined, it takes a certain amount of time in the course of evolution to reach the point at which the emergence of "human intelligence" becomes possible.

If this period is relatively short (e.g., one billion years), then the cosmos should be teeming with extraterrestrial civilizations, which would contradict the Fermi paradox (point 492). However, if this evolutionary path is very long on average - i.e., longer than the dwell time of a suitable star in the main sequence stage (perhaps 50 billion years) - then extraterrestrial civilizations would be extremely rare. The example of Earth rather points to the second case, if one understands "suitable stars" to be those of the type of our Sun (F and G stars). Without the Chicxulub impact some 66 million years ago, evolution might have taken much longer to produce a civilization, or a civilization with a cultural-technological character might never have emerged. It is not mere intelligence that makes the crucial difference; rather, it takes a variety of complex and fortunate circumstances to enable the emergence of a technological civilization. In fact, these circumstances could be so rare that even in a vast universe, civilizations are an extreme exception.

495. Copernican principle and Drake formula

If, according to the "Copernican principle"[82], humanity occupies a typical middle position in the cosmos, then the potentially remaining 15% "lifetime" also means that the probability of different "civilizations" existing in the Milky Way in parallel (and not one after the other) cannot be particularly high.

An estimate of how many extraterrestrial civilizations could exist "out there" has of course been attempted many times. A simple mathematical relationship named after the radio astronomer Frank Drake (1930-2022) is used for this purpose and is easy to understand:

$$N = R_* \cdot f_P \cdot n_e \cdot f_l \cdot f_i \cdot f_c \cdot \frac{L}{L_S}$$

The meaning of this equation is quickly explained. The starting point is a huge number R_*, the number of all stars in our Milky Way system (between 200 and 400 billion). The result is the number N of stars that host a planet with a developed (i.e. communicative) civilization. In our Milky Way, this number is in any case >= 1. Each additional factor in this equation determines a probability (in the form of a percentage) for a certain condition that such a civilization can arise and exist today (in the sense of being reachable by radio signals):

---

[82] The Copernican Principle states that the Earth does not occupy a special or central place in the universe, but that the physical laws and conditions are essentially the same at eve-ry point in the universe.

f_p Proportion of stars that have planets

n_e Number of Earth-like planets that are located within the habitable zone of a star with planets

f_l Proportion of planets on which some form of life has developed

f_i Proportion of planets with intelligent life

f_c Proportion of planets with a civilization capable of interstellar communication

L Average lifespan of a technical civilization

L_S Average time required for the emergence of a technical civilization (calculated from the formation of the parent star).

The mathematical simplicity of this formula, which was first established in 1960, is offset by a remarkable lack of knowledge of most of the numerical values in the product of the right-hand side. Various authors have therefore made more or less plausible estimates of the individual factors over the course of time, with the tendency for the result, the number N, to have rapidly decreased since 1960. There are several reasons for this. During the time when Frank Drake first wrote this formula on the blackboard of the Green Bank Observatory seminar room (at that time N was still in the order of a million), astronomical and biological research had developed many times over. The discovery of planets around stars is now part of everyday business and astrobiology has developed into a multidisciplinary field that investigates the conditions for the formation and existence of life in detail - research results that are directly incorporated into an estimate of the Drake numbers f_p, n_e and f_l. And the "astronomical" results alone are already sobering. This is not f_p. The number of planets in our Milky Way will almost certainly reach the number of all stars, and in all probability even exceed it. This positive result is somewhat relativized by the fact that the size of habitable zones around suitable stars (which are characterized, among other things, by long-term stability with regard to their luminosity) has had to be drastically reduced. Also, not all zones of a galaxy are equally suitable for the development of life, which is known to have led to the term "galactic habitable zone". This already reduces the number of stars on whose planets life can at least potentially develop undisturbed over several billion years. Nevertheless, this number is still "astronomically large" for our imagination.

Before discussing the Drake numbers, which biologists are responsible for evaluating (i.e. f_l and f_i), it is worth pointing out an extremely important aspect of the mathematical structure of the Drake equation: The equation consists of factors, most of which are probabilities, i.e. numbers between 0 and 1. Only a single probability that is very small, i.e. close to 0, can N be extremely degrading. And two numbers smaller than 1, which are multiplied together, result in an even smaller number ... This means that this formula is practically useless for realistic estimates, as there are not even approximately reliable values for most factors. And theoretically, the Drake equation can be extended by further factors, for example by the proportion of planets with a large moon (stabilization of the rotation axis), the proportion of planets with plate tectonics (planetary thermostat), the proportion of planetary systems with a Jupiter-like gas planet in the outer system (stabilization of planetary orbits in the habitable zone), and, and, and. And each new factor causes N further shrinkage. And these probabilities initially only relate to the astronomical prerequisites for life.

So what about the "biological" Drake numbers? These figures are usually grossly overestimated by astronomers and physicists (but also by laypeople), i.e. they are far too optimistic. This also applies to the Number L, about which it is not really possible to say anything precise, even with the experience of advanced civilizations on Earth, except that its value probably exceeds 200 years - but by how much is completely uncertain. Because one thing is clear: technical civilizations can fail in many different ways, be it through overpopulation (lack of resources), climate change, self-extinction or cosmic catastrophes – L it may well be smaller than generally assumed (a few thousand to ten thousand years). Even f_c does not necessarily have to be 1, considering how many advanced civilizations on earth have found their way to science and technology on their own over the last 6000 years ...

From the point of view of renowned evolutionary biologists, the number f_i is very, if not exceptionally, small in comparison with f_l. Microbial life is probably even quite common in the cosmos, so there is a non-negligible probability of finding such creatures or traces of them elsewhere in the solar system (Mars, Jupiter's moon Europa). But the probability of complex life developing from them is already lower (for a billion years there were only prokaryotes on earth - and they still dominate the earth today in terms of the number of individuals and biomass). And finally, the emergence of a living being such as humans with the potential for technological development represents the pinnacle of the improbable. This realization shrinks the number N to such an extent that it no longer seems inconceivable that Earth is the only planet in our galaxy that is currently home to a technological civilization. This is one of the truly "dangerous ideas" as it implicitly challenges a reassessment of humanity's place in the cosmos.

"And even if something were to happen somewhere in the infinite universe that could be compared to the origin of human intelligence, the chance that we could enter into an exchange must be considered non-existent. Yes, for all practical purposes, man is alone in the universe."

writes the old master of all evolutionary biologists, Ernst Walter Mayr (1904-2005), in his highly readable book *"What Evolution is"*.

In this respect, he is undoubtedly right, because even if the old SETI (*Search for Extraterrestrial Intelligence*) project should be successful at some point, no exchange of information is possible in view of the cosmic distance scales.

496. Jet through the galaxies with WARP drive...

As nice as it may sound to jet from galaxy to galaxy with warp drive, it is not practically possible according to current and most likely future knowledge. But as I said, many things are conceivable, including warp drive, of course. Or, to put it scientifically: With a little skill, one can assign values (functions) to the energy-momentum tensor of Einstein's gravitational field equations in such a way that space-time assumes local properties that enable a quasi-instantaneous change of location - as can be admired in Star Trek. In practice, however, this is impossible due to the unavailability of the necessary "ingredients" (i.e., matter with negative energy density, so-called "exotic matter"). And this will not change in the future. A warp drive may be patentable, but you can't build it, even if you were to order it "from above"...

497. The "SETI" project and its prospects of success

If you can't "fly" to exoplanets that are probably inhabited, then perhaps you can "listen" to see if any extraterrestrial civilizations are emitting signals that can be used to identify them. This is the idea behind the "SETI" project, the "Search for Extraterrestrial Intelligence".

This project, which is now - and to a large extent privately funded - coordinated and continued by the University of Berkeley, goes back to an initiative by Frank Drake at the beginning of the 1960s. The "BigEar Project" at Ohio State University (named after the "Great Listener" from Lyman Frank Baum's (1856–1919) The Wizard of Oz) was characterized by great optimism, as many millions of civilizations were expected in our Milky Way alone. For this reason, the Russian astronomer Nikolai Semyonovich Kardashev (1932–2019) even developed a classification system to categorize them if necessary. Since then, observation possibilities have, of course, improved rapidly. In particular, simultaneous measurement in many hundreds of individual channels provides data volumes that are no longer manageable under reasonable financial and technical conditions. But thanks to the Internet and the "SETI@home" project made possible by it, the computing power of several million home computers could be pooled to sift through the data generated worldwide by radio telescopes during their normal observation operations to find out any peculiarities. There has still been no new "Wow!" signal. But the search continues, and that's a good thing. Because the theory that extraterrestrial civilizations are very rare is, as I said, initially speculation - even if it is well founded. But in science, facts count. Even if the chances of discovering any kind of artificial signal from the vastness of cosmic space in the sense of SETI are very slim, there may be other exciting results that are definitely worth the effort. Just think of the discovery of pulsars by Jocelyn Bell-Burnell and Antony Hewish (1924–2021) in 1967 (LGM 1 - "Little Green Man 1"), where the scientists involved did not immediately rule out "extraterrestrials" as the sender.

Fig. 230: The special antennas of the Allen Telescope Array in southern California are hoped to detect radio signals from extraterrestrial civilizations (Wikimedia)

The feeling that technical civilizations are probably something extremely rare is of course not necessarily true. It is first and foremost a speculation or hypothesis - albeit a well-founded one - that people would naturally like to have resolved. Hence all the money for SETI and exoplanet research. The feeling of being lonely, as a human being or as a civilization, is not a pleasant one. Unfortunately, the question "Are we alone in space" or - better - "... alone in the Milky Way" is a question that can only reasonably be answered with "no" by a positive example. As long as there is no evidence of "aliens," it can always be claimed that they are hiding in one of the many places in the Milky Way that are inaccessible to us or that they have no interest in drawing attention to them-selves.

The example of "Earth" also leads to a narrow view that projects the singular terrestrial evolutionary history of life onto other planets as the only representative one *("Rare Earth Hypothesis"* by Peter Ward and Donald E. Brownlee). This entails the danger of overesti-mating this evolutionary history because we do not know (and cannot know) whether certain developments could have taken place quite differently under different condi-tions - even on a temporal scale. In short, we lack the imagination and, in some cases, the knowledge to even consider other scenarios in this context. In other words, the ques-tion of the existence of extraterrestrial civilizations remains open to speculation.

All attempts to calculate a realistic figure from the Drake equation may have practical value (for example, to justify budgets for new space telescopes or radio antennas), but scientifically they are pure gimmickry. However, this does not apply to the determination of the size and error range of some factors of this equation (point 495), which are acces-sible to observation, such as (the census of all stars in our Milky Way, in progress) and (the task of exoplanet research, unfortunately still quite uncertain due to methodologi-cal measurement artifacts). These factors could perhaps be estimated empirically in the future as soon as it becomes possible to sort out Earth-like planets that probably have a biosphere using bioindicators. But it is impossible to predict whether this endeavor will be successful. All other factors (still) defy objective delimitation, as they can only be statistically determined on the basis of a single example, Earth - and for a statistician, that's really not much...

Let's assume that the thesis is correct that developed civilizations capable of at least making themselves known in the cosmos (for example through radio waves or laser ra-diation) are a rare phenomenon. Would this not then contradict the philosophical *"prin-ciple of mediocrity"* (Copernican principle), which casually states, *"You, the reader, I, the author, we together, all humans, our Earth, our Sun, our Milky Way, are in themselves nothing special in the cosmos"*? Certainly not. You just have to enlarge the radius of vi-sion. In addition to our Milky Way, there are well over 1 trillion galaxies in the (probably infinitely large) universe (diameter of the Hubble bubble ~93 billion light years, depend-ing on the model). Of these, ~34% are spiral galaxies similar to our Milky Way. If we now assume that there is on average one technologically highly developed civilization per spiral galaxy (which is also the assumption of the "Rare Earth" hypothesis), then there are still several hundred billion in the manageable cosmos... This should satisfy the Co-pernican principle. Ironically, however, nature has arranged it so that these many civili-zations can never come into contact with each other...

So no more Klingons, Nausicaans, Talaxians, and whatever they're all called. A pity, re-ally...

498. The Star Trek universe

The Klingons are known to be a more combative race, whose warriors are characterized by a strong stature and a distinctive forehead structure ("forehead bulge"). They have a darker skin color, ranging from bronze to dark brown, and long, thick hair that they often wear loose. However, they do not inhabit "our" universe but the "Star Trek" universe, together with many other species. This fictional universe is notably different from our own; it is remarkably densely populated with humanoid beings who have developed the ability to travel faster than light, forge interstellar alliances, and colonize a multitude of planets. In contrast to our world, which is often characterized by conflict and inequality, humanity in the "Star Trek" universe has overcome many of these problems. Poverty, hunger, and disease have been all but eradicated, and society has evolved into a peaceful and cooperative federation based on the principles of equality, respect, and the pursuit of knowledge. The protagonists include a variety of intelligent species, each with different philosophies, lifestyles, and technologies. These species, such as the aforementioned Klingon warriors, the logical Vulcans, the enigmatic Romulans, and the menacing Borg, contribute significantly to the complexity and depth of the universe.

Fig. 231: The Star Trek spaceship is highly recognizable and boasts some incredible abilities (Creative Common, Kurumi Morishita)

Of course, this (fictional) world is not free of conflict. While in the real world, as we know, military solutions and power struggles often take center stage, the "Star Trek" universe places great emphasis on diplomacy, negotiations, and the pursuit of peaceful solutions. Starfleet, the central organization of the Federation, is a symbol of these values. It serves not only as a defense force but also as a scientific and diplomatic institution dedicated to the exploration of the universe and the protection of peace.

In short, the "Star Trek" universe is a place of hope and progress, where humanity, together with other intelligent species, strives to understand the universe and build a better future. It is a universe that shows us what is possible when we overcome our differences, work together, and use our knowledge and skills for the benefit of all. It is a place where diversity is celebrated, conflicts are resolved peacefully, and infinite space awaits us as the last great adventure.

And it is, unfortunately, a utopia...

The series "Star Trek" was created in the 1960s, when screenwriter and producer Gene Roddenberry (1921-1991) had a visionary idea: a science fiction series that not only entertains but also reflects social issues and shows a positive vision of the future of humanity. Inspired by western series such as "Wagon Train," Roddenberry wanted to create a "train to space" in which a motley crew explores alien worlds and has to overcome moral and ethical dilemmas in the process.

"Star Trek" was broadcast in 1966 despite initial reservations on the part of the television stations and quickly developed into a true cult phenomenon. The series was successful because it dealt with issues such as racism, war, peace, and human rights in a futuristic context and presented a progressive, united, and peaceful vision of the future. The concept of a mixed-race cast, at a time when racial segregation still prevailed in the USA, was simply revolutionary.

Furthermore, "Star Trek" helped bring the science fiction genre into the mainstream by pushing the boundaries of what could be shown on television. The series depicted a future in which humanity has overcome its differences and works together towards a better world. This utopian vision, which stood in sharp contrast to the political tensions and social unrest of the time, offered viewers some much-needed hope and inspiration.

The series not only offered viewers exciting adventures but also a hopeful message that still inspires many people today.

499. The place of education and knowledge in the Star Trek universe

The "Star Trek" universe is deeply connected to the theme of "education and knowledge," both in the stories it tells and in the way it has influenced viewers in the real world. At the center of the universe are the values of knowledge, science, and curiosity. Starfleet, the central organization in the "Star Trek" universe, is not only a military arm but, above all, a scientific and diplomatic institution whose main mission is to explore space, gather knowledge, and promote exchange between different cultures. This mission is reflected in its famous motto: "To boldly go where no one has gone before." It is about discovering new worlds, understanding the unknown, and using the knowledge gained for the benefit of all.

Education and technological achievements play a central role in "Star Trek." The series depicts a future in which education and science enable humanity to develop advanced technologies such as the warp drive, replicators, and transporters. These technologies are not only fascinating but are also used to solve complex problems, avoid conflicts, and improve life for everyone. "Star Trek" shows how important it is to use knowledge responsibly and to consider the ethical implications of every scientific discovery.

Fig. 232: The legendary crew of the starship Enterprise were not only extremely clever and likeable, but also exemplary in their courage and determination. (Creative Commons)

The series also uses its futuristic setting to explore deeper philosophical and ethical questions that otherwise always get short shrift. It encourages the audience to think about important issues such as racism, war, peace, environmental degradation, and human nature. These issues are not only relevant to the fictional world of "Star Trek" but are also exceptionally important in the real world. In this way, the series encourages critical thinking and prompts viewers to look at the world around them from new angles.

Beyond pure entertainment, "Star Trek" has inspired many people in the real world to pursue careers in science, technology, engineering, and math (the so-called STEM subjects). Numerous scientists, engineers, and even astronauts report that the series had a decisive influence on their career choice. The characters in "Star Trek," regardless of their species or origin, share a common goal: the pursuit of knowledge, progress, and personal development. This vision of an educated and diverse society has not only inspired viewers but has also shown that knowledge and education are the keys to a better and fairer future.

In a technologically advanced society like that of "Star Trek," education is crucial. General education enables people to better understand the world and make informed decisions. STEM subjects, on the other hand, are crucial for tackling the technological challenges of the future. They provide the foundations for innovations that can improve our lives. "Star Trek" shows that it is only through sound education and the constant pursuit of knowledge that a society can develop and successfully master the challenges that lie ahead. The series reminds us that education and science are not only tools of progress but also the basis for a peaceful, just, and prosperous future.

EPILOG

500. Knowledge and the pursuit of knowledge as an end in itself

Well, it may not be vital to know about the planet Pluto, or about the ferryman across the river Acheron called "Charon" (*Don't Pay the Ferryman!*) or about Chris de Burgh. But the world becomes more interesting when you do. We should therefore once again recognize knowledge as an end in itself, something that helps us to develop as a person, to understand and classify things in the world, and thus to lead a self-determined life.

An individual can only understand and critically reflect on the central problems of humanity with a certain degree of general education and the ability to criticize based on this. A person who knows something about the French Revolution can use this knowledge to recognize and realistically categorize a series of social problems and human abysses that occupy daily politics, for example. A person who has internalized some of the basic ideas of the natural sciences will, for example, recognize the smoke and mirrors that are presented to them every day by the media in relation to climate policy and "renewable" energies. And one thing should not be forgotten when it comes to acquiring specialist knowledge, as is expected in any demanding profession these days. Only those who have a strong general knowledge are usually able to assess things and issues themselves and critically, as they have looked at various facets and thus gained the necessary vision.

However, "knowing" something is undoubtedly also an intellectual pleasure. It may seem useless to most people to be able to name flowers by the wayside or point out the most important constellations in the sky. But such knowledge is part of a huge network of facts and connections that are interconnected in many different ways. And that is exactly what this short excursion into the "panopticon of fascinating things and events" should convey to you. Everything is somehow connected - and I haven't told you anything about "artificial intelligence," "quantum mechanical entanglement," and the "many-worlds" interpretation of quantum mechanics... What do you think? Should I?

P.S. Point 18: As you probably already know, exactly nine green hunkis fit into one Hemputi...

Inhalt

1.   Topics.. 1

2.   Cabinets of curiosities.. 2

3.   Nothing new under the sun 3

4.   The mysterious number Pi 4

5.   Every book in the world is included in Pi............................... 6

6.   ISBN and Pi.. 6

7.   Snow White and Rose Red 6

8.   The Brothers Grimm and the "German Dictionary 7

9.   Who or what is an idiot?.. 8

10.   Etymological dictionary 8

11.   Hans Albers as "Der Greifer" 9

12.   Munchausen's ride on the cannonball 9

13.   Lies, logic, paradoxes 10

14.   Mr. Shanks has made a mistake 11

15.   Loose and fallacies 12

16.   Gourmet tip: Herring with whipped cream 12

17.   Risk assessments 13

18.   About Knaffs, Plautze and Hunkies as well as Hemputis,
      which are predominantly red 14

19.   Fuzzy logic 14

20.   Advice literature: Ask Mummy! 15

21.   Screws and threads..15

22.   The fight against insufficient depth of field - focus stacking.......16

23.   Blowflies (Schmeißfliegen) 17

24.   Gourmet tip: Witch's eggs 18

25.   Egg cartons and breakfast eggs 19

26.   Non-Newtonian liquids...19

27.   Sinking into quicksand?...20

28.   Navier-Stokes - a millennium problem..................................21

29.   Why mathematics is important 22

30.   Plastic and mother-of-pearl egg spoon 24

31.	Shells and mother-of-pearl	24
32.	Knight Runkel and the black pearl	26
33.	Captain Nearchos and the Diadochi	27
34.	Perseus, the last Antigonide	27
35.	Perseus and 6000 talents of gold	27
36.	Greek historiography	28
37.	Why it's good to study universal history	28
38.	Weimar Classics	29
39.	The Bell (by Schiller)	29
40.	Goethe, Schiller, Arthur Schramm are the best we have.	30
41.	A zeppelin for catching flies	30
42.	Fly agaric and fly death	31
43.	Caesar's mushroom	31
44.	Money doesn't stink	32
45.	Urinals and fine blue fabrics	33
46.	Friedrich Wöhler and the urea	34
47.	What is "life"?	35
48.	The riding primordial dwarf	35
49.	When is life life at all?	36
50.	The mule paradox	36
51.	Stupidity as a World Power	37
52.	Mass stupefaction	38
53.	Is Stupidity Harmful?	41
54.	Then Why Don't You Move to Fukushima!	41
55.	Democracy	42
56.	Democracy and the Pitfalls of Electoral Law	42
57.	The Condorcet Paradox	43
58.	Astrologers, Horoscopes, and Scientific Education	44
59.	Barnum Statements	46
60.	Chang and Eng Bunker from Siam	47
61.	Separation of Siamese Twins	48
62.	Daily Newspapers and Journals	49
63.	Boulevard and Public Dumbing Down	50

64.	„Quality Press"	51
65.	Sapere aude! - The Best Defense Against Propaganda	52
66.	Self-Fulfilling Prophecies	53
67.	Oedipus and the Oracle of Delphi	53
68.	Jim Morrison and the Oedipus Complex	54
69.	At the Foot of Mount Parnassus...	55
70.	The Misfortune of Croesus	55
71.	The Gold of Midas	56
72.	Programming with Delphi	56
73.	Blaise Pascal and his Pascaline	57
74.	Invention of the Omnibus	58
75.	Why Paris is Called "Paris"	59
76.	Trojan Troubles	59
77.	The Hunchback of Notre-Dame	60
78.	Death in the Middle Ages	62
79.	Saint Christopher Sheets	63
80.	Indulgences	64
81.	Martin Luther and his 95 Theses	65
82.	Martin Luther and the German Language	66
83.	Martin Luther and the Reformation	68
84.	Prague in Bohemia	68
85.	The Prague St. Wenceslas Treaty	69
86.	Krušovice Dark	70
87.	Braunschweig Mumme	70
88.	Beer with a Head	71
89.	Not Always Are Dreams Foam – Carrie Jane Everson	72
90.	Gold Extraction	73
91.	„Biogold" and the Golden Fleece	75
92.	The Argonaut Saga	75
93.	Medea's Revenge	76
94.	Gold, Gold, Gold...	77
95.	How is Gold Created in the Cosmos?	77
96.	Life History of a Star with Eight Solar Masses	78

97. Heavy elements are built from neutrons 79

98. When Neutron Stars Collide 79

99. A Supernova and the Early History of the Solar System 81

100. Is Reality Real? 81

101. The Brain in the Nutrient Broth 82

102. Is there an external world? 83

103. Solipsism 84

104. Claudia Brücken and Propaganda 84

105. Dr. Mabuse 85

106. Those Magnificent Men in Their Flying Machines 87

107. Gert (Karl-Gerhart) Fröbe 88

108. Why Don't Ton-Heavy Airplanes Fall from the Sky? 88

109. Airspeed Measurement with the Pitot Tube 89

110. The Bumblebee Paradox 89

111. The Buff-tailed Bumblebee as a Rodent 90

112. Spring Aspect 90

113. Bee flies 90

114. Why are butterflies called "Schmetterlinge" in German? 93

115. Molkadiebschnopper, Mottakeenich and Tuud 93

116. The Book- and Clothes-Eating Moth 94

117. Naphthalene 94

118. The Skirt and What's Underneath... 95

119. The Massacre on the Champ de Mars 96

120. Women turn into Hyenas 97

121. Decisive Moments in History 97

122. "... if not, what remains of glory?" - Robert Falcon Scott 98

123. The Whaler Terra Nova 99

124. Sonar 100

125. Claustrophobia 101

126. Assassination of the President 101

127. The Mary Surratt Justice Scandal 102

128. Selbstmordattentäter, bomb ein bischen später... 102

129. Sunnis and Shiites 102

130.	Tyrannicide	103
131.	The older and younger tyranny	103
132.	„Die Bürgschaft" - by Schiller	104
133.	Jurassic Park	104
134.	The big brother of Sue	105
135.	How big can an Animal get?	106
136.	The Tower of Babel	107
137.	The Entropie Theorem	109
138.	What is Energy?	109
139.	Energy cannot be renewed	110
140.	There is no energy consumption	111
141.	The value of energy	111
142.	What is Entropie?	112
143.	Heat death of the universe	113
144.	The "measured Variable" Energy	114
145.	Primary energy and useful energy	115
146.	Primary Sources of Energy	116
147.	Entropy export and structure formation	117
148.	Primary energy consumption	117
149.	Biogas power plants	117
150.	"Energiewende" and castles in the air	120
151.	Why is the night sky black?	120
152.	Olbers Paradox	121
153.	Curved spaces and the parallelism problem	121
154.	The Spacetime	124
155.	The expansion of cosmic space - galaxy escape	125
156.	The cosmic background radiation	126
157.	What is "time"?	128
158.	Motion and the Zeno arrow paradox	129
159.	The "flow" of time	130
160.	The thermodynamic and cosmological arrow of time	131
161.	Thomas Mann and his "Zauberberg"	131
162.	„Als wär's ein Stück von mir"	133

163.	The job description of the censor	134
164.	The business of "censorship"	135
165.	The practice of censorship	137
166.	"Censorship does not take place" - political correctness	138
167.	The euphemism treadmill	139
168.	Biologically speaking, there are no human races	139
169.	Fools and jokers	140
170.	Dyl Ulenspegel	140
171.	Is mirror milk poisonous?	142
172.	Das Chiralitätsproblem	142
173.	The Ozma Problem	143
174.	Buridan's donkey	144
175.	Dilemma and Trilemma	145
176.	Ultimate truths	146
177.	Cosmological proof of God	146
178.	All about the devil	147
179.	Hell topography, demonology and torture science	148
180.	Black cat - white spot	153
181.	Unsolved problems in cat research	154
182.	Dead cats don't purr	154
183.	The "seven lives" of the cat	155
184.	Pisans can't even build straight towers	155
185.	Friesland - the land of leaning towers	156
186.	The Tower of Hanoi	158
187.	Mersenne numbers and prime numbers	159
188.	Citizen science	160
189.	Computer networks	160
190.	How does the Internet work? - TCP/IP	161
191.	A sufficiently advanced technology...	162
192.	Solaris	162
193.	For friends of Soviet film	163
194.	And it went "boom"...	164
195.	We haven't gotten away with it yet	166

196.	Self-made disasters	167
197.	Disturbed risk perception	168
198.	Pisa sends its regards	168
199.	Chernobyl, Fukushima and their nuclear deaths	169
200.	The dead fish from Fukushima	170
201.	Earth radiation and earth radiation detection	173
202.	The eco power filter	174
203.	Decadence indicator esotericism	174
204.	Water revitalization according to Grander	174
205.	Stupidity is not a disease	176
206.	The opponents of vaccination - bankrupt	176
207.	Miracle Mineral Supplements	177
208.	Quackery	179
209.	Diagnostics of ineffective alternative remedies	180
210.	The boom in pseudoscience	180
211.	Placebo effect and spiritual healing methods	181
212.	Why do glues stick?	181
213.	Climate glue protesters	182
214.	Miracle glue „gecko feet"	183
215.	Bird glue and bird murder	185
216.	Wind turbines and bats	185
217.	Zapping electricity from wind power	186
218.	How the German Renewable Energy Sources Act (EEG) works...	187
219.	Wind turbines, crazy minks and infrasound	187
220.	Thunderstorm	188
221.	Heinz Erhardt and the thunderstorm	189
222.	Lightning and thunder	190
223.	The lightning rod	191
224.	What is Ball lightning?	191
225.	Sprites, blue jets and elves	193
226.	Ionosphere and shortwave broadcasting	193
227.	Meteor scatter	194
228.	Laurentian tears - or the invention of barbecuing	194

229.	Shooting stars, bolides and meteorites	196
230.	Meteorite impacts	196
231.	The Yucatan dinosaur killer	198
232.	Mass extinctions	199
233.	The oxygen catastrophe	199
234.	Supervolcanoes	200
235.	Yellowstone	201
236.	The genetic bottleneck of mankind	202
237.	We haven't escaped yet	203
238.	It rumbles ominously under the Campi Flegrei	203
239.	The mini super volcano in the Eifel	204
240.	Louis-Auguste Cyparis and the Montagne Pelée	206
241.	Fever flies and building the Panama Canal	207
242.	Corpses preservation	208
243.	The Mausoleum of Maussolos II.	209
244.	The "genius" of the Carpathians	211
245.	Dracula	212
246.	The Fearless Vampire Killers	213
247.	Vampire bats	214
248.	The rabies	215
249.	Homeopathy	215
250.	Potentiate	216
251.	Homeopathic potency experiment	216
252.	Effect without active ingredients	217
253.	Island talents - Savants	217
254.	Kim Peek - the "Rain man"	218
255.	Photographic memory	219
256.	Geniuses	220
257.	Deficits in the education system	221
258.	Abitur oder Matura	221
259.	Noether's theorem	223
260.	Symmetry and beauty	224
261.	The golden ratio	225

262. Kirkwood gaps and KAM theorem 227

263. The golden ratio in photography and painting 228

264. Albrecht Dürer 229

265. The Shroud of Turin..231

266. Sindonology...232

267. Radiocarbon dating 233

268. Dendrochronology 235

269. Regional climate history 236

270. Milankovic cycles 236

271. Martians 240

272. War of the worlds 241

273. Life on Mars? 243

274. Earth - the planet of bacteria......................................244

275. Deadly bacteria 245

276. Penicillins...245

277. Multi-resistant germs 246

278. Orphan drugs and Ebola 246

279. Plague pandemics...248

280. Isaac Newton and his axiomatic mechanics 249

281. The Higgs boson..251

282. Reality and imagination 251

283. The (failed) Hilbert program 252

284. Maurits Cornelis Escher 252

285. Euler's identity 254

286. Exponential growth and the Club of Rome 254

287. Chess and rice 255

288. Limited growth 255

289. The ecological footprint 256

290. The logistic equation 256

291. Big numbers - googol and googolplex..............................259

292. The Observable Universe 259

293. Big Bang and Hubble bubble.......................................260

294. The Multiverse of Doppelgangers 262

295.	Big Bang theories	262
296.	Theory and empiricism	263
297.	The demon theory of friction	264
298.	Quaestio disputata	265
299.	The scholastic problem of Almighty	268
300.	Unity of teaching and research	269
301.	The vagabond literature	269
302.	Carmina Burana	269
303.	Codex Manesse	270
304.	The Nibelungenlied	273
305.	Humboldt's ideal of education	275
306.	Knowledge-based society	277
307.	Internet	278
308.	People and books	278
309.	Bibliomaniac murderer	280
310.	The Pitaval	282
311.	Books can change the world	282
312.	Copernican revolution as a paradigm shift	283
313.	Industrial revolution - digital revolution	285
314.	LCD - Liquid crystal displays	286
315.	Liquid crystals	287
316.	The Schadt-Helfrich cell	288
317.	Lippmann's color photography	289
318.	Even technical devices can die out	291
319.	The ancestry of the automobile	291
320.	The mitochondrial Eve	295
321.	Colonization of America	296
322.	Amerigo Vespucci	297
323.	Cortés and Pizarro	299
324.	Aguirre, the wrath of God	299
325.	Arquebus	300
326.	Gunpowder	301
327.	Powder mills	301

328. Sitting in the ink...303

329. Codex Argenteus...304

330. Inkjet printer printheads...306

331. Printing banknotes..307

332. Real blossoms...308

333. Tulip mania...310

334. Tulip mosaic virus...312

335. Poisonous plants...313

336. Poisoners and poison murders.......................................314

337. The Marsh sample...316

338. Poisonous plants from the garden and from fields and meadows..............317

339. Ricin...320

340. The umbrella assassination...321

341. Jonah and the castor bean worm.....................................321

342. Taxonomy and taxonomists...323

343. Famous entomologists...323

344. Lolita..325

345. Kubrick, Nixon and the man in the moon..........................326

346. Conspiracy theories..328

347. Why the manned moon landings are not fakes..................329

348. A small panopticon of conspiracy theories.......................331

349. "True conspiracy theories"..332

350. A pending case - the laboratory hypothesis on the origin of the coronavirus...333

351. Propaganda in the media and how to recognize it............334

352. Image forgers and image retouching...............................336

353. The stinky finger or can you insult a speed trap?.............337

354. Noah and the Flood...338

355. Gilgamesh and the Atrahasis epic...................................338

356. The Ararat anomaly...341

357. Dam burst on the Bosporus...341

358. Great Spokane Flood...342

359. Jökulhlaups...343

360.	Doomsday scenarios	344
361.	The cosmic year	345
362.	The next cosmic year - what will it bring?	347
363.	Man - a cosmic mayfly	347
364.	Doomsday scenarios	348
365.	The "gray goo scenario"	349
366.	Designer viruses	350
367.	Brave new world	351
368.	"Intelligence" as a limiting factor of technical civilizations	351
369.	The end of the world is near - prophets and charlatans	352
370.	Nostradamus	352
371.	Exegesis and hermeneutics	354
372.	The Revelation of John	354
373.	The Omega Point Theory	355
374.	Christian fundamentalism	358
375.	Intelligent Design	358
376.	Basic type theory	360
377.	Irreducible complexity	360
378.	The Flying Spaghetti Monster (FSM) doctrine	363
379.	Islamic fundamentalism	364
380.	Progressive role of Islam in history	364
381.	Iglauer Kompaktat	366
382.	Letter of Majesty from Emperor Rudolf II	367
383.	Bohemian brothers in Herrnhut	367
384.	Wool and plucking science	369
385.	Dichlorodiphenyltrichloroethane (DDT) and birdlife	370
386.	The silent spring	371
387.	Ozone destruction by CFCs	373
388.	Catastrophic chemical accidents in Seveso and Bhopal	374
389.	Ecotoxicology	375
390.	The Seveso poison dioxin	376
391.	Environmental protection - an idea of the 20th century	378
392.	Humans as the ultimate natural disaster	379

393. The thunder of guns can already be heard 380

394. Pandemics 380

395. Ebola 382

396. COVID-19 383

397. The threat of new viruses and the limits of global pandemic
preparedness 384

398. Man-made climate change - the new warning sign 385

399. Little ice age 387

400. Greenhouse effect 388

401. Anthropogenic greenhouse effect 389

402. Can climate simulations be trusted? 390

403. Climate simulations can neither be validated nor verified 391

404. Usurpation of climate change by politics 392

405. Positive effects of an increase in carbon dioxide concentration 392

406. Photosynthesis 393

407. Georg Imbert and the wood gasifier 395

408. The gyro car 398

409. The potter's wheel 399

410. Ceramics 400

411. Porcelain 401

412. Ehrenfried Walther von Tschirnhaus and Johann Friedrich Böttger 402

413. Meissen porcelain 405

414. High-temperature superconductivity 406

415. What exactly is superconductivity? 406

416. Superfluidity 408

417. Wonderful world of quantum physics 409

418. Fermions and bosons - the spin makes the difference... 409

419. Quantum mechanical state 410

420. Pauli exclusion principle 411

421. Cooper pairs and superconductivity 412

422. Magnetic resonance tomography 413

423. Computer tomography 414

424. The pharaonic murder - finally solved 415

425.	Seismic tomography	416
426.	Johann Radon from Tetschen on the River Elbe	418
427.	Mathematics is not a natural science, but it is indispensable	419
428.	Hilbert's problems	420
429.	Millennium problems	421
430.	Joseph Weizenbaum's "dunghill"	422
431.	AI - phrase threshers and bullshit generators	423
432.	Bicycles and cycling	424
433.	Electromobility	427
434.	Battery technology: the key to electromobility	428
435.	Lithium	430
436.	Big Bang Theory	430
437.	Why is there anything at all and not rather nothing?	433
438.	The ominous "nothing"	434
439.	The "nothing" as "emptiness"	436
440.	The vacuum	436
441.	The "world ether"	438
442.	Overcoming the "world ether"	440
443.	The Michelson-Morley experiment	440
444.	The quantum vacuum	441
445.	The Unruh effect	442
446.	The emergence of our universe from the quantum vacuum	444
447.	Cargo Cult Sciences	445
448.	Cargo cult	445
449.	The Piltdown Man	447
450.	Lysenkoism	448
451.	Mitschurin hat festgestellt...	451
452.	Epigenetic processes	451
453.	Great little queen - Matilda of Flanders	452
454.	The Bayeux Tapestry	453
455.	Halley's comet	454
456.	Comet pamphlets	455
457.	Comet research	459

458.   Problem Progress 460

459.   Belief in progress..461

460.   Sustainability..462

461.   Fears for the future 463

462.   Global conflicts and the need for a paradigm shift.....................464

463.   Bloch's concrete utopias 465

464.   Romantic utopianism 467

465.   Brave new world and "Big Brother is watching you"....................468

466.   Thomas Hobbes' theory of the state 470

467.   Liberty, equality, fraternity and the planet Neptune....................471

468.   Galileo Galilei was the first person to see the planet Neptune 473

469.   The real discovery of Neptune 474

470.   Talented ladies as "human calculators"......................................476

471.   A clever woman corrects Newton 476

472.   Diderot and his encyclopedia...477

473.   Encyclopædia Britannica...480

474.   Wrynecks and weathercocks 482

475.   Paradigm shift in science 484

476.   Nicolaus Copernicus as a great economist..................................485

477.   What did Copernicus look like in reality? 486

478.   Facial reconstructions of historical personalities 489

479.   A painting in the Seine... 490

480.   Make lenses to see the moon big...491

481.   Camera obscura and Renaissance painting 493

482.   The secret of Jan Vermeer's "Letter Reader at the Open Window".............494

483.   Canaletto view from Dresden...496

484.   Bomber Harris 497

485.   "World peace" - the utopia of modernity.......................................499

486.   The Pax Romana 500

487.   The lingering lessons of the "Pax Romana".................................502

488.   Roman decadence 503

489.   The Decline of the West..504

490.   The Clash of Civilizations...505

491. How long will humanity continue to exist? 506
492. The Fermi paradox 507
493. Yes, where are they, the aliens... 508
494. The emergence of intelligent life... 510
495. Copernican principle and Drake formula 513
496. Jet through the galaxies with WARP drive... 516
497. The "SETI" project and its prospects of success 516
498. The Star Trek universe 519
499. The place of education and knowledge in the Star Trek universe................. 520
500. Knowledge and the pursuit of knowledge as an end in itself 523

Panoptikum - A journey of discovery through the universe of knowledge

Do you fancy a journey through the wonderful world of knowledge? "Panoptikum of fascinating things and events" takes you on a fascinating collection of facts, curious stories and philosophical musings. In this book, Mathias Scholz brings together topics ranging from mathematics to fairy tales, from historical cabinets of curiosities to the secrets of science - all presented with a pinch of humor and a keen eye for the unexpected.

Whether you are looking for new insights for your next small talk or just want to be surprised, this book will captivate you, make you laugh and make you think. "Panoptikum" is an invitation to experience the world with open eyes and an alert mind.